Quantitative Finance

WILEY SERIES IN STATISTICS IN PRACTICE

Advisory Editor, Marian Scott, University of Glasgow, Scotland, UK

Founding Editor, Vic Barnett, Nottingham Trent University, UK

Statistics in Practice is an important international series of texts which provide detailed coverage of statistical concepts, methods, and worked case studies in specific fields of investigation and study. With sound motivation and many worked practical examples, the books show in down-to-earth terms how to select and use an appropriate range of statistical techniques in a particular practical field within each title's special topic area.

The books provide statistical support for professionals and research workers across a range of employment fields and research environments. Subject areas covered include medicine and pharmaceutics; industry, finance, and commerce; public services; the earth and environmental sciences, and so on.

The books also provide support to students studying statistical courses applied to the above areas. The demand for graduates to be equipped for the work environment has led to such courses becoming increasingly prevalent at universities and colleges.

It is our aim to present judiciously chosen and well-written workbooks to meet everyday practical needs. Feedback of views from readers will be most valuable to monitor the success of this aim.

A complete list of titles in this series appears at the end of the volume.

Quantitative Finance

Maria C. Mariani
University of Texas at El Paso
Texas, United States

Ionut Florescu
Stevens Institute of Technology
Hoboken, United States

This edition first published 2020
© 2020 John Wiley & Sons, Inc.

All rights reserved. No part of this publication may be reproduced, stored in a retrieval system, or transmitted, in any form or by any means, electronic, mechanical, photocopying, recording or otherwise, except as permitted by law. Advice on how to obtain permission to reuse material from this title is available at http://www.wiley.com/go/permissions.

The right of Maria C. Mariani and Ionut Florescu to be identified as the authors of this work has been asserted in accordance with law.

Registered Office
John Wiley & Sons, Inc., 111 River Street, Hoboken, NJ 07030, USA

Editorial Office
111 River Street, Hoboken, NJ 07030, USA

For details of our global editorial offices, customer services, and more information about Wiley products visit us at www.wiley.com.

Wiley also publishes its books in a variety of electronic formats and by print-on-demand. Some content that appears in standard print versions of this book may not be available in other formats.

Limit of Liability/Disclaimer of Warranty
While the publisher and authors have used their best efforts in preparing this work, they make no representations or warranties with respect to the accuracy or completeness of the contents of this work and specifically disclaim all warranties, including without limitation any implied warranties of merchantability or fitness for a particular purpose. No warranty may be created or extended by sales representatives, written sales materials or promotional statements for this work. The fact that an organization, website, or product is referred to in this work as a citation and/or potential source of further information does not mean that the publisher and authors endorse the information or services the organization, website, or product may provide or recommendations it may make. This work is sold with the understanding that the publisher is not engaged in rendering professional services. The advice and strategies contained herein may not be suitable for your situation. You should consult with a specialist where appropriate. Further, readers should be aware that websites listed in this work may have changed or disappeared between when this work was written and when it is read. Neither the publisher nor authors shall be liable for any loss of profit or any other commercial damages, including but not limited to special, incidental, consequential, or other damages.

Library of Congress Cataloging-in-Publication Data

Names: Mariani, Maria C., author. | Florescu, Ionut, 1973– author.
Title: Quantitative finance / Maria C. Mariani, University of Texas at El Paso, Texas,
    United States, Ionut Florescu, Stevens Intistute of Technology, Hoboken, United States.
Description: Hoboken, NJ : Wiley, 2020. | Series: Wiley series in
    statistics in practice | Includes index.
Identifiers: LCCN 2019035349 (print) | LCCN 2019035350 (ebook) | ISBN
    9781118629956 (hardback) | ISBN 9781118629963 (adobe pdf) | ISBN
    9781118629888 (epub)
Subjects: LCSH: Business mathematics. | Finance–Mathematical models. |
    Finance–Econometric models.
Classification: LCC HF5691 .M29 2020 (print) | LCC HF5691 (ebook) | DDC
    332.01/5195–dc23
LC record available at https://lccn.loc.gov/2019035349
LC ebook record available at https://lccn.loc.gov/2019035350

Cover Design: Wiley
Cover Image: Courtesy of Maria C. Mariani

Set in 9.5/12.5pt STIXTwoText by SPi Global, Pondicherry, India
10 9 8 7 6 5 4 3 2 1

Contents

List of Figures *xv*
List of Tables *xvii*

Part I   Stochastic Processes and Finance *1*

1   Stochastic Processes *3*
1.1   Introduction *3*
1.2   General Characteristics of Stochastic Processes *4*
1.2.1   The Index Set I *4*
1.2.2   The State Space S *4*
1.2.3   Adaptiveness, Filtration, and Standard Filtration *5*
1.2.4   Pathwise Realizations *7*
1.2.5   The Finite Dimensional Distribution of Stochastic Processes *8*
1.2.6   Independent Components *9*
1.2.7   Stationary Process *9*
1.2.8   Stationary and Independent Increments *10*
1.3   Variation and Quadratic Variation of Stochastic Processes *11*
1.4   Other More Specific Properties *13*
1.5   Examples of Stochastic Processes *14*
1.5.1   The Bernoulli Process (Simple Random Walk) *14*
1.5.2   The Brownian Motion (Wiener Process) *17*
1.6   Borel—Cantelli Lemmas *19*
1.7   Central Limit Theorem *20*
1.8   Stochastic Differential Equation *20*
1.9   Stochastic Integral *21*
1.9.1   Properties of the Stochastic Integral *22*
1.10   Maximization and Parameter Calibration of Stochastic Processes *22*
1.10.1 Approximation of the Likelihood Function (Pseudo Maximum Likelihood Estimation) *24*
1.10.2 Ozaki Method *24*

vi Contents

1.10.3 Shoji-Ozaki Method 25
1.10.4 Kessler Method 25
1.11 Quadrature Methods 26
1.11.1 Rectangle Rule: ($n = 1$) (Darboux Sums) 27
1.11.2 Midpoint Rule 28
1.11.3 Trapezoid Rule 28
1.11.4 Simpson's Rule 28
1.12 Problems 29

2 Basics of Finance 33
2.1 Introduction 33
2.2 Arbitrage 33
2.3 Options 35
2.3.1 Vanilla Options 35
2.3.2 Put–Call Parity 36
2.4 Hedging 39
2.5 Modeling Return of Stocks 40
2.6 Continuous Time Model 41
2.6.1 Itô's Lemma 42
2.7 Problems 45

Part II Quantitative Finance in Practice 47

3 Some Models Used in Quantitative Finance 49
3.1 Introduction 49
3.2 Assumptions for the Black–Scholes–Merton Derivation 49
3.3 The B-S Model 50
3.4 Some Remarks on the B-S Model 58
3.4.1 Remark 1 58
3.4.2 Remark 2 58
3.5 Heston Model 60
3.5.1 Heston PDE Derivation 61
3.6 The Cox–Ingersoll–Ross (CIR) Model 63
3.7 Stochastic α, β, ρ (SABR) Model 64
3.7.1 SABR Implied Volatility 64
3.8 Methods for Finding Roots of Functions: Implied Volatility 65
3.8.1 Introduction 65
3.8.2 The Bisection Method 65
3.8.3 The Newton's Method 66
3.8.4 Secant Method 67
3.8.5 Computation of Implied Volatility Using the Newton's Method 68
3.9 Some Remarks of Implied Volatility (Put–Call Parity) 69

3.10 Hedging Using Volatility *70*
3.11 Functional Approximation Methods *73*
3.11.1 Local Volatility Model *74*
3.11.2 Dupire's Equation *74*
3.11.3 Spline Approximation *77*
3.11.4 Numerical Solution Techniques *78*
3.11.5 Pricing Surface *79*
3.12 Problems *79*

4 Solving Partial Differential Equations *83*
4.1 Introduction *83*
4.2 Useful Definitions and Types of PDEs *83*
4.2.1 Types of PDEs (2-D) *83*
4.2.2 Boundary Conditions (BC) for PDEs *84*
4.3 Functional Spaces Useful for PDEs *85*
4.4 Separation of Variables *88*
4.5 Moment-Generating Laplace Transform *91*
4.5.1 Numeric Inversion for Laplace Transform *92*
4.5.2 Fourier Series Approximation Method *93*
4.6 Application of the Laplace Transform to the Black–Scholes PDE *96*
4.7 Problems *99*

5 Wavelets and Fourier Transforms *101*
5.1 Introduction *101*
5.2 Dynamic Fourier Analysis *101*
5.2.1 Tapering *102*
5.2.2 Estimation of Spectral Density with Daniell Kernel *103*
5.2.3 Discrete Fourier Transform *104*
5.2.4 The Fast Fourier Transform (FFT) Method *106*
5.3 Wavelets Theory *109*
5.3.1 Definition *109*
5.3.2 Wavelets and Time Series *110*
5.4 Examples of Discrete Wavelets Transforms (DWT) *112*
5.4.1 Haar Wavelets *112*
5.4.2 Daubechies Wavelets *115*
5.5 Application of Wavelets Transform *116*
5.5.1 Finance *116*
5.5.2 Modeling and Forecasting *117*
5.5.3 Image Compression *117*
5.5.4 Seismic Signals *117*
5.5.5 Damage Detection in Frame Structures *118*
5.6 Problems *118*

6	**Tree Methods** *121*	
6.1	Introduction *121*	
6.2	Tree Methods: the Binomial Tree *122*	
6.2.1	One-Step Binomial Tree *122*	
6.2.2	Using the Tree to Price a European Option *125*	
6.2.3	Using the Tree to Price an American Option *126*	
6.2.4	Using the Tree to Price Any Path-Dependent Option *127*	
6.2.5	Using the Tree for Computing Hedge Sensitivities: the Greeks *128*	
6.2.6	Further Discussion on the American Option Pricing *128*	
6.2.7	A Parenthesis: the Brownian Motion as a Limit of Simple Random Walk *132*	
6.3	Tree Methods for Dividend-Paying Assets *135*	
6.3.1	Options on Assets Paying a Continuous Dividend *135*	
6.3.2	Options on Assets Paying a Known Discrete Proportional Dividend *136*	
6.3.3	Options on Assets Paying a Known Discrete Cash Dividend *136*	
6.3.4	Tree for Known (Deterministic) Time-Varying Volatility *137*	
6.4	Pricing Path-Dependent Options: Barrier Options *139*	
6.5	Trinomial Tree Method and Other Considerations *140*	
6.6	Markov Process *143*	
6.6.1	Transition Function *143*	
6.7	Basic Elements of Operators and Semigroup Theory *146*	
6.7.1	Infinitesimal Operator of Semigroup *150*	
6.7.2	Feller Semigroup *151*	
6.8	General Diffusion Process *152*	
6.8.1	Example: Derivation of Option Pricing PDE *155*	
6.9	A General Diffusion Approximation Method *156*	
6.10	Particle Filter Construction *159*	
6.11	Quadrinomial Tree Approximation *163*	
6.11.1	Construction of the One-Period Model *164*	
6.11.2	Construction of the Multiperiod Model: Option Valuation *170*	
6.12	Problems *173*	
7	**Approximating PDEs** *177*	
7.1	Introduction *177*	
7.2	The Explicit Finite Difference Method *179*	
7.2.1	Stability and Convergence *180*	
7.3	The Implicit Finite Difference Method *180*	
7.3.1	Stability and Convergence *182*	
7.4	The Crank–Nicolson Finite Difference Method *183*	
7.4.1	Stability and Convergence *183*	

7.5	A Discussion About the Necessary Number of Nodes in the Schemes *184*
7.5.1	Explicit Finite Difference Method *184*
7.5.2	Implicit Finite Difference Method *185*
7.5.3	Crank–Nicolson Finite Difference Method *185*
7.6	Solution of a Tridiagonal System *186*
7.6.1	Inverting the Tridiagonal Matrix *186*
7.6.2	Algorithm for Solving a Tridiagonal System *187*
7.7	Heston PDE *188*
7.7.1	Boundary Conditions *189*
7.7.2	Derivative Approximation for Nonuniform Grid *190*
7.8	Methods for Free Boundary Problems *191*
7.8.1	American Option Valuations *192*
7.8.2	Free Boundary Problem *192*
7.8.3	Linear Complementarity Problem (LCP) *193*
7.8.4	The Obstacle Problem *196*
7.9	Methods for Pricing American Options *199*
7.10	Problems *201*
8	**Approximating Stochastic Processes** *203*
8.1	Introduction *203*
8.2	Plain Vanilla Monte Carlo Method *203*
8.3	Approximation of Integrals Using the Monte Carlo Method *205*
8.4	Variance Reduction *205*
8.4.1	Antithetic Variates *205*
8.4.2	Control Variates *206*
8.5	American Option Pricing with Monte Carlo Simulation *208*
8.5.1	Introduction *209*
8.5.2	Martingale Optimization *210*
8.5.3	Least Squares Monte Carlo (LSM) *210*
8.6	Nonstandard Monte Carlo Methods *216*
8.6.1	Sequential Monte Carlo (SMC) Method *216*
8.6.2	Markov Chain Monte Carlo (MCMC) Method *217*
8.7	Generating One-Dimensional Random Variables by Inverting the cdf *218*
8.8	Generating One-Dimensional Normal Random Variables *220*
8.8.1	The Box–Muller Method *221*
8.8.2	The Polar Rejection Method *222*
8.9	Generating Random Variables: Rejection Sampling Method *224*
8.9.1	Marsaglia's Ziggurat Method *226*
8.10	Generating Random Variables: Importance Sampling *236*
8.10.1	Sampling Importance Resampling *240*

8.10.2 Adaptive Importance Sampling *241*
8.11 Problems *242*

9 Stochastic Differential Equations *245*

9.1 Introduction *245*
9.2 The Construction of the Stochastic Integral *246*
9.2.1 Itô Integral Construction *249*
9.2.2 An Illustrative Example *251*
9.3 Properties of the Stochastic Integral *253*
9.4 Itô Lemma *254*
9.5 Stochastic Differential Equations (SDEs) *257*
9.5.1 Solution Methods for SDEs *259*
9.6 Examples of Stochastic Differential Equations *260*
9.6.1 An Analysis of Cox–Ingersoll–Ross (CIR)-Type Models *263*
9.6.2 Moments Calculation for the CIR Model *265*
9.6.3 Interpretation of the Formulas for Moments *267*
9.6.4 Parameter Estimation for the CIR Model *267*
9.7 Linear Systems of SDEs *268*
9.8 Some Relationship Between SDEs and Partial Differential Equations (PDEs) *271*
9.9 Euler Method for Approximating SDEs *273*
9.10 Random Vectors: Moments and Distributions *277*
9.10.1 The Dirichlet Distribution *279*
9.10.2 Multivariate Normal Distribution *280*
9.11 Generating Multivariate (Gaussian) Distributions with Prescribed Covariance Structure *281*
9.11.1 Generating Gaussian Vectors *281*
9.12 Problems *283*

Part III Advanced Models for Underlying Assets *287*

10 Stochastic Volatility Models *289*

10.1 Introduction *289*
10.2 Stochastic Volatility *289*
10.3 Types of Continuous Time SV Models *290*
10.3.1 Constant Elasticity of Variance (CEV) Models *291*
10.3.2 Hull–White Model *292*
10.3.3 The Stochastic Alpha Beta Rho (SABR) Model *293*
10.3.4 Scott Model *294*
10.3.5 Stein and Stein Model *295*
10.3.6 Heston Model *295*

10.4	Derivation of Formulae Used: Mean-Reverting Processes *296*
10.4.1	Moment Analysis for CIR Type Processes *299*
10.5	Problems *301*

11 Jump Diffusion Models *303*

11.1	Introduction *303*
11.2	The Poisson Process (Jumps) *303*
11.3	The Compound Poisson Process *304*
11.4	The Black–Scholes Models with Jumps *305*
11.5	Solutions to Partial-Integral Differential Systems *310*
11.5.1	Suitability of the Stochastic Model Postulated *311*
11.5.2	Regime-Switching Jump Diffusion Model *312*
11.5.3	The Option Pricing Problem *313*
11.5.4	The General PIDE System *314*
11.6	Problems *322*

12 General Lévy Processes *325*

12.1	Introduction and Definitions *325*
12.2	Lévy Processes *325*
12.3	Examples of Lévy Processes *329*
12.3.1	The Gamma Process *329*
12.3.2	Inverse Gaussian Process *330*
12.3.3	Exponential Lévy Models *330*
12.4	Subordination of Lévy Processes *331*
12.5	Rescaled Range Analysis (Hurst Analysis) and Detrended Fluctuation Analysis (DFA) *332*
12.5.1	Rescaled Range Analysis (Hurst Analysis) *332*
12.5.2	Detrended Fluctuation Analysis *334*
12.5.3	Stationarity and Unit Root Test *335*
12.6	Problems *336*

13 Generalized Lévy Processes, Long Range Correlations, and Memory Effects *337*

13.1	Introduction *337*
13.1.1	Stable Distributions *337*
13.2	The Lévy Flight Models *339*
13.2.1	Background *339*
13.2.2	Kurtosis *343*
13.2.3	Self-Similarity *345*
13.2.4	The H - α Relationship for the Truncated Lévy Flight *346*
13.3	Sum of Lévy Stochastic Variables with Different Parameters *347*

13.3.1 Sum of Exponential Random Variables with Different Parameters *348*
13.3.2 Sum of Lévy Random Variables with Different Parameters *351*
13.4 Examples and Applications *352*
13.4.1 Truncated Lévy Models Applied to Financial Indices *352*
13.4.2 Detrended Fluctuation Analysis (DFA) and Rescaled Range Analysis Applied to Financial Indices *357*
13.5 Problems *362*

14 Approximating General Derivative Prices *365*
14.1 Introduction *365*
14.2 Statement of the Problem *368*
14.3 A General Parabolic Integro-Differential Problem *370*
14.3.1 Schaefer's Fixed Point Theorem *371*
14.4 Solutions in Bounded Domains *372*
14.5 Construction of the Solution in the Whole Domain *385*
14.6 Problems *386*

15 Solutions to Complex Models Arising in the Pricing of Financial Options *389*
15.1 Introduction *389*
15.2 Option Pricing with Transaction Costs and Stochastic Volatility *389*
15.3 Option Price Valuation in the Geometric Brownian Motion Case with Transaction Costs *390*
15.4 Stochastic Volatility Model with Transaction Costs *392*
15.5 The PDE Derivation When the Volatility is a Traded Asset *393*
15.5.1 The Nonlinear PDE *395*
15.5.2 Derivation of the Option Value PDEs in Arbitrage Free and Complete Markets *397*
15.6 Problems *400*

16 Factor and Copulas Models *403*
16.1 Introduction *403*
16.2 Factor Models *403*
16.2.1 Cross-Sectional Regression *404*
16.2.2 Expected Return *406*
16.2.3 Macroeconomic Factor Models *407*
16.2.4 Fundamental Factor Models *408*
16.2.5 Statistical Factor Models *408*
16.3 Copula Models *409*
16.3.1 Families of Copulas *411*
16.4 Problems *412*

Part IV Fixed Income Securities and Derivatives *413*

17   Models for the Bond Market *415*
17.1   Introduction and Notations *415*
17.2   Notations *415*
17.3   Caps and Swaps *417*
17.4   Valuation of Basic Instruments: Zero Coupon and Vanilla Options on Zero Coupon *419*
17.4.1   Black Model *419*
17.4.2   Short Rate Models *420*
17.5   Term Structure Consistent Models *422*
17.6   Inverting the Yield Curve *426*
17.6.1   Affine Term Structure *427*
17.7   Problems *428*

18   Exchange Traded Funds (ETFs), Credit Default Swap (CDS), and Securitization *431*
18.1   Introduction *431*
18.2   Exchange Traded Funds (ETFs) *431*
18.2.1   Index ETFs *432*
18.2.2   Stock ETFs *433*
18.2.3   Bond ETFs *433*
18.2.4   Commodity ETFs *433*
18.2.5   Currency ETFs *434*
18.2.6   Inverse ETFs *435*
18.2.7   Leverage ETFs *435*
18.3   Credit Default Swap (CDS) *436*
18.3.1   Example of Credit Default Swap *437*
18.3.2   Valuation *437*
18.3.3   Recovery Rate Estimates *439*
18.3.4   Binary Credit Default Swaps *439*
18.3.5   Basket Credit Default Swaps *439*
18.4   Mortgage Backed Securities (MBS) *440*
18.5   Collateralized Debt Obligation (CDO) *441*
18.5.1   Collateralized Mortgage Obligations (CMO) *441*
18.5.2   Collateralized Loan Obligations (CLO) *442*
18.5.3   Collateralized Bond Obligations (CBO) *442*
18.6   Problems *443*

Bibliography *445*

Index *459*

List of Figures

1.1	An example of three paths corresponding to three ω's for a certain stochastic process.	7
6.1	One-step binomial tree for the return process.	123
6.2	The basic successors for a given volatility value. Case 1.	165
6.3	The basic successors for a given volatility value. Case 2.	168
8.1	Rejection sampling using a basic uniform.	226
8.2	The Ziggurat distribution for $n = 8$.	227
8.3	Candidate densities for the importance sampling procedure as well as the target density.	239
13.1	Lévy flight parameter for City Group.	353
13.2	Lévy flight parameter for JPMorgan.	354
13.3	Lévy flight parameter for Microsoft Corporation.	355
13.4	Lévy flight parameter for the DJIA index.	356
13.5	Lévy flight parameter for IBM, high frequency (tick) data.	358
13.6	Lévy flight parameter for Google, high frequency (tick) data.	359
13.7	Lévy flight parameter for Walmart, high frequency (tick) data.	360
13.8	Lévy flight parameter for the Walt Disney Company, high frequency (tick) data.	361
13.9	DFA and Hurst methods applied to the data series of IBM, high frequency data corresponding to the crash week.	363
13.10	DFA and Hurst methods applied to the data series of Google, high frequency data corresponding to the crash week.	364

List of Tables

1.1	Sample outcome	15
2.1	Arbitrage example	34
2.2	Binomial hedging example	39
3.1	Continuous hedging example	50
3.2	Example of hedging with actual volatility	71
3.3	Example of hedging with implied volatility	72
5.1	Determining p and q for $N = 16$	113
6.1	An illustration of the typical progression of number of elements in lists a, d, and b	173
8.1	Calibrating the delta hedging	206
8.2	Stock price paths	212
8.3	Cash flow matrix at time $t = 3$	212
8.4	Regression at $t = 2$	213
8.5	Optimal early exercise decision at time $t = 2$	213
8.6	Cash flow matrix at time $t = 2$	214
8.7	Regression at time $t = 1$	214
8.8	Optimal early exercise decision at time $t = 1$	215
8.9	Stopping rule	215
8.10	Optimal cash flow	216
8.11	Average running time in seconds for 30 runs of the three methods	235
11.1	Moments of the poisson distribution with intensity λ	304
12.1	Moments of the $\Gamma(a, b)$ distribution	330
13.1	Lévy parameter for the high frequency data corresponding to the crash week	362

Part I

Stochastic Processes and Finance

1

Stochastic Processes

A Brief Review

1.1 Introduction

In this chapter, we introduce the basic mathematical tools we will use. We assume the reader has a good understanding of probability spaces and random variables. For more details we refer to [67, 70]. This chapter is not meant to be a replacement for a book. To get the fundamentals please consult [70, 117]. In this chapter, we are reviewing fundamental notions for the rest of the book.

So, *what is a stochastic process?* When asked this question, *R.A. Fisher* famously replied, "What is a stochastic process? Oh, it's just one darn thing after another." We hope to elaborate on Fisher's reply in this introduction.

We start the study of stochastic processes by presenting some commonly assumed properties and characteristics. Generally, these characteristics simplify analysis of stochastic processes. However, a stochastic process with these properties will have simplified dynamics, and the resulting models may not be complex enough to model real-life behavior. In Section 1.6 of this chapter, we introduce the simplest stochastic processes: the coin toss process (also known as the Bernoulli process) which produces the simple random walk.

We start with the definition of a stochastic process.

Definition 1.1.1 Given a probability space $(\Omega, \mathcal{F}, \mathbf{P})$, a stochastic process is any collection $\{X(t) : t \in \mathcal{I}\}$ of random variables defined on this probability space, where $\mathcal{I}$ is an index set. The notations X_t and $X(t)$ are used interchangeably to denote the value of the stochastic process at index value t.

Specifically, for any fixed t the resulting X_t is just a random variable. However, what makes this index set $\mathcal{I}$ special is that it confers the collection of random variables a certain structure. This will be explained next.

Quantitative Finance, First Edition. Maria C. Mariani and Ionut Florescu.
© 2020 John Wiley & Sons, Inc. Published 2020 by John Wiley & Sons, Inc.

1.2 General Characteristics of Stochastic Processes

1.2.1 The Index Set $\mathcal{I}$

The set $\mathcal{I}$ indexes and determines the type of stochastic process. This set can be quite general but here are some examples:

- If $\mathcal{I} = \{0, 1, 2 \ldots\}$ or equivalent, we obtain the so-called discrete-time stochastic processes. We shall often write the process as $\{X_n\}_{n\in\mathbb{N}}$ in this case.
- If $\mathcal{I} = [0, \infty)$, we obtain the continuous-time stochastic processes. We shall write the process as $\{X_t\}_{t\geq 0}$ in this case. Most of the time t represents time.
- The index set can be multidimensional. For example, with $\mathcal{I} = \mathbb{Z} \times \mathbb{Z}$, we may be describing a discrete random field where at any combination $(x, y) \in \mathcal{I}$ we have a value $X(x, y)$ which may represent some node weights in a two-dimensional graph. If $\mathcal{I} = [0, 1] \times [0, 1]$ we may be describing the structure of some surface where, for instance, $X(x, y)$ could be the value of some electrical field intensity at position (x, y).

1.2.2 The State Space $\mathcal{S}$

The state space is the domain space of all the random variables X_t. Since we are discussing about random variables and random vectors, then necessarily $\mathcal{S} \subseteq \mathbb{R}$ or $\mathbb{R}^n$. Again, we have several important examples:

- If $\mathcal{S} \subseteq \mathbb{Z}$, then the process is integer valued or a process with discrete state space.
- If $\mathcal{S} = \mathbb{R}$, then X_t is a real-valued process or a process with a continuous state space.
- If $\mathcal{S} = \mathbb{R}^k$, then X_t is a k-dimensional vector process.

The state space $\mathcal{S}$ can be more general (for example, an abstract Lie algebra), in which case the definitions work very similarly except that for each t we have X_t measurable functions.

We recall that a real-valued function f defined on Ω is called measurable with respect to a sigma algebra $\mathcal{F}$ in that space if the inverse image of set B, defined as $f^{-1}(B) \equiv \{\omega \in E : f(\omega) \in B)\}$ is a set in sigma algebra $\mathcal{F}$, for all Borel sets B of $\mathbb{R}$.

A sigma algebra $\mathcal{F}$ is a collection of sets F of Ω satisfying the following conditions:

1) $\emptyset \in \mathcal{F}$.
2) If $F \in \mathcal{F}$ then its complement $F^c \in \mathcal{F}$.
3) If $F_1, F_2, \ldots$ is a countable collection of sets in $\mathcal{F}$ then their union $\cup_{n=1}^{\infty} F_n \in \mathcal{F}$

Suppose we have a random variable X defined on a space Ω. The sigma algebra generated by X is the smallest sigma algebra in Ω that contains all the pre images of sets in $\mathbb{R}$ through X. That is,

$$\sigma(X) = \sigma\left(\{X^{-1}(B) \mid \text{for all } B\text{Borel sets in } \mathbb{R}\}\right)$$

This abstract concept is necessary to make sure that we may calculate any probability related to the random variable X.

1.2.3 Adaptiveness, Filtration, and Standard Filtration

In the special case when the index set I possesses a total order relationship,[1] we can discuss about the information contained in the process $X(t)$ at some moment $t \in I$. To quantify this information we generalize the notion of sigma algebras by introducing a sequence of sigma algebras: the filtration.

Definition 1.2.1 (Filtration). A probability space $(\Omega, \mathcal{F}, \mathbf{P})$ is a filtered probability space if and only if there exists a sequence of sigma algebras $\{\mathcal{F}_t\}_{t \in I}$ included in $\mathcal{F}$ such that $\mathcal{F}$ is an increasing collection i.e.:

$$\mathcal{F}_s \subseteq \mathcal{F}_t, \quad \forall s \leq t, \quad s, t \in I.$$

The filtration is called *complete* if its first element contains all the null sets of $\mathcal{F}$. If, for example, 0 is the first element of the index set (the usual situation) then $\forall N \in \mathcal{F}$, with $\mathbf{P}(N) = 0 \Rightarrow N \in \mathcal{F}_0$. This particular notion of a complete filtration is not satisfied and may lead to all sorts of contracdictions and counterexamples. To avoid any such case we shall assume that any filtration defined in this book is complete and all filtered probability spaces are complete.

In the particular case of continuous time (i.e. $I = [0, \infty)$), it makes sense to discuss about what happens with the filtration when two consecutive times get close to one another. For some specific time $t \in I$ we define the left and right sigma algebras:

$$\mathcal{F}_{t+} = \bigcap_{u > t} \mathcal{F}_u = \lim_{u \downarrow t} \mathcal{F}_u,$$

$$\mathcal{F}_{t-} = \sigma\left(\bigcup_{u > t} \mathcal{F}_u\right).$$

The countable intersection of sigma algebras is always a sigma algebra [67], but a union of sigma algebras is not necessarily a sigma algebra. This is why we modified the definition of $\mathcal{F}_{t-}$ slightly. The notation used $\sigma(\mathcal{C})$ represents the smallest sigma algebra that contains the collection of sets $\mathcal{C}$.

---

[1] i.e. for any two elements $x, y \in I$, either $x \leq y$ or $y \leq x$.

Definition 1.2.2 (Right and Left Continuous Filtrations). A filtration $\{F_t\}_{t \in I}$ is right continuous if and only if $\mathscr{F}_t = \mathscr{F}_{t+}$ for all t, and the filtration is left continuous if and only if $\mathscr{F}_t = \mathscr{F}_{t-}$ for all t.

In general we shall assume throughout (if applicable) that any filtration is right continuous.

Definition 1.2.3 (Adapted Stochastic Process). A stochastic process $\{X_t\}_{t \in I}$ defined on a filtered probability space $(\Omega, \mathscr{F}, \mathbf{P}, \{\mathscr{F}_t\}_{t \in I})$ is called adapted if and only if X_t is $\mathscr{F}_t$-measurable for any $t \in I$.

This is an important concept since in general, $\mathscr{F}_t$ quantifies the flow of information available at any moment t. By requiring that the process be adapted, we ensure that we can calculate probabilities related to X_t based solely on the information available at time t. Furthermore, since the filtration by definition is increasing, this also says that we can calculate the probabilities at any later moment in time as well.

On the other hand, due to the same increasing property of a filtration, it may not be possible to calculate probabilities related to X_t based only on the information available in $\mathscr{F}_s$ for a moment s earlier than t (i.e. $s < t$). This is the reason why the conditional expectation is a crucial concept for stochastic processes. Recall that $\mathbf{E}[X_t | F_s]$ is $\mathscr{F}_s$-measurable. Suppose we are sitting at time s and trying to calculate probabilities related to the random variable X_t at some time t in the future. Even though we may not calculate the probabilities related to X_t directly (nobody can since X_t will be in the future), we can still calculate its distribution according to its best guess based on the current information. That is precisely $\mathbf{E}[X_t | \mathscr{F}_s]$.

Definition 1.2.4 (Standard Filtration). In some cases, we are only given a standard probability space (without a separate filtration defined on the space). This typically corresponds to the case where we assume that all the information available at time t comes from the stochastic process X_t itself. No external sources of information are available. In this case, we will be using the standard filtration generated by the process $\{X_t\}_{t \in I}$ itself. Let

$$\mathscr{F}_t = \sigma(\{X_s : s \leq t, s \in I\}),$$

denote the sigma algebra generated by the random variables up to time t. The collection of sigma algebras $\{\mathscr{F}_t\}_t$ is increasing and obviously the process $\{X_t\}_t$ is adapted with respect to it.

Notation In the case when the filtration is not specified, we will always construct the standard filtration and denote it with $\{F_t\}_t$.

In the special case when $I = \mathbb{N}$, the set of natural numbers, and the filtration is generated by the process, we will sometimes substitute the notation $X_1, X_2, \ldots, X_n$ instead of $\mathcal{F}_n$. For example we may write

$$E[X_T^2 | \mathcal{F}_n] = E[X_T^2 | X_n, \ldots, X_1]$$

1.2.4 Pathwise Realizations

Suppose a stochastic process X_t is defined on some probability space $(\Omega, \mathcal{F}, \mathbf{P})$. Recall that by definition for every $t \in I$ fixed, X_t is a random variable. On the other hand, for every fixed $\omega \in \Omega$ we shall find a particular realization for any time t's, this outcome is typically denoted $X_t(\omega)$. Therefore, for each ω we can find a collection of numbers representing the realization of the stochastic process. That is a path. This realization may be thought of as the function $t \mapsto X_t(\omega)$.

This pathwise idea means that we can map each ω into a function from I into $\mathbb{R}$. Therefore, the process X_t may be identified as a subset of all the functions from I into $\mathbb{R}$.

In Figure 1.1 we plot three different paths each corresponding to a different realization ω_i, $i \in \{1, 2, 3\}$. Due to this pathwise representation, calculating probabilities related to stochastic processes is equivalent with calculating

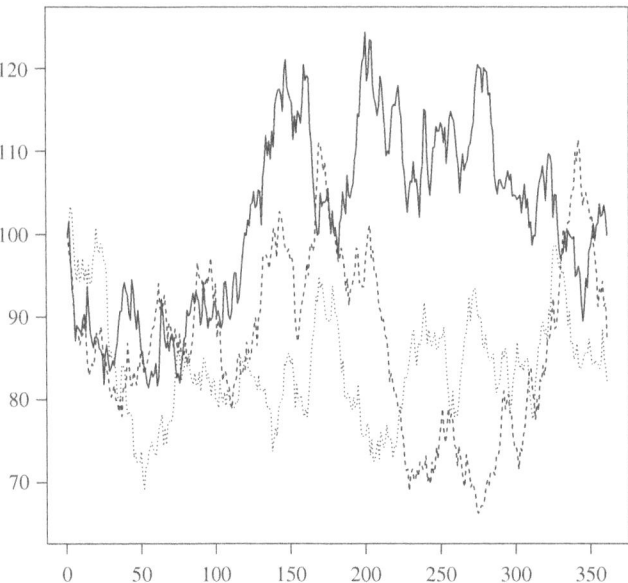

Figure 1.1 An example of three paths corresponding to three ω's for a certain stochastic process.

the distribution of these paths in subsets of the two-dimensional space. For example, the probability

$$P(\max_{t\in[0,1]} X_t \leq 1 \text{ and } \min_{t\in[0,1]} X_t \geq 0)$$

is the probability of the paths being in the unit square. However, such a calculation is impossible when the state space is infinite or when the index set is uncountable infinite such as the real numbers. To deal with this problem we need to introduce the concept of finite dimensional distribution.

1.2.5 The Finite Dimensional Distribution of Stochastic Processes

As we have seen, a stochastic process $\{X_t : t \geq 0\}_{t\in\mathcal{I}}$ is a parametrized collection of random variables defined on a probability space and taking values in $\mathbb{R}$. Thus, we have to ask: what quantities characterize a random variable? The answer is obviously its distribution. However, here we are working with a lot of variables. Depending on the number of elements in the index set $\mathcal{I}$, the stochastic process may have a finite or infinite number of components. In either case we will be concerned with the joint distribution of a finite sample taken from the process. This is due to practical consideration and the fact that in general we cannot study the joint distribution of a continuum of random variables. The processes that have a continuum structure on the set $\mathcal{I}$ serve as subject for a more advanced topic in stochastic differential equations (SDE). However, even in that more advanced situation, the finite distribution of the process still is the primary object of study.

Next, we clarify what we mean by finite dimensional distribution. Let $\{X_t\}_{t\in\mathcal{I}}$ be a stochastic process. For any $n \geq 1$ and for any subset $\{t_1, t_2, \ldots, t_n\}$ of $\mathcal{I}$, we denote with $F_{X_{t_1}, X_{t_2}, \ldots, X_{t_n}}$ the joint distribution function of the variables $X_{t_1}, X_{t_2}, \ldots, X_{t_n}$. The statistical properties of the process X_t are completely described by the family of distribution functions $F_{X_{t_1}, X_{t_2}, \ldots, X_{t_n}}$ indexed by the n and the t_i's. This is a famous result due to Kolmogorov in the 1930's. Please refer to [119] and [159] for more details.

If we can describe these finite dimensional joint distributions for all n and t's, we completely characterize the stochastic process. Unfortunately, in general this is a complicated task. However, there are some properties of the stochastic processes that make this calculation task much easier. Figure 1.1 has three different paths. It is clear that every time the paths are produced they are different but the paths may have common characteristics. In the example plotted, the paths tend to keep coming back to the starting value, and they seem to have large oscillations when the process has large values and small oscillations when the process is close to 0. These features, if they exist, will help us calculate probabilities related to the distribution of the stochastic processes. Next we discuss the most important types of such features.

1.2.6 Independent Components

This is one of the most desirable properties of stochastic processes, however, no reasonable real life process has this property. For any collection $\{t_1, t_2, \ldots, t_n\}$ of elements in $\mathcal{I}$, the corresponding random variables $X_{t_1}, X_{t_2}, \ldots, X_{t_n}$ are independent. Therefore, the joint distribution $F_{X_{t_1}, X_{t_2}, \ldots, X_{t_n}}$ is just the product of the marginal distributions $F_{X_{t_i}}$. Thus, it is very easy to calculate probabilities using such a process. However, every new component being random implies no structure. In fact, this is the defining characteristic of a noise process.

Definition 1.2.5 (White Noise Process). A stochastic process $\{X_t\}_{t \in \mathcal{I}}$ is called a white noise process if it has independent components. That is, for any collection $\{t_1, t_2, \ldots, t_n\}$ of index elements, the corresponding random variables $X_{t_1}, X_{t_2}, \ldots, X_{t_n}$ are independent. Additionally, for any time t, the random variables X_t have the same distribution $F(x)$, with the expected value $E[X_t] = 0$.

The process is called a Gaussian white noise process if it is a white noise process, and in addition the common distribution of the stochastic process X_t is a normal with mean 0.

Please note that independent components do not require the distribution to be the same for all variables. In practical applications, modeling a signal often means eliminating trends until eventually reaching this noise process. At that time the process does not expose any more trends since only "noise" remains. Typically the modeling process is complete at that point.

1.2.7 Stationary Process

A stochastic process X_t is said to be *strictly stationary* if the joint distribution function of the vectors

$$(X_{t_1}, X_{t_2}, \ldots, X_{t_n}) \quad \text{and} \quad (X_{t_1+h}, X_{t_2+h}, \ldots, X_{t_n+h})$$

are the same for all $h > 0$ and all arbitrary selection of index points $\{t_1, t_2, \ldots, t_n\}$ in $\mathcal{I}$. In particular, the distribution of X_t is the same for all t. Note that this property simplifies the calculation of the joint distribution function. The condition implies that the process is in equilibrium. The process will behave the same regardless of the particular time at which we examine it.

A stochastic process X_t is said to be *weak stationary* or *covariance stationary* if X_t has finite second moments for any t and if the covariance function $\text{Cov}(X_t, X_{t+h})$ depends only on h for all $t \in \mathcal{I}$. Note that this is a weaker version than the notion of strict stationarity. A strictly stationary process with finite second moments (so that covariance exists) is going to be automatically covariance stationary. The reverse is not true. Indeed, examples of processes which are covariance stationary but are not strictly stationary include autoregressive conditionally heteroscedastic (ARCH) processes. ARCH processes are knows as discrete-time stochastic variance processes.

The notion of weak stationarity was developed because of the practical way in which we observe stochastic processes. While strict stationarity is a very desirable concept, it is not possible to test it with real data. To prove strict stationarity means we need to test all joint distributions. In real life the samples we gather are finite so this is not possible. Instead, we can test the stationarity of the covariance matrix which only involves bivariate distributions.

Many phenomena can be described by stationary processes. Furthermore, many classes of processes eventually become stationary if observed for a long time. The white noise process is a trivial example of a strictly stationary process.

However, some of the most common processes encountered in practice – the Poisson process and the Brownian motion – are not stationary. However, they have stationary and independent **increments**. We define this concept next.

1.2.8 Stationary and Independent Increments

In order to discuss the increments for stochastic processes, we need to assume that the set $\mathcal{I}$ has a total order, that is, for any two elements s and t in $\mathcal{I}$ we have either $s \leq t$ or $t \leq s$. As a clarifying point a two-dimensional index set, for example, $\mathcal{I} = [0, 1] \times [0, 1]$ does not have this property.

A stochastic process X_t is said to have *independent increments* if the random variables

$$X_{t_2} - X_{t_1}, \ X_{t_3} - X_{t_2}, \ \ldots, \ X_{t_n} - X_{t_{n-1}}$$

are independent for any n and any choice of the sequence $\{t_1, t_2, \ldots, t_n\}$ in $\mathcal{I}$ with $t_1 < t_2 < \cdots < t_n$.

A stochastic process X_t is said to have *stationary increments* if the distribution of the random variable $X_{t+h} - X_t$ depends only on the length $h > 0$ of the increment and not on the time t.

Notice that this is not the same as stationarity of the process itself. In fact, with the exception of the constant process, there exists no process with stationary and independent increments *which is also* stationary. This is proven in the next proposition.

Proposition 1.2.1 *If a process $\{X_t, t \in [0, \infty)\}$ has stationary and independent increments then,*

$$E[X_t] = m_0 + m_1 t$$
$$\mathrm{Var}[X_t - X_0] = \mathrm{Var}[X_1 - X_0]t,$$

where $m_0 = E[X_0]$, and $m_1 = E[X_1] - m_0$.

Proof. We present the proof for the variance, and the result for the mean is entirely similar (see [119]). Let $f(t) = \text{Var}[X_t - X_0]$. Then for any t, s we have

$$\begin{aligned}
f(t+s) &= \text{Var}[X_{t+s} - X_0] = \text{Var}[X_{t+s} - X_s + X_s - X_0] \\
&= \text{Var}[X_{t+s} - X_s] + \text{Var}[X_s - X_0] \text{ (because the increments are independent)} \\
&= \text{Var}[X_t - X_0] + \text{Var}[X_s - X_0] \text{ (because of stationary increments)} \\
&= f(t) + f(s)
\end{aligned}$$

that is, the function f is additive (the above equation is also called Cauchy's functional equation). If we assume that the function f obeys some regularity conditions,[2] then the only solution is $f(t) = f(1)t$ and the result stated in the proposition holds. □

1.3 Variation and Quadratic Variation of Stochastic Processes

The notion of the variation of a stochastic process is originated from deterministic equivalents. We recall these deterministic equivalents.

Definition 1.3.1 (Variation for Deterministic Functions). Let $f : [0, \infty) \to \mathbb{R}$ be a deterministic function. Let $\pi_n = (0 = t_0 < t_1 < \ldots t_n = t)$ be a partition of the interval $[0, t]$ with n subintervals. Let $\| \pi_n \| = \max_i (t_{i+1} - t_i)$ be the length of the largest subinterval in the partition. We define the first order variation as

$$FV_t(f) = \lim_{\|\pi_n\| \to 0} \sum_{i=0}^{n-1} |f(t_{i+1}) - f(t_i)|.$$

We define the quadratic variation as

$$[f, f]_t = \lim_{\|\pi_n\| \to 0} \sum_{i=0}^{n-1} |f(t_{i+1}) - f(t_i)|^2.$$

In general, the d-order variation is defined as

$$\lim_{\|\pi_n\| \to 0} \sum_{i=0}^{n-1} |f(t_{i+1}) - f(t_i)|^d.$$

Next, we remark why we have not used a notation to the higher order variations (that is for orders three or more).

---

2 These regularity conditions are either (i) f is continuous, (ii) f is monotone, and (iii) f is bounded on compact intervals. In particular the third condition is satisfied by any process with finite second moments. The linearity of the function under condition (i) was first proven by [34].

Lemma 1.3.1 The first order variation at point t of a differentiable function $f(t)$ with continuous derivative is the length of the curve from 0 to t, that is,

$$FV_t(f) = \int_0^t |f'(s)| ds$$

Proof. This lemma is easy to prove using the mean value theorem. Recall that for any differentiable function f with continuous derivative ($f \in \mathscr{C}^1([0, \infty))$), the mean value theorem states that

$$f(t_{i+1}) - f(t_i) = f'(t_i^*)(t_{i+1} - t_i),$$

where t_i^* is some point between t_i and t_{i+1}. Hence, we obtain

$$\begin{aligned} FV_t(f) &= \lim_{\|\pi_n\| \to 0} \sum_{i=0}^{n-1} |f(t_{i+1}) - f(t_i)| \\ &= \lim_{\|\pi_n\| \to 0} \sum_{i=0}^{n-1} |f'(t_i^*)|(t_{i+1} - t_i) \\ &= \int_0^t |f'(s)| ds, \end{aligned}$$

recognizing that the last sum is just a Darboux sum which converges to the integral. $\square$

Lemma 1.3.2 For a deterministic function f which is differentiable with continuous first order derivative, all d-order variations with $d \geq 2$ are zero. We denote $\mathscr{C}^1([0, \infty))$ the collection of all functions with first derivative continuous.

Proof. This lemma is the reason why we do not need to discuss about higher order variations for deterministic function, since they are all 0. To prove the lemma, we look at the formula for the quadratic variation. All higher d-orders ($d > 2$) use the same reasoning. We have

$$\begin{aligned} [f,f]_t &= \lim_{\|\pi_n\| \to 0} \sum_{i=0}^{n-1} |f(t_{i+1}) - f(t_i)|^2 \\ &= \lim_{\|\pi_n\| \to 0} \sum_{i=0}^{n-1} |f'(t_i^*)|^2 (t_{i+1} - t_i)^2 \\ &\leq \lim_{\|\pi_n\| \to 0} \|\pi\| \sum_{i=0}^{n-1} |f'(t_i^*)|^2 (t_{i+1} - t_i) \\ &= \lim_{\|\pi_n\| \to 0} \|\pi\| \lim_{\|\pi_n\| \to 0} \sum_{i=0}^{n-1} |f'(t_i^*)|^2 (t_{i+1} - t_i). \end{aligned} \quad (1.1)$$

The second term in 1.1 is the integral $\int_0^t |f'(s)|^2 ds$. This integral is bounded, since the function has a continuous first order derivative and furthermore the first term converges to 0. Therefore, the product goes to 0. □

We note that the only way the product at the end of the above proof does not equal 0 is when the integral is infinite. However, as we know the integral of any derivable function of finite intervals is finite. Therefore, it must be that the functions with finite quadratic variation on $[0, t]$ have to be non-derivable. In fact, for any point t we may repeat this argument for an arbitrary interval $[t - \Delta t, t + \Delta t]$; thus we can easily conclude that the functions with finite quadratic variation are not derivable at any point in $\mathbb{R}$.

The notion of a function which is continuous but not derivable at any point is very strange. However, it is this strange behavior that is the defining characteristic for stochastic processes.

Definition 1.3.2 (Quadratic Variation for Stochastic Processes). Let X_t be a stochastic process on the probability space $(\Omega, \mathcal{F}, \mathbf{P})$ with filtration $\{\mathcal{F}_t\}_t$. Let $\pi_n = (0 = t_0 < t_1 < \ldots t_n = t)$ be a partition of the interval $[0, t]$. We define the quadratic variation process

$$[X, X]_t = \lim_{\|\pi_n\| \to 0} \sum_{i=0}^{n-1} |X_{t_{i+1}} - X_{t_i}|^2,$$

where the limit of the sum is defined in probability.

The quadratic variation process is a stochastic process. The quadratic variation may be calculated explicitly only for some classes of stochastic processes. The stochastic processes used in finance have finite second order variation. The third and higher order variations are all zero while the first order is infinite. This is the fundamental reason why the quadratic variation has such a big role for stochastic processes used in finance.

1.4 Other More Specific Properties

- **Point Processes.** These are special processes that count rare events. They are very useful in practice due to their frequent occurrence. For example, consider the process that gives at any time t the number of buses passing by a particular point on a certain street, starting from an initial time $t = 0$. This is a typical rare event ("rare" here does not refer to the frequency of the event, rather to the fact that there are gaps between event occurrence). Or, consider the process that counts the number of defects in a certain area of material (say 1 cm^2). Two particular cases of such a process (and the most important) are the Poisson and jump diffusion processes which will be studied in Chapter 11.

- **Markov Processes.** In general terms this is a process with the property that at time s and given the process value X_s, the future values of the process (X_t with $t > s$) only depend on this X_s and not any of the earlier X_r with $r < s$. Or equivalently the behavior of the process at any future time when its present state is exactly known is not modified by additional knowledge about its past. The study of Markov processes constitutes a big part of this book. The finite distribution of such a process has a much simplified structure. Using conditional distributions, for a fixed sequence of times $t_1 < t_2 < \cdots < t_n$ we may write

$$\begin{aligned} F_{X_{t_1},X_{t_2},\ldots,X_{t_n}} &= F_{X_{t_n}|X_{t_{n-1}},\ldots,X_{t_1}} F_{X_{t_{n-1}}|X_{t_{n-2}},\ldots,X_{t_1}} \cdots F_{X_{t_2}|X_{t_1}} F_{X_{t_1}} \\ &= F_{X_{t_n}|X_{t_{n-1}}} F_{X_{t_{n-1}}|X_{t_{n-2}}} \cdots F_{X_{t_2}|X_{t_1}} F_{X_{t_1}} \\ &= F_{X_{t_1}} \prod_{i=2}^{n} F_{X_{t_i}|X_{t_{i-1}}} \end{aligned}$$

which is a much simpler structure. In particular it means that we only need to describe one-step transitions.

- **Martingales.** Let $(\Omega, \mathcal{F}, P)$ be a probability space. A martingale sequence of length n is a set of variables $X_1, X_2, \ldots, X_n$ and corresponding sigma algebras $\mathcal{F}_1, \mathcal{F}_2, \ldots, \mathcal{F}_n$ that satisfy the following relations:
  1) Each X_i is an integrable random variable adapted to the corresponding sigma algebra $\mathcal{F}_i$.
  2) The $\mathcal{F}_i$'s form a filtration.
  3) For every $i \in [1, 2, \ldots, n-1]$, we have

  $$X_i = \mathbf{E}[X_{i+1}|\mathcal{F}_i].$$

  This process has the property that the expected value of the future given the information we have today is going to be equal to the known value of the process today. These are some of the oldest processes studied in the history of probability due to their tight connection with gambling. In fact in French (the origin of the name is attributed to Paul Lévy) a martingale means a winning strategy (winning formula).

Examples of martingales include the standard Brownian motion, Brownian motion with drift, Wald's martingale and several others.

In the next section, we present some examples of stochastic processes.

1.5 Examples of Stochastic Processes

1.5.1 The Bernoulli Process (Simple Random Walk)

We will start the study of stochastic processes with a very simple process – tosses of a (not necessarily fair) coin. Historically, this is the first stochastic process ever studied.

1.5 Examples of Stochastic Processes

Table 1.1 Sample outcome.

Y_i	0	0	1	0	0	1	0	0	0	0	1	1	1
N_i	0	0	1	1	1	2	2	2	2	2	3	4	5

Let $Y_1, Y_2, \ldots$ be independent and identically distributed (iid) Bernoulli random variables with parameter p, i.e.

$$Y_i = \begin{cases} 1 & \text{with probability } p \\ 0 & \text{with probability } 1-p \end{cases}$$

To simplify the analogy, let $Y_i = 1$ when a head appears and a tail is obtained at the i-th toss if $Y_i = 0$. Let

$$N_k = \sum_{i=1}^{k} Y_i,$$

be the number of heads up to the k-th toss, which we know is distributed as a Binomial (k, p) random variable ($N_k \sim \text{Binom}(k, p)$). An example of the above analogy is presented in Table 1.1

A sample outcome may look like the following:

Let S_n be the time at which the n-th head (success) occurred. Then mathematically,

$$S_n = \inf\{k : N_k = n\}$$

Let $X_n = S_n - S_{n-1}$ be the number of tosses to get the n-th head starting from the $(n-1)$-th head. We will present some known results about these processes.

Proposition 1.5.1

1) "Waiting times" $X_1, X_2 \ldots$ are independent and identically distributed "trials" $\sim \text{Geometric}(p)$ random variables.
2) The time at which the n-th head occurs is negative binomial, i.e. $S_n \sim \text{negative binomial}(n, p)$.
3) Given $N_k = n$ the distribution of $(S_1, \ldots, S_n)$ is the same as the distribution of a random sample of n numbers chosen without replacement from $\{1, 2, \ldots, k\}$.
4) Given $S_n = k$ the distribution of $(S_1, \ldots, S_{n-1})$ is the same as the distribution of a random sample of $n-1$ numbers chosen without replacement from $\{1, 2, \ldots, k-1\}$.
5) We have as sets:

$$\{S_n > k\} = \{N_k < n\}.$$

6) Central limit theorems (CLT):
$$\frac{N_k - \exp[N_k]}{\sqrt{\operatorname{Var}[N_k]}} = \frac{N_k - kp}{\sqrt{kp(1-p)}} \xrightarrow{D} N(0,1).$$

7)
$$\frac{S_n - \exp[S_n]}{\sqrt{\operatorname{Var}[S_n]}} = \frac{S_n - n/p}{\sqrt{n(1-p)/p}} \xrightarrow{D} N(0,1).$$

8) As $p \downarrow 0$
$$\frac{X_1}{\exp[X_1]} = \frac{X_1}{1/p} \xrightarrow{D} \text{Exponential}(\lambda = 1).$$

9) As $p \downarrow 0$
$$\mathbf{P}\left\{N_{[\frac{t}{p}]} = j\right\} \to \frac{t^j}{j!}e^{-t}.$$

We will prove several of these properties. The rest are assigned as exercises.

For 1) and 2) The distributional assertions are easy to prove; we may just use the definition of geometric (p) and negative binomial random variables. We need only to show that the X_i's are independent. See problem 1.

Proof for 3). We take $n = 4$ and $k = 100$ and prove this part only for this particular case. The general proof is identical to problem 2. A typical outcome of a Bernoulli process looks like as follows:

$$\omega : 0010010100010111 0000100$$

In the calculation of probability we have $1 \leq s_1 < s_2 < s_3 < s_4 \leq 100$. Using the definition of the conditional probability we can write:

$$\mathbf{P}(S_1 = s_1 \ldots S_4 = s_4 | N_{100} = 4)$$
$$= \frac{\mathbf{P}(S_1 = s_1 \ldots S_4 = s_4 \text{ and } N_{100} = 4)}{\mathbf{P}(N_{100} = 4)}$$
$$= \frac{\mathbf{P}\left(\overbrace{00\ldots0}^{s_1-1}1\overbrace{00\ldots0}^{s_2-s_1-1}1\overbrace{00\ldots0}^{s_3-s_2-1}1\overbrace{00\ldots0}^{s_4-s_3-1}1\overbrace{00\ldots0}^{100-s_4}\right)}{\binom{100}{4}p^4(1-p)^{96}}$$
$$= \frac{(1-p)^{s_1-1}p(1-p)^{s_2-s_1-1}p(1-p)^{s_3-s_2-1}p(1-p)^{s_4-s_3-1}p(1-p)^{100-s_4}}{\binom{100}{4}p^4(1-p)^{96}}$$
$$= \frac{(1-p)^{96}p^4}{\binom{100}{4}p^4(1-p)^{96}} = \frac{1}{\binom{100}{4}}.$$

The result is significant since it means that if we only know that there have been 4 heads by the 100-th toss, then any 4 tosses among these 100 are equally likely to contain the heads. □

Proof for 8).

$$P\left(\frac{X_1}{1/p} > t\right) = P\left(X_1 > \frac{t}{p}\right) = P\left(X_1 > \left[\frac{t}{p}\right]\right)$$

$$= (1-p)^{\left[\frac{t}{p}\right]} = \left[(1-p)^{-\frac{1}{p}}\right]^{-p\left[\frac{t}{p}\right]} \to e^{-t},$$

since

$$\lim_{p \to 0} -p\left[\frac{t}{p}\right] = \lim_{p \to 0} -p\left(\frac{t}{p} + \left[\frac{t}{p}\right] - \frac{t}{p}\right)$$

$$= -t + \lim_{p \to 0} p\underbrace{\left(\frac{t}{p} - \left[\frac{t}{p}\right]\right)}_{\in [0,1]} = -t$$

Therefore $P\left(\frac{X_1}{1/p} \leq t\right) \to 1 - e^{-t}$ and the proof is complete. □

The next example of a stochastic process is the Brownian motion.

1.5.2 The Brownian Motion (Wiener Process)

Let $(\Omega, \mathcal{F}, P)$ be a probability space. A Brownian motion is a stochastic process Z_t with the following properties:

1) $Z_0 = 0$.
2) With probability 1, the function $t \to Z_t$ is continuous in t.
3) The process Z_t has stationary and independent increments.
4) The increment $Z_{t+s} - Z_s$ has a $N(0, t)$ distribution.

The Brownian motion, also known as the Wiener process, may be obtained as a limit of a random walk. Assuming a random walk with probability $\frac{1}{2}$, we will have two variables: the time $t \in [0, T]$ and the position $x \in [-X, X]$.

For each $j = 1, 2, ..., n$, consider

$$u_{k,j} = P(x_k = j),$$

where k represents the time and j the position.

For $P(B) \neq 0$ we can define the conditional probability of the event A given that the event B occurs as

$$P(A|B) = \frac{P(A \cap B)}{P(B)}.$$

1 Stochastic Processes

To the position j in time $k+1$ we can arrive only from the position $j-1$, or j at time k, so we have:

$$u_{k+1,j} = \frac{1}{2}\left(u_{k,j} + u_{k,j-1}\right). \tag{1.2}$$

We can rewrite (6.9) as

$$u_{k+1,j} = \frac{1}{2}\left[\left(u_{k,j+1} - u_{k,j}\right) - \left(u_{k,j} - u_{k,j-1}\right)\right] + u_{k,j}$$

or

$$u_{k+1,j} - u_{k,j} = \frac{1}{2}\left[\left(u_{k,j+1} - u_{k,j}\right) - \left(u_{k,j} - u_{k,j-1}\right)\right].$$

Using the notation,

$$u_{k,j} = u\left(t_k, x_j\right),$$

we obtain

$$u\left(t_{k+1}, x_j\right) - u\left(t_k, x_j\right) = \frac{1}{2}\left[\left(u\left(t_k, x_{j+1}\right) - u\left(t_k, x_j\right)\right) - \left(u\left(t_k, x_j\right) - u\left(t_k, x_{j-1}\right)\right)\right]. \tag{1.3}$$

Now let Δt and Δx be such that $X = n\Delta x$, $T = n\Delta t$ and then $\frac{X}{\Delta x} = \frac{T}{\Delta t}$. Then multiplying (6.10) by $\frac{1}{\Delta t}$ we obtain

$$\frac{1}{\Delta t}\left(u\left(t_{k+1}, x_j\right) - u\left(t_k, x_j\right)\right)$$
$$= \frac{1}{2\Delta t}\left[\left(u\left(t_k, x_{j+1}\right) - u\left(t_k, x_j\right)\right) - \left(u\left(t_k, x_j\right) - u\left(t_k, x_{j-1}\right)\right)\right]. \tag{1.4}$$

If we take $\Delta t \to 0$ the first term in (6.11) converges to $\partial_t(u)$. For the second term, if we assume that

$$\Delta t \approx (\Delta x)^2$$

taking into account that $\Delta t = \frac{T\Delta x}{X}$,

$$\frac{1}{2\Delta t} = \frac{X}{2T\Delta x},$$

we can conclude that the second term converges to $\partial_{xx}(u)$. So, from the random walks we get a discrete version of the heat equation

$$\partial_t(u) = \frac{1}{2}\partial_{xx}(u)$$

As an example, consider the random walk with step $\Delta x = \sqrt{\Delta t}$, that is,

$$\begin{cases} x_{j+1} = x_j \pm \sqrt{\Delta t} \\ x_0 = 0 \end{cases}$$

We claim that the expected value after n steps is zero. The stochastic variables $x_{j+1} - x_j = \pm\sqrt{\Delta t}$ are independents and so

$$E(x_n) = VE\left(\sum_{j=0}^{n-1}(x_{j+1} - x_j)\right) = \sum_{j=0}^{n-1} VE(x_{j+1} - x_j) = 0$$

and

$$\text{Var}(x_n) = \sum_{j=0}^{n-1} \text{Var}(x_{j+1} - x_j) = \sum_{j=0}^{n-1} \Delta t = n\Delta t = T.$$

If we interpolate the points $\{x_j\}_{1 \leq j \leq n-1}$ we obtain

$$x(j) = \frac{x_{j+1} - x_j}{\Delta t}(t - t_j) + x_j \tag{1.5}$$

for $t_j \leq t \leq t_{j+1}$. Equation (6.12) is a *Markovian process*, because:

1) $\forall a > 0$, $\{x(t_k + a) - x(t_k)\}$ is independent of the history $\{x(s) : s \leq t_k\}$
2) $E(x(t_k + a) - x(t_k)) = 0$
   $\text{Var}(x(t_k + a) - x(t_k)) = a$

Remark 1.5.1 If $\Delta t \ll 1$ then $x \approx N(0, \sqrt{a})$ (i.e. normal distributed with mean 0 and variance a) and then,

$$P(x(t+a) - x(t) \geq y) \approx \frac{1}{\sqrt{2\pi}} \int_y^\infty e^{-\frac{t^2}{2a}} dt.$$

This is due to the CLT that guarantees that if N is large enough, the distribution can be approximated by Gaussian.

The Brownian motion will be discussed in details later in the book.
In the next sections, we recall some known results that would be used throughout this book.

1.6 Borel–Cantelli Lemmas

In probability theory, the Borel—Cantelli lemmas are statements about sequences of events. The lemmas state that, under certain conditions, an event will have probability either zero or one. We formally state the lemmas as follows: Let $\{A_n : n \neq 1\}$ be a sequence of events in a probability space. Then the event $A(i.o.) = \{A_n$ occurs for infinitely many $n\}$ is given by

$$A(i.o.) = \cap_{k=1}^\infty \cup_{n=k}^\infty A_n,$$

where *i.o.* stands for "infinitely often."

1 Stochastic Processes

Lemma 1.6.1 If $\{A_n\}$ is a sequence of events and $\sum_{n=1}^{\infty} P(A_n) < \infty$, then $P(\{A_n \ \ i.o\}) = 0$.

Lemma 1.6.2 Suppose that $\{A_n\}$ is a sequence of independent events and $\sum_{n=1}^{\infty} P(A_n) = \infty$, then $P(\{A_n \ \ i.o\}) = 1$.

The problems at the end of this chapter involves applications of the Borel–Cantelli lemmas to the Bernoulli process. Please refer to [67]. Section 1.4 for more details of the lemmas.

1.7 Central Limit Theorem

The CLT is the second fundamental theorem in probability which states that if S_n is the sum of n mutually independent random variables, then the distribution function of S_n is well approximated by a certain type of continuous function known as a normal density function, which is given by the formula

$$f_{\mu,\sigma}(x) = \frac{1}{\sqrt{2\pi}\sigma} e^{-\frac{(x-\mu)^2}{2\sigma^2}} \tag{1.6}$$

where μ and σ are the mean and standard deviation, respectively. The CLT gives information on what happens when we have the sum of a large number of independent random variables each of which contributes a small amount to the total.

1.8 Stochastic Differential Equation

The theory of differential equations is the origin of classical calculus and motivated the creation of differential and integral calculus. A differential equation is an equation involving an unknown function and its derivative. Typically, a differential equation is a functional relationship

$$f(t, x(t), x'(t), x''(t), \ldots) = 0, \quad 0 \leq t \leq T \tag{1.7}$$

involving the time t, an unknown function $x(t)$, and its derivative. The solution of 1.7 is to find a function $x(t)$ which satisfies Equation 1.7.

Now consider the deterministic differential equation:

$$dx(t) = a(t, x(t))dt, \quad x(0) = x_0. \tag{1.8}$$

The easiest way to introduce randomness in this equation is to randomize the initial condition. The solution $x(t)$ then becomes a stochastic process $(X_t, t \in [0, T])$ defined as

$$dX_t = a(t, X_t)dt, \quad X_0(\omega) = Y(\omega). \tag{1.9}$$

Equation 1.9 is called a random differential equation. Random differential equations can be considered as deterministic equations with a perturbed initial condition. Note this is not a full SDE.

For a complete introduction and study of stochastic differential equations we refer to [159]. For our purpose, we introduce a simplified definition.

Definition 1.8.1 An SDE is defined as a deterministic differential equation which is perturbed by random noise.

Note this is very different from a random differential equation because the randomness is now included in the dynamics of the equation. Examples of SDE's include:

1) $dX_t = adt + bdB_t$.
2) $dX(t) = -\lambda X(t)dt + dB_t$.

and many, many others. dB_t is a notation for the derivative of the Brownian motion – a process that does not exist also called the white noise process.

A comprehensive discussion of SDE's will be presented in Chapter 9 of this book.

1.9 Stochastic Integral

Let $(\Omega, \mathcal{F}, \mathbf{P})$ be a probability space and $\{\mathcal{F}_t\}$ a filtration on this space. Define for some fixed $S \leq T$ a class of functions $v = v(S, T)$:

$$f(t, \omega) : [0, \infty) \times \Omega \to \mathbb{R}$$

such that:

1) $(t, \omega) \mapsto f(t, \omega)$ is a $\mathcal{B} \times \mathcal{F}$ measurable function, where $\mathcal{B} = \mathcal{B}[S, T]$ is the Borel sigma algebra on that interval.
2) $\omega \mapsto f(t, \omega)$ is $\mathcal{F}_t$-adapted for all t.
3) $\mathrm{E}[\int_S^T f^2(t, \omega)dt] < \infty$

Then for every such $f \in v$ we can construct

$$\int_S^T f_t dB_t = \int_S^T f(t, \omega) dB_t(\omega),$$

where B_t is a standard Brownian motion with respect to the same filtration $\{\mathcal{F}_t\}$. This quantity is called a stochastic integral with respect to the Brownian motion B_t. We note that the stochastic integral is a random quantity.

1.9.1 Properties of the Stochastic Integral

- Linearity:

$$\int_S^T (af_t + bg_t)dB_t = a\int_S^T f_t dB_t + b\int_S^T g_t dB_t, \quad a.s.$$

-
$$\int_S^T f_t dB_t = \int_S^U f_t dB_t + \int_U^T f_t dB_t, \quad a.s., \forall S < U < T$$

-
$$E\left[\int_S^T f_t dB_t\right] = 0$$

- Itô Isometry:

$$E\left[\left(\int_S^T f_t dB_t\right)^2\right] = E\left[\int_S^T f_t^2 dt\right]$$

- If $f \in v(0, T)$ for all T then $M_t(\omega) = \int_0^t f(s, \omega)dB_s(\omega)$ is a martingale with respect to $\mathscr{F}_t$ and

$$P\left(\sup_{0 \leq t \leq T} |M_t| \geq \lambda\right) \leq \frac{1}{\lambda^2} E\left[\int_0^T f_t^2 dt\right], \lambda, T > 0 \text{(Doob's inequality)}$$

A detailed discussion on the construction of stochastic integrals will be presented in Chapter 9 of this book.

1.10 Maximization and Parameter Calibration of Stochastic Processes

There are primarily two methods to estimate parameters for a stochastic process in finance. The difference is in the data that is available.

The first method uses a sample of observations of the underlying process. It recovers the parameters of the stochastic process under the objective probability measure P. The second method uses the particular data specific to finance. The input is derivative data (such as options futures) and it estimates the parameters under the equivalent martingale measure Q. The first method is appropriate if we want to identify features of the underlying process or we want to construct financial instruments based on it. For example, in portfolio optimization, assessing the risk of a portfolio are problems that need to use parameters estimated under the probability P. On the other hand estimating parameters under Q is needed if we want to price other derivatives which are not regularly traded on the market.

1.10 Maximization and Parameter Calibration of Stochastic Processes

Method 1: Given a history of a stochastic process X_t: $x_0, x_1, \ldots, x_n$ estimate the parameters of the process. To clarify we assume the process follows a stochastic differential equation:

$$dX_t = f(X_t, \theta)dt + g(X_t, \theta)dW_t, t \geq 0, X_0 = x0, \quad (1.10)$$

Here, the functions f, g are given and the problem is to estimate the vector of parameters θ. For example, to consider the Black–Scholes model, we take $f(x, \theta) = ux$, $g(x, \theta) = \sigma x$ and $\theta = (u, \sigma)$. For the Cox–Ingersoll–Ross (CIR) model, we take $f(x, \theta) = k(\bar{x} - x)$, $g(x, \theta) = \sigma \sqrt{x}$ and $\theta = (k, \bar{x}, \sigma)$. Almost all stochastic models used in the financial markets may be written in this form.

The classical approach is to write the likelihood function and maximize it to produce the maximum likelihood estimator (MLE). If $f(x_1, \ldots, x_n | \theta)$ is the density of the probability

$$P(X_{t_1} \leq x_1, \ldots, X_{t_n} \leq x_n) = \int_{-\infty}^{x_1} \cdots \int_{-\infty}^{x_n} f(x_1, \ldots, x_n | \theta) dx_1 \ldots dx_n$$

then we maximize the likelihood function:

$$L(\theta) = f(x_1, \ldots, x_n | \theta),$$

for the observed $x_1, x_2, \ldots, x_n$, as a function of θ.

Typically this distribution is hard to find. We can however write

$$f(x_0, x_1, \ldots, x_n | \theta) = f(x_n | x_{n-1}, \ldots, x_0, \theta) f(x_{n-1} | x_{n-2}, \ldots, x_0) \ldots f(x_0 | \theta).$$

When X_t is a Markov process (any solution to the SDE 1.10 is) we can reduce the density:

$$f(x_0, \ldots, x_n | \theta) = f(x_n | x_{n-1}, \theta) f(x_{n-1} | x_{n-2}, \theta) \ldots f(x_0 | \theta).$$

This can be calculated for diffusions of the type in Eq. (1.10) using the Kolmogorov backward equation (Fokker–Planck equation). Specifically, if f and g do not depend on time (i.e. $f(x, t, \theta) = f(x, \theta)$) as in the specifications of the Eq. (1.10), the Markov process is homogeneous and the transition density reduces to $p(t - s, x, y)$, which is the density of

$$X_t = y | X_s = x.$$

This density satisfies the equation

$$\frac{\partial p}{\partial t} = f(x) \frac{\partial p}{\partial x} + \frac{1}{2} g(x)^2 \frac{\partial^2 p}{\partial x^2},$$

where $p(t, x, y)$ is the transition density function to be calculated. If we can somehow calculate or estimate p, we can express the maximum likelihood function as

$$L(\theta) = \Pi_{i=1}^n p_\theta(\Delta t, x_{i-1}, x_i) p_\theta(x_0)$$

where $p_\theta(x_0)$ is the initial distribution at time $t = 0$, $x_0, x_1, \ldots, x_n$ are the observations at times $t_i = i\Delta t$ and θ is the vector of parameters. Since this function is hard to optimize we typically maximize a log transformation:

$$l(\theta) = \log L(\theta) = \sum_{i=1}^{n} \log p_\theta(\Delta t, x_{i-1}, x_i) + \log p_\theta(x_0)$$

since the logarithm is an increasing function and it will not change the location of the maximum. This function is called the *score function*.

The major issue with this approach is that the function $p(t, x, y)$ may only be calculated exactly in very few cases. We present a way to deal with this issue next.

1.10.1 Approximation of the Likelihood Function (Pseudo Maximum Likelihood Estimation)

We adapt the method in [3] for the probability density function of V_T. The main idea is to transform the variance process into an equivalent process, which has a density function close to the normal density. Then carry out the series expansion around that normal density. Finally, we invert the transformation to obtain the density expansion for the original process.

In this instance, we replace p_θ with a density h_θ that has a simple functional form and same moments as p_θ.

Euler Method We discretize 1.10 using Euler scheme as follows:

$$X_{t+\Delta t} - X_t = f(X_t, \theta)\Delta t + g(X_t, \theta)\Delta W_t \tag{1.11}$$

Conditioned by $\mathscr{F}_t$ this is Gaussian so we use a transition density (approximate). Therefore $X_{t+\Delta} - X_t | \mathscr{F}_t$ is approximately normal with mean $f(X_t, \theta)\Delta t$ and variance $g(X_t, \theta)^2 \Delta t$ which implies that

$$h(\Delta t, x, y) = \frac{1}{\sqrt{2\pi g^2(x, \theta)\Delta t}} e^{\frac{-(y-x-f(x,\theta)\Delta t)^2}{2\Delta t g^2(x,\theta)}}$$

is a good approximation. Thus the approximate log likelihood is

$$\ln(\theta)(x_1, \ldots, x_n) = \frac{-1}{2}\left[\sum_{i=1}^{n} \frac{(x_i - x_{i-1} - f(x_{i-1}, \theta)\Delta t)^2}{2\Delta t g^2(x_{i-1}, \theta)} + \sum_{i=1}^{n} \log(2\pi g^2(x_{i-1}, \theta)\Delta t).\right]$$

This approximate may be maximized easily.

1.10.2 Ozaki Method

The second approach we present is the Ozaki method, and it works for homogeneous stochastic differential equations. Given the following SDE,

$$dX_t = f(X_t, \theta)dt + \sigma dW_t$$

one can show that

$$X_{t+\Delta t} | X_t = x \sim N(E_x, V_x);$$

where

$$E_x = x + \frac{f(x)}{\partial f / \partial x} \left(e^{\frac{\partial f}{\partial x} \Delta t} - 1 \right)$$

$$V_x = \sigma^2 \frac{e^{2K_x \Delta t} - 1}{2K_x}$$

and

$$K_x = \frac{1}{\Delta t} \log \left(1 + \frac{f(x)}{x \partial f / \partial x} \left(e^{\frac{\partial f}{\partial x} \Delta t} - 1 \right) \right)$$

Also note that the general SDE may be transformed to a constant σ SDE using the Lamperti transform. For example, consider the Vasicek model: $f(x) = \theta_1(\theta_2 - x)$, $g(x) = \theta_3$. Take $\theta_1 = 3, \theta_2 = 0.7$ and $\theta_3 = 0.5$. Use pmle = "Ozaki" option in SimDiff.Proc in R.

1.10.3 Shoji-Ozaki Method

The Shoji–Ozaki method is an extension of the method to Ozaki. It is a more general case where the drift is allowed to depend on the time variable t, and also the diffusion coefficient can be varied. See [170] for more details of this methodology.

1.10.4 Kessler Method

Kessler [120] proposed to use a higher order Itô-Taylor expansion to approximate the mean and variance in a conditional Gaussian density. For example, consider the Hull White model:

$$dX_t = a(t)(b(t) - X_t)dt + \sigma(t)dW_t$$

For this example, take $a(t) = \theta_1 t$, $b(t) = \theta_2 \sqrt{t}$, and $\sigma(t) = \theta_3 t$ where $\theta_1 = 2, \theta_2 = 0.7, \theta_3 = 0.8$, and $\Delta t = 0.001$. We refer to [170] for more details. An implementation of all these methods may be found in the Sim.DiffProc package in R [26].

Now we briefly discuss the second method of estimating parameters for a stochastic process by using option data. For this method, we obtain the option data $C_1, \ldots, C_n$ and assume a model (CIR, …). In this model, we calculate a formula for the option price $C(K, T, \theta)$, i.e.

$$\min_\theta \sum_{i=1}^n (C(K, T, \theta) - C_i)^2$$

to obtain θ. Please refer to [83] for more details.

We now present some numerical approximation methods that are used to evaluate well defined integrals.

1.11 Quadrature Methods

Quadrature methods allow one to evaluate numerically an integral which is well defined (i.e. has a finite value).

To motivate the need of such methods suppose the price process follows a general dynamics of the form

$$dX_t = \mu(t, X_t)dt + \sigma(t, X_t)dW_t, \quad X_0 = x.$$

The general formula for the price of a European-type derivative at time t is

$$F(t, X_t) = \mathbb{E}\left[e^{-\int_t^T r(s)ds} F(T, X_T) | F_t\right]$$

where $F(T, X_T)$ is the terminal payoff of the option at maturity time $t = T$. Suppose we can find the transition probability density for the diffusion process $X_t : f(\Delta, x, t, y)$. That is the density of the probability of going from $X_0 = x$ at time Δ to $X_t = y$ at time t. For the geometric Brownian motion, that is, the process $\frac{dS_t}{S_t} = rdt + \sigma dWt$, we can calculate this transition explicitly:

$$f(\Delta, x, t, y) = \frac{1}{y\sqrt{2\pi\sigma^2(t-\Delta)}} e^{\frac{-1}{2}\left(\frac{\ln y - \ln x - (r - \frac{\sigma^2}{2})(t-\Delta)}{\sigma\sqrt{t-\Delta}}\right)^2}.$$

In general if we know this function $f(\Delta, x, t, y)$ and we use the notation $P(t, T) = e^{-\int_t^T r(s)ds}$, we may rewrite the expectation as

$$F(t, X_t) = F(t, x) = \int_0^\infty P(t, T) F(T, y) f(\Delta, x, t, y) dy.$$

To now compute the option price we need to calculate this integral. This is usually hard and we typically do it in a numerical way. This is where the quadrature methods become useful.

The Problem Suppose we have a real function f. Consider

$$I(f) = \int_A f(x)dx$$

where $A \subset \mathbb{R}$ is some subset of the real axis.

Definition 1.11.1 A quadrature rule of order n is an expression of the form

$$I_n(f) = \sum_{i=1}^n w_i^n f(x_i^n)$$

where w_i^n are called weights and x_i^n are called abscissa or quadrature nodes such that $I_n(f) \to I(f)$ as $n \to \infty$.

1.11 Quadrature Methods

The basic rules of constructing a quadrature approximation I_n are:

1) Approximate f by some interpolating function (polynomials) of order n such that $P_n(x_i^n) = f(x_i^n)$ for all i.
2) Integrate P_n and return $I(P_n)$ as an approximation to $I(f)$.

1.11.1 Rectangle Rule: ($n = 1$) (Darboux Sums)

For the rectangle rule, the main idea is to approximate f with a piece-wise constant function. Suppose we integrate on an interval $[a, b]$: take

$$x_0 = a,$$
$$x_1 = x_0 + \left(\frac{b-a}{n}\right)$$
$$x_i = x_0 + \frac{b-a}{n}i$$
$$x_n = b.$$

Note that we are using equidistant points but we may use any points in the interval as long as the maximum interval $\max_i |x_i - x_{i-1}|$ is going to 0 with n.

Next we calculate $f(x_i)$ and we set

$$I_n(f) = \sum_{i=0}^{n-1} hf(x_i) \text{ where } h = \frac{b-a}{n}.$$

Here is a pseudocode that accomplishes the task:

- Input a, b, n, f
- Set $h = \frac{b-a}{n}$
- Set sum $= 0$
- For $i = 0$ to $n - 1$
  sum $=$ sum $+ h * f(a + ih)$
  $i \to i + 1$
  Return sum

If $a = \infty$ or $b = \infty$ the algorithm needs to be suitably modified to avoid calculating $f(a)$ or $f(b)$.

Since

$$I(f) = \lim_{||\pi|| \to 0} \sum_{i=1}^{n} |x_i - x_{i-1}| f(\xi_i),$$

where $\pi = x_0, x_1, \ldots, x_n$, $||\pi|| = \min_i |x_i - x_{i-1}|$ where ξ_i is any number in the interval $[x_{i-1}, x_i]$, then in fact the algorithm works not only with the endpoints but also with any point in the interval $\xi_i \in [x_{i-1}, x_i]$. In particular the next rule uses the midpoint in the interval.

1.11.2 Midpoint Rule

The midpoint rule is the same as the rectangular rule but in this case, $I_n(f)$ is defined as

$$I_n(f) = \sum_{i=1}^{n} hf\left(\frac{x_i + x_{i-1}}{2}\right)$$

where $h = \frac{b-a}{n}$.

1.11.3 Trapezoid Rule

The idea is to use both endpoints of the interval to create a trapezoid. This is a linear approximation with a polynomial degree 1. Specifically, $I_n(f)$ is defined as

$$I_n(f) = \sum_{i=1}^{n} h\left(\frac{f(x_i) + f(x_{i-1})}{2}\right) = h\left(\frac{1}{2}f(x_0) + f(x_1) + \ldots + \frac{1}{2}f(x_n)\right).$$

The error of approximation is often quoted as $O(h^2)$ but this is an upper bound. Depending on the function the actual error may be quite a bit better.

1.11.4 Simpson's Rule

For this quadrature method, we approximate the function using a quadratic polynomial. We use two intervals to construct a second order polynomial so that $f(x_{i-1}) = P(x_{i-1}), f(x_i) = P(x_i)$ and $f(x_{i+1}) = P(x_{i+1})$.

1.11.4.1 Lagrange Interpolating Polynomial

Given $(x_1, y_1), \ldots, (x_k, y_k)$ all different, let

$$l_j(x) = \prod_{\substack{i=1 \\ i \neq j}}^{n} \frac{(x - x_i)}{(x_j - x_i)} = \frac{(x - x_1)}{(x_j - x_1)} \frac{(x - x_{j-1})}{(x_j - x_{j-1})} \frac{(x - x_{j+1})}{(x_j - x_{j+1})} \cdots \frac{(x - x_n)}{(x_j - x_n)}$$

Then the polynomial $L(x) = \sum_{j=1}^{k} y_j l_j(x)$ passes through each of the points $(x_1, y_1), \ldots, (x_n, y_n)$, i.e. $L(x_j) = y_j$.

Going back and translating to the Simpson rule notation the quadrinomial interpolating polynomial is

$$P(x) = f(x_{i-1})\frac{(x - x_i)}{(x_{i-1} - x_i)} \frac{(x - x_{i+1})}{(x_{i-1} - x_{i+1})} + f(x_i)\frac{(x - x_{i-1})}{(x_i - x_{i-1})} \frac{(x - x_{i+1})}{(x_i - x_{i+1})}$$
$$+ f(x_{i+1})\frac{(x - x_{i-1})}{(x_{i+1} - x_{i-1})} \frac{(x - x_i)}{(x_{i+1} - x_i)},$$

So now we approximate $\int_{x_i}^{x_{i+1}} f(x)dx$ with $\int_{x_i}^{x_{i+1}} P(x)dx$. After integrating, we obtain assuming equidistant points $x_i - x_{i-1} = x_{i+1} - x_i$:

$$\int_{x_i}^{x_{i+1}} f(x)dx \simeq \frac{x_{i+1} - x_{i-1}}{6} \left(f(x_{i-1}) + 4f(x_i) + f(x_{i+1}) \right).$$

In this exposition we used two consecutive intervals to calculate the approximating polynomial. However, we do not need to do this if we simply use the midpoint in each interval as the third point. That is, if we take the points x_i, $\frac{x_i + x_{i+1}}{2}$, and x_{i+1} to replace the points x_{i-1}, x_i, and x_{i+1} in the previous expression and we consider $x_{i+1} - x_{i-1} = h$ we obtain

$$I(f) \simeq \frac{h}{6} \sum_{i=1}^{n} \left(f(x_{i-1}) + 4f\left(\frac{x_i + x_{i-1}}{2}\right) + f(x_{i+1}) \right).$$

There are many other quadrature rule variants; we stop here from the exposition and refer the reader for example to [78].

1.12 Problems

1. Prove that the random variables X_i's in Proposition 1.5.0 are in fact independent.
2. Give a general proof of parts 3) and 4) in Proposition 1.5.0 for any $n, k \in \mathbb{N}$.
3. Show that the equality of sets in part 5) of Proposition 1.5.0 holds by double inclusion.
4. Prove parts 6) and 7) of Proposition 1.5.0 by applying the CLT.
5. Prove part 9) of Proposition 1.5.0.
6. These exercises are due to [54]. Consider an infinite Bernoulli process with $p = 0.5$, that is, an infinite sequence of random variables $\{Y_i, i \in \mathbb{Z}\}$ with $\mathbf{P}(Y_i = 0) = \mathbf{P}(Y_i = 1) = 0.5$, for all $i \in \mathbb{Z}$. We would like to study the length of the maximum sequence of 1's. Let

$$l_m = \max\{i \geq 1 : X_{m-i+1} = \cdots = X_m = 1\},$$

be the length of the run of 1's up to the m-th toss and including it. Obviously, l_m will be 0 if the m-th toss is a tail. We are interested in the asymptotic behavior of the longest run from 1 to n for large n.
That is we are interested in the behavior of L_n where

$$L_n = \max_{m \in \{1,\ldots,n\}} l_m$$
$$= \max\{i \geq 1 : X_{m-i+1} = \cdots = X_m = 1, \text{ for some } m \in \{1, \ldots, n\}\}$$

(a) Explain why $\mathbf{P}(l_m = i) = 2^{-(i+1)}$, for $i = 0, 1, 2, \ldots$ and for any m.
(b) Apply the first Borel–Cantelli lemma to the events

$$A_n = \{l_n > (1+\varepsilon)\log_2 n\}.$$

Conclude that for each $\varepsilon > 0$, with probability one, $l_n \leq (1+\varepsilon)\log_2 n$ for all n large enough.

Taking a countable sequence $\varepsilon_k \downarrow 0$ deduce that
$$\limsup_{n \to \infty} \frac{L_n}{\log_2 n} \leq 1, \quad \text{a.s.}$$

(c) Fixing $\varepsilon > 0$ and letting $A_n = \{L_n < k_n\}$ for $k_n = (1-\varepsilon)\log_2 n$. Explain why
$$A_n \subseteq \bigcap_{i=1}^{m_n} B_i^c,$$
where $m_n = [n/k_n]$ (integer part) and $B_i = \{X_{(i-1)k_n+1} = \ldots = X_{ik_n} = 1\}$ are independent events.

Deduce that $\mathbf{P}(A_n) \leq \mathbf{P}(B_i^c)^{m_n} \leq \exp(-n^\varepsilon/(2\log_2 n))$, for all n large enough.

(d) Apply the first Borel–Cantelli for the events A_n defined in problem 6(c), followed by $\varepsilon \downarrow 0$, to conclude that
$$\liminf_{n \to \infty} \frac{L_n}{\log_2 n} \geq 1 \quad \text{a.s.}$$

(e) Combine problems 6(b) and 6(d) together to conclude that
$$\frac{L_n}{\log_2 n} \to 1 \quad \text{a.s.}$$

Therefore the length of the maximum sequence of heads is approximately equal to $\log_2 n$ when the number of tosses n is large enough.

7. Let Z be a Brownian motion defined in $[0, T]$. Given a partition $\mathscr{P}$ such that $0 = t_0 < t_1 < \ldots < t_n = T$, we define
$$V_{\mathscr{P}}(Z) = \sum_{j=0}^{n-1} (Z(t_{j+1}) - Z(t_j))^2$$
and the quadratic variation of Z as the limit (when it exists)
$$VC(Z) = \lim_{|\mathscr{P}| \to 0} V_{\mathscr{P}}(Z)$$

Prove that:
(a) $E[(Z(t_{j+1}) - Z(t_j))^2] = t_{j+1} - t_j$. Conclude that $E(V_{\mathscr{P}}(Z)) = T$.
(b) $\mathrm{Var}[(Z(t_{j+1}) - Z(t_j))^2] = 2(t_{j+1} - t_j)^2$, and then
$$\mathrm{Var}(V_{\mathscr{P}}(Z)) = \sum_{j=0}^{n-1} 2(t_{j+1} - t_j)^2 \to 0 \quad \text{for } |\mathscr{P}| \to 0.$$

(c) Tchebycheff inequality: if X is an stochastic variable with $E(X) = \mu$ and $\mathrm{Var}(X) = \sigma^2$, then for all $\varepsilon > 0$ we have that
$$P(|X - \mu| \geq \varepsilon) \leq \left(\frac{\sigma}{\varepsilon}\right)^2$$

Deduce that $VC(Z) = T$ with probability 1, i.e.:

$$P(|VC(Z) - T| \geq \varepsilon) = 0$$

for all $\varepsilon > 0$. Conclude that with probability 1, Z is not differentiable in any interval $[t, t+a]$.

8. Suppose that the price of an asset follows a Brownian motion :

$$dS = \mu S dt + \sigma S dz.$$

  (a) What is the stochastic process for S^n?
  (b) What is the expected value for S^n?

9. The Hull White model is

$$dX_t = a(t)(b(t) - X_t)dt + \sigma(t)dW_t.$$

In this problem take $a(t) = \theta_1 t$, $b(t) = \theta_2 \sqrt{t}$ and $\sigma(t) = \theta_3 t$ where $\theta_1 = 2, \theta_2 = 0, 7, \theta_3 = 0.8$, and $\Delta t = 0.001$.

  (a) Generate a single path of the process by choosing a $\Delta t = 0.001$ from $t = 0$ to $t = 1$.
  (b) Use the Sim.DiffProc package in R to estimate the parameters $\theta_1, \theta_2, \theta_3$. Use the Euler method and compare with the known values of the parameters.
  (c) Repeat part (b) with all the methods available in the package. Write a conclusion based on the results obtained.

10. The variance process in the Heston model satisfy a CIR process:

$$dV_t = \kappa(\overline{V} - V_t) + \sigma\sqrt{V_t}dW_t.$$

Use Ito to calculate the dynamics of the volatility process $U_t = \sqrt{V_t}$. Under which conditions on the parameters $\kappa, \overline{V}$, and σ the process becomes an Ørnstein–Uhlenbeck process, i.e. of the form

$$dU_t = \gamma X_t dt + \delta dW_t$$

for some parameters δ and γ?

2
Basics of Finance

2.1 Introduction

The value of money over time is a very important concept in financial mathematics. A dollar that is earned today is valued higher than a dollar that will be earned in a year's time. This is because a dollar earned today can be invested today and accrue interest, making this dollar worth more in a year. We can then calculate how much money we will need to invest now to have k dollars in a certain time frame. For example, if we invest $100 at 5% interest for a year, we will have 100(5%) + 100 = $105 at the end of the year. A question that often arises is how can one model interest mathematically? This can be modeled by an amount $M(t)$ in the bank at time t. The amount of interest gained during an elapsed time of dt is represented as $M(t+dt) - M(t) = \frac{dM}{dt}dt + \cdots$. Since interest is proportional to the amount of money in the account, given the interest rate r and time step dt, we can then write $\frac{dM}{dt}dt = rM(t)dt$. This is now a differential equation that can be solved to obtain $M(t) = M(0)e^{rt}$. As regards to pricing options, which we will study in Section 2.3, we are concerned with the present valuation of an option. We will use the time value of money to answer the question how much we would be willing to pay now in order to receive an exercise amount E at some time T in the future?

In the subsequent sections, we introduce some concepts that will be used throughout this book.

2.2 Arbitrage

Arbitrage is one of the fundamental concepts of mathematical finance. The easiest way to understand arbitrage is through the phrase "free lunch." Arbitrage is basically investing nothing and being able to get a positive return without undergoing any risk at all, like getting a free meal without having to pay for it. This is a very crucial principle for the mathematical modeling of option pricing.

Quantitative Finance, First Edition. Maria C. Mariani and Ionut Florescu.
© 2020 John Wiley & Sons, Inc. Published 2020 by John Wiley & Sons, Inc.

2 Basics of Finance

Table 2.1 Arbitrage example.

Holding	Value today (time t)	Value upon exercise (T)
Forward	0	$S(T) - F$
-Stock	$-S(t)$	$-S(T)$
Cash	$S(t)$	$S(t)e^{r(T-t)}$
Cashflow	0	$S(t)e^{r(T-t)} - F$

We will present an example of an arbitrage opportunity, but first we will define a forward contract.

Definition 2.2.1 A forward contract is an agreement between two parties to buy and sell specified asset from the other for a specified price, known as the forward price, on a specified date in the future (the delivery date or maturity date).

The agreed upon price of the underlying asset at the delivery date is called the delivery price, which we will denote by F. In a forward contract, there is no choice to buy or sell the underlying asset; the underlying asset must be paid for and delivered, unlike "options" which we will define in the next section. In a forward contract there is no exchange of money until the delivery date. The objective is to find an appropriate value of the delivery price F. If $S(T)$ represents the stock price at time T, then the buyer of the forward is entitled to an amount $S(T) - F$ at expiry, and we do not know this value. In Table 2.1, we present an example of pricing F using an arbitrage argument.

For this particular example of arbitrage opportunity, we will enter into a forward contract. At the same time we will sell the asset or "go short." It costs nothing to enter into a forward contract and we will sell our stock that is worth $S(t)$. In order to make the market arbitrage free, we must have $S(t)e^{r(T-t)} - F = 0$, that is, $F = S(t)e^{r(T-t)}$. This is called the arbitrage-free payoff for a forward contract.

Consider the case where $S(t)e^{r(T-t)} - F > 0$, and we enter into the situation explained earlier. At the end of the contract we will have $S(t)e^{r(T-t)}$ in a bank, a shorted asset, and a long forward. The position you take on the asset then washes out when you give back F, so we have obtained a profit of $S(t)e^{r(T-t)} - F$ without spending any money or taking any risk! This is an arbitrage opportunity. Considering the case where $S(t)e^{r(T-t)} - F < 0$, we may construct a similar arbitrage opportunity as well.

Therefore, according to this argument the price of the forward at time t is exactly

$$F = S(t)e^{r(T-t)}.$$

However, note that the rate $r(T - t)$ is typically unknown and estimated. It is supposed to be the expected value of the future rate, and much of quantitative finance is about finding this value.

In practice, traders profit from finding and exploiting more complex opportunities than the simple one described earlier. They will act quickly on these opportunities, and the prices will adjust.

2.3 Options

Definition 2.3.1 An option is a financial contract that gives the holder the right to trade in the future a specified underlying asset at an agreed and specified price. The details of the underlying asset (S), the calculation of the settlement price (strike K), and the time at which the transaction is to take place (maturity or exercise time) are typically written in the contract. The sum of money received at maturity of the contract is called the Payoff. The payoff is typically positive.

The party that gives the right to trade is called the option underwriter or short the option writer. The writer can either own the particular underlying asset in which case the option is called "covered" or provide means to purchase the asset in which case the option is called "naked."

The party that buys the option becomes the holder of the contract. The holder has the option to exercise the trade – this gives the name of the contract. In return for this option right the holder must pay a certain amount called *the option premium*. Most of quantitative finance is dedicated to calculate this option premium.

There are two fundamental option types. A Call type option gives the holder the right to **buy** the underlying asset. A Put type option gives the holder the right to **sell** the underlying.

In the following subsection, we briefly describe the most commonly encountered option types: the so called vanilla options. In English language (owing we assume to England and the United States), apparently the most commonly encountered flavor of ice cream is vanilla; thus the most basic types of anything are typically called vanilla. This moniker is most certainly not true in other languages.

2.3.1 Vanilla Options

The most basic type of option is a *European option contract*. At a time specified in the option contract called the maturity time or the expiration date and which we will denote with T, the holder of the option may exercise the right to purchase or sell the underlying asset, for a specified amount, called the strike price and which will be denoted with K. The option to buy is a European Call and the option to sell is a European Put.

The writer is of course obligated by the contract to provide the trade counter party should the option holder decide to exercise it.

Consider a Call option. If at time T the price of the underlying asset is $S > K$, then the Call option is worth exercising. The holder would buy the asset from the writer for the specified amount K and then sell it immediately for the market price S. Thus the option's payoff in this case is $S - K$. On the other hand if $S < K$ at maturity, the option is not worth exercising as the holder can purchase the asset cheaper by paying its market price. Thus the value of the Call option at maturity can be written as

$$V(S, T) = \max(S - K, 0). \tag{2.1}$$

Using a similar argument the Put option payoff is

$$V(S, T) = \max(K - S, 0).$$

An American option contract is similar with a European option except that the option to purchase or sell the asset may be exercised at any time prior to the contract maturity date T. For this reason, the underlying asset path is very important for American options, as opposed to European where only the final value at time T is important.

As we shall see next if the underlying asset does not pay dividends for the lifetime of the option, then an American Call has the same value as a European Call option. Dividend is defined as payments made to shareholders out of the profits made by the company. This is very counterintuitive but logically correct. The same is not true for the Put.

2.3.2 Put–Call Parity

This parity refers to the relation between the price of European Put and Call Contracts. The equation derived is valid regardless of the dynamics of the stock which may be following any complicated model.

Suppose first that the asset pays no dividend. We denote the price of the asset at time t with S_t. Construct a portfolio by buying a Call and selling a Put with the same exact maturity and strike price. At time T the payoff of the portfolio is

$$(S_T - K)_+ - (K - S_T)_+ = S_T - K$$

regardless of whether S_T is above or below K. But this payoff may be replicated by buying a unit of stock and buying a bond that pays exactly K at time T.

A zero coupon bond is the price at time t of an instrument that pays \$1 at some specified time in the future T. We denote this number with $P(t, T)$. A zero coupon bond assumes no default (i.e. we always get the dollar at time T), and in the simple case where the interest rate stays constant at r between t and T, its price is simply

$$P(t, T) = e^{-r(T-t)}$$

Coming back to our portfolio story, since the two portfolios have the same exact payoff, using standard no arbitrage arguments, their value must be the same at all times. Thus, at any time t we must have the following relationship between the price of a European Call and Put:

$$C(t) - P(t) = S_t - KP(t, T),$$

where $C(t)$ is the European Call price and $P(t)$ is the European Put both at t. As mentioned, $P(t, T)$ is the price of a zero coupon bond that pays \$1 at time T. This is known as the Put–Call parity.

In the special case when the risk-free interest rate r is assumed constant, we obtain the more familiar form of the Put–Call parity:

$$C(t) - P(t) = S_t - Ke^{-r(T-t)},$$

If the asset pays dividends then we simply add all dividend values paid over the option lifetime into some quantity $D(T)$. The options have nothing to do with this dividend, and having an option will not mean receiving any dividend as only the asset holder does. Thus, at time t, if we held the stock minus the present value of K, we will produce at time T S_T PLUS all the dividends received in the meantime (recall that owning the stock receives dividends). Therefore, for a dividend paying asset if $D(t)$ is the present value of dividends received between time t and maturity T, then the Put–Call parity in this situation is

$$C(t) - P(t) = S_t - KP(t, T) - D(t), \quad \forall t \leq T$$

2.3.2.1 For American Options

For clarity let us denote $C_A(t)$ and $P_A(t)$ American Call and Put prices and with $C_E(t)$ and $P_E(t)$ European option prices. For simplicity, we are also going to use $P(t, T) = e^{-r(T-t)}$. If one prefers just replace the more general term for all the following formulas.

Clearly, since American options are European with added bonus of being able to exercise anytime, we must have

$$C_A(t) \geq C_E(t) \text{ and } P_A(t) \geq P_E(t)$$

Furthermore, we must have

$$C_A(t) \geq (S_t - K)_+ \text{ and } P_A(t) \geq (K - S_t)_+$$

If one of these relations is not true, for example, $C_A(t) < (S_t - K)_+$, then buy an American Call for $C_A(t)$ and immediately exercise to receive $(S_t - K)_+$. Thus, we can obtain a profit $(S_t - K)_+ - C(t) > 0$ and thus immediate arbitrage opportunity. Therefore, the inequalities as stated must be true at all times.

A slightly more complex argument shows that

$$C_A(t) \geq S_t - Ke^{-r(T-t)}.$$

Once again assume that this isn't true. Thus, there must be a t at which we have $C_A(t) < S_t - Ke^{-r(T-t)}$. At that exact time form a portfolio by short selling 1 share of the stock for S_t, buying one Call at $C_A(t)$ and putting $Ke^{-r(T-t)}$ into a bank account at interest rate r. Total balance at t is thus

$$S_t - Ke^{-r(T-t)} - C_A(t)$$

which by assumption is strictly positive so at t we make a profit. Then at time T:

- We need to buy back the stock for S_T.
- Exercise the option and receive $(S_T - K)_+$.
- Get the money back from the bank K.

Balance at time T is

$$(S_T - K)_+ + K - S_T \geq 0$$

which is positive for all situations. Thus again, we made money with no risk and therefore it must be that our assumption is false and that

$$C_A(t) \geq S_t - Ke^{-r(T-t)} \text{ at all times} t.$$

Now, note that for any $t < T$ the term $e^{-r(T-t)} < 1$; therefore

$$C_A(t) \geq S_t - Ke^{-r(T-t)} > S_t - K, \quad \forall t$$

But on the right-hand side is the value received if the call option is exercised at time t. Therefore, IT IS NEVER OPTIMAL TO EXERCISE AN AMERICAN OPTION EARLY in the case when THE UNDERLYING PAYS NO DIVIDENDS.

This is not always true. For example, when the underlying pays dividends or the process modeling the underlying has jumps, it isn't true anymore.

Using similar no arbitrage arguments, we may show that for the case of no dividends, the following Put–Call inequalities hold:

$$S_t - K \leq C_A(t) - P_A(t) \leq S_t - Ke^{-r(T-t)}$$

for American type options.

For dividend paying stock let $D(t)$ denote the total value of dividends paid between t and T and discounted back at time t. The following Put–Call inequality holds in this case:

$$S_t - K - D(t) \leq C_A(t) - P_A(t) \leq S_t - Ke^{-r(T-t)}$$

2.4 Hedging

When investing in the stock market and developing a portfolio, there is always a chance that the stock market or the assets invested in will drop in value. Hedging is the term used when protecting the investment from such unfavorable events.

Definition 2.4.1 Hedging is the process of reducing the exposure of a portfolio to the movement of an underlying asset.

Mathematically, sensitivity is defined as the derivative of the portfolio value with respect to the particular asset value we want to hedge. Reducing the exposure means setting the derivative equal to zero. This is a process known as *delta hedging*.

To better understand the concept of delta hedging, we will use a simple binomial model. In this example, we will start with a Call option on an asset worth $100 at time t. The contract maturity time is $T = t = \Delta t$. If we decide to exercise the option, in the case when the value of the asset increases, we will receive $1, and if the option decreases we will not exercise the option, and we will make no profit.

Suppose the price of the stock tomorrow is unknown, but the probability that the stock rises is 0.6, and the probability that the stock falls is 0.4. The expected value of V is $E[V] = 1(0.6) + 0(0.4) = 0.6$. This is actually not the correct value of V, as we will see in the binomial hedging example presented in Table 2.2.

To hedge we are in some sense "going short" or selling a quantity Δ of our stock to protect against unfavorable outcomes in the market. Our goal is to make our portfolio insensitive to market movement. The question is how much should we sell.

For simplicity, we will ignore the concept of time value of money in this example. To calculate the value of Δ that does not make us dependent on the market in Table 2.2, set the totals in the last two columns equal:

$$1 - \Delta(101) + 100\Delta - V = -99\Delta + 100\Delta - V. \tag{2.2}$$

Table 2.2 Binomial hedging example.

	Value today (time t)	Value tomorrow ($t + dt$)	Value tomorrow ($t + dt$)
Option	V	1	0
$-\Delta$ Stock	$-\Delta 100$	$-\Delta(101)$	$-\Delta(99)$
Cash	$100\Delta - V$	$100\Delta - V$	$100\Delta - V$
Total	0	$1 - \Delta(101) + 100\Delta - V$	$-99\Delta + 100\Delta - V$

Solving in terms of Δ in (2.2), we obtain $\Delta = \frac{1}{2}$. Therefore if we short a half share of the stock, our portfolio will be insensitive to the market movement. Since we have correctly hedged, we are not concerned whether or not the stock rises or falls; the portfolio is now insensitive to market movement.

More generally, assume that at time t the value of our asset is S and in some elapsed time dt, the price of the asset can either rise or fall. We will denote the value risen as uS and the value fallen as vS, where $0 < v < 1 < u$ and $u = 1 + \sigma\sqrt{dt}, v = 1 - \sigma\sqrt{dt}$. At time of expiry, $t + dt$, the option takes on the value of either V^+ or V^- if the asset price rises or falls, respectively. Then $\Delta = \frac{V^+ - V^-}{(u-v)S}$. If we let r to be the interest rate, the price of the option is given today by

$$V = \frac{1}{1 + rdt}(p'V^+ + (1 - p')V^-),$$

where

$$p' = \frac{1}{2} + \frac{r\sqrt{dt}}{2\sigma},$$

is known as the risk-neutral probability. This is the general case for the binomial model that will be studied in Chapter 6. Next, we study the modeling of returns in order to proceed to continuous hedging.

2.5 Modeling Return of Stocks

When investing in the stock market, it is very important to model the money you make from a stock over a certain time period, or the return.

Definition 2.5.1 A return is a percentage growth in the value of an asset, together with accumulated dividends, over some period of time.

Mathematically the return is given as

$$\text{Return} = \frac{\text{Change in the value of the asset} + \text{accumulated cashflows}}{\text{Original value of the asset}}.$$

For example, suppose we invest in two different stocks A and B. Stock A is worth $10, and stock B is worth $100. Now suppose that both stocks rise by $10, so that stock A is worth $20 and stock B is worth $110. Since both stocks have risen $10, they have the same absolute growth. Applying our definition of return, we observe that stock A grew 100% and stock B by 10%. When comparing the two stocks, we observe that we had much more return on stock A. Letting S_i denote the asset value on the ith day, the return from day i to day $i + 1$ is calculated as

$$\frac{S_{i+1} - S_i}{S_i} = R_i.$$

If we suppose that the empirical returns are close enough to normal (for a good approximation, then ignoring dividends), we can model the return by

$$R_i = \frac{S_{i+1} - S_i}{S_i} = \mu + \sigma\phi, \qquad (2.3)$$

where μ represents the mean, σ represents the standard deviation, and ϕ represents the probability density function (pdf) for a standard normal distribution, i.e.

$$\phi(x) = \frac{1}{\sqrt{2\pi}} e^{-\frac{1}{2}x^2}.$$

Suppose we let the time step from the asset price S_i to S_{i+1} be denoted δt. Thus, we can write the mean of the return model as

$$\text{mean} = \mu \delta t,$$

where μ is assumed to be constant. The standard deviation can also be written as

$$\text{standard deviation} = \sigma(\delta t)^{\frac{1}{2}},$$

where σ measures the amount of randomness. Large values of σ suggest a more uncertain return. Therefore, we can now model the return as

$$R_i = \frac{S_{i+1} - S_i}{S_i} = \mu \delta t + \sigma(\delta t)^{\frac{1}{2}} \phi. \qquad (2.4)$$

2.6 Continuous Time Model

From the previous section, we deduced that the return model is given as

$$R_i = \mu \delta t + \sigma(\delta t)^{\frac{1}{2}} \phi. \qquad (2.5)$$

If we take $\delta t \to 0$ in (2.5) and define a process $dX = \phi(dt)^{\frac{1}{2}}$, so that $E[dX] = 0$ and $E[(dX)^2] = dt$, we can model the continuous model of an asset price by using the equation

$$\frac{dS}{S} = \mu dt + \sigma dX. \qquad (2.6)$$

This model for the corresponding return on the asset dS/S decomposes this return into two parts. The first part is a predictable deterministic return, μdt, where μ could be a constant or a function, i.e. $\mu = \mu(S, t)$. μ is a measure of the average rate of growth of the asset price also known as the drift. The second part, $\sigma(S, t)dX$, models the random change in the asset price in response to external effects. Here σ is called the volatility. The volatility is a measure of the fluctuation (risk) in the asset prices, and it corresponds to the diffusion coefficient, dX. X is known as the Brownian motion, which is a random variable drawn from the normal distribution with mean 0 and variance dt.

Equation (2.6) models what is known as a geometric Brownian motions and has been used to model currency, equities, indices, and commodities.

In classical calculus, given $F = X^2$, the derivative is given as $dF = 2XdX$. Finding the derivatives of stochastic variables does not follow classical calculus. The most important rule for stochastic calculus is known as the Itô lemma which will be explained in Section 2.7. In order to present Itô's lemma, we first need to introduce the concept of Itô integral.

Definition 2.6.1 (Itô Integral). Let $f = f(t, X)$ be a differentiable function with $t \in [0, T]$ and X a Brownian motion in $[0, T]$. Consider the partition

$$\pi = \{t_0, t_1, \ldots, t_n\}, 0 = t_0 < t_1 < \ldots < t_n = T,$$

with $\pi = \max\{|t_{i+1} - t_i| : 0 \leq i \leq n-1\}$. Associated to each partition π, we define the sum of stochastic variables

$$S_\pi = \sum_{i=0}^{n-1} f\left(t_i, X\left(t_i\right)\right) \Delta X_i = \sum_{i=0}^{n-1} f\left(t_i, X_i\right) \Delta X_i$$

with

$$\Delta X_i = X\left(t_{i+1}\right) - X\left(t_i\right) = X_{i+1} - X_i.$$

When the limit

$$\lim_{\pi \to 0} \sum_{i=0}^{n-1} f\left(t_i, X_i\right) \Delta X_i$$

exists, we say that f is an Itô-integrable function and

$$\lim_{\pi \to 0} \sum_{i=0}^{n-1} f\left(t_i, X_i\right) \Delta X_i = \int_0^T f(t, X(t)) \, dX.$$

2.6.1 Itô's Lemma

The Itô's lemma [111] is a Taylor series expansion for stochastic calculus and has very important results. It serves as the stochastic calculus counterpart of the chain rule. In this subsection, we will present the heuristic derivation showing the details that we have to take into account for the proof of the Itô's lemma. For a rigorous proof see [159]. In order to derive Itô's lemma, we will first need to introduce various timescales. This derivation can be found in [210]. Suppose we have a stochastic process $X(t)$ and an arbitrary function $F(x)$. Itô lemma is

2.6 Continuous Time Model

concerned with deriving an expression for the dynamics of $F(X(t))$. We define a timescale

$$\frac{\delta t}{n} = h.$$

This timescale is so small that $F(X(t+h))$ can be approximated by a Taylor series:

$$F(X(t+h)) - F(X(t)) = (X(t+h) - X(t))\frac{\partial F}{\partial X}(X(t))$$
$$+ \frac{1}{2}(X(t+h) - X(t))^2 \frac{\partial^2 F}{\partial X^2}(X(t)) + \cdots .$$

It follows that

$$(F(X(t+h)) - F(X(t))) + (F(X(t+2h)) - F(X(t+h))) + \cdots$$
$$+ (F(X(t+n)) - F(X(t+(n-1)h)))$$
$$= \sum_{j=1}^{n}(X(t+jh) - X(t+(j-1)h))\frac{\partial F}{dX}(X(t+(j-1)h))$$
$$+ \frac{1}{2}\frac{\partial^2 F}{\partial X^2}(X(t))\sum_{j=1}^{n}(X(t+jh) - X(t+(j-1)h))^2 + \cdots$$

The above equation uses the approximation

$$\frac{\partial^2 F}{\partial X^2}(X(t+(j-1)h)) = \frac{\partial^2 F}{\partial X^2}(X(t)).$$

The first line then becomes

$$F(X(t+nh)) - F(X(T)) = F(X(t+\delta t)) - F(X(T)). \tag{2.7}$$

The second line is the definition of

$$\int_{t}^{t+\delta t} \frac{\partial F}{\partial X} dX \tag{2.8}$$

and the last line becomes

$$\frac{1}{2}\frac{\partial^2 F}{\partial X^2}(X(t))\delta t, \tag{2.9}$$

in the mean squared sense. Combining (2.7)–(2.9) we obtain

$$F(X(t+\delta t)) - F(X(t)) = \int_{t}^{t+\delta t} \frac{\partial F}{\partial X}(X(\tau))dX(\tau) + \frac{1}{2}\int_{t}^{t+\delta t} \frac{\partial^2 F}{\partial X^2}(X(\tau))d\tau. \tag{2.10}$$

Lemma 2.6.1 (Itô's Lemma). Given a stochastic process $X(t)$ and a twice differentiable function $F(x)$, the dynamics of the process $F(X(t))$ can be written as

$$F(X(t)) = F(X(0)) + \int_{0}^{t} \frac{\partial F}{\partial X}(X(\tau))dX(\tau) + \frac{1}{2}\int_{0}^{t} \frac{\partial^2 F}{\partial X^2}(X(\tau))d\tau.$$

2 Basics of Finance

This is the integral version of Itô's lemma which is conventionally written as follows:

$$dF = \frac{\partial F}{\partial X}dX + \left(\frac{\partial F}{\partial t} + \frac{1}{2}\frac{\partial^2 F}{\partial X^2}\right)dt. \tag{2.11}$$

This is the most famous version of the Itô's lemma.

Remarks 2.6.1

1) In (2.11), the Itô's lemma is the "stochastic version" of the fundamental theorem of calculus.
2) In (2.11), the first term is a stochastic component and the second one is a deterministic component.
   The stochastic component is due to the presence of dX. We recall that X is a Brownian motion.

Next we present examples and applications of the Itô's lemma.

Example 2.6.1 (The return process). We will state basic facts about the return equation. More details about the model can be found in [38, 101]. In a general setting the stock price follows the equation

$$dS = \mu(S, t)dt + \sigma(S, t)dX \tag{2.12}$$

where μ and σ are general functions of class $\mathscr{C}^2$ (i.e. first and second derivatives are continuous) and X is a regular one dimensional Brownian motion.

The Black–Scholes–Merton model is a particular form described earlier where the stock follows a geometric Brownian motion (2.12). Specifically in this case $\mu(S, t) = \mu S$ and $\sigma(S, t) = \sigma S$, i.e. the functions are linear. The process $R_t = \log S_t$ increment is $R_t - R_{t-\Delta t} = \log S_t - \log S_{t-\Delta t} = \log(S_t/S_{t-\Delta t})$, the continuously compounded return over the interval $(t - \Delta t, t)$. We can obtain an equation for the continuously compounded return applying Itô's lemma to the function $g(t, x) = \log x$. This is given as

$$dR_t = \frac{\partial g}{\partial t}(t, S_t)dt + \frac{\partial g}{\partial x}(t, S_t)dS_t + \frac{1}{2}\frac{\partial^2 g}{\partial x^2}(t, S_t)(dS_t)^2$$

$$= \frac{1}{S_t}(\mu S_t dt + \sigma S_t dB_t) + \frac{1}{2}\left(-\frac{1}{S_t^2}\right)(\mu S_t dt + \sigma S_t dB_t)^2$$

$$= \mu dt + \sigma dB_t - \frac{1}{2S_t^2}\left(\mu^2 S_t^2 dt^2 + 2\mu\sigma S_t^2 dt dB_t + \sigma^2 S_t^2 dB_t^2\right)$$

$$= \left(\mu - \frac{\sigma^2}{2}\right)dt + \sigma dB_t,$$

The integral form of the last equation is given as

$$R_t - R_0 = \int_0^t \left(\mu - \frac{\sigma^2}{2}\right) ds + \int_0^t \sigma dB_s = \left(\mu - \frac{\sigma^2}{2}\right) t + \sigma B_t. \quad (2.13)$$

We recall that $R_t = \log S_t$. Therefore substituting R_t into (2.13) and solving for S_t we obtain an explicit formula for the stock dynamics:

$$S_t = S_0 e^{\left(\mu - \frac{\sigma^2}{2}\right) t + \sigma B_t}. \quad (2.14)$$

Example 2.6.2 Many other stock dynamics may be considered. For example:

- Ornstein–Uhlenbeck process:

$$dS_t = \mu S_t dt + \sigma dB_t.$$

- Mean reverting Ornstein–Uhlenbeck process:

$$dS_t = \alpha(m - S_t)dt + \sigma dB_t.$$

Both of these stochastic differential equation's (SDE's) have explicit solutions that may be found using Itô's lemma. See problem 8.

2.7 Problems

1. Prove the following version of the fundamental theorem of arbitrage (the fundamental theorem of arbitrage provides the necessary and sufficient conditions for a market to be arbitrage free and for a market to be complete).

    Let $\mathcal{M}$ be a market with N securities, one of them a risk-free security and M futures at time T. Then the following statements are equivalent.
    (a) There are no strict arbitrage opportunities.
    (b) There exists a probability vector $(\hat{p}_1, \cdots, \hat{p}_M)$ so that the present value of each security is its expected value at time T discounted with the risk-free interest rate factor.

2. Show that the previous theorem does not hold if the market can take infinitely many future states. Consider a market with three securities, where the third one has a bond with return 0, and the other two with present values $S_1 = 0$ and $S_2 = 2$. Suppose that the future states vectors for $(S_1(T), S_2(T))$ are given by the set

$$\Omega = \{(a_1, a_2) \in \mathbb{R}^2 : a_1, a_2 > 0, a_1 + a_2 > 1\} \cup \{(0, 1)\}$$

    Show that:
    i) There are no strict arbitrage opportunities.
    ii) The analogous to problem 1(b) does not hold. What can we conclude about the original version of the fundamental theorem?

3. Consider a market with two securities, one of them risk free with rate of return R, and three future states. Suppose that the risky security has a present value S_1 and can take three future values $S_1^1 > S_1^2 > S_1^3$.
   i) Give an "if and only if" condition for no arbitrage.
   ii) Show an example of an option O that can not be replicated (i.e.: there is no hedging strategy). Why is it possible to find O?
   iii) Show that the present value of O is not uniquely determined. More precisely, verify that the set of possible present values for O is an interval.
4. Explain carefully the difference between the European option and the American option. Give an example for each type of option.
5. Suppose a trader buys a European Put on a share for $4. The stock price is $53 and the strike price is $49.
   (a) Under what circumstances will the trader make a profit?
   (b) Under what circumstances will the option be exercised?
   (c) Draw a diagram showing the variation of the trader's profit with the stock price at the maturity of the option.
6. Suppose that an August Call option to buy a share for $600 costs $60.5 and is held until August.
   (a) Under what circumstances will the holder of the option make a profit?
   (b) Under what circumstances will the option be exercised?
   (c) Draw a diagram showing how the profit from a long position in the option depends on the stock price at maturity of the option.
7. Use Itô's lemma to express dF given that $F(x) = x^{1/2}$, where the stochastic process $\{S_t, t \geq 0\}$ satisfies the stochastic differential equation
$$dS_t = \alpha(\beta - S_t)dt + \sigma\sqrt{S_t}dZ_t$$
where α, β and σ are positive constants and $\{Z_t, t \geq 0\}$ a standard Brownian motion.
8. Find the explicit solutions of the following stochastic differential equations:
   (a) Ornstein–Uhlenbeck process:
$$dX_t = \mu X_t dt + \sigma dB_t.$$
   (b) Mean reverting Ornstein–Uhlenbeck process:
$$dX_t = \alpha(m - X_t)dt + \sigma dB_t.$$
   Hint: Apply Itô's lemma to the function $g(t,x) = xe^{-\mu t}$ (Ornstein–Uhlenbeck process) and $g(t,x) = xe^{\alpha t}$ (mean reverting Ornstein–Uhlenbeck process).

Part II

Quantitative Finance in Practice

3

Some Models Used in Quantitative Finance

3.1 Introduction

The Black-Scholes (B–S) model (see e.g. [24, 57, 101, 104, 152, 211]) is one of the most important models in the evolution of the quantitative finance. The model is used extensively for pricing derivatives in financial markets. As we shall see, using a backward parabolic differential equation or a probabilistic argument, it produces an analytical formula for European type options.

In this chapter, we start by deriving the B-S differential equation for the price of an option. The boundary conditions for different types of options and their analytical solutions will also be discussed. We present the assumptions for the derivation of the B-S equation. We then present a series of other models used in finance. Finally, we introduce the concept of volatility modeling.

3.2 Assumptions for the Black–Scholes–Merton Derivation

1) We assume that the risk-free interest rate r is constant and the same for all maturities. In practice, r is actually stochastic and this will lead to complex formulas for fixed rate instruments. It is also assumed that the lending and borrowing rates the same.
2) The next assumption is that trading and consequently delta hedging is done continuously. Obviously, we cannot hedge continuously, portfolio rebalancing must be done at discrete time intervals.
3) We assume that there are no transaction costs on the underlying asset in the market. This is done for simplicity, much like friction due to air resistance is neglected in simple physics equations. In reality, there are transaction costs, and in fact the hedging strategy often depends on these costs for the underlying asset in the market. We will obviously rehedge more often when transaction costs for the underlying are inexpensive.

Quantitative Finance, First Edition. Maria C. Mariani and Ionut Florescu.
© 2020 John Wiley & Sons, Inc. Published 2020 by John Wiley & Sons, Inc.

4) The asset price follows a log-normal random walk, as described in the next section. There are other models, but most of these models lack formulae.
5) Another assumption is that there are no arbitrage opportunities. This is an interesting assumption, since arbitrage opportunities actually do exist, and traders typically search for these opportunities. However, the market dynamics today is such that if such opportunities exist, they will be exploited and they will disappear very fast.
6) Finally, we assume that short selling is permitted and that assets are infinitely divisible.

3.3 The B-S Model

In Table 3.1, we present an example of a continuous hedging. We will apply Itô's lemma for this example. To begin, we let V be the value of an option at time t. Suppose we want to go short some amount of stock, say ΔS. Then when we look at the total cash amount; we have $-V + \Delta S$, since we bought the option V and sold some Δ quantity of stock. This is all done at time t, so we can sum these values to get $V + (-\Delta S) + (-V + \Delta S) = 0$. An interesting time to consider in the future to make profit is time T. Then the value of the stock at time T is given as $V + dV$, and the amount that we went short is now worth $-\Delta(S + dS)$. The cash we had after borrowing and going short is now worth $-(V + \Delta S)(1 + rdt)$, where r is the risk-free interest rate. Then if we sum both columns we get

$$dV - \Delta dS - (V - \Delta S)rdt = 0. \qquad (3.1)$$

Applying Itô's lemma to (3.1) we obtain the equation

$$\left(\frac{\partial V}{\partial S} - \Delta\right) dS + \left(\frac{\partial V}{\partial t} + \frac{1}{2}\sigma^2 S^2 \frac{\partial^2 V}{\partial S^2} + rS\Delta - rV\right) dt = 0.$$

Finally, we hedge with $\Delta = \frac{\partial V}{\partial S}$ to eliminate fluctuations in the stock and we obtain

$$\frac{\partial V}{\partial t} + \frac{1}{2}\sigma^2 S^2 \frac{\partial^2 V}{\partial S^2} + rS\frac{\partial V}{\partial S} - rV = 0. \qquad (3.2)$$

Table 3.1 Continuous hedging example.

Holding	Value today(time t)	Value tomorrow($t + dt$)
Option	V	$V + dV$
$-\Delta$ Stock	$-\Delta S$	$-\Delta(S + dS)$
Cash	$-V + \Delta S$	$(-V + \Delta S)(1 + rdt)$
Total	0	$dV - \Delta dS - (V - \Delta S)rdt$

This Eq. (3.2) is called the B-S differential equation. Any European type option written on the stock which follows the geometric Brownian motion dynamics will solve this equation.

In practice one hedges against risky market movements. This is done using derivatives of the option price with respect to various parameters in the model. These derivatives are denoted with Greek letters and are thus known as the Greeks:

$$\Theta = \frac{\partial V}{\partial t}, \quad \Gamma = \frac{\partial^2 V}{\partial S^2}, \quad \Delta = \frac{\partial V}{\partial S}, \quad \rho = \frac{\partial V}{\partial r}, \quad \text{Vega} = \frac{\partial V}{\partial \sigma}.$$

Note that "Vega" is not a Greek letter. In the B-S model the volatility is constant; therefore a derivative with respect to a constant is nonsense (it is always zero). Nevertheless, the volatility moves and this was observed long time ago. To cope with this movement produced the Greek which is not an actual Greek.

Substituting these values into the B-S equation, we obtain the following equation:

$$\Theta + \frac{1}{2}\sigma^2 S^2 \Gamma + rS\Delta - rV = 0. \tag{3.3}$$

Remarks 3.3.1

1) $\Theta = \frac{\partial V}{\partial t}$ measures the variation of V with time, keeping the asset price constant.
2) $\Gamma = \frac{\partial^2 V}{\partial S^2}$ measures the delta variation with respect to the asset price.
3) The delta, $\Delta = \frac{\partial V}{\partial S}$ for a whole portfolio is the rate of change of the value of the portfolio with respect to changes in the underlying asset.
4) $\rho = \frac{\partial V}{\partial r}$ measures the variation of the asset price with respect to the risk-free interest rate.
5) Vega $= \frac{\partial V}{\partial \sigma}$ represents the variation of the price with the volatility.

Next consider the problem of pricing a European Call option in the B-S equation. Denote with $C(S, t)$ the value of this option.

$$\frac{\partial C}{\partial t} + \frac{1}{2}\sigma^2 S^2 \frac{\partial^2 C}{\partial S^2} + rS\frac{\partial C}{\partial S} - rC = 0. \tag{3.4}$$

The payoff for the Call at the final time T is

$$C(S, T) = \max(S - K, 0),$$

where K is the option's strike. If the asset price S reaches 0 at any time, the geometric Brownian motion becomes 0, and thus the boundary conditions are

$$C(0, t) = 0, \quad C(S, t) \sim S \quad \text{as } S \to \infty.$$

We will now look at a more rigorous version of this idea, which involves continuously rebalancing our portfolio in order to eliminate risk. This will allow us to value European Puts and Calls.

3 Some Models Used in Quantitative Finance

Theorem 3.3.1 (B-S formula for European Call options). The solution to the B-S equation for European Call options is given by

$$C(S,t) = SN(d_1) - Ee^{-r(T-t)}N(d_2), \quad (3.5)$$

where

$$d_1 = \frac{\log(\frac{S}{E}) + (r + \frac{1}{2}\sigma^2)(T-t)}{\sigma\sqrt{(T-t)}}$$

$$d_2 = \frac{\log(\frac{S}{E}) + (r - \frac{1}{2}\sigma^2)(T-t)}{\sigma\sqrt{(T-t)}},$$

and

$$N(x) = \frac{1}{\sqrt{2\pi}} \int_{-\infty}^{x} e^{-\frac{1}{2}s^2} ds.$$

Proof. Equation (3.4) is a backward equation, and we want to transform it into a forward equation. Using the following change of variables,

$$S = Ee^x, \quad t = T - \frac{\tau}{\frac{1}{2}\sigma^2}, \quad C = Ec(x, \tau),$$

Equation (3.4) becomes

$$\frac{\partial v}{\partial \tau} = \frac{\partial^2 v}{\partial x^2} + (k-1)\frac{\partial v}{\partial x} - kv, \quad (3.6)$$

where we set $k = \frac{r}{\frac{1}{2}\sigma^2}$. Hence the initial condition is now

$$v(x, 0) = \max(e^x - 1, 0).$$

To transform a backward equation into a forward equation, we need to use the diffusion equation. This can be done by making a change of variable

$$v = e^{\alpha x + \beta \tau} u(x, \tau)$$

for some α and β that are to be determined. After the change of variable we differentiate the equation to obtain

$$\beta u + \frac{\partial u}{\partial \tau} = \alpha^2 u + 2\alpha \frac{\partial u}{\partial x} + \frac{\partial^2 u}{\partial x^2} + (k-1)(\alpha u + \frac{\partial u}{\partial x}) - ku.$$

In order to eliminate the u terms, we choose

$$\beta = \alpha^2 + (k-1)\alpha - k$$

and to eliminate the $\frac{\partial u}{\partial x}$ terms, we note that

$$0 = 2\alpha + (k-1).$$

3.3 The B-S Model

Then solving for α and β, we get

$$\alpha = -\frac{1}{2}(k-1), \quad \beta = -\frac{1}{4}(k+1)^2.$$

Plugging the values of α and β into our change of variable, we obtain,

$$v = e^{\frac{-1}{2}(k-1)x - \frac{1}{4}(k+1)^2\tau} u(x,\tau),$$

where

$$\frac{\partial u}{\partial \tau} = \frac{\partial^2 u}{\partial x^2} \quad \text{for } -\infty < x < \infty, \tau > 0,$$

with

$$u(x,0) = u_0(x) = \max(e^{\frac{1}{2}(k+1)x} - e^{\frac{1}{2}(k-1)x}, 0).$$

This is now the diffusion equation, and the solution to this equation is given by

$$u(x,\tau) = \frac{1}{2\sqrt{\pi\tau}} \int_{-\infty}^{\infty} u_0(s) e^{-\frac{(x-s)^2}{4\tau}} ds.$$

We can then evaluate this integral, making a change of variable $x' = (s-x)\sqrt{2\tau}$:

$$u(x,\tau) = \frac{1}{\sqrt{2\pi}} \int_{-\infty}^{\infty} u_0(x'\sqrt{2\tau} + x) e^{-\frac{1}{2}x'^2} dx'$$

$$= \frac{1}{\sqrt{2\pi}} \int_{-x/\sqrt{2\tau}}^{\infty} e^{\frac{1}{2}(k+1)(x+x'\sqrt{2\tau})} e^{-\frac{1}{2}x'^2} dx'$$

$$- \frac{1}{\sqrt{2\pi}} \int_{-x/\sqrt{2\tau}}^{\infty} e^{\frac{1}{2}(k-1)(x+x'\sqrt{2\tau})} e^{-\frac{1}{2}x'^2} dx' = I_1 - I_2.$$

Next, by completing the square in the exponent, we can solve for I_1, i.e.

$$I_1 = \frac{1}{\sqrt{2\pi}} \int_{-x/\sqrt{2\tau}}^{\infty} e^{\frac{1}{2}(k+1)(x+x'\sqrt{2\tau})} e^{-\frac{1}{2}x'^2} dx'$$

$$= \frac{e^{\frac{1}{2}(k+1)x}}{\sqrt{2\pi}} \int_{-x/\sqrt{2\tau}}^{\infty} e^{\frac{1}{4}(k+1)^2\tau} e^{-\frac{1}{2}(x' - \frac{1}{2}(k+1)\sqrt{2\tau})^2} dx'$$

$$= \frac{e^{\frac{1}{2}(k+1)x + \frac{1}{4}(k+1)^2\tau}}{\sqrt{2\pi}} \int_{\frac{-x}{\sqrt{2\tau}} - \frac{1}{2}(k+1)\sqrt{2\tau}}^{\infty} e^{-\frac{1}{2}\rho^2} d\rho$$

$$I_1 = e^{\frac{1}{2}(k+1)x + \frac{1}{4}(k+1)^2\tau} N(d_1). \tag{3.7}$$

Wherein the above equation,

$$d_1 = \frac{x}{\sqrt{2\tau}} + \frac{1}{2}(k+1)\sqrt{2\tau},$$

and

$$N(d_1) = \frac{1}{\sqrt{2\pi}} \int_{-\infty}^{d_1} e^{-\frac{1}{2}s^2} ds.$$

The last equation is the distribution function (CDF) for a normal distribution. Now that we have calculated I_1, we can similarly calculate I_2 by replacing the $(k+1)$ in (3.7) by $(k-1)$. Thus we obtain

$$v(x,\tau) = e^{-\frac{1}{2}(k-1)x - \frac{1}{4}(k+1)^2\tau} u(x,\tau).$$

Then setting $x = \log(\frac{S}{E})$, $\tau = \frac{1}{2}\sigma^2(T-t)$, and $C = Ev(x,\tau)$, we get back to

$$C(S,t) = SN(d_1) - Ee^{-r(T-t)}N(d_2),$$

and with the change of variables, we obtain

$$d_1 = \frac{\log(\frac{S}{E}) + (r + \frac{1}{2}\sigma^2)(T-t)}{\sigma\sqrt{T-t}}$$

$$d_2 = \frac{\log(\frac{S}{E}) + (r - \frac{1}{2}\sigma^2)(T-t)}{\sigma\sqrt{T-t}}.$$

Hence the proof. □

We state a similar theorem for European Put options.

Theorem 3.3.2 (B-S formula for European Put options). *The solution to the B-S equation for European Put options is given by*

$$P(S,t) = e^{-r(T-t)}KN(-d_2) - S_0N(-d_1),$$

where d_1, d_2 and $N(x)$ are given in Theorem 3.3.1.

Proof. The B-S formula can also be obtained by solving Eq. (3.4) with the specific boundary conditions for the Put. However, we shall use [88] results to get to the formula using a probabilistic approach. Suppose we wish to obtain the formula for the Put option denoted $p(t)$. We know that at time T (maturity) the value is $p(T) = (K - S_T)_+$, where K is the strike price of the option. Using Harrison and Pliska [88], the discounted price is a martingale, therefore,

$$e^{-rt}p(t) = \mathbf{E}\left[e^{-rT}P(T)|\mathcal{F}_t\right].$$

Furthermore, under this martingale measure the equation of the stock has drift r and thus the formula (2.14) becomes

$$S_t = S_0 e^{\left(r - \frac{\sigma^2}{2}\right)t + \sigma B_t} \tag{3.8}$$

For simplicity of notation, we set $t = 0$. We use the fact that the conditional expectation becomes the regular expectation if the variable is independent of the conditional filtration. This is our case since S_T only depends on the

3.3 The B-S Model

increment $B_T - B_0$ and that is independent of $\mathcal{F}_0$. Taking this observation into account, the price of the Put is

$$p(0) = \mathbf{E}\left[e^{-rT}(K - S_T)_+ | \mathcal{F}_0\right]$$

$$= \mathbf{E}\left[e^{-rT}(K - S_0 e^{\left(r-\frac{\sigma^2}{2}\right)T + \sigma B_T})_+ | \mathcal{F}_0\right]$$

$$= \mathbf{E}\left[e^{-rT}\left(K - S_0 e^{\left(r-\frac{\sigma^2}{2}\right)T + \sigma \sqrt{T}\frac{B_T}{\sqrt{T}}}\right)_+ \Big| \mathcal{F}_0\right].$$

Now we recognize that $B_T/\sqrt{T}$ is distributed as a normal distribution with mean zero and variance 1 (i.e. $N(0, 1)$) and thus we obtain

$$p(0) = e^{-rT} \int_{-\infty}^{\infty} \left(K - S_0 e^{\left(r-\frac{\sigma^2}{2}\right)T + \sigma\sqrt{T}z}\right)_+ \frac{1}{\sqrt{2\pi}} e^{-\frac{z^2}{2}} dz.$$

The question that arises is when exactly is the expression inside positive? We note that

$$K - S_0 e^{\left(r-\frac{\sigma^2}{2}\right)T + \sigma\sqrt{T}z} \geq 0 \quad \Leftrightarrow \quad e^{\left(r-\frac{\sigma^2}{2}\right)T + \sigma\sqrt{T}z} \leq \frac{K}{S_0}$$

$$\Leftrightarrow \left(r - \frac{\sigma^2}{2}\right)T + \sigma\sqrt{T}z \leq \log\frac{K}{S_0}$$

$$\Leftrightarrow z \leq \frac{\log\frac{K}{S_0} - \left(r - \frac{\sigma^2}{2}\right)T}{\sigma\sqrt{T}} = -\frac{\log\frac{S_0}{K} + \left(r - \frac{\sigma^2}{2}\right)T}{\sigma\sqrt{T}} = -d_2$$

using the notation given in Eq. (3.5). We also use the same notation $N(x) = \int_{-\infty}^{x} \frac{1}{\sqrt{2\pi}} e^{-z^2/2} dz$ for the CDF of the normal distribution with mean 0 and variance 1. Thus we obtain

$$p(0) = e^{-rT} \int_{-\infty}^{-d_2} \left(K - S_0 e^{\left(r-\frac{\sigma^2}{2}\right)T + \sigma\sqrt{T}z}\right) \frac{1}{\sqrt{2\pi}} e^{-\frac{z^2}{2}} dz$$

$$= e^{-rT} K \int_{-\infty}^{-d_2} \frac{1}{\sqrt{2\pi}} e^{-\frac{z^2}{2}} dz - e^{-rT} S_0 e^{\left(r-\frac{\sigma^2}{2}\right)T} \int_{-\infty}^{-d_2} e^{\sigma\sqrt{T}z} \frac{1}{\sqrt{2\pi}} e^{-\frac{z^2}{2}} dz$$

$$= e^{-rT} KN(-d_2) - S_0 e^{-\frac{\sigma^2 T}{2}} \int_{-\infty}^{-d_2} \frac{1}{\sqrt{2\pi}} e^{-\frac{z^2}{2} + \sigma\sqrt{T}z - \frac{\sigma^2 T}{2} + \frac{\sigma^2 T}{2}} dz$$

$$= e^{-rT} KN(-d_2) - S_0 e^{-\frac{\sigma^2 T}{2} + \frac{\sigma^2 T}{2}} \int_{-\infty}^{-d_2} \frac{1}{\sqrt{2\pi}} e^{-\frac{(z-\sigma\sqrt{T})^2}{2}} dz$$

$$= e^{-rT} KN(-d_2) - S_0 \int_{-\infty}^{-d_2 - \sigma\sqrt{T}} \frac{1}{\sqrt{2\pi}} e^{-\frac{u^2}{2}} du,$$

where we have completed the square in the second integral and made the change of variables $z - \sigma\sqrt{T} = u$. Now recognizing that $d_2 + \sigma\sqrt{T} = d_1$ we finally obtain

$$p(0) = e^{-rT}KN(-d_2) - S_0N(-d_1),$$

which is exactly the formula for the Put option in the statement of the theorem.

□

In the following remark, we discuss another way to obtain the B-S equations. It is derived by constructing a portfolio that replicates an option.

Remark 3.3.2 We say that a portfolio "replicates an option" when it has the value of the option. Let π be defined as

$$\begin{cases} \pi = \pi_1 S + \pi_2 B \\ \pi_u = V_u \\ \pi_d = V_d \end{cases}$$

so that

$$\begin{array}{c} S_u \\ (V_u) \\ \nearrow^p \\ S \\ (V) \\ \searrow_{1-p} \quad S_d \\ (V_d) \end{array}$$

Then from the system we have

$$\pi_1 Su + \pi_2 e^{rT} B = V_u$$
$$\pi_1 Sd + \pi_2 e^{rT} B = V_d$$

Next solving for π_1 and π_2 we obtain

$$\pi_1 = \frac{V_u - V_d}{Su - Sd} = \Delta$$

and

$$\pi_2 B = e^{-rT}\left(V_u - \frac{V_u - V_d}{u - d}u\right).$$

Thus the value for this portfolio is given as

$$\pi = \frac{V_u - V_d}{u - d} + e^{-rT}\left(V_u - \frac{V_u - V_d}{u - d}u\right)$$
$$= e^{-rT}\left(pV_u + (1-p)V_d\right)$$

where

$$p = \frac{e^{-rT} - d}{u - d}.$$

3.3 The B-S Model

For example considering the continuous case

$$dS = \mu S dt + \sigma S dZ,$$

we want to construct a portfolio that finances itself, i.e.

$$\pi = \pi_1 S + \pi_2 B$$

and

$$d\pi = \pi_1 dS + \pi_2 dB.$$

Even if this is a risky portfolio (i.e. there are chances that combination of assets within individual group of investments fail to meet financial objectives) and we cannot proceed as before, using Itô's 's lemma, we get the same value. We want to find the value of $V(S,t)$, i.e.

$$dV = \sigma S \frac{\partial V}{\partial S} dZ + \left\{ \frac{\partial V}{\partial t} + \mu S \frac{\partial V}{\partial S} + \frac{1}{2}\sigma^2 S^2 \frac{\partial^2 V}{\partial S^2} \right\} dt.$$

If we want the portfolio to replicate the option, we need

$$\begin{cases} \pi = V \\ d\pi = dV \end{cases}$$

Then, we have that

$$d\pi = \pi_1 dS + \pi_2 dB$$
$$= \pi_1 (\mu S dt + \sigma S dZ) + \pi_2 dB = dV.$$

Since the stochastic components are equal, so we conclude that

$$\pi_1 \sigma S = \sigma S \frac{\partial V}{\partial S}.$$

Thus:

$$\pi_1 = \frac{\partial V}{\partial S}.$$

On the other hand,

$$\pi_2 dB = \left(\frac{\partial V}{\partial t} + \frac{1}{2}\sigma^2 S^2 \frac{\partial^2 V}{\partial S^2} \right) dt.$$

We recall that if B is a risk-free bond, i.e. $dB = rBdt$, then,

$$\pi_2 rB = \frac{\partial V}{\partial t} + \frac{1}{2}\sigma^2 S^2 \frac{\partial^2 V}{\partial S^2}. \tag{3.9}$$

Adding $r\pi_1 S$ to (3.9) and taking account of the condition that $\pi = V$, we obtain

$$rV = r\pi = r\pi_1 S + r\pi_2 B = \frac{\partial V}{\partial t} + \frac{1}{2}\sigma^2 S^2 \frac{\partial^2 V}{\partial S^2} + r\pi_1 S$$
$$= \frac{\partial V}{\partial t} + \frac{1}{2}\sigma^2 S^2 \frac{\partial^2 V}{\partial S^2} + r\frac{\partial V}{\partial S} S$$

which is the the B-S equation.

3.4 Some Remarks on the B-S Model

3.4.1 Remark 1

Many assets, such as equities, pay dividends. As mentioned in Section 2.3, dividends are payments to shareholders out of the profits made by the respective company. The future dividend stream of a company is reflected in today's share price. Suppose that we have a European Call option on an asset that does not pay dividends. Then the B-S model to evaluate is

$$\frac{\partial V}{\partial t} + \frac{1}{2}\sigma^2 S^2 \frac{\partial^2 C}{\partial V^2} + rS\frac{\partial V}{\partial S} = rV.$$

In this case,

$$V(S,T) = (S-K)^+ = \max\{S-K, 0\}$$

and we can assume that V is a function of S, t, K and T:

$$V = V(S, t, K, T).$$

In general we assume that the market "knows" the price of the option – more precisely, the "fair price" of an option. The market keeps the prices at equilibrium. Assuming that the present values given in the market are "true values," we can obtain a function $V^*(K,T)$. The "true market values" are the highest estimated price that a buyer would pay and a seller would accept for an item in an open and competitive market. If the B-S model is correct, we obtain

$$V(S, t, K, T) = V^*(K, T).$$

We can assume that we know the risk-free interest rate r (rate corresponding to a risk-free bond). If we know r and σ we can solve the previous equation. In order to estimate σ so that the B-S model holds, we need to compute the implied volatility. That is, using market option values, estimate the value of σ implied by these market prices. The estimated parameter $\sigma = \sigma(S, t)$ will be a function of time.

3.4.2 Remark 2

We recall from Chapter 2 that an American option is the type of option that may be exercised at any time prior to expiry. Suppose we construct the portfolio:

$$\pi = \begin{cases} 1 & \text{option in long} \\ 1 & \text{unit of share in short} \\ ke^{-r(T-t)} & \text{bonds in long} \end{cases}$$

If we exercise at time t, for a Call option $S > K$ (otherwise we do not exercise) we have that

$$S - K + Ke^{-r(T-t)} - S < 0,$$

3.4 Some Remarks on the B-S Model

that is, the portfolio exercised is negative. If we keep the option without exercising it until $t = T$, we have that

$$\max\{S_T - K, 0\} + K - S = \begin{cases} 0 & \text{if } S \geq K \\ (K - S) > 0 & \text{if } S < K \end{cases}$$

So it is more convenient to exercise at the final time T. We remark that this result is valid for a Call option on an asset without dividends (see [210]). However, an American Put option has a higher price than a European Put option. In this case the payoff is

$$(K - S)^+ = \max\{K - S; 0\}.$$

In order to proceed, we assume a claim. Without this claim we obtain a contradiction.

Claim: We need the inequality

$$V(S, t) \geq \max\{K - S; 0\}$$

This is related to the free boundary obstacle problem, and we use the non-arbitrage condition if the condition does not hold. Suppose that there exists t so that $0 \leq V(S, t) < \max\{K - S; 0\}$. In this case $\max\{K - S; 0\} = K - S$. Then we consider the following strategy where we buy a share and an option that we will exercise immediately (i.e. we buy the option to exercise it). In this case we are paying $S + V$ and we receive K. Then the net profit is $K - (S + V) > 0$ (assumption). Then there is arbitrage and the proof is done! The obstacle problem [93] assumes that:

1) V has a lower constraint,

$$V(S, t) \geq \max\{K - S; 0\}$$

2) V and $\frac{dV}{dS}$ are continuous (i.e. the contact between V and the obstacle is of first order). In this case the B-S model (PDE) is replaced by a differential.

Heuristically, we can assume that when $V > \max\{K - S; 0\}$ the B-S model holds, and we have a European Call option. But when we have that $V(S, t) = \max\{K - S; 0\}$ in an interval $[A, B]$, it is better to exercise. In this case, the B-S model does not hold because $V(S, t) = \max\{K - S; 0\}$. At what time will this change take effect? There exists a free boundary condition $S_f(t)$ such that if $S < S_f(t)$ we have to exercise. We can conclude that if P is the price of an American Put option, $P \in C^1$ (i.e. P is continuous and the first derivative is also continuous) and also

$$P(S, t) \geq \max\{K - S; 0\}$$

then if $S < S_f(t)$ it is better to exercise. If the boundary satisfies that

$$0 \leq P(S, t) = \max\{K - S; 0\}$$

and $S < S_f$, then the following equation must hold:

$$P(S_f(t), t) = K - S_f(t)$$

Intuitively, taking the derivative with respect to S, we obtain:

$$\frac{\partial P}{\partial S}(S_f(t), t) = -1$$

This is a classical first-order contact-free-boundary conditions for the obstacle problem. The equality holds only when V "does not have contact" with the obstacle.

In the next section, we will discuss a mathematical model used to describe the evolution of the volatility of an underlying asset.

3.5 Heston Model

One of the main assumptions in the B-S Merton model is that the volatility is constant. However, judging from observed market prices, the volatility is not constant but rather changing in time. The Heston model is a stochastic volatility model that assumes that the volatility of the asset is non-constant and follows a random process.

In this model the underlying stock price S_t follows:

$$dS_t = \mu S_t dt + \sqrt{V_t} S_t dW_t^1 \tag{3.10}$$

$$dV_t = \kappa(\theta - V_t)dt + \sigma\sqrt{V_t} dW_t^2$$

where the two Brownian motions are correlated with coefficients ρ, i.e.

$$\mathbb{E}^P[dW_t^1, dW_t^2] = \rho dt,$$

with parameters defined as follows: μ is the stock drift, $\kappa > 0$ is the mean reversion speed for variance, $\theta > 0$ is the mean reversion level for variance, $\sigma > 0$ is the volatility of the variance, V_0 is the initial (time zero) variance level, and $\rho \in [-1, 1]$ is the correlation between the two Brownian motions. The initial variance level, V_0 is often treated as a constant.

Girsanov's theorem ([158]) states that there exists a measure Q under which $\tilde{W}_t^1$ is a Brownian motion. $\frac{\mu-r}{\sqrt{V_t}}$ is known as the market price of risk. If we define

$$\tilde{W}_t^1 = W_t^1 + \frac{\mu - r}{\sqrt{V_t}} t,$$

Girsanov's theorem ([159]) states that there exists a measure Q under which $\tilde{W}_t^1$ is a Brownian motion. $\frac{\mu-r}{\sqrt{V_t}}$ is known as the market price of risk. then Eq. (3.10) becomes

$$dS_t = rS_t dt + \sqrt{V_t} S_t \tilde{W}_t^1.$$

3.5 Heston Model

We can also write an equation for $\log(S_t)$ and using Itô's lemma we obtain:

$$d \ln S_t = (\mu - \frac{V_t}{2})dt + dW_t^1$$

or under the risk-neutral measure Q or simply measure Q (a probability measure such that each share price is exactly equal to the discounted expectation of the share price under this measure), we obtain

$$d \ln S_t = (r - \frac{V_t}{2})dt + dW_t^1.$$

For the variance, we take

$$\tilde{W}_t^2 = W_t^2 + \frac{\lambda(S_t, V_t, t)}{\sigma\sqrt{V_t}}t$$

and we obtain

$$dV_t = \left[\kappa(\theta - V_t) - \lambda(S_t, V_t, t)\right]dt + \sigma\sqrt{V_t}\tilde{W}_t^2,$$

which is the dynamics under measure Q, and also $\lambda(S_t, V_t, t)$ is the volatility risk premium (a measure of the extra amount investors demand in order to hold a volatile security, above what can be computed based on expected returns). If we take $\lambda(S_t, V_t, t) = \lambda V_t$, we obtain $dV_t = (\kappa\theta - (\kappa + \lambda)V_t)dt + \sigma\sqrt{V_t}\tilde{W}_t^2$ (see [95] for more details). If we take $\kappa^* = \kappa + \lambda$ and $\theta^* = \frac{\kappa\theta}{\kappa+\theta}$, we can rewrite the Heston model under risk-neutral measure Q as

$$dS_t = rS_t dt + \sqrt{V_t}S_t d\tilde{W}_t^1$$

$$dV_t = \kappa^*(\theta^* - V_t)dt + \sigma\sqrt{V_t}d\tilde{W}_t^2,$$

with

$$\mathbb{E}^Q[d\tilde{W}_t^1, d\tilde{W}_t^2] = \rho dt.$$

When $\lambda = 0$, $\kappa^* = \kappa$ and $\theta^* = \theta$. Henceforth, we will assume $\lambda = 0$, but this is not necessarily needed. Thus, we will always assume that the Heston model is specified under the risk-neutral problem measure Q. We shall not use the asterisk or the tilde notation from now on.

3.5.1 Heston PDE Derivation

This derivation of the Heston differential equation is similar to the B-S PDE derivation. Before we derive the Heston PDE, we need another derivation to hedge volatility. So we form a portfolio made of $V(S, v, t)$– option, Δ– units of stock, μ– units of another option, and $U(s, v, t)$– used for the volatility hedge. Therefore the portfolio value is $\Pi = V + \Delta S + \mu U$. Assuming the self-financing condition we have $d\Pi = dV + \Delta dS + \mu dU$. The idea of the derivation is to apply Itô's lemma for V and U and then find Δ and μ that make the portfolio riskless.

Using the two-dimensional Itô's lemma,

$$dV = \frac{\partial V}{\partial t}dt + \frac{\partial V}{\partial S}dS + \frac{\partial V}{\partial v}dv + \frac{1}{2}vS^2\frac{\partial^2 V}{\partial S^2}dt + \frac{1}{2}v\sigma^2\frac{\partial^2 V}{\partial v^2}dt + \rho\sigma vS\frac{\partial^2 V}{\partial S\partial v}dt \quad (3.11)$$

because $(dS_t)^2 = v_t S_t^2 (dW_t^1)^2 = v_t S_t dt$, $(dv)^2 = \sigma_t^2 v_t dt$, $dW_t^1 dt = dW_t^2 dt = 0$. Substituting (3.11) in $d\Pi$, we get

$$d\Pi = \left[\frac{\partial V}{\partial t} + \frac{1}{2}vS^2\frac{\partial^2 V}{\partial S^2} + \rho\sigma vS\frac{\partial^2 V}{\partial S\partial v} + \frac{1}{2}\sigma^2 v\frac{\partial^2 V}{\partial S^2}\right]dt$$

$$+ \mu\left[\frac{\partial U}{\partial t} + \frac{1}{2}vS^2\frac{\partial^2 U}{\partial S^2} + \rho\sigma vS\frac{\partial^2 U}{\partial S\partial v} + \frac{1}{2}\sigma^2 v\frac{\partial^2 U}{\partial S^2}\right]dt$$

$$+ \left[\frac{\partial V}{\partial S} + \mu\frac{\partial U}{\partial S} + \Delta\right]dS + \left[\frac{\partial V}{\partial v} + \mu\frac{\partial U}{\partial v}\right]dv$$

So in order to make the portfolio risk free, we need to set $\frac{\partial V}{\partial S} + \mu\frac{\partial U}{\partial S} + \Delta$ and $\frac{\partial V}{\partial v} + \mu\frac{\partial U}{\partial v}$ to zero. Therefore the hedge parameters are $\mu = -(\frac{\partial V}{\partial v})/(\frac{\partial U}{\partial v})$ and $\Delta = -\mu\frac{\partial U}{\partial S} - \frac{\partial V}{\partial S}$. If we substitute these values back into the portfolio dynamics, we get

$$d\Pi = \left[\frac{\partial V}{\partial t} + \frac{1}{2}vS^2\frac{\partial^2 V}{\partial S^2} + \rho\sigma vS\frac{\partial^2 U}{\partial S\partial v} + \frac{1}{2}\sigma^2 v\frac{\partial^2 V}{\partial v^2}\right]dt$$

$$+ \left[\frac{\partial U}{\partial t} + \frac{1}{2}vS^2\frac{\partial^2 U}{\partial S^2} + \rho\sigma vS\frac{\partial^2 U}{\partial S\partial v} + \frac{1}{2}\sigma^2 v\frac{\partial^2 U}{\partial v^2}\right]dt$$

since the portfolio earns the risk-free rate:

$$d\Pi = r\Pi dt = r(V + \Delta S + \mu U)dt.$$

Now we set the two terms equal and substitute μ and Δ and rearrange the terms as follows:

$$\frac{\left[\frac{\partial V}{\partial t} + \frac{1}{2}vS^2\frac{\partial^2 V}{\partial S^2} + \rho\sigma vS\frac{\partial^2 V}{\partial S\partial v} + \frac{1}{2}\sigma^2 v\frac{\partial^2 V}{\partial v^2}\right] - rV + rS\frac{\partial V}{\partial S}}{\frac{\partial V}{\partial v}}$$

$$= \frac{\left[\frac{\partial U}{\partial t} + \frac{1}{2}vS^2\frac{\partial^2 U}{\partial S^2} + \rho\sigma vS\frac{\partial^2 U}{\partial S\partial v} + \frac{1}{2}\sigma^2 v\frac{\partial^2 U}{\partial v^2}\right] - rU + rS\frac{\partial U}{\partial S}}{\frac{\partial U}{\partial v}}.$$

However since the left side of the above equation is a function of V while the right side is a function of U only, it implies we can write both sides as function of S, v, and t only, i.e. $f(S, v, t)$. Now let $f(S, v, t) = -\kappa(\theta - v) + \lambda(S, v, t)$ and take $\lambda(S, v, t) = \lambda v$, then the Heston PDE is given as

$$\frac{\partial U}{\partial t} + \frac{1}{2}vS^2\frac{\partial^2 U}{\partial S^2} + \rho\sigma vS\frac{\partial^2 U}{\partial S\partial v} + \frac{1}{2}\sigma^2 v\frac{\partial^2 U}{\partial v^2}$$

$$- rU + rS\frac{\partial U}{\partial S} + (\kappa(\theta - v) - \lambda(S, v, t))\frac{\partial U}{\partial v} = 0 \quad (3.12)$$

with boundary conditions:

$U(S, v, T) = \max(S - k, 0)$ for the Call option

$U(0, v, t) = 0$ (as stock price gets to 0, Call is worthless)

$\frac{\partial U}{\partial S}(\infty, v, t) = 1$ (when stock gets to ∞, change in stock implies change in option value)

$U(S, \infty, t) = S$ (when volatility is ∞, option price is S)

Taking $x = \ln S$ in (3.12) and simplifying, we obtain

$$\frac{\partial U}{\partial t} + \frac{1}{2}v\frac{\partial^2 U}{\partial x^2} + \left(v - \frac{1}{2}\right)\frac{\partial U}{\partial x} + \rho\sigma v\frac{\partial^2 U}{\partial v \partial x} + \frac{1}{2}\sigma^2 v\frac{\partial^2 U}{\partial v^2}$$
$$- rU + [\kappa(\theta - v) - \lambda v]\frac{\partial U}{\partial v} = 0. \qquad (3.13)$$

Example 3.5.1 Consider the volatility process in the Heston model

$$U_t = \sqrt{V_t}.$$

Use Itô to calculate the dynamic for the volatility process U_t. Under what conditions on the parameters κ, θ, σ the process becomes an Ornstein–Uhlenbeck (O-U) process (i.e. $dX_t = X_t dt + \sigma dW_t$)?

The solution of the example is as follows:
The volatility process $\sqrt{V_t}$ follows an O-U process. Let $U_t = \sqrt{V_t}$. Applying Itô's lemma to $f(x, t) = \sqrt{x}$, we obtain

$$dU_t = d\sqrt{V_t} = \frac{1}{2\sqrt{V_t}} dV_t - \frac{1}{2} \cdot \frac{1}{4} \cdot \frac{1}{\sqrt{V_t^3}} d\langle V, V \rangle_t$$

$$= \frac{1}{2\sqrt{V_t}} \left(\kappa(\theta - V_t)dt + \sigma\sqrt{V_t} dW_t^2 \right) - \frac{1}{8} \cdot \frac{1}{\sqrt{V_t^3}} \sigma^2 V_t dt$$

$$= \frac{\kappa\theta}{2\sqrt{V_t}} dt - \frac{\kappa\sqrt{V_t}}{2} dt - \frac{\sigma^2}{8\sqrt{V_t}} dt + \frac{\sigma}{2} dW_t^2$$

If we take $\frac{\kappa\theta}{2} = \frac{\sigma^2}{8}$, we obtain the O-U process.

3.6 The Cox–Ingersoll–Ross (CIR) Model

The CIR model is typically used to describe the evolution of interest rates. It describes the interest rate movements as driven by only one source of market risk. The model is also relevant providing the dynamics of the variance process in the Heston model. The model follows the SDE:

$$dV_t = \kappa(\theta - V_t)dt + \sigma\sqrt{V_t} dW_t \qquad (3.14)$$

If (3.14) is satisfied then $V_t|V_s = 0$ for $s < t$ has a scaled non-central χ^2 distribution ([29]). Specifically,

$$2C_t V_t | V_\Delta \sim \chi_d^2 = \frac{4\kappa\theta}{\sigma^2} \left(2C_t V_\Delta \exp(-\kappa(t - \Delta)) \right)$$

where $C_t = \frac{2\kappa}{\sigma^2(1 - e^{\kappa(t-\Delta)})}$ and $\frac{4\kappa\theta}{\sigma^2}$ denotes the degrees of freedom.

We can also calculate the conditional mean and variance as follows:

$$\mathbb{E}[V_t | V_\Delta] = \theta + (V_\Delta - \theta) e^{-\kappa(t-\Delta)}$$

and

$$\text{Var}[V_t | V_\Delta] = V_\Delta \frac{\sigma^2 e^{\kappa(t-\Delta)}}{\kappa} \left(1 - e^{-\kappa(t-\Delta)}\right) + \frac{\theta \sigma^2}{2\kappa} \left(1 - e^{-\kappa(t-\Delta)}\right)^2.$$

Therefore as $k \to \infty$, the conditional expectation goes to θ and the variance goes to 0. Also as $k \to 0$ the conditional expectation goes to V_Δ and the variance goes to $\sigma^2 V_t(t - \Delta)$. If $2\kappa\theta > \sigma^2$ (Feller condition) then the variance process will not go negative.

3.7 Stochastic α, β, ρ (SABR) Model

The SABR model is a stochastic volatility model, which is used to capture the volatility smile in derivatives markets and also used in the forward rate modeling of fixed income instruments. SABR is a dynamic model in which both S and α are represented by stochastic state variables whose time evolution is given by the following system of stochastic differential equations:

$$dS_t = \alpha_t S_t^\beta dW_t^1$$
$$d\alpha_t = v\alpha_t dW_t^2$$

with $\mathbb{E}[dW_t^1, dW_t^2] = \rho dt$, where S_t is a typical forward rate, α_0- is the initial variance, v is the volatility of variance, β is the exponent for the rate, and ρ is the correlation coefficient. The constant parameters β, v satisfy the conditions $0 \leq \beta \leq 1$, $v \geq 0$.

Remarks 3.7.1 If $\beta = 0$ the SABR model reduces to the stochastic normal model; if $\beta = 1$, we have the stochastic log normal model; and if $\beta = 1/2$, we obtain the stochastic CIR model.

3.7.1 SABR Implied Volatility

In order to obtain the option price, one uses the usual B-S model. For example a European Call with forward rate S_0, strike price K, maturity T and implied volatility σ_B is

$$C_B(S_0, K, \sigma_B, T) = e^{-rT(S_0 N(d_1) - KN(d_2))} \tag{3.15}$$

where

$$d_{1,2} = \frac{\ln \frac{S_0}{K} \pm \frac{1}{2}\sigma_B^2 T}{\sigma_B \sqrt{T}}.$$

Equation (3.15) is the same as the B-S model. However, σ_B is the implied volatility from the SABR model.

In the next section, we elaborate on the concept of implied volatility introduced in Section 3.4.1 and then we discuss some methods used to compute implied volatility.

3.8 Methods for Finding Roots of Functions: Implied Volatility

3.8.1 Introduction

This section addresses the following problem. Assume that we have a function $f : \mathbb{R} \to \mathbb{R}$ which perhaps has a very complicated form. We wish to look at the equation $f(x) = 0$. If f would be invertible it would be easy to find the roots as the set $\{x : x = f^{-1}(0)\}$. We also assume that we can calculate $f(x)$ for all x. More specifically, this is needed in the context when f is the value given by the B-S formula. Suppose that we look at a Call option whose value is denoted by $C_{BS} = C_{BS}(S_0, K, T, r, \sigma)$. Note that S_0 and r are values that are known at the current time (stock value and the short term interest rate). K and T are strike price and maturity (in years) which characterize each option. Therefore the unknown in this context is the volatility σ. What one does is observe the bid-ask spread, (denoted B and A) for the specific option value (characterized by K, T earlier) from the market. Then one considers the function

$$f(x) = C_{BS}(S_0, K, T, r, x) - (B + A)/2.$$

The value x for which the previous function is zero is called the volatility implied by the marked or simply the *implied volatility*.[1] Note that this value is the root of the function f. Thus finding the implied volatility is the same as finding the roots of a function. Below we present several methods of finding roots of a function. For more details of the root finding methods, see [11, 122].

3.8.2 The Bisection Method

The bisection method is a very simple method. It is based on the fact that if the function f has a single root at x_0, then the function changes the sign from

---

[1] Note that the value $(B + A)/2$ is just a convention; one could also use them separately and obtain two implied volatility values and infer that the real implied volatility value is somewhere inside the interval.

negative to positive or vice versa. Suppose there exists a neighborhood of x_0 which contains only that one root x_0. Take any two points a and b in the neighborhood. Then since the function changes sign at x_0, if x_0 is between a and b, the values $f(a)$ and $f(b)$ will have opposite signs. Therefore, the product $f(a)f(b)$ will be negative. However, if x_0 is not between a and b, then $f(a)$ and $f(b)$ will have the same sign (be it negative or positive), and the product $f(a)f(b)$ will be positive. This is the main idea of the bisection method. The algorithm starts by finding two points a and b such that the product $f(a)f(b)$ is negative. In the case of implied volatility the suggested starting points are 0 and 1. *Pseudo-Algorithm:*

Step 1: Check if the distance between a and b (i.e. $b - a$) is less than some tolerance level ε. If Yes STOP report either point you want (a or b). If No step further.

Step 2: Calculate $f(\frac{a+b}{2})$. Evaluate $f(a)f(\frac{a+b}{2})$ and $f(\frac{a+b}{2})f(b)$.

Step 3: If $f(a)f(\frac{a+b}{2}) < 0$ make $b \leftarrow \frac{a+b}{2}$. Go to Step 1.

Step 4: If $f(\frac{a+b}{2})f(b) < 0$ make $a \leftarrow \frac{a+b}{2}$. Go to Step 1.

If there are more than one root in the starting interval $[a, b]$ or the root is a multiple root, the algorithm fails. There are many improvements that can be found in [10, 11, 122]. Examples of these improvements are presented in the following subsequent subsections.

3.8.3 The Newton's Method

This method assumes the function f has a continuous derivative. It also assumes that the derivative $f'(x)$ can be calculated easily (fast) at any point x. Newton's method may not converge if you start too far away from a root. However, if it does converge, it is faster than the bisection method since its convergence is quadratic rather than linear. Newton's method is also important because it readily generalizes to higher-dimensional problems[2] see [122]. Suppose we have a function f whose zeros are to be determined numerically. Let r be a zero of f and let x be an approximation to r. If $f''(x)$ exists and is continuous, then by Taylor's Theorem,

$$0 = f(r) = f(x + h) = f(x) + hf'(x) + \mathcal{O}(h^2) \tag{3.16}$$

where $h = r - x$. If h is small, then it is reasonable to ignore the $\mathcal{O}(h^2)$ term and solve the remaining equation for h. Ignoring the $\mathcal{O}(h^2)$ term and solving for h in (3.16), we obtain $h = -f(x)/f'(x)$. If x is an approximation to r, then $x - f(x)/f'(x)$ should be a better approximation to r. Newton's method begins with an estimate x_0 of r and then defines inductively

$$x_{n+1} = x_n - \frac{f(x_n)}{f'(x_n)}, \quad (n \geq 0). \tag{3.17}$$

---

[2] The bisection method does not generalize.

In some cases the conditions on the function that are necessary for convergence are satisfied, but the initial point x_0 chosen is not in the interval where the method converges. This can happen, for example, if the function whose root is sought approaches zero asymptotically as x goes to $-\infty$ or ∞. In some cases, Newton's method is combined with other slower methods such as the bisection method in a hybrid method that is numerically globally convergent (i.e. the iterative method converges for any arbitrary initial approximation).

3.8.4 Secant Method

In (3.18), we defined the Newton iteration:

$$x_{n+1} = x_n - \frac{f(x_n)}{f'(x_n)}, \quad (n \geq 0). \tag{3.18}$$

An issue with the Newton's method is that it involves the derivative of the function whose zero is sought. Notice that in (3.18) we need to know the derivative value at x_n to calculate the next point. In most cases we cannot compute this value so instead we need to estimate it. A number of methods have been proposed. For example, Steffensen's iteration

$$x_{n+1} = x_n - \frac{[f(x_n)]^2}{f(x_n + f(x_n)) - f(x_n)} \tag{3.19}$$

gives one approach to this problem. Another method is to replace $f'(x)$ in (3.18) by a difference quotient, such as

$$f'(x) \approx \frac{f(x_n) - f(x_{n-1})}{x_n - x_{n-1}}. \tag{3.20}$$

The approximation given in (3.20) comes directly from the definition of $f'(x)$ as a limit. When we replace $f'(x)$ in (3.18) with (3.20), the resulting algorithm is called the **secant method**. Its formula is

$$x_{n+1} = x_n - f(x_n)\left[\frac{x_n - x_{n-1}}{f(x_n) - f(x_{n-1})}\right] \quad (n \geq 1). \tag{3.21}$$

Since the calculation of x_{n+1} requires x_n and x_{n-1}, two initial points must be prescribed at the beginning. However, each new x_{n+1} requires only one new evaluation of f.

Remark 3.8.1 Other methods used in finding roots of a function are mentioned in the following. The respective formulas have been omitted since they are a bit less elegant than the few presented so far.

- **Muller method.** This method is similar to the secant method except that the Muller method uses a second order approximation (using second derivative).
- **Inverse quadratic interpolation.** This method uses second order approximation for the inverse function f^{-1}.

- **Brent's method.** This method is the most powerful deterministic algorithm since it combines three of the methods presented thus far to find the root faster. The three basic algorithms used are bisection, secant, and inverse quadratic interpolation methods.

Besides these methods, new random algorithms exist where perturbations are introduced for a more certain convergence. These algorithms are better known for finding extreme points of functions (the best known one is called simulated annealing), but there are some variations used to find roots of functions as well. Please refer to [22] for details and for details of these methods, please refer to [10, 153].

We now present an example of calculating the implied volatility using the Newton's method.

3.8.5 Computation of Implied Volatility Using the Newton's Method

The inversion of the B-S formula to get the implied volatility is done by using the root finding methods described in the previous subsections. In this subsection, we consider the Newton's method for computing the implied volatility. The method works very well for a single option.

Let us recall the B-S formula for the price of a European Call option.

$$C(S,t) = SN(d_1) - Ke^{-r(T-t)}N(d_2), \tag{3.22}$$

where,

$$d_1 = \frac{\ln\left(\frac{S}{K}\right) + \left(r + \frac{1}{2}\sigma^2\right)(T-t)}{\sigma\sqrt{T-t}}$$

$$d_2 = d_1 - \sigma\sqrt{T-t} = \frac{\ln\left(\frac{S}{K}\right) + \left(r - \frac{1}{2}\sigma^2\right)(T-t)}{\sigma\sqrt{T-t}}$$

and N is the distribution function of the standard normal distribution.

The procedure for the "daily" computation of the implied volatility of an underlying asset is as follows. We compute the volatility $\sigma := \sigma_{K,T-t}$ by solving the equation

$$C(S_*, t_*; K, T-t, r, \sigma) = v \tag{3.23}$$

where t_* is time, S_* is the price of the underlying asset, r is the interest rate constant in the future, K is the strike price of a European Call option, T is the maturity, $T - t$ is the term to maturity, and $v := C_{market}$ is the observed market price of the option. In order to simplify the notation, we set

$$f(\sigma) := C(S_*, t_*; K, T-t, r, \sigma), \quad d_2 := d_2(\sigma, S, K, T-t, r). \tag{3.24}$$

As a result, from (3.23) we have to compute a solution of

$$f(\sigma) - v = 0. \tag{3.25}$$

The function $f(\sigma)$ is a smooth function that depends on σ in a highly nonlinear manner. Since f is differentiable, we can apply the Newton's method as follows:

1) Choose an initial guess σ_0 for the implied volatility and a stopping-bound $N \in \mathbb{N}$ for the iteration steps.
2) Set $f(\sigma) := C(S_*, t_*; K, T-t, r, \sigma)$ and $f'(\sigma) := \frac{dC}{d\sigma}(S_*, t_*; K, T-t, r, \sigma)$.
3) Compute for $n = 0, \ldots, N-1$

$$\sigma_{n+1} = \sigma_n - \frac{f(\sigma_n) - v}{f'(\sigma_n)} \tag{3.26}$$

4) Approximate the value σ_n for the implied volatility $\sigma_{implied}(K, T-t)$.

3.9 Some Remarks of Implied Volatility (Put–Call Parity)

Put–Call parity which we explained in Chapter 2 provides a good starting point for understanding volatility smiles. It is an important relationship between the price C of a European Call and the price P of a European Put:

$$P + S_0 e^{-qT} = C + K e^{-rT}. \tag{3.27}$$

The Call and the Put have the same strike price, K, and time to maturity, T. The variable S_0 is the price of the underlying asset today, r is the risk-free interest rate for the maturity T, and q is the yield of the asset.

An essential feature of the Put–Call parity relationship is that it is based on a relatively simple non-arbitrage argument.

Suppose that for a particular value of the volatility, P_α and C_α are the values of the European Put and Call options calculated using the B-S model. Suppose further that P_β and C_β are the market values of these options. Since the Put–Call parity holds for the B-S model, we must have

$$P_\alpha + S_0 e^{-qT} = C_\alpha + K e^{-rT}. \tag{3.28}$$

Similarly, since the Put–Call parity holds for the market prices, we have

$$P_\beta + S_0 e^{-qT} = C_\beta + K e^{-rT}. \tag{3.29}$$

Subtracting (3.29) from (3.28) gives

$$P_\alpha - P_\beta = C_\alpha - C_\beta. \tag{3.30}$$

This shows that the pricing error when the B-S model is used to price the European Put option should be exactly the same as the pricing error when it is used to price a European Call option with the same strike price and time to maturity. In the next section, we discuss an important concept known as hedging using volatility.

3.10 Hedging Using Volatility

Definition 3.10.1 Volatility is a measure for the variation of price of a financial instrument over time. It is usually denoted by σ and it corresponds to the squared root of the quadrative variance of a stochastic process.

There are two types of volatility that we are concerned with, actual volatility and implied volatility. Actual volatility is the unknown parameter or process. Implied volatility is a market-suggested value for volatility. We can then take historical values of stock prices and other parameters to figure out the market's opinion on future volatility values. The next example presents a way to use volatility in trading.

Example 3.10.1 Suppose we calculate the implied volatility to be 40% using the price of the stock, but we think that the actual volatility is 50%. If our assumption is correct, how can we make money, and which delta should we use?

It is unclear which delta we should hedge with, so in this section we will use both and analyze the results. For a European Call option in the B-S model we have

$$\Delta = N(d_1),$$

and we saw earlier that

$$N(x) = \frac{1}{\sqrt{2\pi}} \int_{-\infty}^{x} e^{\frac{1}{2}x^2} dx$$

and

$$d_1 = \frac{\log(S/E) + (r + \frac{1}{2}\sigma^2)(T-t)}{\sigma\sqrt{T-t}}.$$

We still do not know which volatility value to use for σ. The other variables are easily determined. We will denote actual volatility as σ and $\tilde{\sigma}$ as implied volatility. In Table 3.2, we present an example of hedging with actual volatility. We will also use Δ^a to represent the delta using the actual volatility, and V^i is the theoretical value of the option found by plugging implied volatility in the formula. So we will buy an option for V^i and hedge with Δ^a.

3.10 Hedging Using Volatility

Table 3.2 Example of hedging with actual volatility.

Holding	Value today(time t)	Value tomorrow($t + dt$)
Option	V^i	$V^i + dV^i$
$-\Delta^a S$	$-\Delta^a S$	$-\Delta^a(S + dS)$
Cash	$-V^i + \Delta^a S$	$(-V^i + \Delta^a S)(1 + rdt)$
Total	0	$dV^i - \Delta^a dS - (V^i - \Delta^a S)rdt$

Therefore we get

$$\text{Profit} = dV^i - \Delta^a dS - (V^i - \Delta^a S)rdt.$$

Then the B-S equation yields

$$dV^a - \Delta^a dS - (V^a - \Delta^a s)rdt = 0.$$

Hence the profit for a time step is given by

$$= dV^i - dV^a + (V^a - \Delta^a S)rdt - (V^i - \Delta^a S)rdt$$
$$= dV^i - dV^a - (V^i - V^a)rdt$$
$$= e^{rt}d(e^{-rt(V^i-V^a)}).$$

Then the profit at time t_0 is

$$e^{-r(t-t_0)}e^{rt}d(e^{-rt}(V^i - V^a)) = e^{rt_0}d(e^{-rt}(V^i - V^a))$$

and the total profit is given by

$$e^{rt_0}\int_{t_0}^{T} d(e^{-rt}(V^i - V^a)) = V^a(t_0) - V^i(t_0).$$

Thus, if you are good at approximating actual volatility, you can hedge in such a way that there will always be arbitrage. The profit in this case is

$$\text{Profit} = V(S, t, \sigma) - V(S, t, \tilde{\sigma}).$$

With this approach, we can determine the profit from time t_0 to time T, but the profit at each step is random. In Table 3.3, an example of hedging with implied volatility is presented. If we hedge with implied volatility $\tilde{\sigma}$ we have

Then profit is given from [210] by

$$\text{Profit} = dV^i - \Delta^i dS - r(V^i - \Delta^i S)dt$$
$$\text{Profit} = \Theta^i dt + \frac{1}{2}\sigma^2 S^2 \Gamma^i dt - r(V^i - \Delta^i S)dt$$
$$\text{Profit} = \frac{1}{2}(\sigma^2 - \tilde{\sigma}^2)S^2 \Gamma^i dt.$$

Table 3.3 Example of hedging with implied volatility.

Holding	Value today(time t)	Value tomorrow($t + dt$)
Option	V^i	$V^i + dV^i$
$-\Delta^i S$	$-\Delta^i S$	$-\Delta^i(S + dS)$
Cash	$-V^i + \Delta^i S$	$(-V^i + \Delta^i S)(1 + rdt)$
Total	0	$dV^i - \Delta^i dS - (V^i - \Delta^i S)rdt$

Then profit present value is

$$\frac{1}{2}(\sigma^2 - \tilde{\sigma}^2)e^{-r(t-t_0)}S^2\Gamma^i dt.$$

Unlike hedging with actual volatility, we can determine profit at every step. In [162, 210] the authors have shown that if you hedge using σ_h, then the total profit is modeled by

$$V(S, t; \sigma_h) - V(S, t, \tilde{\sigma}) + \frac{1}{2}(\sigma^2 - \sigma_h^2)\int_{t_0}^{T} e^{-r(t-t_0)}S^2\Gamma^h dt,$$

where the superscript h on the Γ means that it uses the B-S formula using volatility of σ_h. We have now modeled profit for $\sigma, \tilde{\sigma}$, and σ_h. We will now examine the advantages and disadvantages of using these different volatilities.

We discuss the pros and cons of hedging with actual volatility and implied volatility, and then finally we discuss hedging using other volatilities.

When using actual volatility, the main advantage is that we know exactly what profit we will have at expiration. This is a very desirable trait, as we can better see the risk and reward before hand. A drawback is that the profit and loss can fluctuate drastically during the life of an option. We are also going to have a lack of confidence when hedging with this volatility, since the number we have chosen is likely to be wrong. On the other hand, we can play around with numbers, and since profit is deterministic in this case, we can see what will happen if we do not have the correct figure for actual volatility. We can then decide if there is a small margin for success not to hedge.

Unlike hedging with actual volatility, hedging with implied volatility gives three distinct advantages. First and foremost, there are not any local fluctuations with profit and loss; instead you are continually making a profit. The second advantage is that we only need to be on the right side of a trade to make a profit. We can simply buy when the actual volatility is higher than implied and sell when lower. The last advantage is that we are using implied volatility for delta, which is easily found. The disadvantage is that we do not know how much we will make, but we do know that this value will be positive.

We can balance these advantages and disadvantages by choosing another volatility entirely. This is an open problem and is beyond the scope of this book. When using this in the real world, applied and implied volatility have their purposes. If we are not concerned with the day-to-day profit and loss, then actual volatility is the natural choice. However, if we are expected to report profits or losses on a regular basis, then we will need to hedge with implied volatility, as profit or loss can be measured periodically.

3.11 Functional Approximation Methods

An option pricing model establishes a relationship between several variables. They are the traded derivatives, the underlying asset, and the market variables, e.g. volatility of the underlying asset. These models are used in practice to price derivative securities given knowledge of the volatility and other market variables.

The classical B-S model assumes that the volatility is constant across strikes and maturity dates. However it is well known in the world of option pricing that this assumption is very unrealistic. Option prices for different maturities change significantly, and option prices for different strikes also experience significant variations. In this section we consider the numerical problems arising in the computation of the implied volatilities and the implied volatility surface.

Volatility is one of the most basic market variables in financial practice and theory. However, it is not directly observable in the market. The B-S model can be used to estimate volatility from observable option prices. By inverting the B-S formula with option market data, we obtain estimates which are known as implied volatility. These estimates show strong dependence of volatility values on strike price and time to maturity. This dependence (popularly called volatility smile) cannot be captured by the B-S model.

The constant implied volatility approach, which uses different volatilities for options with different strikes and maturities, works well for pricing simple European options, but it fails to provide adequate solutions for pricing exotic or American options. This approach also produces incorrect hedge factors even for simple options. We already discussed several stochastic models that do not have this problem (e.g. Heston SABR). However, we will present next a very popular practitioner approach. In this approach, we use a one-factor diffusion process with a volatility function depending on both the asset price and time, i.e. $\sigma(S_t, t)$. This is a deterministic approach and the function $\sigma(S_t, t)$ is also known as the "local volatility" or the "forward volatility." We discuss the local volatility model next.

3.11.1 Local Volatility Model

We recall from Section 2.6 that if the underlying asset follows a continuous one-factor diffusion process for some fixed time horizon $t \in [0, T]$, then we can write

$$dS_t = \mu(S_t, t)S_t dt + \sigma(S_t, t)S_t dW_t, \qquad (3.31)$$

where W_t is a Brownian notion, $\mu(S_t, t)$ is the drift, and $\sigma(S_t, t)$ is a local volatility function. The local volatility function is assumed to be continuous and sufficiently smooth so that (3.31) with corresponding initial and boundary conditions have a unique solution.

The Local volatility models are widely used in the financial industry to hedge barrier options (see [61]). These models try to stay close to the B-S model by introducing more flexibility into the volatility. In the local volatility model, the only stochastic behaviour comes from the underlying asset price dynamics. Thus there is only one source of randomness, ensuring that the completeness of the B-S model is preserved. Completeness guarantees that the prices are unique. Please refer to [115] for more details about local volatility modeling.

Next we present the derivation of local volatility from B-S implied volatility.

3.11.2 Dupire's Equation

The Dupire's Eq. [60] is a forward equation for the Call option price C as a function of the strike price K and the time to maturity T. According to standard financial theory, the price at time t of a Call option with strike price K and maturity time T is the discounted expectation of its payoff, under the risk-neutral measure. Let

$$D_{0,T} = \exp\left(-\int_{t_0}^{T} r_s ds\right) \qquad (3.32)$$

denote the discount rate from the current time t_0 to maturity T and $\phi(T, s)$ denote the risk-neutral probability density of the underlying asset at maturity. More accurately the density should be written as $\phi(T, s; t_0, S_0)$, since it is the transition probability density function of going from state (t_0, S_0) to (T, s). However, since t_0 and S_0 are considered to be constants, for brevity of notation, it is written as $\phi(T, s)$. It is assumed that the term structure for the short rate r_t is a deterministic function known at the current time t_0.

The risk-neutral probability density distribution of the underlying asset price at maturity is known to be $\phi(T, s)$. Therefore since this is a probability density function, its time evolution satisfies the equation

$$\frac{\partial \phi(t, s)}{\partial t} = -(r_t - q_t)\frac{\partial}{\partial s}[s\phi(t, s)] + \frac{1}{2}\frac{\partial^2}{\partial s^2}[\sigma(t, s)^2 s^2 \phi(t, s)]. \qquad (3.33)$$

Next, let the price of a European Call with strike price K be denoted $C = C(S_t, K)$ such that

$$C = E[D_{0,T}(S_T - K)^+ | F_0]$$
$$= D_{0,T} \int_K^\infty (s - K)\phi(T, s)ds. \qquad (3.34)$$

Taking the first derivative of (3.34) with respect to strike K, we obtain

$$\frac{\partial C}{\partial K} = D_{0,T} \frac{\partial}{\partial K} \int_K^\infty (s - K)\phi(T, s)ds$$
$$= D_{0,T} \left[-(K - K)\phi(T, K) - \int_K^\infty \phi(T, s)ds \right]$$
$$= -D_{0,T} \int_K^\infty \phi(T, s)ds, \qquad (3.35)$$

and the second derivative with respect to the strike price yields

$$\frac{\partial^2 C}{\partial K^2} = -D_{0,T} \frac{\partial}{\partial K} \int K^\infty \phi(T, s)ds$$
$$= D_{0,T}\phi(T, K). \qquad (3.36)$$

Assuming that $\lim_{s \to \infty} \phi(T, s) = 0$.

Taking the first derivative of (3.34) with respect maturity T and using the chain rule, we obtain

$$\frac{\partial C}{\partial T} + r_T C = D_{0,T} \int_K^\infty (s - K) \frac{\partial \phi(T, s)}{\partial T} ds. \qquad (3.37)$$

Substituting (3.33) into (3.37) yields

$$\frac{\partial C}{\partial T} + r_T C$$
$$= D_{0,T} \int_K^\infty (s - K) \left(\frac{1}{2} \frac{\partial^2}{\partial s^2} [\sigma(T, s)^2 s^2 \phi(T, s)] - (r_T - q_T) \frac{\partial}{\partial S_T} [S_T \phi(T, s)] \right) ds$$
$$= -\frac{1}{2} D_{0,T} [\sigma(T, s)^2 S_T^2 \phi(T, s)]_{s=K}^\infty + (r_T - q_T) D_{0,T} \int_K^\infty s\phi(T, s)ds$$
$$= \frac{1}{2} D_{0,T} \sigma(T, K)^2 K^2 \phi(T, K) + (r_T - q_T) \left(C + K D_{0,T} \int_K^\infty \phi(T, s)ds \right). \qquad (3.38)$$

Finally, substituting (3.35) and (3.36) into the (3.38), we get

$$\frac{\partial C}{\partial T} + r_T C = \frac{1}{2} \sigma(T, K)^2 K^2 \frac{\partial^2 C}{\partial K^2} + (r_T - q_T) \left(C + K \frac{\partial C}{\partial K} \right) \qquad (3.39)$$

In this context, it is assumed that $\phi(T, S_T)$ behaves appropriately at the boundary condition of $S_t = \infty$ (for instance, this is the case when ϕ decays exponentially fast for $S_T \to \infty$). Therefore

$$\frac{\partial C}{\partial T} = \frac{1}{2} \sigma(T, K)^2 K^2 \frac{\partial^2 C}{\partial K^2} - q_T C - (r_T - q_T) K \frac{\partial C}{\partial K}.$$

Thus

$$\sigma(T,K)^2 = 2\frac{\frac{\partial C}{\partial T} + (r_T - q_T)K\frac{\partial C}{\partial K} + q_T C}{K^2 \frac{\partial^2 C}{\partial K^2}}. \tag{3.40}$$

Equation (3.40) is commonly known as *Dupire formula* [60]. Since at any point in time the value of Call options with different strikes and times to maturity can be observed in the market, the local volatility is a deterministic function, even when the dynamics of the spot volatility is stochastic.

Given a certain local volatility function, the price of all sorts of contingent claims on that the underlying can be priced. By using the B-S model described in Section 3.3, this process can be inverted, by extracting the local volatility surface from option prices given as a function of strike and maturity. The hidden assumption is that the option price is a continuous $C^{2,1}$ function (i.e. twice continuously differentiable function), known over all possible strikes and maturities. Even if this assumption holds, a problem arises in the implementation of Eq. (3.40). In general the option price function will never be known analytically, neither will its derivatives. Therefore numerical approximations for the derivatives have to be made, which are by their very nature approximations. Problems can arise when the values to be approximated are very small, and small absolute errors in the approximation can lead to big relative errors, perturbing the estimated quantity. When the disturbed quantity is added to other values, the effect will be limited. This is not the case in Dupire's formula where the second derivative with respect to the strike in the denominator stands by itself. This derivative will be very small for options that are far in or out of the money (i.e. the effect is particularly large for options with short maturities). Small errors in the approximation of this derivative will get multiplied by the strike value squared, resulting in big errors at these values, sometimes even giving negative values that result in negative variances and complex local volatilities.

Another problem is the derivability assumption for option prices. Using market data option prices are known for only discrete points. Beyond the current month, option maturities correspond to the third Friday of the month, thus the number of different maturities is always limited with large gaps in time especially for long maturities. The same holds to a lesser degree for strikes. As a result, in practice the inversion problem is ill-posed: the solution is not unique and is unstable. This is a known problem when dealing with Dupire's formula in practice. One can smooth the option price data using Tikhonov regularization (see [87]) or by minimizing the functions entropy (see [13]). Both of these methods try to estimate a stable option price function. These methods need the resulting option price function to be convex in the strike direction at every point to avoid negative local variance. This guarantees the positivity of the second derivative in the strike direction. This seems sensible since the nonconvexity of the option prices leads to arbitrage opportunities; however this add a considerable amount of complexity to the model. An easier and inherently more stable method is to use implied volatilities and implied volatility surface to obtain the local volatility surface (see [115]). In the subsections that follow, we discuss methods used to represent the local volatility function.

3.11.3 Spline Approximation

In this method the local volatility function is obtained using a bicubic spline approximation which is computed by solving an inverse box-constrained non-linear optimization problem.

Let $\{\overline{V}_j\}, j = 1, \ldots, m$ denote the m given market option prices. Given $\{K_i, T_i\}$ (where $i = 1, \ldots, p, p \leq m$) spline knots with corresponding local volatility values $\sigma_i = \sigma(K_i, T_i)$, an interpolating cubic spline $c(K, T, \sigma)$ with a fixed end condition (e.g. the natural spline end condition) is uniquely defined by setting $c(K_i, T_i) = \sigma_i, i = 1, \ldots, p$. The local volatility values σ_i at a set of fixed knots are determined by solving the following minimization problem:

$$\min_{\overline{\sigma} \in \mathbb{R}^p} f(\overline{\sigma}) = \frac{1}{2} \sum_{j=1}^{m} w_j (V_j(\overline{\sigma}) - \overline{V}_j)^2, \tag{3.41}$$

subject to $l_i \leq \sigma_i \leq u_i$, for $i = 1, \ldots, p$. Where $V_j(\overline{\sigma}) = V_j(c(K_j, T_j), K_j, T_j), \overline{V}_j$'s are the given option prices at given strike price and expiration time $(K_j, T_j), j = 1, \ldots, m$ pairs, w_j's are weights, $\sigma_i, i = 1, \ldots, p$ are the model parameters, and l's and u's are lower and upper bounds, respectively. The $\{V_j(\overline{\sigma})\}$'s are the model option values.

To guarantee an accurate and appropriate reconstruction of the local volatility, the number of spline knots p should not be greater than the number of available market option prices m.

The bicubic spline is defined over a rectangle R in the (K, T) plane where the sides of R are parallel to the K and T axes. R is divided into rectangular panels by lines parallel to the axes. Over each panel the bicubic spline is a bicubic polynomial which can be presented in the form

$$spline(K, T) = \sum_{i=0}^{3} \sum_{j=0}^{3} a_{ij} K^i T^j.$$

Each of these polynomials joins the polynomials in adjacent panels with continuity up to the second derivative. The constant K-values of the dividing lines parallel to the T-axis form the set of interior knots for the variable K, corresponding to the set of interior knots of a cubic spline. Similarly, the constant T-values of dividing lines parallel to the K-axis form the set of interior knots for the variable T. Instead of representing the bicubic spline in terms of the previous set of bicubic polynomials, in practice it is represented for the sake of computational speed and accuracy in the form

$$spline(K, T) = \sum_{i=1}^{p} \sum_{j=1}^{q} c_{ij} M_i(K) N_j(T),$$

where $M_i(K)$, $i = 1, \ldots, p$, $N_j(T)$, and $j = 1, \ldots, q$ are normalized B-splines. Please refer to [90] for further details of normalized B-splines and [91] for further details of bicubic splines.

In the next subsection, we briefly discuss a general numerical solution technique for the Dupire's equation.

3.11.4 Numerical Solution Techniques

In the previous subsection, we discussed the bicubic approximation for the volatility surface. In this subsection, both the B-S equations and the Dupire equations are solved by the finite difference method using the implicit time integration scheme. Please refer to Sections 7.2–7.4 for more details of the finite difference approximation scheme.

A uniform grid with $N \times M$ points in the region $[0, 2S_{init}] \times [0, \tau]$, where S_{init} is the initial money region and τ is the maximum maturity in the market option data used for solving the discretized B-S and the Dupire equations. The spline knots are also chosen to be on a uniform rectangular mesh covering the region Ω in which volatility values are significant in pricing the market options (the region Ω covers not far out of the money and in the money region of S). This region is not known explicitly; therefore a grid $[\delta_1 S_{init}, \delta_2 S_{init}] \times [0, \tau]$ is used. The parameters δ_1 and δ_2 are suitably chosen.

The general scheme of the volatility calibration algorithm are as follows when we consider the case of the Dupire equation:

1) Set the uniform grid for solving the discretized Dupire equation.
2) Set the spline knots.
3) Set the initial values for the local volatility at chosen spline knots $\{\sigma_i^{(0)}\}$, $i = 1, \ldots, p$.
4) Solve the discretized Dupire equation with the local volatility function approximated by spline with the local volatility values at chosen spline knots $\{\sigma^{(k)}\}$, $i = 1, \ldots, p$. Obtain the set of model-based option values $\{V_j(\overline{\sigma})\}$ and the values of the objective function $f(\overline{\sigma})$ for (3.41).
5) Using the objective function $f(\overline{\sigma})$ (3.41) and the chosen minimization algorithm, find the undated values of the local volatility at spline knots $\{\sigma_i^{(k+1)}\}$.
6) Repeat steps 4–5 until a local minimum is found. The local minimizer $\{\sigma_i^*\}$, $i = 1, \ldots, p$ defines the local volatility function.

The algorithm remains the same for the case of the B-S equation.

In the next subsection, we discuss the problem of determining a complete surface of implied volatilities and of option prices.

3.11.5 Pricing Surface

Given a set of market prices, we consider the problem of determining a complete surface of the implied volatilities of the option prices. By varying the strike price K and the term to maturity T, we can create a table whose elements represent volatilities for different strikes and maturities. Under this practice, the implied volatility parameters will be different for options with different time to maturity T and strike price K. This collection of implied volatilities for different strikes and maturities is called the implied volatility surface, and the market participants use this table as the option price quotation. For many applications (calibration, pricing of nonliquid or nontraded options, etc.) we are interested in an implied volatility surface which is complete, i.e. which contains an implied volatility for each pair (K, T) in a reasonable large enough set $[0, K_{max}] \times [0, T_{max}]$.

However, in a typical option market, one often observes the prices of a few options with the same time to maturity but different strike levels only. Some of these option contracts are not liquid at all, i.e. they are not traded at an adequatable extent. Therefore, we are faced with the problem of how to interpolate or extrapolate the table of implied volatilities. Some well known methods for completing the table of implied volatilities includes polynomials in two variables; linear, quadratic, or cubic splines in one or two variables; and parametrization of the surface and fitting of the parameters (See [80]). However, it is appropriate to complete the pricing table of a pricing surface instead of completing the volatilities table. This is due to the fact that the properties of the pricing surface are deeply related to the assumptions concerning the market.

In order to ensure a pricing (using the implied volatility surface) which is arbitrage free, we have to find sufficient conditions for the arbitrage condition to hold. These conditions are related to the pricing surface. In the case of a Call option, there exists a mapping $C : (0, \infty) \times (0, \infty) \to [0, \infty)$, such that $(K, T) \mapsto C(K, T)$. The Call price surface defined by the mapping is called free of (static) arbitrage if and only if there exists a nonnegative martingale, say S, such that $C(K, T) = E((S_T - K)^+)$ for all $(K, T) \in (0, \infty) \times [0, \infty)$. The price surface is called free of (static) arbitrage if there are no arbitrage opportunities. Please refer to [80] and references therein for more details.

3.12 Problems

1. Suppose $V = V(S)$. Find the most general solution of the Black–Scholes equation.

2. Suppose $V = \Lambda_1(t)\Lambda_2(S)$. Find the most general solution of the Black–Scholes equation.
3. Prove that for a European Call option on an asset that pays no dividends the following relations hold:

$$C \leq S, \quad C \geq S - E\exp(-r(T-t)).$$

4. Given the formulation of the free boundary problem for the valuation of an American Put option,

$$\frac{\partial P}{\partial t} + \frac{1}{2}\sigma^2 S^2 \frac{\partial^2 P}{\partial S^2} + rS\frac{\partial P}{\partial S} - rP \leq 0$$

$$\frac{\partial P}{\partial t} + \frac{1}{2}\sigma^2 S^2 \frac{\partial^2 P}{\partial S^2} + rS\frac{\partial P}{\partial S} - rP = 0 \quad \text{if } S < S_f$$

$$P(S_f(t), t) = K - S_f(t)$$

$$\frac{\partial P}{\partial S}(S_f(t), t) = -1$$

using a suitable change of variable, transform the free boundary problem into the modified problem

$$\frac{\partial u}{\partial \tau} = \frac{\partial^2 u}{\partial x^2} \quad \text{for } x > x_f(\tau)$$

$$u(x, \tau) = e^{\frac{1}{2}(k+1)^2\tau}\{e^{\frac{1}{2}(k-1)x} - e^{\frac{1}{2}(k+1)x}\} \quad \text{for } x \leq x_f(\tau)$$

with initial condition

$$u(x, 0) = g(x, 0) = \{e^{\frac{1}{2}(k-1)x} - e^{\frac{1}{2}(k+1)x}\}$$

where $k = r/\frac{1}{2}\sigma^2$ (Hint: Use the change of variable, $S = Ke^x$ and $t = T - \tau/\frac{1}{2}\sigma^2$.)
This modified problem can be solved numerically using the finite difference method (See Chapter 7 for more details).

5. In a market with N securities and M futures, where $(S_1, \cdots, S_N)$ is the present values vector and $(S_1^j(T), \cdots, S_N^j(T))$ the future values vector ($j = 1, \cdots, M$), we say that a portfolio $\pi = (\pi_1, \cdots, \pi_N)$ produces *strict arbitrage* if one of the following conditions hold:
   i) $\pi.S < 0$ and $\pi.S^j(T) \geq 0$ for all $j = 1, \cdots, M$.
   ii) $\pi.S = 0$ and $\pi.S^j(T) > 0$ for all $j = 1, \cdots, M$.
   If R is the risk-free interest rate, prove that the following statements are equivalent:
   (a) There exists a portfolio that produces strict arbitrage.
   (b) There exists a portfolio satisfying (i).
   (c) There exists a portfolio satisfying (ii).

6. Consider the case of a European Call option whose price is $1.875 when

$$S = 21, K = 20, r = 0.2, T - t = 0.25.$$

Compute the implied volatility using the Newton's method by choosing $\sigma_0 = 0.20$.

7. Using the same information in problem 6, compute the implied volatility using the bisection method in the interval $[0.20, 0.30]$.

8. Use provided data to construct a pricer for Call options using SABR model.

9. Given the general form of Dupire's equation, show that

$$\sigma^2 = \frac{\frac{\partial C}{\partial T}}{\frac{1}{2} K^2 \frac{\partial^2 C}{\partial K^2}}$$

asssuming that the risk rate r_T and the dividend yield q_T each equal to zero.

10. Explain the reason why it is convenient to represent the bicubic spline in the form

$$spline(K, T) = \sum_{i=1}^{p} \sum_{j=1}^{q} c_{ij} M_i(K) N_j(T),$$

where $M_i(K)$, $i = 1, \ldots, p$, $N_j(T)$, and $j = 1, \ldots, q$ are normalized B-splines instead of

$$spline(K, T) = \sum_{i=0}^{3} \sum_{j=0}^{3} a_{ij} K^i T^j.$$

4
Solving Partial Differential Equations

4.1 Introduction

Many problems in mathematical finance result in partial differential equations (PDEs) that need to be solved. The best way to solve these equations is to find the exact solution using analytical methods. This is however a hard task and only the simplest PDEs have exact analytical formulas for the solution. In this chapter, we talk about some PDEs that provide exact solution. We will also present transformation methods that may produce exact analytical solutions. When applied in practice these are in fact approximations as well, but they are theoretical ones as opposed to the numerical approximations presented in later chapters.

We begin the chapter with a discussion of the three main types of PDEs and some important properties.

4.2 Useful Definitions and Types of PDEs

PDEs describe the relationships among the derivatives of an unknown function with respect to different independent variables, such as time and position. PDEs are used to describe a wide variety of phenomena such as sound, heat, electrostatics, electrodynamics, fluid flow, and several others.

4.2.1 Types of PDEs (2-D)

A two-dimensional linear PDE is of the form

$$a\frac{\partial^2 U}{\partial t^2} + b\frac{\partial^2 U}{\partial x \partial t} + c\frac{\partial^2 U}{\partial x^2} + d\frac{\partial U}{\partial t} + e\frac{\partial U}{\partial x} + fU = g \quad (4.1)$$

where a, b, c, d, e, f, and g are constants. The problem is to find the function U.

Equation (4.1) can be rewritten as a polynomial in the form

$$P(\alpha, \beta) = a\alpha^2 + b\alpha\beta + c\beta^2 + d\alpha + e\beta + j. \quad (4.2)$$

Quantitative Finance, First Edition. Maria C. Mariani and Ionut Florescu.
© 2020 John Wiley & Sons, Inc. Published 2020 by John Wiley & Sons, Inc.

The nature of the PDE (4.2) is determined by the properties of the polynomial $P(\alpha, \beta)$. Looking at the values of the discriminant, i.e $\Delta = b^2 - 4ac$ of (4.2), we can classify the PDE (4.1) as follows:

1) If $\Delta < 0$, we have an elliptic equation. An example is the Laplace equations which is of the form

$$\frac{\partial^2 U}{\partial t^2} + \frac{\partial^2 U}{\partial x^2} = 0.$$

The solutions of Laplace's equation are the harmonic functions, which have wide applications in electromagnetism, astronomy, and fluid dynamics, since they can be used to accurately describe the behavior of electric, gravitational, and fluid potentials.

2) If $\Delta = 0$, we have the parabolic equation. An example is the diffusion (heat) equation, i.e.:

$$\frac{\partial U}{\partial t} - \frac{\partial^2 U}{\partial x^2} = 0.$$

The diffusion equation arises in the modeling of a number of phenomena that describes the transport of any quantity that diffuses spreads out as a consequence of the spatial gradients in its concentration and is often used in financial mathematics in the modeling of options. The heat equation was used by Black and Scholes to model the behavior of the stock market.

3) If $\Delta > 0$, we have the hyperbolic equation. An example includes the wave equation, i.e.

$$\frac{\partial^2 U}{\partial t^2} - \frac{\partial^2 U}{\partial x^2} = 0.$$

This type of equation appears often in wave propagation (acoustic, seismic waves, sound waves, wind waves on seas, lakes and rivers etc.)

Please note that the nature of the PDE is classified based on the coefficients a, b, c, d, e, f, and g being constant. When dealing with variable coefficients (that is, when the coefficients depend on the state variables x and t), the character of the PDE may switch between these three basic types.

4.2.2 Boundary Conditions (BC) for PDEs

PDEs are solved on a domain in $\mathbb{R}^2$. This domain could be bounded or unbounded. An example is $x, t \in [0, t] \times [0, \infty)$. The PDEs specified without boundary conditions (BC) have typically an infinite number of solutions. Once the behavior of the solution is specified on the boundaries, the solution becomes unique.

There are two types of boundary conditions normally used in practice: the Dirichlet- and the Newman-type BC. The Dirichlet-type conditions specify the actual value of the function on the boundary, e.g. $U(T,x) = 4$. The Newman-type BC specify the partial derivatives of the function on the boundaries, e.g. $\frac{\partial U}{\partial t}(T,x) = 2x$.

There are certainly other type of boundary conditions, for instance, the "mixed boundary condition" which is a weighted sum of the Dirichlet- and the Newman-type BC. Nevertheless, the basic ones are the two mentioned.

In the next section, we introduce some notations that will be used throughout the chapter and the rest of the book. We also give a brief account of few functional spaces that are commonly used when studying the PDEs.

4.3 Functional Spaces Useful for PDEs

For $d \geq 1$, let $\Omega \subset \mathbb{R}^d$ be an open set. $Q_T = \Omega \times (0, T)$ is called a parabolic domain for any $T > 0$. If the set Ω has a boundary, denoted by $\partial \Omega$, then $\partial \Omega$ will be smooth. We shall use $x = (x_1, x_2, \ldots, x_d) \in \Omega$ and $t \in (0, T)$.

We denote with $C^{i,j}(Q_T)$ the space of all functions which are derivable i times in the first variable and j times in the second variable, and both partial derivatives are continuous. We denote with $C_0^{i,j}(Q_T)$ the set of functions discussed earlier with additional property that they have compact support (i.e. they are zero outside a compact set in Q_T). The set of functions $C_0^\infty(Q_T)$ are infinitely derivable on the domain. These functions are also known as test functions.

The space $L^1_{loc}(Q_T)$ is the set of all functions that are locally integrable (i.e. integrable on every compact subset of its domain of definition Q_T).

A multi-index of nonnegative integers will be denoted by $\alpha = (\alpha_1, \ldots, \alpha_d)$ with $|\alpha| = \alpha_1 + \ldots + \alpha_d$. We will let k be a nonnegative integer and δ be a positive constant with $0 < \delta < 1$. Unless otherwise indicated, we will always use the standard Lebesgue integrals.

Definition 4.3.1 Let $u, v \in L^1_{loc}(Q_T)$. For a nonnegative integer ρ, we say v is the $\alpha \rho^{th}$ weak partial derivative of u of order $|\alpha| + \rho$, $D^\alpha \partial_t^\rho u = v$, provided that

$$\iint_{Q_T} u\, D^\alpha \partial_t^\rho \phi \; dx\, dt = (-1)^{|\alpha|+\rho} \iint_{Q_T} v\, \phi \; dx\, dt,$$

for any test function $\phi \in C_0^\infty(Q_T)$.

It can be shown that weak derivatives are unique up to a set of zero measure. In order to proceed we present several important definitions.

Definition 4.3.2 A function $u \in L^1_{loc}(\Omega)$ is weakly differentiable with respect to x_i if there exists a function $g_i \in L^1_{loc}(\Omega)$ such that

$$\int_\Omega f \partial_i \phi\, dx = -\int_\Omega g_i \phi\, dx \quad \text{for all} \quad \phi \in C_c^\infty(\Omega). \tag{4.3}$$

Definition 4.3.3 Let $1 \leq p \leq \infty$ and $k \in \mathbb{N} = \{1, 2, 3, \ldots\}$. We define the following Sobolev spaces,

$$W_p^k(\Omega) := \{u \in L^p(\Omega) \mid D^\alpha u \in L^p(\Omega), \ 1 \leq |\alpha| \leq k\}, \tag{4.4}$$

$$W_p^{2k,k}(Q_T) := \{u \in L^p(Q_T) \mid D^\alpha \partial_t^\rho u \in L^p(Q_T), 1 \leq |\alpha| + 2\rho \leq 2k\}. \tag{4.5}$$

The spaces discussed earlier become Banach spaces if we endow them with the respective norms

$$|u|_{W_p^k(\Omega)} = \sum_{0 \leq |\alpha| \leq k} |D^\alpha u|_{L^p(\Omega)}, \tag{4.6}$$

$$|u|_{W_p^{2k,k}(Q_T)} = \sum_{0 \leq |\alpha| + 2\rho \leq 2k} |D^\alpha \partial_t^\rho u|_{L^p(Q_T)}. \tag{4.7}$$

For more on the theory of Sobolev spaces, we refer the reader to [2].

Next, we discuss spaces with classical derivatives, known as Hölder spaces. We will follow the notation and definitions given in the books [129] and [205]. We define $C_{loc}^k(\Omega)$ to be the set of all real-valued functions $u = u(x)$ with continuous classical derivatives $D^\alpha u$ in Ω, where $0 \leq |\alpha| \leq k$. Next, we set

$$|u|_{0;\Omega} = [u]_{0;\Omega} = \sup_{\Omega} |u|,$$

$$[u]_{k;\Omega} = \max_{|\alpha|=k} |D^\alpha u|_{0;\Omega}.$$

Definition 4.3.4 The space $C^k(\Omega)$ is the set of all functions $u \in C_{loc}^k(\Omega)$ such that the norm

$$|u|_{k;\Omega} = \sum_{j=0}^{k} [u]_{j;\Omega}$$

is finite. With this norm, it can be shown that $C^k(\Omega)$ is a Banach space.

If the seminorm

$$[u]_{\delta;\Omega} = \sup_{\substack{x,y \in \Omega \\ x \neq y}} \frac{|u(x) - u(y)|}{|x - y|^\delta}$$

is finite, then we say the real-valued function u is *Hölder continuous in Ω with exponent δ*. For a k-times differentiable function, we will set

$$[u]_{k+\delta;\Omega} = \max_{|\alpha|=k} [D^\alpha u]_{\delta;\Omega}.$$

Definition 4.3.5 The Hölder space $C^{k+\delta}(\overline{\Omega})$ is the set of all functions $u \in C^k(\Omega)$ such that the norm

$$|u|_{k+\delta;\Omega} = |u|_{k;\Omega} + [u]_{k+\delta;\Omega}$$

is finite. With this norm, it can be shown that $C^{k+\delta}(\overline{\Omega})$ is a Banach space.

For any two points $P_1 = (x_1, t_1)$ and $P_2 = (x_2, y_2) \in Q_T$, we define the parabolic distance between them as

$$d(P_1, P_2) = \left(|x_1 - x_2|^2 + |t_1 - t_2|\right)^{1/2}.$$

For a real-valued function $u = u(x, t)$ on Q_T, let us define the seminorm

$$[u]_{\delta,\delta/2;Q_T} = \sup_{\substack{P_1, P_2 \in Q_T \\ P_1 \neq P_2}} \frac{|u(x_1, t_1) - u(x_2, t_2)|}{d^\delta(P_1, P_2)}.$$

If this seminorm is finite for some u, then we say u is Hölder continuous with exponent δ. The maximum norm of u is given by

$$|u|_{0;Q_T} = \sup_{(x,t) \in Q_T} |u(x, t)|.$$

Definition 4.3.6 The space $C^{\delta,\delta/2}\left(\overline{Q}_T\right)$ is the set of all functions $u \in Q_T$ such that the norm

$$|u|_{\delta,\delta/2;Q_T} = |u|_{0;Q_T} + [u]_{\delta,\delta/2;Q_T}$$

is finite. Furthermore, we define

$$C^{2k+\delta,k+\delta/2}\left(\overline{Q}_T\right) = \{u \,:\, D^\alpha \partial_t^\rho u \in C^{\delta,\delta/2}\left(\overline{Q}_T\right),\ 0 \leq |\alpha| + 2\rho \leq 2k\}.$$

We define a seminorm on $C^{2k+\delta,k+\delta/2}\left(\overline{Q}_T\right)$ by

$$[u]_{2k+\delta,k+\delta/2;Q_T} = \sum_{|\alpha|+2\rho=2k} [D^\alpha \partial_t^\rho u]_{\delta,\delta/2;Q_T},$$

and a norm by

$$|u|_{2k+\delta,k+\delta/2;Q_T} = \sum_{0 \leq |\alpha|+2\rho \leq 2k} |D^\alpha \partial_t^\rho u|_{\delta,\delta/2;Q_T}.$$

Using this norm, it can be shown that $C^{2k+\delta,k+\delta/2}\left(\overline{Q}_T\right)$ is a Banach space.

Next we discuss a classical solution method to PDEs using the method of separation of variables.

4.4 Separation of Variables

Solutions to many PDEs can be obtained using the technique known as separation of variables. This solution technique is based on the fact that if $f(x)$ and $g(t)$ are functions of independent variables x and t, respectively, and if $f(x) = g(t)$, then there must be a constant λ for which $f(x) = \lambda$ and $g(t) = \lambda$. Applying the method of separation of variables enables one to obtain two ordinary differential equations (ODE). To illustrate the method of separation of variables for an initial-boundary value problem, we consider the one dimensional wave equation

$$\frac{\partial^2 u}{\partial t^2} = \frac{\partial^2 u}{\partial x^2}, \tag{4.8}$$

$$u(0, t) = u(L, t) = 0 \quad \text{for all} \quad t$$

$$u(x, 0) = f(x)$$

$$\frac{\partial u}{\partial t}(x, 0) = g(x)$$

where L is a constant and $f(x)$ and $g(x)$ are given functions. In order to solve (4.8) using the method of separation of variables, we assume a solution of the form

$$u(x, t) = f(x)g(t) \tag{4.9}$$

By inserting (4.9) into the wave equation, we obtain

$$f(x)g''(t) = f(x)''g(t) \tag{4.10}$$

and dividing by the product $f(x)g(t)$ (4.10) becomes

$$\frac{g''(t)}{g(t)} = \frac{f(x)''}{f(x)} \tag{4.11}$$

Since the left-hand side of (4.11) is a function of x alone and the right-hand side is a function of t alone, both expressions must be independent of x and t. Therefore,

$$\frac{g''(t)}{g(t)} = \frac{f(x)''}{f(x)} = \lambda. \tag{4.12}$$

for a suitable $\lambda \in \mathbb{R}$. Thus, we obtain

$$g''(t) - \lambda g(t) = 0, \tag{4.13}$$

$$f''(x) - \lambda f(x) = 0. \tag{4.14}$$

We allow λ to take any value and then show that only certain values allow you to satisfy the boundary conditions. This gives us three distinct types of solutions that are restricted by the initial and boundary conditions.

Case 1: $\lambda = 0$
In this case, Eqs. (4.13) and (4.14) become

$$g'' = 0 \quad \Rightarrow \quad g(t) = At + B,$$

and

$$f'' = 0 \quad \Rightarrow \quad f(x) = Cx + D,$$

for constants A, B, C, and D. However, the boundary conditions in (4.8) imply that

$$C = 0 \quad \text{and} \quad D = 0.$$

Thus, the only solution, with $\lambda = 0$, is a trivial solution.

Case 2: $\lambda = p^2$
When λ is positive, the separable equations are

$$g'' - p^2 g = 0, \qquad (4.15)$$

and

$$f'' - p^2 f = 0. \qquad (4.16)$$

Equations (4.15) and (4.16) have constant coefficients and so can be solved. The solution to (4.16) is

$$f(x) = Ae^{px} + Be^{-px}. \qquad (4.17)$$

To satisfy the boundary conditions in x, we find that the only solution is the trivial solution with

$$A = 0 \quad \text{and} \quad B = 0.$$

Case 3: $\lambda = -p^2$
This case when the separation constant, λ, is negative is interesting and generates a nontrivial solution. The two equations are now

$$g'' + p^2 g = 0, \qquad (4.18)$$

and

$$f'' + p^2 f = 0. \qquad (4.19)$$

The solution to (4.19) is

$$f(x) = A \cos px + B \sin px. \qquad (4.20)$$

The boundary condition at $x = 0$ implies that

$$u(0, t) = f(0)g(t) = 0 \quad \Rightarrow \quad f(0) = 0, \qquad (4.21)$$

and using this in (4.21) gives

$$A = 0.$$

The condition at $x = L$ gives

$$f(L) = B \sin pL = 0. \tag{4.22}$$

From (4.22), if $B = 0$ we have the trivial solution, but there is a nontrivial solution if

$$\sin pL = 0 \quad \Rightarrow \quad pL = n\pi \quad \Rightarrow \quad p = \frac{n\pi}{L}, \tag{4.23}$$

where n is an integer. The solution for (4.21) follows in a similar manner and we obtain

$$g(t) = C \cos pt + D \sin pt, \tag{4.24}$$

where $p = \frac{n\pi}{L}$. Thus, a solution for $u(x, t)$ is

$$u(x, t) = \sin \frac{n\pi}{L} x \left(C \cos \frac{n\pi}{L} t + D \sin \frac{n\pi}{L} t \right). \tag{4.25}$$

where C and D are constants. The solution of (4.25) satisfy the initial conditions:

$$u(x, 0) = C \sin \frac{n\pi}{L} x$$

and

$$\frac{\partial u}{\partial t}(x, 0) = D \frac{n\pi}{L} \sin \frac{n\pi}{L} x.$$

In order to obtain the general solution, we consider the superposition of solutions of the form (4.25) and obtain

$$u(x, t) = \sum_{n=1}^{\infty} \sin \frac{n\pi}{L} x \left(C_n \cos \frac{n\pi}{L} t + D_n \sin \frac{n\pi}{L} t \right). \tag{4.26}$$

with initial conditions

$$u(x, 0) = f(x), \text{ for } 0 \leq x \leq L$$

and

$$\frac{\partial u}{\partial t}(x, 0) = g(x), \text{ for } 0 \leq x \leq L.$$

The first initial condition

$$u(x, 0) = \sum_{n=1}^{\infty} C_n \sin \frac{n\pi}{L} x = f(x)$$

is satisfied by choosing the coefficients of C_n to be

$$C_n = \frac{2}{L} \int_0^L f(x) \sin \frac{n\pi}{L} x \, dx,$$

the Fourier sine coefficients for $f(x)$. The second initial condition

$$\frac{\partial u}{\partial t}(x, 0) = \sum_{n=1}^{\infty} D_n \frac{n\pi}{L} \sin \frac{n\pi}{L} x = g(x)$$

is satisfied by choosing the coefficients of D_n to be

$$D_n = \frac{2}{n\pi} \int_0^L g(x) \sin \frac{n\pi}{L} x \, dx,$$

the Fourier sine coefficient for $g(x)$. Inserting the values of C_n and D_n into (4.26) gives the complete solution to the wave equation.

Next, we discuss a transformation methodology that is very useful in finance. The Laplace transform is a very important tool in finance since it can produce quasi-analytical formulas delivering real-time option pricing [95].

4.5 Moment-Generating Laplace Transform

The kth moment of a distribution is the expected value of the kth power of the random variable. Generally, moments are hard to compute. Sometimes this calculation may be simplified using a moment generating function (M.G.F).

Definition 4.5.1 (M.G.F). The M.G.F. of a random variable X with probability density function $f(x)$ is a function $M : \mathbb{R} \to [0, \infty)$ given by

$$M_X(t) = E[e^{tX}] = \int_{-\infty}^{\infty} e^{tx} f(x) dx$$

Definition 4.5.2 (Laplace transform). If f is a given positive function, the Laplace transform of f is

$$\mathcal{L}f(t) = \int_{-\infty}^{\infty} e^{-tx} f(x) dx$$

for all t for which the integral exists and is finite.

We note that the two definitions are very much related. In fact we can write $M_X(t) = \int_{-\infty}^{\infty} e^{tx} f(x) dx = \mathcal{L}f_-(t) 1_{[0,\infty)}(t) + \mathcal{L}f_+(t) 1_{(-\infty,0]}(t)$ where

$$f_-(t) = \begin{cases} f(-x), & x > 0 \\ 0, & \text{if otherwise.} \end{cases}$$

$$f_+(t) = \begin{cases} f(x), & x > 0 \\ 0, & \text{if otherwise.} \end{cases}$$

Therefore, since we already know about the Laplace transform of the density function, we may find the M.G.F. of the variable with that density.

We can calculate moments of random variables using M.G.F as follows:

$$M_X(t) = E[e^{tX}] = \sum_{k=0}^{\infty} \frac{E(X^k)}{k!} t^k,$$

using the exponential function expansion. Then the moments are

$$E(X^k) = M_X^{(k)}(0)$$

i.e. the k-th derivative of the M.G.F. calculated at 0.

We remark that if two random variable X and Y are independent, then we can write

$$M_{X+Y}(t) = M_X(t) + M_Y(t).$$

This simplifies greatly calculating M.G.F. for sums (and averages) of independent random variables.

Some properties of the M.G.F are presented below:

1) If X and Y have the same M.G.F, they have the same distribution.
2) If $X_1, \ldots, X_n$, is a sequence of random variable with M.G.F $M_{X_1}(t), \ldots, M_{X_n}(t)$, and

$$M_{X_n}(t) \to M(t)$$

for all t, then $X_n \to X$ in distribution and M.G.F of X is $M(t)$.

For more properties please consult [67].

Theorem 4.5.1 (Inversion theorem). If f has a Laplace transform $\mathcal{L}f(t)$,

$$\mathcal{L}f(t) = \int_0^t e^{-tx} f(x) dx$$

then we must have

$$f(x) = \frac{1}{2\pi i} \lim_{T \to \infty} \int_{C-iT}^{C+iT} e^{tx} \mathcal{L}f(t) dt$$

where C is any constant greater than the real part of all irregularities of $\mathcal{L}f(t)$.

4.5.1 Numeric Inversion for Laplace Transform

In general, finding the inverse Laplace transform theoretically is complicated unless we are dealing with simple functions.

Numerically, it seems like a simple problem. We can potentially approximate the integral using standard quadrature methods. Specifically,

$$\mathcal{L}f(t) = \sum_{i=1}^n w_i e^{-t\zeta_i} f(\zeta_i)$$

where ζ_i are quadrature nodes and w_i are weights.

The idea is to use $t_1, \ldots, t_n$ and write a system of n equations:

$$\mathcal{L}f(t_j) = \sum_{i=1}^n w_i e^{-t_j \zeta_i} f(\zeta_i). \tag{4.27}$$

However, the system (4.27) thus formed is ill conditioned (unstable) in $\mathbb{R}$. Specifically, small changes in values of $\mathcal{L}f(t_j)$ lead to very different solutions. The main reason for this behavior is that the exponential term in (4.27) dominates the values of the function. Thus, this method generally fails.

The proper way is to use complex integrals. By the inversion theorem we have to calculate

$$f(x) = \frac{1}{2\pi i} \int_{C-i\infty}^{C+i\infty} e^{tx} \mathcal{L}f(t) dt. \tag{4.28}$$

In this instance, we take a contour and perform a complex integration. We set $t = C + iu$ which implies that

$$f(x) = \frac{1}{2\pi} \int_{-\infty}^{\infty} e^{Cx} e^{iux} \mathcal{L}f(C+iu) du \tag{4.29}$$

where $e^{iux} = \cos(ux) + i\sin(ux)$.

Transforming (4.29) to $[0, \infty)$, we get

$$f(x) = \frac{2e^{Cx}}{\pi} \int_0^\infty \text{Re}(\mathcal{L}f(C+iu)) \cos(ux) dx \tag{4.30}$$

and

$$f(x) = -\frac{2e^{Cx}}{\pi} \int_0^\infty \text{Im}(\mathcal{L}f(C+iu)) \sin(ux) dx. \tag{4.31}$$

The next step is to approximate these integrals.

4.5.2 Fourier Series Approximation Method

The main idea of the Fourier series approximation method is to apply the trapezoidal rule (see Section 1.11) to Eq. (4.30) and obtain

$$f(x) \simeq \frac{\Delta e^{Cx}}{\pi} \text{Re}(\mathcal{L}f(C)) + \frac{2\Delta e^{Cx}}{\pi} \sum_{k=1}^{\infty} \text{Re}(\mathcal{L}f(C+ik\Delta)) \cos(k\Delta x) \tag{4.32}$$

where Δ is the step in the variable u ($u = k\Delta$). If we set $\Delta = \frac{\pi}{2x}$ and $C = \frac{A}{2x}$ (for some constant A), then we have that $\frac{\Delta}{\pi} = \frac{1}{2x}$ and $Cx = \frac{A}{2}$. This implies that $k\Delta x = k\frac{\pi}{2}$, and therefore $\cos(k\Delta x) = 0$ if k is odd. The Eq. (4.32) becomes

$$f(x) \simeq \frac{\Delta e^{A/2}}{2x} \text{Re}\left(\mathcal{L}f\left(\frac{A}{2x}\right)\right) + \frac{e^{A/2}}{x} \sum_{k=1}^{\infty} (-1)^k \text{Re}\left(\mathcal{L}f\left(\frac{A+2k\pi i}{2x}\right)\right) \tag{4.33}$$

Generally, if f is a probability density function, $C = 0$. We can show that the discretization error if $|f(x)| < M$ is bounded by

$$|f(x) - f_\Delta(x)| < M \frac{e^{-A}}{1 - e^{-A}} \simeq Me^{-A};$$

therefore A should be large. In practice, A is chosen to be approximately 18.4. Next we briefly describe a methodology for calculating the infinite series in Eq. (4.33). The idea is to apply the so-called Euler's algorithm. The idea of the algorithm to use explicitly the first n terms and then a weighted average of next m terms as follows:

$$f_A(x) \simeq E(x, n, m) = \sum_{k=0}^{m} \binom{m}{k} 2^{-m} \Delta_{n+k}(x)$$

where

$$\Delta_n(x) = \frac{Ae^{A/2}}{2x} \operatorname{Re}\left(\mathcal{L}f\left(\frac{A}{2x}\right)\right) + \frac{e^{A/2}}{x} \sum_{k=1}^{\infty} (-1)^k \operatorname{Re}\left(\mathcal{L}f\left(\frac{A + 2k\pi i}{2x}\right)\right).$$

In order for this to be effective, the term

$$a_k = \operatorname{Re}\left(\mathcal{L}f\left(\frac{A + 2k\pi i}{2x}\right)\right)$$

needs to satisfy the following three properties:

1) Has to be the same sign for all k.
2) Has to be monotone in k.
3) The higher order differences $(-1)^m \Delta^m a_{n+k}$ are monotone.

However, during practical implementation, we rarely check these properties. Usually, $E(x, n, m)$ is within 10^{-13} of the integral for $n = 38$ and $m = 11$. The test algorithm for series

$$\sum_{k=1}^{\infty} \frac{(-1)^k}{k}$$

converges to $-\ln 2$. Please see [1] for more details.

Next we present some properties of the Laplace transform methodology. This transformation is well known for its applications to differential equations, especially ODE. Laplace transforms can also be used to compute option prices.

We recall that the Laplace transform function $\mathcal{L}f$ of a function $f(x)$ is defined by

$$\mathcal{L}f(s) = \int_0^{\infty} f(x) e^{-sx} dx,$$

for all numbers s for which the integral converges.

For example, if $f(x) = e^{3x}$, then its Laplace transform $\mathcal{L}f(s)$ is determined by

$$\mathcal{L}f(s) = \int_0^{\infty} e^{3x} e^{-sx} dx$$
$$= \int_0^{\infty} e^{(3-s)x} dx.$$

Clearly $\mathcal{L}f(s)$ is undefined for $s \leq 3$ and $\mathcal{L}f(s) = \frac{1}{s-3}$, when $s > 2$.
We list some properties of the Laplace transform.

4.5 Moment-Generating Laplace Transform

1) Given a function $y(t)$ with Laplace transform $\mathcal{L}(y)$, the Laplace transform of dy/dt is given by

$$\mathcal{L}(\frac{dy}{dt}) = s\mathcal{L}(y) - y(0).$$

2) Given functions f and g and a constant c,

$$\mathcal{L}(f+g) = \mathcal{L}(f) + \mathcal{L}(g),$$

and

$$\mathcal{L}(cf) = c\mathcal{L}(f).$$

Next we define inverse Laplace transform. The notation for this inverse transformation is $\mathcal{L}^{-1}$, that is,

$$\mathcal{L}^{-1}(F) = f,$$

if and only if

$$\mathcal{L}(f) = F.$$

The inverse Laplace transform is also a linear operator.

Example 4.5.1 Compute the inverse Laplace transform of

$$y = \mathcal{L}^{-1}\left[\frac{1}{s-1} - \frac{4}{(s-1)(s+1)}\right].$$

By linearity we have

$$y = \mathcal{L}^{-1}\left[\frac{1}{s-1}\right] - \mathcal{L}^{-1}\left[\frac{4}{(s-1)(s+1)}\right].$$

But we observe $\mathcal{L}(e^t) = \frac{1}{s-1}$. Hence

$$\mathcal{L}^{-1}(\frac{1}{s-1}) = e^t.$$

By the partial fraction decomposition $\frac{4}{(s-1)(s+1)} = \frac{2}{s-1} - \frac{2}{s+1}$. Thus,

$$\mathcal{L}^{-1}\left[\frac{4}{(s-1)(s+1)}\right] = \mathcal{L}^{-1}\left[\frac{2}{s-1}\right] - \mathcal{L}^{-1}\left[\frac{2}{s+1}\right]$$

$$= 2\mathcal{L}^{-1}\left[\frac{1}{s-1}\right] - 2\mathcal{L}^{-1}\left[\frac{1}{s+1}\right]$$

$$= 2e^t - 2e^{-t}.$$

Hence

$$y = \mathcal{L}^{-1}\left[\frac{1}{s-1} - \frac{4}{(s-1)(s+1)}\right] = e^t - (2e^t - 2e^{-t}) = -e^t + 2e^{-t}.$$

Example 4.5.2 Consider the initial value problem

$$\frac{dy}{dt} = y - 4e^{-t}, \quad y(0) = 1.$$

Taking the Laplace transform of both sides, we obtain

$$\mathcal{L}(\frac{dy}{dt}) = \mathcal{L}(y - 4e^{-t}).$$

Therefore, using the properties of Laplace transform, we obtain,

$$s\mathcal{L}(y) - y(0) = \mathcal{L}(y) - 4\mathcal{L}(e^{-t}).$$

This implies (using $y(0) = 1$),

$$s\mathcal{L}(y) - 1 = \mathcal{L}(y) - 4\mathcal{L}(e^{-t}).$$

Now $\mathcal{L}(e^{at}) = \frac{1}{(s-a)}$. Therefore the previous equation becomes

$$s\mathcal{L}(y) - 1 = \mathcal{L}(y) - \frac{4}{s+1}.$$

Hence

$$\mathcal{L}(y) = \frac{1}{s-1} - \frac{4}{(s-1)(s+1)}.$$

Thus

$$y = \mathcal{L}^{-1}\left[\frac{1}{s-1} - \frac{4}{(s-1)(s+1)}\right].$$

This is solved in Example 4.5.1. We obtained

$$y = \mathcal{L}^{-1}\left[\frac{1}{s-1} - \frac{4}{(s-1)(s+1)}\right] = e^t - (2e^t - 2e^{-t}) = -e^t + 2e^{-t}.$$

4.6 Application of the Laplace Transform to the Black–Scholes PDE

We recall the forward time form of the Black–Scholes PDE,

$$\frac{\partial F}{\partial t} + rS\frac{\partial F}{\partial S} + \frac{1}{2}\sigma^2 S^2 \frac{\partial^2 F}{\partial S^2} = rF$$

with the boundary condition $F(T, S) = \Phi(S)$ (this is $(S - K)_+$) for a Call option, for example).

Define $\zeta = (T - t)\frac{\sigma^2}{2}$, $x = \ln S$ and f such that $F(t, S) = f(\zeta, x)$; then using this change of variables the Black–Scholes PDE becomes

$$-\frac{\partial f}{\partial \zeta} + \left(\frac{r}{\sigma^2/2} - 1\right)\frac{\partial f}{\partial x} + \frac{\partial^2 f}{\partial x^2} - \frac{r}{\sigma^2/2}f = 0 \qquad (4.34)$$

where $f(0, x) = F(T, e^x) = \Phi(e^x)$ (which is $(e^x - K)_+$) for a Call option).

4.6 Application of the Laplace Transform to the Black–Scholes PDE

Applying Laplace transform to $f(\zeta, x)$ in ζ, we obtain

$$(\mathcal{L}f)(u,x) = \int_0^\infty e^{-u\zeta} f(\zeta, x) d\zeta$$

$$\left(\mathcal{L}\frac{\partial f}{\partial \zeta}(\zeta, x)\right)(u, x) = u\mathcal{L}f(u, x) - f(0, x)$$

$$\left(\mathcal{L}\frac{\partial f}{\partial \zeta}(\zeta, x)\right)(u, x) = \int_0^\infty e^{-u\zeta} \frac{\partial f}{\partial \zeta}(\zeta, x) d\zeta = \frac{\partial}{\partial x} \int_0^\infty e^{-u\zeta} f(\zeta, x) d\zeta$$

since $\mathcal{L}\left(\frac{\partial f}{\partial x}\right)(u, x) = \frac{\partial \mathcal{L}f}{\partial x}(u, x)$ and similarly, $\mathcal{L}\left(\frac{\partial^2 f}{\partial x^2}\right)(u, x) = \frac{\partial^2 \mathcal{L}f}{\partial x^2}(u, x)$.

For simplicity of notations we use $\mathcal{L}f(u, x) = \hat{f}(u, x)$, and the Eq. (4.34) reduces to

$$-\left(u\hat{f} - \Phi(e^x)\right) + \left(\frac{r}{\sigma^2/2} - 1\right)\frac{\partial \hat{f}}{\partial x} + \frac{\partial^2 \hat{f}}{\partial x^2} - \frac{r}{\sigma^2/2}\hat{f} = 0. \tag{4.35}$$

We observe that (4.35) is now a second order ODE. Let's denote $m = \frac{r}{\sigma^2/2}$ and for the Call option $\Phi(e^x) = (e^x - e^k)_+$ where $K = e^k$; then ODE (4.35) becomes

$$\frac{\partial^2 \hat{f}}{\partial x^2} + (m-1)\frac{\partial \hat{f}}{\partial x} - (m+u)\hat{f} + (e^x - e^k)_+ = 0. \tag{4.36}$$

The boundary values for Eq. (4.36) are obtained by applying the Laplace transform as follows:

$$\hat{f}(u, x) \to \mathcal{L}(e^z - e^{mt}e^k) = \frac{e^x}{u} - \frac{e^k}{u+m} \quad \text{as } x \to \infty$$

$$\hat{f}(u, x) \to \mathcal{L}(0) = 0 \quad \text{as } x \to \infty$$

In order to solve the ODE (4.36) for $\hat{f}$, we multiply through the equation with an integration factor $e^{-\alpha x}$ where $\alpha = \frac{1-m}{2}$ and specifically choose $\hat{g}$ such that $\hat{f}(u, x) = e^{\alpha x}\hat{g}(u, x)$. Thus the first derivative will cancel out and obtain

$$\frac{\partial^2 \hat{g}}{\partial x^2} - (b+u)\hat{g} + e^{-\alpha x}(e^x - e^k)_+ = 0 \tag{4.37}$$

where $b = \alpha^2 + m = \frac{(m-1)^2}{4} + m$. We solve (4.37) on $x > k$ and on $x \le k$ to get

$$\hat{g}(u, x) = \begin{cases} \frac{e^{-(\alpha-1)x}}{u} - \frac{e^{-\alpha x+k}}{u+m} + h_1(u, x)A_1 + h_2(u, x)A_2 & \text{if } x > k \\ h_1(u, x)B_1 + h_2(u, x)B_2 & \text{if } x < k \end{cases}$$

where $h_1(u, x) = e^{-\sqrt{b+ux}}$, $h_2(u, x) = e^{\sqrt{b+ux}}$ and $A_1, A_2, B_1,$ and B_2 are constants to be determined from the boundary conditions. We have that

$$\lim_{x \to \infty} e^{\alpha x} h_1(u, x) = \lim_{x \to \infty} e^{\alpha x - \sqrt{b+ux}} = 0 \text{ when } u > 0$$

and also

$$\lim_{x \to \infty} e^{ax} h_2(u, x) = \infty \text{ when } u > 0.$$

We must have $A_2 = 0$; otherwise the solution will explode, i.e. it will diverge. Similarly when $x < k$ the same reasoning for $x \to -\infty$ gives $B_1 = 0$. Therefore

$$\hat{g}(u, x) = \begin{cases} \dfrac{e^{-(a-1)x}}{u} - \dfrac{e^{-ax+k}}{u+m} + h_1(u, x) A_1 \text{ if } x > k \\ h_2(u, x) B_2 \text{ if } x < k \end{cases}$$

Using the relation

$$\lim_{\substack{x \to k \\ x > k}} \hat{f}(u, x) = \lim_{\substack{x \to k \\ x < k}} \hat{f}(u, x),$$

$$\lim_{\substack{x \to k \\ x > k}} \frac{\partial \hat{f}}{\partial x} = \lim_{\substack{x \to k \\ x < k}} \frac{\partial \hat{f}}{\partial x}$$

and after performing some algebra, we get

$$A_1(u) = \frac{e^{(1-a+\sqrt{b+u})k}(u - (a - 1 + \sqrt{b+u})m)}{2u\sqrt{b+u}(u+m)}$$

$$B_2(u) = \frac{e^{(1-a-\sqrt{b+u})k}(u - (a - 1 - \sqrt{b+u})m)}{2u\sqrt{b+u}(u+m)}$$

Thus,

$$\hat{f}(u, x) = e^{ax} \left(\frac{e^{-(a-1)x}}{u} - \frac{e^{-ax+k}}{u+m} \right) 1_{z > k}$$
$$+ \frac{e^{\sqrt{b+u}(x-k)+(1-a)k}(u - (a - 1 + \sqrt{b+u}\operatorname{sgn}(x-k))m)}{2u\sqrt{b+u}(u+m)}$$

where

$$\operatorname{sgn}(x) = \begin{cases} 1 \text{ if } x \geq 0 \\ -1 \text{ if } x < 0 \end{cases},$$

and the indicator function $1_{z > k}$ is one if the condition is true and 0 otherwise.

Finally, in order to obtain the solution for the original PDE, we need to numerically invert this exact form of the Laplace transform.

Finite difference methods and Monte Carlo methods can also be used to approximate a PDE. These methods will be discussed in Chapters 7 and 9, respectively. In those two chapters, we will also discuss how to apply the methods to option pricing.

4.7 Problems

1. (a) Find the solution to the following PDE:
$$\frac{\partial^2 u}{\partial t^2} = \frac{\partial^2 u}{\partial x^2}$$
$$u(0, t) = u(L, t) = 0$$
$$u(x, 0) = 0$$
$$\frac{\partial u}{\partial t}(x, 0) = f(x)$$
   where L is a constant and f is a given function.
   (b) Find the solution when $f(x) = \sin x$.

2. Find the solution to the following PDE:
$$\frac{\partial^2 u}{\partial t^2} = \frac{\partial^2 u}{\partial x^2}$$
$$u(0, t) = u(\pi, t) = 0$$
$$u(x, 0) = 0$$
$$\frac{\partial u}{\partial t}(x, 0) = g(x)$$

3. Find the solution to the Laplace equation in R^3:
$$\frac{\partial^2 u}{\partial x^2} + \frac{\partial^2 u}{\partial y^2} + \frac{\partial^2 u}{\partial z^2} = 0$$
   assuming that
$$u(x, y, z) = f(x)g(y)h(z).$$

4. Solve the following initial value problem using Laplace transform:
   (a)
$$\frac{dy}{dt} = -y + e^{-3t}, \quad y(0) = 2.$$
   (b)
$$\frac{dy}{dt} + 11y = 3, \quad y(0) = -2.$$
   (c)
$$\frac{dy}{dt} + 2y = 2e^{7t}, \quad y(0) = 0$$

5. Explain what is a linear second order differential equation and a nonlinear second order differential equation, and give examples.

6. Solve the following boundary value problem:
$$y''(x) = f(x), \quad y(0) = 0, \quad y(1) = 0.$$
   Hence solve $y''(x) = x^2$ subject to the same boundary conditions.

5
Wavelets and Fourier Transforms

5.1 Introduction

In this chapter we continue studying transformation methods. The main idea behind the techniques presented is to transform data from the spatial domain to a frequency domain. We start by introducing the Fourier transform since it is simpler and historically older.

5.2 Dynamic Fourier Analysis

A physical process can be described either in the time domain, by the values of some quantity f as a function of time t, for example, $f[t]$, or in the frequency domain, where the process is specified by giving its amplitude H (generally a complex number) as a function of frequency n, that is, $F[n]$, with $-\infty < n < \infty$. For many applications, it is useful to think of $f[t]$ and $F[n]$ as being two different representations of the same function. Generally, these two representations are connected using Fourier transform formulas:

$$F[n] = \int_{-\infty}^{\infty} f[t]e^{2\pi int} dt = \int_{-\infty}^{\infty} f[t] (\cos(2\pi nt) + i\sin(2\pi nt)) dt \quad (5.1)$$

$$f[t] = \int_{-\infty}^{\infty} F[n]e^{-2\pi int} dn,$$

where $F[n]$ represents the Fourier transform and $f[t]$ represents the inverse Fourier transform. The Fourier transform $F[n]$ converts the data from the time domain into the frequency domain. The inverse Fourier transform $f[t]$ converts the frequency-domain components back into the original time-domain signal. A frequency-domain plot shows how much of the signal lies within each given frequency band over a range of frequencies.

In order to analyze a statistical time series using Fourier transforms, we need to assume that the structure of the statistical or stochastic process which generates the observations is essentially invariant through time. This assumption is

Quantitative Finance, First Edition. Maria C. Mariani and Ionut Florescu.
© 2020 John Wiley & Sons, Inc. Published 2020 by John Wiley & Sons, Inc.

summarized in the condition of stationarity which states that its finite dimensional distribution remains the same throughout time.

This condition is hard to verify and we generally enforce a weak stationarity condition. For example, a time series x_t is weak stationary, if its second order behavior remains the same, regardless of the time t. Looking at the Fourier transform representation (5.1), we see that a stationary series is represented as a superposition of sines and cosines that oscillate at various frequencies. Therefore, a stationary time series can be matched with its sine and cosine series representation (these are the representative frequencies).

Next we introduce some definitions that will be used when applying the dynamic Fourier analysis.

5.2.1 Tapering

Generally, calculating the continuous Fourier transform and its inverse (5.1) is a hard problem in practice. For this reason, a discrete Fourier transform (DFT) is used when we observe discrete time series with a finite number of samples from a process that is continuous in time. We will discuss the DFT in Section 5.2.3; however applying it requires an extra step known as tapering.

When the original function (process) is discontinuous, the corresponding signal value abruptly jumps, yielding spectral leakage (that is, the input signal does not complete a whole number of cycles within the DFT time window). In this case to perform DFT, we need to multiply the finite sampled times series by a windowing function or "a taper." The taper is a function that smoothly decays to zero near the ends of each window, and it is aimed at minimizing the effect of the discontinuity by decreasing the time series magnitude, so it approaches zero at the ends of the window. Although, spectral leakage cannot be prevented, it can be significantly reduced by changing the shape of the taper function in a way to minimize strong discontinuities close to the window edges. In seismic data analysis cosine tapers are often used since it is both effective and easy to calculate. The cosine taper can be written as

$$c(t) = \begin{cases} \frac{1}{2}\left(1 - \cos\frac{\pi t}{a}\right) & 0 \leq t \leq a \\ 1 & a \leq t \leq (1-a) \\ \frac{1}{2}\left(1 - \cos\frac{\pi}{a}(1-t)\right) & (1-a) \leq t \leq 1 \end{cases}$$

where t is time and a is the taper ratio. The cosine window represents an attempt to smoothly set the data to zero at the boundaries while not significantly reducing the level of the values within the window. This form of tapering reduces the leakage of the spectral power from a spectral peak to frequencies far away, and it coarsens the spectral resolution by a factor $1/(1-a)$ for the aforementioned cosine tapers.

In general, the effects of applying a taper are:

1) Decrease the time series magnitude to zero or near zero at its start and end so that there is no sharp discontinuity between the first and last point in the periodic time series.
2) Change the weighting of samples in the time series so that those near the middle contribute more to the Fourier transform.
3) Reduce the resolution of the spectrum by averaging adjacent samples.

For nonstationary time series signals, tapering may bias the spectral amplitudes even if the taper is normalized. However, for these we should not use Fourier transform in the first place.

Generally, it is difficult to give precise recommendations on which tapers to use in all specific situation. The work of [23] recommends to reduce the leakage by increasing the length N of the time window and at the same time decreasing the taper ratio a, such that the length (Na) of the cosine half-bells is kept constant.

In practice, a value of $a = 5\%$ for window lengths of 30 seconds or 60 seconds is good enough for frequencies down to 0.2 Hz. See [167] for more details.

5.2.2 Estimation of Spectral Density with Daniell Kernel

In this section we present a probabilistic approach to estimate the magnitudes and frequency coefficients in a Fourier transform. A stationary process X_t may be defined by taking linear combinations of the form

$$X_t = \sum_{j=1}^{m}(A_j \cos(2\pi\lambda_j t) + B_j \sin(2\pi\lambda_j t)) \tag{5.2}$$

where $0 \leq \lambda \leq \frac{1}{2}$ is a fixed constant and $A_1, B_1, A_2, B_2, \ldots A_m$, and B_m are all uncorrelated random variables with mean zero and

$$Var(A_j) = Var(B_j) = \sigma_j^2.$$

We assume $\sum_{j=1}^{m} \sigma_j^2 = \sigma^2$ so that the variance of the process X_t is σ^2, and let the spectral density $f(\lambda)$ satisfy the equation

$$\int_{-\frac{1}{2}}^{\frac{1}{2}} f(\lambda)d\lambda = \sigma^2.$$

Then, the process given in (5.2) converges to a stationary process with spectral density f as $m \to \infty$.

In order to estimate the spectral density f, we define the estimators as a weighted average of periodogram values (I) for frequencies in the range $(j-m)/n$ to $(j+m)/n$. In particular we define

$$\hat{f}(j/n) = \sum_{k=-m}^{m} W_m(k) I\left(\frac{j+k}{n}\right)$$

The set of weights $\{W_m(k)\}$ sum to one and the set is often referred to as a kernel or a spectral window. Essentially, this kernel with parameter m is a centered moving average which creates a smoothed value at time t by averaging all values between and including times $t - m$ and $t + m$.

We define

$$W_m(k) = \frac{1}{2m+1} \quad \text{for } -m \le k \le m, \quad \sum_k W_m(k) = 1 \quad \text{and} \quad \sum_k k W_m(k) = 0$$

The smoothing formula $\{u_t\}$ for a Daniell kernel with $m = 1$ corresponds to the three weights $(\frac{1}{3}, \frac{1}{3}, \frac{1}{3})$ and is given by

$$\hat{u}_t = \frac{1}{3}(u_{t-1} + u_t + u_{t+1})$$

Applying the Daniell kernel again on smoothed values $\{\hat{u}_t\}$ produces a more extensive smoothing by averaging over a wider time interval.

$$\hat{\hat{u}}_t = \frac{\hat{u}_{t-1} + \hat{u}_t + \hat{u}_{t+1}}{3} = \frac{1}{9}u_{t-2} + \frac{1}{9}u_{t-2} + \frac{3}{9}u_t + \frac{2}{9}u_{t+1} + \frac{1}{9}u_{t+2}. \quad (5.3)$$

Consequently applying the Daniell kernel transforms the spectral windows into a form of Gaussian probability density function.

5.2.3 Discrete Fourier Transform

The DFT is the equivalent of the continuous Fourier transform for signals known only at N instants separated by sample times T (i.e. a finite sequence of data). It is commonly used in practical applications since we generally cannot observe a signal continuously. The DFT converts a finite sequence of equally spaced samples of a function into a sequence of equally spaced samples of the discrete-time Fourier transform (DTFT), which is a complex-valued function of the frequency.

The DFT is the most important discrete transform, used to perform Fourier analysis in many practical applications. In image processing, the samples can be the values of pixels along a row or column of an image. The DFT can also be used to efficiently solve partial differential equations and to perform other operations such as convolutions or multiplying large integers.

The DFT of the sequence $f[k]$ is defined as

$$F[n] = \sum_{k=0}^{N-1} f[k] e^{-j\frac{2\pi}{N}nk}, n = 0, \ldots, N-1. \quad (5.4)$$

We observe that Eq. (5.4) can be written in matrix form as

$$\begin{bmatrix} F[0] \\ F[1] \\ F[2] \\ F[3] \\ \vdots \\ F[N-1] \end{bmatrix} = \begin{bmatrix} 1 & 1 & 1 & 1 & \cdots & 1 \\ 1 & W & W^2 & W^3 & \cdots & W^{N-1} \\ 1 & W^2 & W^4 & W^6 & \cdots & W^{N-2} \\ 1 & W^3 & W^6 & W^9 & \cdots & W^{N-3} \\ \vdots & & & & & \\ 1 & W^{N-1} & W^{N-2} & W^{N-3} & \cdots & W \end{bmatrix} \begin{bmatrix} f[0] \\ f[1] \\ f[2] \\ f[3] \\ \vdots \\ f[N-1] \end{bmatrix}$$

where $W = \exp(-j2\pi/N)$. Since $W^N = 1$ note that the columns are powers of the initial column. In linear algebra this is a particular type of a Vandermonde matrix and multiplying vectors with it is a very fast operation.

The inverse DFT corresponding to (5.4) is

$$f[k] = \frac{1}{N} \sum_{n=0}^{N-1} F[n] e^{j\frac{2\pi}{N}nk}, k = 0, \ldots, N-1. \tag{5.5}$$

Thus the inverse matrix is $1/N$ times the complex conjugate of the original matrix.

In the process of taking the inverse transform, the terms $F[n]$ and $F[N-n]$ combine to produce two frequency components, only one of which is considered to be valid (the one at the lower of the two frequencies). From (5.5), the contribution to $f[k]$ of $F[n]$ and $F[N-n]$ is

$$f_n[k] = \frac{1}{N} \left\{ F[n] e^{j\frac{2\pi}{N}nk} + F[N-n] e^{j\frac{2\pi}{N-n}k} \right\} \tag{5.6}$$

For all $f[k]$ real, $F[N-n] = \sum_{k=0}^{N-1} f[k] e^{-j\frac{2\pi}{N}(N-n)k}$.

The time taken to evaluate a DFT on a digital computer depends principally on the number of multiplications involved, since these are the slowest operations. With the DFT, this number is directly related to N^2 (matrix multiplication of a vector), where N is the length of the transform. For most problems, N is chosen to be at least 256 in order to get a reasonable approximation for the spectrum of the sequence under consideration; hence computational speed becomes a major consideration. Efficient computer algorithms for estimating DFT are known as fast Fourier transforms (FFT) algorithms (Section 5.2.4). These algorithms rely on the fact that the standard DFT involves a lot of redundant calculations.

For instance, we can rewrite

$$F[n] = \sum_{k=0}^{N-1} f[k] e^{-j\frac{2\pi}{N}nk}, n = 0, \ldots, N-1$$

as

$$F[n] = \sum_{k=0}^{N-1} f[k] W_N^{nk}, n = 0, \ldots, N-1. \tag{5.7}$$

We observe that in (5.7), the same values of W_N^{nk} are calculated many times as the computation proceeds.

5.2.4 The Fast Fourier Transform (FFT) Method

The work of Heston [95] in stochastic volatility modeling marks the first time Fourier inversion appears in option pricing literature. Since then, Fourier inversion has become more and more prevalent. This is due to the fact that analytic formulas are hard to calculate and can only be done in the simplest of models.

The FFT allows us to compute efficiently sums of the form

$$W_m = \sum_{j=1}^{N} \exp\left(\frac{2\pi i (j-1)(m-1)}{N}\right) x_k, \quad [28] \qquad (5.8)$$

with a complexity of $O(N \log N)$. The big O notation is a mathematical notation that describes the limiting behavior of a function when the argument tends towards a particular value. It will be described in details in Chapter 6.

Discretizing the semidefinite integral with equidistant quadrature roots $\Delta x = x_{i+1} - x_i$ gives

$$\int_0^\infty e^{-ivk} \psi_\alpha(v) dv \approx \sum_{j=1}^{N} e^{-i(j-1)\Delta x k} \psi_\alpha(x_j). \qquad (5.9)$$

In finance this expression will allow us to simultaneously compute option prices for a range of N strikes. We choose

$$k = -\frac{N\Delta k}{2} + (m-1)\Delta k, \quad m = 1, \ldots, N. \qquad (5.10)$$

Here the discretization in (log)-strike domain and the quadrature discretization have to obey the Nyquist relation:

$$\Delta k \Delta x = \frac{2\pi}{N}, \quad [64] \qquad (5.11)$$

which effectively restricts the choice of the integration domain and the range of strikes to be computed. Therefore, in practice we have to balance the interpolation error in the strike domain against the quadrature error. The quadrature error comes from the numerical integration.

The Fourier transform methods arise naturally with the Levy–Khintchine formula (see [21, 121]). In fact, the characteristic function of a distribution μ corresponds to the Fourier transform of the density function. The characteristic function of a distribution μ on $\mathbb{R}$ is defined as

$$\phi_\mu(u) = \hat{\mu}(u) = \int_\mathbb{R} e^{iux} \mu(dx). \qquad (5.12)$$

In the case where μ is a probability density function, its Fourier transform may be written in term of its moments. If we define the n-th moment of μ by $\mu_n = \int x^n \mu(dx)$, then from (5.12) we obtain

$$\phi_\mu(u) = \int_{\mathbb{R}} \sum_{n=0}^{\infty} \frac{(iux)^n}{n!} \mu(dx) = \sum_{n=0}^{\infty} \frac{(iu)^n}{n!} \mu_n.$$

Fourier transform methods have become popular in finance in the context of exponential Lévy processes primarily due to lack of knowledge of the density functions in closed form and the impossibility to obtain closed form expressions for option prices. We will study Lévy processes in Chapter 12 of this book.

As we stated, the FFT methods have become a very important tool in finance. The fact that many characteristic functions can be expressed analytically using it makes the Fourier transform a natural approach to compute option prices.

Theorem 5.2.1 Let f be a function $f : \mathbb{R} \to \mathbb{R}$, continuous, bounded in $L^1(\mathbb{R})$, so that $F(f) \in L^1(\mathbb{R})$. Then

$$E[f(X_t)] = F^{-1}[\tilde{f}](.)\phi_{X_t}(.)]_0,$$

with F denoting the Fourier transform operator and $\tilde{f}(x) = f(-x)$.

Proof. We observe that

$$\int_{\mathbb{R}} f(x)\mu(dx) = \int_{\mathbb{R}} f(y-x)\mu(dx)|_{y=0},$$

and the convolution product is integrable since f is bounded and continuous. Hence the theorem follows. $\square$

This theorem allows us to compute option values which are typically expressed as expectations of the final payoff f from the Fourier transform of the density of X_t (its characteristic function). When applied to the option pricing problem, the assumptions on f are usually not satisfied since most payoff functions are not in $L^1(\mathbb{R})$. The payoff function is usually replaced with a truncated version to satisfy the assumption.

Two Fourier transformations are possible, one with respect to the log strike price of the option, the other with respect to the log spot price at T. We will consider the first approach.

Theorem 5.2.2 Define the truncated time value option by

$$z_T(k) = E[(e^{X_T} - e^k)_+] - (1 - e^k)_+,$$

Then for an exponential martingale e^{X_T} such that $E[e^{(1+\alpha)X_T}] < \infty$, the Fourier transform of the time value of the option is defined by

$$\zeta_T(v) = \int_{-\infty}^{\infty} e^{ivk} z_T(k) dk = e^{ivrT} \frac{\Phi_T(v-i) - 1}{iv(1+iv)}.$$

The price of the option is then given by

$$C(t,x) = x\left[(1 - e^{k_t}) + \frac{1}{2\pi}\int_{\mathbb{R}}\zeta_T(v)e^{-ivk_t}dv\right].$$

Proof. We sketch an outline of the proof: The discounted stock process $\tilde{S}_t = e^{X_t}$ is a martingale. Therefore

$$\int_{-\infty}^{\infty} e^x q_{X_t}(x) dx = 1.$$

The stock process admits a moment of order $1 + \alpha$ so that there exists $\alpha > 0$, such that

$$\int_{-\infty}^{\infty} e^{(1+\alpha)x} q_{X_t}(x) dx < 0,$$

which can be used to show both integrals $\int_{-\infty}^{\infty} xe^x q_{X_t}(x)$ and $\int_{-\infty}^{\infty} e^x q_{X_t}(x) dx$ are finite. Using Fubini's theorem, we explicitly compute the Fourier transform z_T as follows: We write

$$z_T(k) = E[(e^{X_T} - e^k)_+] - (1 - e^k)_+$$

$$= \int_{-\infty}^{\infty}[(e^x - e^k)1_{x \geq k} - (1 - e^k)1_{0 \geq k}]q_{X_t}(x)dx$$

$$\int_{-\infty}^{\infty}[(e^x - e^k)(1_{x \geq k} - 1_{0 \geq k})]q_{X_t}(x)dx,$$

where in the second term we use the fact that $\tilde{S}_t$ is a martingale. Assuming we can interchange the integrals, the Fourier transform of z_T is given by

$$F(z_T)[v] = \int_{-\infty}^{\infty} e^{ivk} z_T(k) dk$$

$$= \int_{-\infty}^{\infty} e^{ivk} \int_{-\infty}^{\infty}(e^x - e^k)(1_{x \geq k} - 1_{0 \geq k})q_{X_t}(x)dxdk$$

$$= \int_{-\infty}^{\infty} q_{X_t}(x) \int_{-\infty}^{\infty} e^{ivk}(e^x - e^k)(1_{x \geq k} - 1_{0 \geq k})dkdx$$

$$= \int_{-\infty}^{\infty} q_{X_t}(x) \int_0^x e^{ivk}(e^x - e^k)dkdx.$$

When applying the Fubini's theorem, we observe that when $x \geq 0$, $\int_0^x |e^{ivk}(e^x - e^k)| \leq e^x(x - 1)dk$ and $\int_{-\infty}^{\infty} e^x(x-1)q_{X_t}(x)dx < \infty$. Also, when $x < 0$, $\int_0^x |e^{ivk}(e^x - e^k)| \leq (e^x - 1)dk$ and $\int_{-\infty}^{\infty}(e^x - 1)q_{X_t}(x)dx < \infty$. Thus,

$$F(z_t)[v] = \int_{-\infty}^{\infty} q_{X_t}(x) \int_0^x e^{ivk}(e^x - e^k)dkdx$$

$$= \int_{-\infty}^{\infty} q_{X_t}(x) \left[\frac{e^x}{iv}(e^{ivx} - 1) - \frac{1}{iv+1}(e^{(iv+1)x} - 1)\right]dx$$

$$= \int_{-\infty}^{\infty} q_{X_t}(x)\frac{e^{(iv+1)x} - 1}{iv(iv+1)}dx.$$

$\square$

This approach has the advantage of not requiring an exact value of α in order to compute the option prices. The numerical implementation for the use of this truncation function, which is not smooth, has the inconvenience of generating rather large truncation errors, making the convergence slow. Thus we may observe that the Fourier transform of the time value of the option is

$$\zeta_T(v) = e^{ivrT} \frac{\Phi_T(v-i) - 1}{iv(1+iv)} \sim |v|^{-2}.$$

5.3 Wavelets Theory

Recall that in order to perform Fourier decomposition, the time series as well as the process needs to be stationary. From a regression point of view, we may imagine a system responding to various driving frequencies by producing linear combinations of sine and cosine functions. If a time series X_t is stationary, its mean, variance, and covariances remain the same regardless of the time t; therefore we can match a stationary time series with sines and cosines because they behave the same indefinitely.

Nonstationary time series require a deeper analysis. The concept of wavelet analysis generalizes dynamic Fourier analysis, with functions that are better suited to capture the local behavior of nonstationary time series. These functions are called wavelet functions.

Wavelet transform is often compared with the Fourier transform, in the sense that we have a similar decomposition with wavelets replacing the sene functions. In fact, the Fourier transform may be viewed as a special case of a continuous wavelet transform with the choice of the wavelet function

$$\psi(t) = e^{-2\pi it} \psi(t) = e^{-2\pi it}.$$

The main difference between a wavelet transform and the Fourier transform is the fact that the wavelets analysis is a more general method which is localized in both time and frequency and has a faster computation speed; whereas the standard Fourier analysis is only localized in frequency [189].

As a mathematical tool, wavelets can be used to extract information from various kinds of data, including digital signals, images, and several others. Wavelets are well suited for approximating data with sharp discontinuities [82].

5.3.1 Definition

The wavelet transform of a function $f(t)$ with finite energy is defined as the integral transform for a family of functions

$$\psi_{\lambda,t}(u) \equiv \frac{1}{\sqrt{\lambda}} \psi\left(\frac{u-t}{\lambda}\right)$$

and is given as

$$Wf(\lambda, t) = \int_{-\infty}^{\infty} f(u)\psi_{\lambda,t}(u)du \quad \lambda > 0. \tag{5.13}$$

The function ψ is called the wavelet function (mother wavelet). The λ is a scale parameter and t a location parameter, and the generated functions $\psi_{\lambda,t}(u)$ are called wavelets. In the case where $\psi_{\lambda,t}(u)$ is complex, we use complex conjugate $\overline{\psi}_{\lambda,t}(u)$ in the definition (5.13). The normalizing constant $\frac{1}{\sqrt{\lambda}}$ is chosen so that

$$|\psi_{\lambda,t}(u)|^2 \equiv \int |\psi_{\lambda,t}(t)|^2 du = \int |\psi(t)|^2 dt = 1$$

for all the scales λ.

The choice of the mother wavelet $\psi(t)$ is critical. Different choices lead to different decompositions, and the right choice very much depends on the particular data studied. The only property the mother wavelet function $\psi(t)$ has is unit energy (or variance in probability, i.e. $|\psi(t)|^2_{L^2} = \int |\psi(t)|^2 dt = 1$). However, to produce a good decomposition it must possess the following properties:

- Fast decay as $|t| \to \infty$ in order to obtain localization in space
- Zero mean, that is, $\int \psi(t)dt = 0$

These two properties discussed earlier typically make the function look like a wave reflected around 0 which is why the function $\psi(t)$ is called a wavelet. Please refer to [75] for more details.

5.3.2 Wavelets and Time Series

The wavelet function ψ is in effect a band-pass filter. When we scale it using λ for each level, the bandwidth is reduced. This creates the problem that in order to cover the entire spectrum, an infinite number of levels would be required. To deal with this in practice an extra decomposition is used. Wavelets are generated by a scaling function (father wavelet), ϕ, in addition to a mother wavelet function, ψ. The scaling function is used to capture the smooth, low-frequency nature of the data; whereas the wavelet functions are used to capture the detailed and high-frequency nature of the data.

The scaling function integrates to one, and the wavelet function integrates to zero:

$$\int \phi(t)dt = 1 \quad \text{and} \quad \int \psi(t)dt = 0. \tag{5.14}$$

Generally, the analytic form of wavelets does not exist and they are typically generated using numerical schemes. Unlike in the Fourier transform case, here we talk about time and scale, rather than time and frequency. The departure from the periodic functions (sin and cos) means that frequency looses its precise meaning.

5.3 Wavelets Theory

The orthogonal wavelet decomposition of a time series x_t, $t = 1, \ldots n$ is defined as

$$x_t = \sum_k s_{J,k} \phi_{J,k}(t) + \sum_k d_{J,k} \psi_{J,k}(t)$$
$$+ \sum_k d_{J-1,k} \psi_{J-1,k}(t) + \ldots + \sum_k d_{1,k} \psi_{1,k}(t) \tag{5.15}$$

where the J-th level represents the number of scales (frequencies) for the orthogonal wavelet decomposition of the time series x_t. The index k ranges from 1 to the number of coefficients associated with the specified component. The functions $\phi_{J,k}(t), \psi_{J,k}(t), \psi_{J-1,k}(t), \ldots$, and $\psi_{1,k}(t)$ are generated from the scaling function, $\phi(t)$, and the wavelet function, $\psi(t)$, by shifting and scaling based on powers of 2:

$$\phi_{J,k}(t) = 2^{\frac{-J}{2}} \phi\left(\frac{t - 2^J k}{2^J}\right),$$

$$\psi_{j,k}(t) = 2^{\frac{-j}{2}} \psi\left(\frac{t - 2^j k}{2^j}\right), \quad j = 1, \ldots, J,$$

where $2^j k$ is the shift parameter and 2^j is the scale parameter. The wavelet functions are spread out and shorter for larger values of j and narrow and taller for smaller values of the scale ([189]). The reciprocal of the scale parameter ($\frac{1}{2^j}$) in wavelet analysis is analogous to frequency ($\omega_j = \frac{j}{n}$) in Fourier analysis. This is as a result of the fact that larger values of the scale refer to slower, smoother movements of the signal, and smaller values of the scale refer to faster, finer movements of the signal.

The **discrete wavelet transform (DWT)** of a time series data x_t is given by the coefficients $s_{J,k}$ and $d_{J-1,k}$ for $j = J, J-1, \ldots, 1$ in (5.15). To some degree of approximation, they are given by

$$s_{J,k} = n^{-1/2} \sum_{t=1}^{n} x_t \phi_{J,k}(t), \tag{5.16}$$

$$d_{j,k} = n^{-1/2} \sum_{t=1}^{n} x_t \psi_{j,k}(t) \quad j = J, J-1, \ldots, 1. \tag{5.17}$$

The magnitudes of these coefficients measure the importance of the corresponding wavelet term in describing the behavior of the time series x_t. The $s_{J,k}$ are known as the smooth coefficients because they describe the smooth behavior of the data. The $d_{j,k}$ are known as the detail coefficients since they represent the finer and high frequency nature of the time series data.

A way to measure the importance of each level is to evaluate the proportion of the total power or energy explained by the phenomenon under study. The level corresponds to the number of scales (frequencies) for the orthogonal wavelet decomposition of the time series. For example, in seismic studies larger values of the scale correspond to slower and smoother movements of the seismogram; whereas, smaller values of the scale correspond to faster and finer movements of the seismogram.

The total energy P of a time series x_t, for $t = 1, \ldots, n$, is

$$P = \sum_{t=1}^{n} x_t^2. \tag{5.18}$$

The total energy associated to each level of scale is

$$P_J = \sum_{k=1}^{n/2^J} s_{J,k}^2 \tag{5.19}$$

and

$$P_j = \sum_{k=1}^{n/2^j} d_{j,k}^2, \quad j = J, J-1, \ldots, 1. \tag{5.20}$$

Thus, we can rewrite Eq. (5.18) as

$$P = P_J + \sum_{j=1}^{J} P_j. \tag{5.21}$$

The proportion of the total energy explained by each level is the ratios of the total energy associated with each coefficient of detail to the total energy of the time series, or

$$P_j/P$$

for $j = J, J-1, \ldots, 1$. Please refer to [18] for more details.

5.4 Examples of Discrete Wavelets Transforms (DWT)

A DWT is defined as any wavelet transform for which the wavelets are discretely sampled. Equations (5.16) and (5.17) exemplify such DWT. A DWT decomposes a signal into a set of mutually orthogonal wavelet basis functions. One important feature of the DWT over other transforms, such as the Fourier transform, lies in its ability to offer temporal resolution, i.e. it captures both frequency and location (or time) information.

5.4.1 Haar Wavelets

The Haar DWT is one of the simplest possible wavelet transforms. One disadvantage of the Haar wavelet is the fact that it is not continuous and therefore not differentiable. This property can, however, be an advantage for the analysis of

5.4 Examples of Discrete Wavelets Transforms (DWT)

Table 5.1 Determining p and q for $N = 16$.

k	0	1	2	3	4	5	6	7	8	9	10	11	12	13	14	15
p	0	0	1	1	2	2	2	3	3	3	3	3	3	3	3	3
q	0	1	1	2	1	2	3	4	1	2	3	4	5	6	7	8

signals with sudden transitions, such as monitoring of tool failure in machines (see e.g. [131]).

The Haar wavelet function $\psi(t)$ can be described as

$$\psi(t) = \begin{cases} 1, & 0 \leq t < \frac{1}{2} \\ -1, & \frac{1}{2} \leq t < 1 \\ 0, & \text{if otherwise.} \end{cases}$$

and the scaling function $\phi(t)$ can be described as

$$\phi(t) = \begin{cases} 1, & 0 \leq t < 1 \\ 0, & \text{if otherwise.} \end{cases}$$

The Haar functions in fact are very useful for demonstrating the basic characteristics of wavelets (see [189]).

5.4.1.1 Haar Functions

The family of N Haar functions $h_k(t)$, $(k = 0, \ldots, N-1)$ are defined on the interval $0 \leq t \leq 1$. The shape of the specific function $h_k(t)$ of a given index k depends on two parameters p and q:

$$k = 2^p + q - 1 \tag{5.22}$$

For any value of $k \geq 0$, p and q are uniquely determined so that 2^p is the largest power of 2 contained in k ($2^p < k$) and $q - 1$ is the remainder $q - 1 = k - 2^p$.

Example 5.4.1 When $N = 16$, the index k with the corresponding p and q are shown in Table (5.1):

Thus the Haar functions can be defined recursively as:

1) When $k = 0$, the Haar function is defined as a constant

$$h_0(t) = 1/\sqrt{N}.$$

2) When $k > 0$, the Haar function is defined by

$$\psi(t) = \frac{1}{\sqrt{N}} \begin{cases} 2^{p/2}, & (q-1)/2^p \leq t < (q-0.5)/2^p \\ -2^{p/2}, & (q-0.5)/2^p \leq t < (q-0.5)/2^p \\ 0, & \text{if otherwise.} \end{cases}$$

From the definition, it can be seen that p determines the amplitude and width of the nonzero part of the function; while q determines the position of the nonzero part of the function.

5.4.1.2 Haar Transform Matrix

The N Haar functions can be sampled at $t = m/N$, where $m = 0, \cdots, N-1$ to form an N by N matrix for discrete Haar transform. For example, when $N = 2$, we have

$$\mathbf{H}_2 = \frac{1}{\sqrt{2}} \begin{bmatrix} 1 & 1 \\ 1 & -1 \end{bmatrix};$$

when $N = 4$, we have

$$\mathbf{H}_4 = \frac{1}{2} \begin{bmatrix} 1 & 1 & 1 & 1 \\ 1 & 1 & -1 & -1 \\ \sqrt{2} & -\sqrt{2} & 0 & 0 \\ 0 & 0 & \sqrt{2} & -\sqrt{2} \end{bmatrix};$$

and when $N = 8$, we have

$$\mathbf{H}_8 = \frac{1}{\sqrt{8}} \begin{bmatrix} 1 & 1 & 1 & 1 & 1 & 1 & 1 & 1 \\ 1 & 1 & 1 & 1 & -1 & -1 & -1 & -1 \\ \sqrt{2} & -\sqrt{2} & \sqrt{2} & -\sqrt{2} & 0 & 0 & 0 & 0 \\ 0 & 0 & 0 & 0 & \sqrt{2} & -\sqrt{2} & \sqrt{2} & -\sqrt{2} \\ 2 & -2 & 0 & 0 & 0 & 0 & 0 & 0 \\ 0 & 0 & 2 & -2 & 0 & 0 & 0 & 0 \\ 0 & 0 & 0 & 0 & 2 & -2 & 0 & 0 \\ 0 & 0 & 0 & 0 & 0 & 0 & 2 & -2 \end{bmatrix}.$$

One important property of the Haar transform matrix is the fact that it is real and orthogonal, i.e:

$$\mathbf{H} = \mathbf{H}^*, \quad \mathbf{H}^{-1} = \mathbf{H}^T, \quad \text{i.e.} \quad \mathbf{H}^T \mathbf{H} = \mathbf{I} \tag{5.23}$$

where $\mathbf{I}$ is the identity matrix.

Example 5.4.2 When $N = 2$,

$$\mathbf{H}_2^{-1} \mathbf{H}_2 = \mathbf{H}_2^T \mathbf{H}_2 = \frac{1}{2} \begin{bmatrix} 1 & 1 \\ 1 & -1 \end{bmatrix} \begin{bmatrix} 1 & 1 \\ 1 & -1 \end{bmatrix}$$

$$= \begin{bmatrix} 1 & 0 \\ 0 & 1 \end{bmatrix}.$$

5.4.2 Daubechies Wavelets

Daubechies wavelets defining the DWT are characterized by a maximal number of vanishing moments for the given support space. As in the Haar wavelets case, the Daubechies wavelets are defined recursively, with each resolution twice that of the previous scale.

There are many Daubechies transforms, but they are all very similar. In this subsection we shall concentrate on the simplest one, the Daub4 wavelet transform. The Daub4 wavelet transform is defined in essentially the same way as the Haar wavelet transform. If a signal f has an even number N of values, then the 1-level Daub4 transform is the mapping $f \to (a^1|d^1)$ from the signal f to its first trend subsignal a^1 and first fluctuation subsignal d^1. Each value a_m of $a^1 = (a_1, \ldots, a_{N/2})$ is equal to the scalar product

$$a_m = f \cdot V_m^1 \tag{5.24}$$

of f with a 1-level scaling signal V_m^1. Similarly, the value d_m of $d^1 = (d_1, \ldots, d_{N/2})$ is equal to the scalar product

$$d_m = f \cdot W_m^1 \tag{5.25}$$

of f with a 1-level wavelet W_m^1. The Daub4 wavelet transform can be extended to multiple levels as many times as the signal length can be divided by 2. A signal is defined as a function that conveys information about the behavior or attributes of some phenomenon.

The main difference between the Daub4 transform and the Haar transform lies in the way that the wavelets and scaling signals are defined. We shall first discuss the scaling signals. The scaling numbers are defined as follows:

$$\alpha_1 = \frac{1+\sqrt{3}}{4\sqrt{2}}$$

$$\alpha_2 = \frac{3+\sqrt{3}}{4\sqrt{2}}$$

$$\alpha_3 = \frac{3-\sqrt{3}}{4\sqrt{2}}$$

$$\alpha_4 = \frac{1-\sqrt{3}}{4\sqrt{2}}$$

The scaling signals have unit energy. This property is due to the following identity satisfied by the scaling numbers:

$$\alpha_1^2 + \alpha_2^2 + \alpha_3^2 + \alpha_4^2 = 1. \tag{5.26}$$

Another identity satisfied by the scaling function is

$$\alpha_1 + \alpha_2 + \alpha_3 + \alpha_4 = \sqrt{2}. \tag{5.27}$$

(5.27) implies that each 1-level trend value $f \cdot V_m^1$ is an average of four values of f, multiplied by $\sqrt{2}$.

The Daub4 wavelet numbers are defined as follows:

$$\beta_1 = \frac{1-\sqrt{3}}{4\sqrt{2}}$$

$$\beta_2 = \frac{\sqrt{3}-3}{4\sqrt{2}}$$

$$\beta_3 = \frac{3+\sqrt{3}}{4\sqrt{2}}$$

$$\beta_4 = \frac{-1-\sqrt{3}}{4\sqrt{2}}$$

The Daub4 wavelets have unit energy. This is evident in the 1-level wavelets since

$$\beta_1^2 + \beta_2^2 + \beta_3^2 + \beta_4^2 = 1. \tag{5.28}$$

The wavelet numbers and the scaling numbers are related by the equations

$$\alpha_1 = -\beta_4, \alpha_2 = \beta_3, \alpha_3 = -\beta_2, \text{ and } \alpha_4 = \beta_1.$$

Other forms of DWT include the dual-tree complex wavelet transform, the undecimated wavelet transform, and the Newland transform. Please refer to [76, 157, 184] for more details of the other forms of the DWT.

The DWT has practical applications in science, engineering, mathematics, and computer science.

5.5 Application of Wavelets Transform

In this section, we briefly discuss some real life applications of the wavelet transforms.

5.5.1 Finance

The wavelet transform has proved to be a very powerful tool for characterizing behavior, especially self-similar behavior, over a wide range of time scales. In the work of [19], the authors investigated the applicability of wavelet decomposition methods to determine if a market crash may be predictable. The premise of the work was that a predictable event produces a time series signal similar in nature with that recorded before a major earthquake which contains before-shock signals. A nonpredictable event would have a time series signal resembling data produced by an explosive event which is very sudden and not preceded by any prior signals.

The wavelets methodology was very useful in yielding a correct identification of the signal type. Correct identification of signal type using wavelets techniques help mitigate some of the potential effects of the events [18]. Wavelets have also been used to investigate the dependence and interdependence between major stock markets (see [49, 203]).

5.5.2 Modeling and Forecasting

Wavelets trasnforms have been used to investigate the modeling and forecasting of nonstationary time series. Ref. [212] the authors propose a modeling procedure that decomposes the time series as the sum of three separate components: trend, harmonic, and irregular components. The proposed estimators used in their study to detect trends and hidden frequencies are strongly consistent. The ten steps ahead prediction for the time series show that the average percentage error criterion of the wavelet approach is the smallest when compared with other methods. Their results suggest that forecasting based on wavelets is a viable alternative to existing methods such as ARIMA and the seasonal ARIMA model, HoltWinters, econometric models, and several others.

5.5.3 Image Compression

One major application of the wavelet transform is data compression and detection of edges in digital images.

Suppose one wants to send a digital image file to a friend via email. In order to expedite the process, the contents of the file need to be compressed. There are two forms of image compression, lossless and lossy compression. Data compressed by lossless schemes can be recovered exactly without any loss of information. However for the lossy compression schemes, the data is altered. Good results can be obtained using lossy compression schemes, but the savings come at the expense of information lost from the original data ([66]).

The Haar wavelet compression is an efficient way to perform both lossless and lossy image compression. It relies on averaging and differencing values in an image matrix to produce a matrix which is sparse or nearly sparse. A sparse matrix is a matrix in which most of the elements are zero. Please refer to [66] for more details of image compression.

5.5.4 Seismic Signals

Wavelet transforms have been applied to the decomposition of seismic signal energy into the time-scaling plane. For better representation the scale is converted into frequency. The seismic signals are obtained from hitting a metal sphere against the lawn. After a number of experiments, both the expected value of the energy center frequencies and standard deviation are found and given in terms of the distance between the detector and place of excitation. The expected value and standard deviation decrease if the distance increases, so it is possible to estimate the distance using only one detector. These results are very useful for analyzing seismic signal whenever several excitations occur. This is demonstrated in experiments in which seismic signals are obtained from two sources. We refer to [197] for more details of the application of wavelets transform to seismic signals.

5.5.5 Damage Detection in Frame Structures

The wavelet transform can be used to detect cracks in frame structures, such as beams and plane frames. The ability of wavelets to detect crack-like damage in structures can be demonstrated by means of several numerical examples. The method requires the knowledge of only the response of the damaged structure, i.e. no information about the original undamaged structure is required. This procedure can help detect the localization of the crack by using a response signal from static or dynamic loads. Moreover, the response needs to be obtained only at the regions where it is suspected that the damage may be present. The results of the simulation show that if a suitable wavelet is selected, the method is capable to extract damage information from the response signal in a simple, robust, and reliable way. Please refer to [160] and references there in for more details.

Wavelets are known to be strong methodologies because they can capture the nonstationary behavior of the time series and are localized in both frequency and time.

5.6 Problems

1. Consider the vector u that consists of 32 equally spaced samples of the function $f(t) \approx \cos(4\pi t)$ on the interval $[0, 1]$. That is, $u_1 = f(0), u_2 = f(\frac{1}{32}), \ldots, u_3 = f(\frac{31}{32})$. Compute the Haar wavelet transform $y = W_{32}u$ and graph the cumulative energy u and y.

2. Define the Haar's wavelets function $\psi(t)$ and verify that for every t,

$$\psi(t) = \begin{cases} 1, & 0 \leq t < \frac{1}{2} \\ -1, & \frac{1}{2} \leq t < 1 \\ 0, & \text{if otherwise.} \end{cases}$$

3. Calculate the Haar wavelets transform for each pair s_{2k}, s_{2k+1} in the array $\vec{s} = (8, 3, 2, 1)$.

4. Calculate the Haar wavelets transform for the data $\vec{s} = (1, 3, 4, 1)$.

5. Discuss the similarities and differences between Fourier transform and wavelet transform.
6. Calculate the Daubechies wavelet transform of the data
$$\vec{a} = (a_0, a_1, a_2, a_3) = (2, 2, 2, 2).$$
7. For the scaling numbers
$$\alpha_1 = \frac{1+\sqrt{3}}{4\sqrt{2}}$$
$$\alpha_2 = \frac{3+\sqrt{3}}{4\sqrt{2}}$$
$$\alpha_3 = \frac{3-\sqrt{3}}{4\sqrt{2}}$$
$$\alpha_4 = \frac{1-\sqrt{3}}{4\sqrt{2}}$$

verify the following identity:
(a) $\alpha_1^2 + \alpha_2^2 + \alpha_3^2 + \alpha_4^2 = 1$
(b) $\alpha_1 + \alpha_2 + \alpha_3 + \alpha_4 = \sqrt{2}$

8. Show that the following eight vectors are pairwise orthogonal:
$$s1 = (1, 1, 0, 0, 0, 0, 0, 0)^T$$
$$s2 = (0, 0, 1, 1, 0, 0, 0, 0)^T$$
$$s3 = (0, 0, 0, 0, 1, 1, 0, 0)^T$$
$$s4 = (0, 0, 0, 0, 0, 0, 1, 1)^T$$
$$s5 = (1, 1, 1, 1, 0, 0, 0, 0)^T$$
$$s6 = (0, 0, 0, 0, 1, 1, 1, 1)^T$$
$$s7 = (1, 1, 1, 1, 1, 1, 1, 1)^T$$
$$s8 = (1, 1, 1, 1, 1, 1, 1, 1)^T$$

9. Haar transform
   (a) For an $N \times N$ Haar transformation matrix, the Haar basis functions are
   $$\psi_k(t) = \psi_{pq}(t) = \frac{1}{\sqrt{N}} \begin{cases} 2^{p/2}, & (q-1)/2^p \le t < (q-0.5)/2^p \\ -2^{p/2}, & (q-0.5)/2^p \le t < q/2^p \\ 0, & 0 \le t \le 1 \end{cases}$$

   They are defined over the continuous closed interval $t \in [0, 1]$ for $k = 0, 1, 2, \ldots, N-1$. We define the integer k such that $k = 2^p + q - 1$, where $0 \le p \le n-1$, $q = 0$ or 1 for $p = 0$, and $1 \le q \le 2^p$ for $p \ne 0$. The i-th row of an $N \times N$ Haar transformation matrix contains the elements of

$\psi_i(t)$ for $t = 0/N, 1/n, \ldots, (N-1)/N$. Write the 5×5 Haar transformation matrix H_5 using these given conditions.

(b) Compute the Haar transfromation of the 2×2 image f with H_2:

$$f = \begin{bmatrix} 5 & -2 \\ -2 & 3 \end{bmatrix}$$

(c) The inverse Haar transform is $f = H^T T H$ where T is the Haar transform of f and H^T is the inverse of the matrix H. Show that $H_2^{-1} = H_2^T$ and use it to compute the inverse Haar transform of the result in part (a).

10. Consider the periodic signal

$$f(t) = \begin{cases} \cos(t) + D, & -1 \le t \le 0 \\ \sin(t^6)/t^3, & 0 \le t \le 2 \end{cases}$$

where $D = \sin(2^6)/2^3 - \cos(1)$, sampled at 1024 equidistant points in $[-1, 2]$. Use the wavelet transform to compress the signal. Use about 15% of the wavelet coefficients for different wavelets models. Plot the compressed signals together with the original signal using Matlab. Calculate the relative energy errors.

11. Consider the signal $f(t) = \sin(7t)$. Use Matlab to obtain the discrete wavelets transform of this signal.

12. Discuss the two main types of data (image) compression including their advantages and disadvantages.

13. Let $f(\theta)$ be the 2π-periodic function determined by the formula

$$f(\theta) = \theta^2, \quad \text{for } -\pi \le \theta \le \pi.$$

Find the Fourier series for $f(\theta)$.

14. Let $f(\theta)$ be the 2π-periodic function determined by the formula

$$f(\theta) = |\sin \theta|, \quad \text{for } -\pi \le \theta \le \pi.$$

Show that the Fourier series for f is given by

$$\frac{2}{\pi} - \frac{4}{\pi} \sum_{n=1}^{\infty} \frac{\cos 2n\theta}{4n^2 - 1}.$$

From this, show that

$$\sum_{n=1}^{\infty} \frac{(-1)^{n+1}}{4n^2 - 1} = \frac{\pi - 2}{4}.$$

6

Tree Methods

6.1 Introduction

In Chapter 3 we discussed various models where the stock process follows a stochastic differential equation (SDE). We discussed the theoretical solution of such equations, and when the solution exists we derived a formula for the vanilla option price.

In general, finding analytic formulas is complicated and often impossible. In this chapter we will present a general methodology to approximate the solution of an SDE with a discrete time, discrete state space process. We shall refer throughout this chapter to such approximating processes as trees. Specifically, the problem can be formulated in the following way.

Suppose we have a stochastic process S_t solving the SDE:

$$dS_t = \mu(S_t)dt + \sigma(S_t)dB_t. \tag{6.1}$$

where $\mu(\cdot)$ and $\sigma(\cdot)$ are known function. The theory presented at the end of this chapter applies to d-dimensional stochastic processes S_t, not just one dimensional. Later in this chapter we shall present an approximation to a stochastic volatility model which will involve a two-dimensional process.

The stochastic process in (6.1) is a time homogeneous Markov process. A tree approximation will be a Markov chain approximating this Markov process. It is important to understand how we create trees that approximate the diffusion process in equation (6.1). The sections at the end of this chapter are very general and may be skipped for the purpose of simple tree approximations (i.e. Black–Scholes–Merton). However, to approximate more general processes this theory is crucial. The theoretical sections in this chapter are a very simplified summary of a much larger and older theoretical development. We refer to [62, 198] for much more detail and proofs of the results presented here. For more details and examples of Markov processes, we refer to [67].

Quantitative Finance, First Edition. Maria C. Mariani and Ionut Florescu.
© 2020 John Wiley & Sons, Inc. Published 2020 by John Wiley & Sons, Inc.

6.2 Tree Methods: the Binomial Tree

We will start this chapter with the simplest possible tree. The binomial tree was developed during the late 1970's by [48] and perfected in the early 1980's. The binomial tree model generates a pricing tree in which every node represents the price of an underlying financial instrument at a given point in time. The methodology can be used to price options with nonstandard features such as path dependence and barrier options. The main idea behind the binomial tree is to construct a discrete version of the geometric Brownian motion model:

$$dS_t = \mu S_t dt + \sigma S_t dW_t$$

which is a simplified version of the model in (6.1).

The process in (6.1) is a continuous process. Furthermore, since it is stochastic every time one constructs a path, the path would look different. The present value of a stock price is known to be S_0. The binomial tree (and indeed any tree in fact) may be thought of forming a collection of "most probably paths." When the number of steps in the tree approximation increases, the tree contains more and better paths until at the limit we obtain the same realizations as the process in (6.1).

6.2.1 One-Step Binomial Tree

We start the construction by looking at a single step. The idea is that, if this step is done properly, the extension to any number of steps is straightforward.

In order to proceed, we state the Girsanov's theorem [159].

Theorem 6.2.1 (Girsanov, One-Dimensional Case). Let $B(t), 0 \leq t \leq T$ be a Brownian motion on a probability space $(\Omega, \mathcal{F}, \mathbb{P})$. Let $\mathcal{F}(t), 0 \leq t \leq T$ be the accompanying filtration, and let $\theta(t), 0 \leq t \leq T$ be a process adapted to this filtration. For $0 \leq t \leq T$, define

$$\tilde{B}(t) = \int_0^t \theta(u) du + B(t),$$

$$Z(t) = \exp\left\{ -\int_0^t \theta(u) dB(u) - \frac{1}{2} \int_0^t \theta^2(u) du \right\},$$

and define a new probability measure by

$$\tilde{\mathbb{P}}(A) = \int_A Z(T) d\mathbb{P}, \quad \forall A \in \mathcal{F}.$$

Under $\tilde{\mathbb{P}}$, the process $\tilde{B}(t), 0 \leq t \leq T$ is a Brownian motion.

Girsanov's theorem allows us to go from the unknown drift parameter μ to a risk-free rate r which is common to all the assets. Applying Girsanov's theorem

6.2 Tree Methods: the Binomial Tree

to (6.1) and using the Itô's formula on the process $X_t = \log S_t$, we obtain the SDE:

$$dX_t = \left(r - \frac{\sigma^2}{2}\right) dt + \sigma dW_t. \tag{6.2}$$

We note that the stochastic equation of the log process $X_t = \log S_t$ is much simpler than the original SDE for the stock process. This translates into a much simpler construction for the X_t than for the original process S_t. For each choice of the martingale measure (probabilities in the tree), a tree for X_t is going to be perfectly equivalent with a tree for S_t and vice versa.

Specifically, suppose that we have constructed a tree for the process X_t which has $x + \Delta x_u$ and $x + \Delta x_d$ steps up and down from x, respectively. Then an equivalent tree for S_t could be constructed immediately by taking the next steps from $S = e^x$ as $Su = e^{x+\Delta x_u} = Se^{\Delta x_u}$ and $Sd = e^{x+\Delta x_d} = Se^{\Delta x_u}$ and keeping the probabilities identical. Similarly the tree in X_t is constructed from an S_t tree by taking $\Delta x_u = \log u$ for the up node and similarly for the down node.

Due to the obvious way in which the successors in the tree are calculated, the tree for the process X_t is called an *additive tree*, and the tree for the stock process S_t is called a *multiplicative tree*. We shall be focusing on the additive tree in the remainder of this chapter since the construction is simpler.

To have two successors for every node, at any point in the tree, we can go up to $x + \Delta x_u$ with probability p_u or down to $x + \Delta x_d$ with probability $p_d = 1 - p_u$. The one-step tree for this process is presented in Figure 6.1.

To determine the appropriate values for $\Delta x_u, \Delta x_d, p_u,$ and p_d, we use the diffusion approximation that will be presented in Section 6.9. However, the model in (6.2) is very simple, and in this particular case it translates into needing to impose that the difference $(\Delta X_t = X_{t+\Delta t} - X_t)$ between any two times calculated for the discrete tree model and the continuous time model should have the first two moments equal.

Fundamentally, in the simple case of a geometric BM, the process X_t is a normal random variable for any time t. We also know that a normal variable is entirely characterized by its mean and variance. Thus, equating the mean and variance increase for an infinitesimally small time step will be sufficient to

Figure 6.1 One-step binomial tree for the return process.

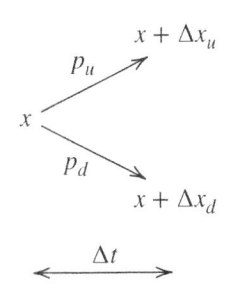

make sure that the binomial tree will converge to the path given by the continuous process. Specifically, the conditions we need to enforce are:

$$\begin{cases} p_u \Delta x_u + p_d \Delta x_d = \left(r - \frac{\sigma^2}{2}\right) \Delta t \\ p_u \Delta x_u^2 + p_d \Delta x_d^2 = \sigma^2 \Delta t + \left(r - \frac{\sigma^2}{2}\right)^2 \Delta t^2 \\ p_u + p_d = 1 \end{cases} \quad (6.3)$$

The system (6.3) has three equations and four unknowns; thus it has an infinite number of solutions. Indeed, the binomial tree approximation is not unique. In practice, we have a choice of a parameter - that choice will create a specific tree.

We note that for any choice of Δx_u and Δx_d the tree recombines. For example, first-step successors are $x + \Delta x_u$ and Δx_d. At the second step their successors will be:

$$x + 2\Delta x_u \text{ and } x + \Delta x_u + \Delta x_d \text{ for the upper node}$$

and $x + \Delta x_d + \Delta x_u$ and $x + 2\Delta x_d$ for the lower node.

Clearly one of the nodes is identical and the tree recombines. The recombining feature in a tree is crucial. For example, consider a binomial tree with $n = 10$ steps that recombines and one that does not recombine. The recombining tree has a total number of nodes: $1 + 2 + 3 + \cdots + 11 = 11 \times 10/2 = 55$ nodes, whereas the one that does not recombine has $1 + 2 + 2^2 + 2^3 + \cdots + 2^{10} = 2^{11} - 1 = 2047$ nodes. The latter has a much bigger number of calculations that quickly become unmanageable.

As long as the resulting probabilities in (6.3) are positive, the corresponding trees constructed with the particular choice of parameters are appropriate. Some popular choices are obtained by taking $\Delta x_u = \Delta x_d$ in the system (6.3). This produces what is called the Trigeorgis tree (see [199]) which supposedly has better approximation power.

Solving the system (6.3) by taking $\Delta x_u = \Delta x_d$ will yield:

$$\begin{cases} \Delta x = \sqrt{\left(r - \frac{\sigma^2}{2}\right)^2 \Delta t^2 + \sigma^2 \Delta t} \\ p_u = \frac{1}{2} + \frac{1}{2} \frac{\left(r - \frac{\sigma^2}{2}\right) \Delta t}{\Delta x} \end{cases} \quad (6.4)$$

The corresponding multiplicative tree for the S_t process obtained by setting the nodes as $S_t = \exp(X_t)$ and keeping all probabilities the same is the famous Cox–Ross–Rubinstein (CRR) tree (see [48]).

Another popular tree is obtained by setting the probabilities of jumps in (6.3) equal (i.e. $p_u = p_d = 1/2$). In this case the tree for S is called Jarrow–Rudd tree (see [113]).

Remarks 6.2.1

- The system (6.3) could be further reduced by subtracting the constant term $\left(r - \frac{\sigma^2}{2}\right)\Delta t$ from each node; the resulting tree will be much simpler. One would have to remember to add the constant back in the final value of the tree at maturity.
- Finding the probabilities p_u and p_d is equivalent to finding the martingale measure under which the pricing is made. For each solution one can construct an additive tree (moving from x to $x + \Delta x$ and $x - \Delta x$ with probabilities p_u and p_d, respectively) or a multiplicative tree (moving from S to Su and Sd with probabilities p_u and p_d, respectively). The two constructed trees are completely equivalent.
- Furthermore any tree works as long as the resulting probabilities are positive. The condition that the parameters need to satisfy to have proper probabilities is different for each type of tree. For example, for the symmetric steps tree ($\Delta x_u = -\Delta x_d = \Delta x$), the condition is $\Delta x > \left|r - \frac{\sigma^2}{2}\right|\Delta t$.

To construct a full tree we need to first solve the one-step tree for a time interval $\Delta t = T/n$ where T is the maturity of the option to be priced and n is a parameter of our choice. We then use this tree to price options as in the next sections.

6.2.2 Using the Tree to Price a European Option

After constructing a binomial tree, we basically have 2^n possible paths of the process X_t or equivalently the process $S_t = \exp(X_t)$. Since these are possible paths, we can use the binomial tree to calculate the present price of any path-dependent option.

We begin this study by looking at the European Call. The European Call is written on a particular stock with current price S_0 and is characterized by maturity T and strike price K.

The first step is to divide the interval $[0, T]$ into $n + 1$ equally spaced points; then we will construct our tree with n steps, $\Delta t = T/n$ in this case. The times in our tree will be $t_0 = 0, t_1 = T/n, t_2 = 2T/n, \ldots, t_n = nT/n = T$.

Next we construct the return tree (X_t) as discussed earlier starting with $X_0 = x_0 = \log(S_0)$ and the last branches of the tree ending in possible values for $X_{t_n} = X_T$. We remark that since we constructed the possible values at time T, we can calculate for every terminal node in the tree the value of the Call option at that terminal node using:

$$C(T) = \left(S_T - K\right)_+ = \left(e^{X_T} - K\right)_+. \tag{6.5}$$

Thus now we know the possible payoffs of the option at time T. Suppose we want to calculate the value of the option at time $t = 0$. Using the Girsanov theorem and the Harrison and Pliska result [88], the discounted price of the process $\{e^{-rt}C(t)\}$ is a continuous time martingale. Therefore we may write that the value at time 0 must be:

$$C(0) = \mathbf{E}\left[e^{-rT}C(T)|\mathcal{F}_0\right].$$

Now we can use the basic properties of conditional expectation and the fact that $T/n = \Delta t$ or $e^{-rT} = \left(e^{-r\Delta T}\right)^n := \delta^n$ to write:

$$\begin{aligned}C(0) &= \mathbf{E}\left[\delta^n C(T)|\mathcal{F}_0\right] = \mathbf{E}\left[\delta^{n-1}\mathbf{E}\left[\delta C(T)|\mathcal{F}_1\right]|\mathcal{F}_0\right] \\ &= \delta \mathbf{E}\left[\delta \mathbf{E}\left[\ldots \delta \mathbf{E}\left[\delta C(T)|F_{n-1}\right]|\ldots\mathcal{F}_1\right]|\mathcal{F}_0\right],\end{aligned}$$

where $\delta = e^{-r\Delta t}$ is a discount factor.

This formula allows us to recursively go back in the tree toward the time $t = 0$. When this happens, we will eventually reach the first node of the tree at C_0 and that will give us the value of the option. More precisely since we know the probabilities of the up and down steps to be p_u and p_d, respectively, and the possible Call values one step ahead to be C^1 and C^2, we can calculate the value of the option at the previous step as:

$$C = e^{-r\Delta t}\left(p_d C^1 + p_u C^2\right). \tag{6.6}$$

What is remarkable about the aforementioned construction is that we constructed the return tree just to calculate the value of the option at the final nodes. Note that when we stepped backwards in the tree, we did not use the intermediate values of the stock; we only used the probabilities. This is due to the fact that we are pricing the European option which only depends on the final stock price. This situation will change when pricing any path-dependent option such as the American, Asian, barrier, etc.

Furthermore, we do not actually need to go through all the values in the tree for this option. For instance, the uppermost value in the final step has probability p_u^n. The next one can only be reached by paths with probability $p_u^{n-1}p_d$ and there are $\binom{n}{1}$ of them. Therefore, we can actually write the final value of the European Call as:

$$C = \sum_{i=0}^{n}\binom{n}{i}p_u^{n-i}p_d^i(e^{x+(n-i)\Delta x_u + i\Delta x_d} - K)_+$$

as an alternative to going through the tree.

6.2.3 Using the Tree to Price an American Option

The American option can be exercised at any time; thus we will need to calculate its value at $t = 0$ and the optimal exercise time τ. Estimating τ is not a simple task, but in principle it can be done. τ is a random variable and therefore it has an expected value and a variance. The expected value can be calculated by looking at all the points in the tree where it is optimal to early exercise the option and the probabilities for all such points. Then we can calculate the expected value and variance.

For the value of the American option, we will proceed in similar fashion as for the European option. We will construct the tree in the same way we did for the European option, and then we will calculate the value of the option at the terminal nodes. For example, for an American Put option we have

$$P(T) = (K - S_T)_+ = (K - e^{X_T})_+ \qquad (6.7)$$

Then we recursively go back in the tree in a similar way as we did in the case of the European option. The only difference is that for the American option, we have the early exercise condition. So at every node we have to decide if it would have been optimal to exercise the option rather than to hold onto it. More precisely using the same notation as before, we again calculate the value of the option today if we would hold onto it as

$$C = e^{-r\Delta t}(p_d C^1 + p_u C^2).$$

But what if we actually exercised the option at that point in time, say t_i? Then we would obtain $(K - S_{t_i})_+$. Since we can only do one of the two, we will obviously do the best thing, so the value of the option at this particular node will be the maximum of the two values. Then again, we recursively work the tree backward all the way to $C(0)$, and that will yield the value of the American Put option. A possible value of the optimal exercise time τ is encountered whenever the value obtained by exercising early $(K - S_{t_i})_+$ is larger than the expectation coming down from the tree.

We now discuss a more general approach of pricing path-dependent options using the binomial tree techniques.

6.2.4 Using the Tree to Price Any Path-Dependent Option

From the previous subsection we see that the binomial tree could be applied to any path-dependent option. The trick is to keep track of the value of the option across various paths of the tree.

Please see [38] for a detailed example of a down and out American option. This is an American option (like before it can be exercised anytime during its lifespan). In addition if the asset price on which the option is dependent falls below a certain level H (down), then the option is worthless; it ceases to exist (out).

Binomial trees for dividend-paying asset are very useful; however they are not sufficiently different from the regular binomial tree construction. The model developed thus far can be modified easily to price options on underlying assets other than dividend and non-dividend-paying assets.

Other applications of the binomial tree methodology is in the computation of hedge sensitivities. A brief introduction is presented in the following. Please refer to [38] for more details. In this application, we discuss how the binomial tree can be used to compute hedge sensitivities.

6.2.5 Using the Tree for Computing Hedge Sensitivities: the Greeks

When hedging an option as we discussed earlier in Section 3.3, it is important to calculate the changes in option price with the various parameters in the model. We also recall in Section 3.3 that these derivatives are denoted with *delta* Δ, *gamma* Γ, *Vega*, *theta* Θ, and *rho* ρ.

Their mathematical definition is given as:

$$\begin{cases} \Delta = \frac{\partial C}{\partial S} \\ \Gamma = \frac{\partial^2 C}{\partial S^2} \\ Vega = \frac{\partial C}{\partial \sigma} \\ \rho = \frac{\partial C}{\partial r} \\ \Theta = \frac{\partial C}{\partial t} \end{cases} \quad (6.8)$$

In order to approximate the Greeks, one can use the method discussed by [38] that uses only one constructed binomial tree to calculate an approximation for Δ and Γ one step in the future.

Alternatively, one can construct two binomial trees in the case of Δ and three in the case of Γ starting from slightly different initial asset prices and use the same formula to approximate Δ and Γ at the present time.

In order to approximate Θ derivative, it requires knowledge of stock prices at two moments in time. It could be calculated using a similar approach as for *Vega* and ρ, but since option price varies in time with the Brownian motion driving the process S_t, it is unnecessary to calculate the slope of the path.

6.2.6 Further Discussion on the American Option Pricing

We consider a multiplicative binomial tree:

```
            Suu ...
      Su
               ...
  S      Sud

      Sd
            Sdd ...
```

In the following examples we work with a probability of an up step:

$$\frac{Se^{r\Delta t} - S_d}{S_u - S_d} = \frac{e^{r\Delta t} - d}{u - d} = p$$

6.2 Tree Methods: the Binomial Tree

Example: European Call option
Suppose a strike price $K = 105$, $u = 1$, $d = 0.9$, and $r = 7\%$ (annual) with T semesters and $\Delta T = \frac{1}{2}$. Suppose that the original price of the asset is $S = 100$. In this case, we have the following tree:

```
                        121
                        (16)
              110
              10,4764
    100                  99
    (6,8597)             (0)
              90
              (0)
                        81
                        (0)
```

The probability is then $p = 0.6781$. Next, when we consider a European Put option, the situation is different. In this case, we have:

```
                        121
                        (0)
              110
              (1,865)
    100                  99
    (4,772)              (6)
              90
              (11,3885)
                        81
                        (24)
```

Remark 6.2.2 We observe that there exists a relation between the values of the Put and the Call options:

$$C_t + Ke^{-r(T-t)} = P_t + S_t.$$

This relation is called the Put–Call parity as discussed in Section 3.9.

We know that it is possible to exercise an American Call option at any time or, in the discrete case, at any node. We compute the maximum between 0 and $K - S_t$, i.e.:

$$P_t = \max\{0, K - S_t\}.$$

We recall that in a Call option, the relation is the reverse of the previous relation, i.e.:

$$C_t = \max\{0, S_t - K\}.$$

We recall in Chapter 2 that without considering dividends, the values of an American and European Call options are the same (i.e. $C_E = C_A$). Suppose that we exercise at time t, (i.e.: $S_t - K > 0$) and at a future time T, it grows with a risk-free rate $e^{r(T-t)}$. As the expected payoff is $E(S(T)) = S_t e^{r(T-t)} > S_t e^{r(T-t)} - K$, then it is not convenient to exercise, because the value will go up. We recall that in Section 3.9, we can write the Put–Call parity as

$$C_t - P_t = S_t - K e^{-r(T-t)}.$$

For $t = T$,

$$C_T = \max\{S_T - K, 0\}$$
$$P_T = \max\{K - S_T, 0\},$$

so that we have the equality

$$C_T - P_T = S_T - K.$$

Now consider the relation

$$\pi_t = \begin{cases} 1\,\text{Call} & \text{in long} \\ 1\,\text{Put} & \text{in short} \\ 1\,\text{share} & \text{in short} \end{cases}$$

Then we have that

$$\pi_t = C_t - P_t - S_t$$
$$\pi_T = C_T - P_T = S_T - K - S_T = -K$$

and assuming no arbitrage,

$$\pi_t = e^{-r(T-t)} \pi_T.$$

In general, suppose that at time T, we have $n+1$ possible states, i.e. $T = n\Delta t$. We assume that

$$S^j(T) = S u^j d^{n-j}$$

with

$$p = \frac{e^{r\Delta t} - d}{u - d}.$$

Then the probability that S is in the state S^j at time T is given as

$$P(S(T) = S^j(T)) = \binom{n}{j} p^j (1-p)^{n-j}.$$

and using the Newton binomial formula, the expected value is

$$E(S(T)) = \sum_{j=0}^{n} \binom{n}{j} p^j (1-p)^{n-j} S u^j d^{n-j}$$
$$= S(pu + (1-p)d)^n.$$

On the other hand,

$$pu + (1-p)d = p(u-d) + d = e^{r\Delta t}$$

and from $T = n\Delta t$ we obtain

$$E(S(T)) = S(pu + (1-p)d)^n = Se^{rn\Delta t} = Se^{rT}.$$

This implies that the expected value is proportional to the basis value and the proportion depends on the interest rate.

Now suppose we have the same situation with a European Call with strike price K and the relation

$$C_j = C_{u^j d^{n-j}} = \max\{S u^j d^{n-j} - K, 0\},$$

then

$$E(C(T)) = \sum_{j=0}^{n} \binom{n}{j} p^j (1-p)^{n-j} C_j$$
$$= \sum_{j=n_0}^{n} \binom{n}{j} p^j (1-p)^{n-j} C_j$$

where

$$n_0 = \min\{j : C_j > 0\}.$$

Then substituting C_j, in the above equation, we have

$$E(C(T)) = \sum_{j=n_0}^{n} \binom{n}{j} p^j (1-p)^{n-j} \left(S u^j d^{n-j} - K\right)$$
$$= S \sum_{j=n_0}^{n} \binom{n}{j} (pu)^j ((1-p)d)^{n-j} - KP(U \geq n_0)$$

where U is a stochastic variable that follows a binomial distribution with mean $E(U) = np$ and variance $Var(U) = np(1-p)$.

Using the central limit theorem, we can approximate the binomial distribution by a normal distribution and obtain

$$P(U \geq n_0) \approx P\left(Z \geq \frac{n_0 - np}{\sqrt{np(1-p)}}\right).$$

6.2.7 A Parenthesis: the Brownian Motion as a Limit of Simple Random Walk

We begin our discussion of the Wiener process (also known as the Brownian motion) with an empirical analysis. Assume we have a random walk, which has a value x at time t. At the next time step $t + \Delta t$, it moves with probability $\frac{1}{2}$, to either $x + \Delta x$ or $x - \Delta x$. Thus the random walk has two variables: the time $t \in [0, T]$ and the position $x \in [-X, X]$. Intuitively, we note that a *Markov process* does not have memory, that is, the history of how the process reached position x is not relevant to where it goes next. Only the present state x is relevant for this.

For each $j = 1, 2, ..., n$, consider

$$u_{k,j} = P\left(x_{k\Delta t} = j\Delta x\right),$$

where k represents the time and j the position.

We recall that for $P(B) \neq 0$ we can define the conditional probability of the event A given that the event B occurs as

$$P(A|B) = \frac{P(A \cap B)}{P(B)}.$$

To the position j in time $k + 1$, we can arrive only from the position $j - 1$, or j at time k, so we have

$$u_{k+1,j} = \frac{1}{2}\left(u_{k,j} + u_{k,j-1}\right). \tag{6.9}$$

We can rewrite (6.9) as

$$u_{k+1,j} = \frac{1}{2}\left[\left(u_{k,j+1} - u_{k,j}\right) - \left(u_{k,j} - u_{k,j-1}\right)\right] + u_{k,j}$$

or

$$u_{k+1,j} - u_{k,j} = \frac{1}{2}\left[\left(u_{k,j+1} - u_{k,j}\right) - \left(u_{k,j} - u_{k,j-1}\right)\right].$$

Using the notation

$$u_{k,j} = u\left(t_k, x_j\right),$$

we obtain

$$u\left(t_{k+1}, x_j\right) - u\left(t_k, x_j\right) = \frac{1}{2}\left[\left(u\left(t_k, x_{j+1}\right) - u\left(t_k, x_j\right)\right) - \left(u\left(t_k, x_j\right) - u\left(t_k, x_{j-1}\right)\right)\right]. \tag{6.10}$$

Now let Δt and Δx be such that $X = n\Delta x$, $T = n\Delta t$, and then $\frac{X}{\Delta x} = \frac{T}{\Delta t}$. Then multiplying (6.10) by $\frac{1}{\Delta t}$ we obtain

$$\frac{1}{\Delta t}\left(u\left(t_{k+1}, x_j\right) - u\left(t_k, x_j\right)\right)$$
$$= \frac{1}{2\Delta t}\left[\left(u\left(t_k, x_{j+1}\right) - u\left(t_k, x_j\right)\right) - \left(u\left(t_k, x_j\right) - u\left(t_k, x_{j-1}\right)\right)\right]. \tag{6.11}$$

If we take $\Delta t \to 0$ the first term in (6.11) converges to $\partial_t(u)$. For the second term, if we assume that

$$\Delta t \approx (\Delta x)^2$$

taking into account that $\Delta t = \frac{T\Delta x}{X}$

$$\frac{1}{2\Delta t} = \frac{X}{2T\Delta x}$$

we can conclude that the second term converges to $\partial_{xx}(u)$. So, from the random walks we get a discrete version of the heat equation

$$\partial_t(u) = \frac{1}{2}\partial_{xx}(u)$$

As an example, consider the random walk with step $\Delta x = \sqrt{\Delta t}$, that is,

$$\begin{cases} x_{j+1} = x_j \pm \sqrt{\Delta t} \\ x_0 = 0 \end{cases}$$

We claim that the expected value after n steps is zero. The stochastic variables $x_{j+1} - x_j = \pm\sqrt{\Delta t}$ are independents and so

$$E(x_n) = VE\left(\sum_{j=0}^{n-1}(x_{j+1}-x_j)\right) = \sum_{j=0}^{n-1} VE(x_{j+1}-x_j) = 0$$

and

$$Var(x_n) = \sum_{j=0}^{n-1} Var(x_{j+1}-x_j) = \sum_{j=0}^{n-1} \Delta t = n\Delta t = T.$$

If we interpolate the points $\{x_j\}_{1 \le j \le n-1}$ we obtain

$$x(j) = \frac{x_{j+1}-x_j}{\Delta t}(t-t_j) + x_j \tag{6.12}$$

for $t_j \le t \le t_{j+1}$. Equation (6.12) is a *Markovian process*, because:

1) $\forall a > 0$, $\{x(t_k + a) - x(t_k)\}$ is independent of the history $\{x(s) : s \le t_k\}$
2)
$$E(x(t_k + a) - x(t_k)) = 0$$
$$Var(x(t_k + a) - x(t_k)) = a$$

Remark 6.2.3 If $\Delta t \ll 1$ then $x \approx N(0, \sqrt{a})$ (i.e normal distributed with mean 0 and variance a) and then,

$$P(x(t+a) - x(t) \ge y) \approx \frac{1}{\sqrt{2\pi}} \int_y^\infty e^{-\frac{t^2}{2a}} dt.$$

This is due to the central limit theorem that guarantees that if N is large enough, the distribution can be approximated by Gaussian.

In the next section, we will study the pricing of options when the volatility varies. We will consider models of the form

$$dS = \mu S dt + \sigma S dZ.$$

The parameter σ is the volatility. When $\sigma = 0$ the model is deterministic. The parameters μ and σ depend on the asset and we have to estimate them. We can estimate them from the historical data for the assets.

We present some examples as follows:

1) Consider the equation

$$\begin{cases} u'' = 1 \\ u(-1) = u(1) = 0 \end{cases}$$

We fix $x \in (-1, 1)$ and define the stochastic process Z (Brownian motion) with $Z(0) = x$; then since the stochastic process Z does not depend on t, $u'' = 1$ and applying Itô's lemma, we obtain

$$du = du(Z(t)) = u'dZ + \frac{1}{2}u''dt = u'dZ + \frac{1}{2}dt.$$

Next, integrating between 0 and T, we obtain

$$u(Z(T)) - u(x) = \int_0^T u'dZ + \int_0^T \frac{1}{2} dt.$$

If we let T_x be the time so that the process arrives at a boundary, then it is possible to prove that this happens with probability one. Thus, using T_x instead of T we have that $u(Z(T_x)) = 0$ and we can therefore conclude that

$$u(x) = -\frac{1}{2} E_x(T_x). \tag{6.13}$$

As a consequence we know the expected value of the time the process arrives at the boundary, because from the solution of our differential equation, we have

$$u = -\frac{1}{2}(1 - x^2). \tag{6.14}$$

Comparing Eqs. (6.13)–(6.14), we observe that

$$E_x(T_x) = 1 - x^2$$

In general, the same computation can be performed for the equation in $\mathbb{R}^n$

$$\begin{cases} \Delta u = 1 \\ u|_{\partial \Omega} = 0 \end{cases}$$

In this case, we define a Brownian motion in $\mathbb{R}^n$ with multivariate normal distribution.

2) Consider the equation
$$\begin{cases} u'' = 0 \\ u(-1) = a; u(1) = b. \end{cases}$$

Let Z be a Brownian motion so that $Z(0) = x$. In the same way as in the previous example, applying Itô's lemma to $u(Z(t))$ we obtain

$$du = u'dZ.$$

Integrating between 0 and T results in

$$u(Z(t)) - u(x) = \int_0^T u' dZ.$$

We recall that $Z(T) \sim N(0, T)$. Therefore the time for arriving at the boundary is T_x; hence $Z(T_x) = \pm 1$. This shows that the solution is linear.

3) Consider the equation

$$u'' = 0$$

with initial point in $x \in (x - r, x + r)$. As in the previous examples, because we are considering a Brownian motion, the probability of arriving at the boundary is $\frac{1}{2}$ and so the solution of the equation is

$$u(x) = \frac{u(x+r) - u(x-r)}{2}.$$

We recall that if we consider a sphere centered at x with radius r, i.e. $S_r(x)$, then the probability of the arriving time is uniform. Therefore since u is represented by its expected value, we obtain the mean value theorem:

$$u(x) = \int_{S_r(x)} u \frac{1}{|S_r(x)|} = \frac{1}{|S_r(x)|} \int_{S_r(x)} u.$$

In the next section we briefly discuss how the tree methodology can be used for assets paying dividends.

6.3 Tree Methods for Dividend-Paying Assets

In this section we discuss tree modifications to accommodate the case when the underlying asset pays continuous dividends, known discrete dividends as well as known cash dividends at a prespecified time point t.

6.3.1 Options on Assets Paying a Continuous Dividend

Suppose an asset pays a dividend rate δ per unit of time. For example, suppose that the equity pays $\delta = 1\%$ annual dividend. If the original model used is geometric Brownian motion, applying Girsanov's theorem to (6.1), we obtain the stochastic equation

$$dS_t = (r - \delta)S_t dt + \sigma S_t dW_t. \tag{6.15}$$

Note that equation (6.15) is exactly a geometric Brownian motion with the risk-free rate r replaced with $r - \delta$. Since everything stays the same, any tree constructed for a stochastic process with r for the drift will work by just replacing it with $r - \delta$. For example, the Trigeorgis tree in this case becomes

$$\Delta x = \sqrt{\sigma^2 \Delta t + (r - \delta - \frac{1}{2}\sigma^2)^2 \Delta t^2}$$

$$p_u = \frac{1}{2} + \frac{1}{2} \frac{(r - \delta - \frac{1}{2}\sigma^2)\Delta t}{\Delta x}$$

$$p_d = 1 - p_u.$$

In our example $\delta = 0.01$. We can obviously write a routine for a general delta and in the case when there are no dividends just set $\delta = 0$. As an observation the unit time in finance is always one year. Therefore in any practical application if for instance expiry is one month, we need to use $T = 1/12$.

6.3.2 Options on Assets Paying a Known Discrete Proportional Dividend

In this instance, the asset pays at some time τ in the future a $\hat{\delta}$ dividend amount which is proportional to the stock value at that time τ. Specifically, the dividend amount is $\hat{\delta} S_\tau$. Due to the way the binomial tree is constructed, we need a very simple modification to accommodate the dividend payment. We only care about the dividend payment if it happens during the option lifetime. If that is the case suppose $\tau \in [(i-1)\Delta t, i\Delta t]$.

When the dividend is paid, the value of the asset drops by that particular amount. The justification is simple – the asset share value represents the value of the company. Since a ceratin amount is paid per share to the shareholders – an entity outside the company – the value of the company drops by that exact amount. To accommodate this we change the value of all nodes at time $i\Delta t$ by multiplying with $1 - \hat{\delta}$ for a multiplicative tree in S_t and by adding $-\hat{\delta}$ for an additive tree. Specifically for multiplicative trees at node $(i,j) = (i\Delta t, S_j)$, the value is $S_0(1-\hat{\delta})u^j d^{i-j}$ where $u = e^{\Delta x_u}$ and $d = e^{\Delta x_d}$.

6.3.3 Options on Assets Paying a Known Discrete Cash Dividend

This is the most complex case and unfortunately this is a realistic case (most assets pay cash dividends). In this case, the major issue is that the tree becomes nonrecombining after the dividend date. Suppose $\tau \in [(i-1)\Delta t, i\Delta t]$ and at that time the stock pays cash dividend D. The value of nodes after this time are supposed to be subtracted by D. Specifically at the node value (i,j) we have

$$S_0(e^{\Delta X_u})^j(e^{\Delta X_d})^{i-j} - D.$$

At the next step the successor down will be $S_0(e^{\Delta X_u})^j(e^{\Delta X_d})^{i-j-1} - De^{\Delta X_d}$. On the other hand the successor up from the lower node $S_0(e^{\Delta X_u})^{j-1}(e^{\Delta X_d})^{i-j-1} - D$ will be $S_0(e^{\Delta X_u})^j(e^{\Delta X_d})^{i-j-1} - De^{\Delta X_u}$. It is easy to see that the two nodes will not match unless $e^{\Delta X_u} = e^{\Delta X_d} = 1$ which makes the tree a straight line. Therefore, after step i the tree becomes nonrecombining and, for example, at time $(i+m)\Delta t$ there will be $m(i+1)$ nodes rather than $i+m+1$ as in regular tree. That is a quadratic number of calculations that quickly become unmanageable.

The trick to deal with this problem is to make the tree nonrecombining in the beginning rather than at the end. Specifically, we assume that S_t has two components, i.e.:

- $\tilde{S}_t$ a random component
- Remainder depending on the future dividend stream:

$$\tilde{S}_t = \begin{cases} S_t & \text{when} \quad t > \tau \\ S_t - De^{-r(\tau-t)} & \text{when} \quad t \leq \tau \end{cases}$$

Suppose $\tilde{S}_t$ follows a geometric Brownian motion with $\tilde{\sigma}$ constant. We calculate $p_u, p_d, \Delta X_u, \Delta X_d$ in the usual way with σ replaced with $\tilde{\sigma}$. The tree is constructed as before; however the tree values are:

at (i,j), when $t = i\Delta t < \tau$: $\tilde{S}_t(e^{\Delta X_u})^j(e^{\Delta X_d})^{i-j} + De^{-r(\tau-t)}$

at (i,j), when $t = i\Delta t \geq \tau$: $\tilde{S}_t(e^{\Delta X_u})^j(e^{\Delta X_d})^{i-j}$.

This tree will be mathematically equivalent with a discretization of a continuous Brownian motion process that suddenly drops at the fixed time τ by the discrete amount D.

6.3.4 Tree for Known (Deterministic) Time-Varying Volatility

Suppose the stochastic process has a known time-varying volatility $\sigma(t)$ and drift $r(t)$. That is, the stock follows

$$dS_t = r(t)S_t dt + \sigma(t)S_t dW_t$$

Fixing a time interval Δt, suppose at times $0, \Delta t, 2\Delta t, \ldots, \mu\Delta t$ the volatility values are

$$\sigma(i\Delta t) = \sigma_i, r(i\Delta t) = r_i.$$

Then, the corresponding drift term for the log process $X_t = \log S_t$ is

$$v_i = r_i - \frac{\sigma_i^2}{2}.$$

We keep the $\Delta X_u, \Delta X_d$ fixed ($\Delta X_u = \Delta X; \Delta X_d = -\Delta X$); this insures that the tree is recombining. However, we vary the probabilities at all steps so that the tree adapts to the time-varying coefficients. We let the probabilities at time step i be denoted using $p_u^i = p_i$ and $p_d^i = 1 - p_i$. Thus, we need to have

$$\begin{cases} p_i \Delta x - (1-p_i)\Delta x = v_i \Delta t_i \\ p_i \Delta x^2 + (1-p_i)\Delta x^2 = \sigma_i^2 \Delta t_i + v_i^2 \Delta t_i^2 \end{cases} \tag{6.16}$$

Simplifying (6.16) we obtain

$$\begin{cases} 2p_i \Delta x - \Delta x = v_i \Delta t_i \\ \Delta x^2 = \sigma_i^2 \Delta t_i + v_i^2 \Delta t_i^2 \end{cases} \tag{6.17}$$

Equation (6.17) can be further simplified to

$$p_i = \frac{1}{2} + \frac{v_i \Delta t_i}{2\Delta x} \tag{6.18}$$

Rearranging terms in (6.18) we obtain

$$v_i^2 \Delta t_i^2 + \sigma_i^2 \Delta t_i - \Delta x^2 = 0 \tag{6.19}$$

Using the quadratic formula we solve for Δt_i to obtain

$$\Delta t_i = \frac{-\sigma_i^2 \pm \sqrt{\sigma_i^4 + 4v_i^2 \Delta x^2}}{2v_i^2} \tag{6.20}$$

From (6.18), we have that

$$\Delta x^2 = \sigma_i^2 \Delta t_i + v_i^2 \Delta t_i^2 \tag{6.21}$$

So now the issue is that Δx must be kept constant but obviously the i's will pose a problem.

So here are two possible approaches to this issue of determining the proper parameter values.

If we sum over i in the last expression, we obtain

$$N\Delta x^2 = \sum_i^N \sigma_i^2 \Delta t_i + \sum_i^N v_i^2 \Delta t_i^2 \tag{6.22}$$

Dividing both sides by N we obtain

$$\Delta x^2 = \frac{1}{N}\sum_i^N \sigma_i^2 \Delta t_i + \frac{1}{N}\sum_i^N v_i^2 \Delta t_i^2$$

Taking $\overline{\Delta t} = \frac{1}{N}\sum_i^N \Delta t_i$ and solving for Δx we obtain

$$\Delta x = \sqrt{\overline{\sigma^2 \Delta t} + \overline{v^2 \Delta t^2}} \tag{6.23}$$

where $\overline{\sigma^2} = \frac{1}{N}\sum_{i=1}^{N} \sigma_i^2$ and $\overline{v^2} = \frac{1}{N}\sum_{i=1}^{N} v_i^2$. These are approximate values. From the practical perspective we just created a circular reasoning (we need the Δt_i values to calculate the Δx to calculate the Δt_i values).

So to actually implement it one needs to initialize the algorithm with $\Delta t_i = T/n$ for all i. Then calculate Δx using (6.23). Then repeat:

1) Calculate all Δt_i's using Eq. (6.20).
2) Recompute Δx using (6.23).
3) Calculate the difference between two consecutive values of Δx. If smaller than ε stop. If larger repeat steps 1–3.

The tree thus constructed is approximately correct. However, the effects of the extra modifications to the value of the option are small.

6.4 Pricing Path-Dependent Options: Barrier Options

Barrier options are triggered by the action of the underlying asset hitting a prescribed value at some time before expiry. For example, as long as the asset remains below a predetermined barrier price during the whole life of the option, the contract will have a Call payoff at expiry. Barrier options are clearly path-dependent options. Path-dependent options are defined as the right, but not the obligation, to buy or sell an underlying asset at a predetermined price during a specified time period; however the exercise time or price is dependent on the underlying asset value during all or part of the contract term. A path-dependent option's payoff is determined by the path of the underlying asset's price.

There are two main types of barrier options:

1) The **In** type option, which exercises as specified in the contract as long as a particular asset level called barrier is reached before expiry. If the barrier is reached, then the option is said to have knocked in. If the barrier level is not reached, the option is worthless.
2) The **Out** type option, which becomes zero and is worthless if the barrier level is not reached. If the barrier is reached then the option is said to have knocked out.

We further characterize the barrier option by the position of the barrier relative to the initial value of the underlying:

1) If the barrier is above the initial asset value, we have an Up option.
2) If the barrier is below the initial asset value, we have a Down option.

We note that if an Out option starts as regular American and barrier hit option becomes zero and similarly if an In option typically starts worthless and if the barrier is hit, the option becomes a regular American option.

Given a barrier B and path $= S_{t1}, \ldots, S_{tN}$, the terminal payoff of the barrier options can be written as

1) Down and Out Call $(S_T - K)_+ 1_{\{\min(S_{t1},\ldots,S_{tN}) > B\}}$
2) Up and Out Call $(S_T - K)_+ 1_{\{\max(S_{t1},\ldots,S_{tN}) < B\}}$
3) Down and In Call $(S_T - K)_+ 1_{\{\min(S_{t1},\ldots,S_{tN}) \leq B\}}$
4) Up and In Call $(S_T - K)_+ 1_{\{\max(S_{t1},\ldots,S_{tN}) \geq B\}}$
5) Down and Out Put $(K - S_T)_+ 1_{\{\min(S_{t1},\ldots,S_{tN}) > B\}}$
6) Up and Out Put $(K - S_T)_+ 1_{\{\max(S_{t1},\ldots,S_{tN}) < B\}}$
7) Down and In Put $(K - S_T)_+ 1_{\{\min(S_{t1},\ldots,S_{tN}) \leq B\}}$
8) Up and In Put $(K - S_T)_+ 1_{\{\max(S_{t1},\ldots,S_{tN}) \geq B\}}$

Remark 6.4.1 Generally, the In type options are much harder to solve than Out type options. However, it is easy to see that we have the following Put–Call parity relations.

Down and Out Call (K, B, T) + Down and In Call (K, B, T) = Call (K, T) for all B.

This is easy to prove by looking at the payoffs and remarking that

$$1_{\min(S_{t1},\ldots,S_{tN}) > B} + 1_{\min(S_{t1},\ldots,S_{tN}) \leq B} = 1$$

and recalling that option price at t is the discounted expectation of the final payoff. Therefore applying expectations will give exactly the relation needed. Thus one just needs to construct a tree method for Down type options and use the In–Out Parity discussed earlier to obtain the price of an In type option.

6.5 Trinomial Tree Method and Other Considerations

The trinomial trees provide an effective method of numerical calculation of option prices within the Black–Scholes model. Trinomial trees can be built in a similar way to the binomial tree. To create the jump sizes u and d and the transition probabilities p_u and p_d in a binomial tree model, we aim to match these parameters to the first two moments of the distribution of our geometric Brownian motion. The same can be done for our trinomial tree for $u, d, p_u, p_m,$ and p_d.

For the trinomial tree, one cannot go up and down by different amounts to keep it recombining. The condition imposed is the fact that the discrete increment needs to match the continuous once, that is,

$$E(\Delta x) = \Delta x p_u + 0 p_u + (-\Delta x) p_d = D dt, \quad D = r - \frac{\sigma^2}{2}$$

$$E(\Delta x^2) = \Delta x^2 p_u + 0 p_u + (-\Delta x)^2 p_d = \sigma^2 \Delta t + D^2 \Delta t^2$$

$$p_u + p_m + p_d = 1$$

6.5 Trinomial Tree Method and Other Considerations

These are three equations and four unknowns. So it has an infinite number of solutions. However we have that

$$p_u = \frac{1}{2}\left(\frac{\sigma^2 \Delta t + D^2 \Delta t^2}{\Delta x^2} + \frac{D\Delta t}{\Delta x}\right)$$

$$p_m = 1 - \frac{\sigma^2 \Delta t + D^2 \Delta t^2}{\Delta x^2}$$

$$p_d = \frac{1}{2}\left(\frac{\sigma^2 \Delta t + D^2 \Delta t^2}{\Delta x^2} - \frac{D\Delta t}{\Delta x}\right)$$

These probabilities need to be numbers between 0 and 1. Imposing this condition we obtain a sufficient condition: $\Delta x \geq \sigma\sqrt{3\Delta t}$. This condition will be explained later in this section. Any Δx with this property produces a convergent tree.

The trinomial tree is an alternate way to approximate the stock price model. The stock price once again follows the equation

$$dS_t = rS_t dt + \sigma S_t dW_t. \tag{6.24}$$

In [38] the authors work with a continuously paying dividend asset and the drift in the equation is replaced by $r - \delta$. All the methods we will implement require an input r. One can easily obtain the formulas for a continuously paying dividend asset by just replacing this parameter r with $r - \delta$. Once again, it is equivalent to work with the return $X_t = \log S_t$ instead of directly with the stock, and we obtain

$$dX_t = v dt + \sigma dW_t, \quad \text{where } v = r - \frac{1}{2}\sigma^2. \tag{6.25}$$

The construction of the trinomial tree is equivalent to the construction of the binomial tree described in previous sections. A one step trinomial tree is presented as follows:

S_u

$S \quad S$

S_d

Trinomial trees allow the option value to increase, decrease, or remain stationary at every time step as illustrated earlier.

Once again we match the expectation and variance. In this case the system contains three equations and three unknowns, so we do not have a free choice as in the binomial tree case. In order to have a convergent tree, numerical experiments have shown that we impose a condition such that

$$\Delta x \geq \sigma\sqrt{3\Delta t} \tag{6.26}$$

Hence for stability there must be restrictions on the relative sizes of Δx and Δt. The condition ensures that our method is stable and converges to the exact solution. Please refer to [38] for details about the stability condition. Once the tree is constructed, we find an American or European option value by stepping back through the tree in a similar manner with what we did for the binomial tree. The only difference is that we calculate the discounted expectation of three node values instead of two as we did for the binomial tree.

The trinomial tree produces more paths (3^n) than the binomial tree (2^n). Surprisingly the order of convergence is not affected for this extra number. In both cases the convergence of the option values is of the order $O(\Delta x^2 + \Delta t)$. Optimal convergence is always guaranteed due to condition (6.26).

Condition (6.26) makes a lot of difference when we deal with barrier options, i.e. options that are path dependent. This is due to the fact that the trinomial tree contains a larger number of possible nodes at each time in the tree. The trinomial tree is capable of dealing with the situations when the volatility changes over time, i.e. is a function of time. For a detailed example of recombining trinomial tree for valuing real options with changing volatility, please refer to [85]. In that study, the trinomial tree is constructed by choosing a parameterization that sets a judicious state space while having sensible transition probabilities between the nodes. The volatility changes are modeled with the changing transition probabilities, while the state space of the trinomial tree is regular and has a fixed number of time and underlying asset price levels.

The meaning of little o and big O Suppose we have two functions f and g. We say that f is of order little o of g at x_0 if

$$f \sim o(g) \Leftrightarrow \lim_{x \to x_0} \frac{f(x)}{g(x)} = 0$$

We say that f is of order big O of g at x_0 if

$$f \sim O(g) \Leftrightarrow \lim_{x \to x_0} \frac{f(x)}{g(x)} = C$$

where C is a constant.

In our context if we calculate the price of an option using an approximation (e.g. trinomial tree) called $\hat{\Pi}$ and the real (unknown) price of the option called Π, then we say that the approximation is of the order $O(\Delta x^2 + \Delta t)$ and we mean that

$$|\hat{\Pi} - \Pi| = C(\Delta x^2 + \Delta t)$$

whenever Δx and Δt both go to zero for some constant C.

6.6 Markov Process

The remaining sections in this chapter are dedicated to a general theory of approximating solutions of SDE's using trees. We shall present in the final section of this chapter what we call the quadrinomial tree approximation for stochastic volatility models. The full version of the presentation may be found in [71].

Markov processes were briefly introduced in Section 1.4 of this book. Here we give a formal definition.

Definition 6.6.1 (Markov Process). Let $\{X_t\}_{t \geq 0}$ be a process on the space $(\Omega, \mathcal{F}, \mathcal{P})$ with values in E and let $\mathcal{G}_t$ a filtration such that X_t is adapted with respect to this filtration. In the most basic case, the filtration is generated by the stochastic process itself: $\mathcal{G}_t = \mathcal{F}_t^X = \sigma(X_s : s \leq t)$.

The process X_t is a Markov process if and only if

$$\mathcal{P}(X_{t+s} \in \Gamma | \mathcal{G}_t) = \mathcal{P}(X_{t+s} \in \Gamma | X_t), \ \forall s, \ t \geq 0 \text{ and } \Gamma \in \mathcal{B}(E). \quad (6.27)$$

Here, the collection $\mathcal{B}(E)$ denote the Borel sets of E.

Essentially, the definition says in order to decide where the process goes next, knowing just the current state is the same as knowing the entire set of past states. The process does not have memory of past states.

The defining Eq. (6.27) is equivalent to

$$E\left[f(X_{t+s}) | \mathcal{F}_t^X\right] = E\left[f(X_{t+s}) | X_t\right],$$

$\forall f$ Borel measurable functions on E.

Now this definition looks very complicated primarily because of those spaces and the daunting probability distribution: $\mathcal{P}(X_{t+s} \in \Gamma | \mathcal{G}_t)$. We wrote it this way to cover any imaginable situation. In practice we use a transition function to define this transition distribution.

6.6.1 Transition Function

Remember that a stochastic process is completely characterized by its distribution. Since knowing the current state is enough to determine the future, the random variable $X_t | X_s$ for any s and t determines the process trajectories. This random variable $X_t | X_s$ has a distribution and in fact that is what we denoted with $\mathcal{P}(X_t \in \Gamma | X_s)$. If this distribution has a density, we will call it **the transition density function**.

Specifically, the random variable $X_t | X_s = x$ has a density denoted by $p(s, t, x, y)$, that is,

$$\mathcal{P}(s, t, x, \Gamma) = \int_\Gamma p(s, t, x, y) dy.$$

This transition density completely characterizes the Markov process. Note that in general the transition probability for a Markov process depends on the time of the transition (s), and it is a four-dimensional function. Such process appears as the solution of a general SDE such as

$$dS_t = \mu(t, S_t)dt + \sigma(t, S_t)dB_t. \tag{6.28}$$

However, in finance we work with simplified SDE's where the drift and diffusion functions $\mu(\cdot, \cdot)$ and $\sigma(\cdot, \cdot)$ are not dependent on time t. The solution of an SDE where the coefficients are not time dependent is a particular type of Markov process.

Definition 6.6.2 (Homogeneous Markov Process). A Markov process is homogeneous if the transition function does not depend on the time when the transition happens, just on the difference between points. Mathematically, the random variable $X_t | X_s$ has the same distribution as $X_{t-s} | X_0$ for all s and t, or written even clearer, $X_{s+h} | X_s$ is the same as distribution of $X_h | X_0$. This should make it clear that the point when the transition is made s is irrelevant for homogeneous processes. In terms of transition function we have

$$p(s, t, x, y) = p(t - s, x, y)$$

or in the other notation,

$$p(s, s + h, x, y) = p(h, x, y).$$

As we mentioned, all solutions to diffusion equations of the type

$$dS_t = \mu(S_t)dt + \sigma(S_t)dB_t \tag{6.29}$$

where μ and σ are functions that do not have a t argument are homogeneous Markov processes.

Formally, a function $\mathscr{P}(t, x, \Gamma) = \int_\Gamma p(t, x, y) dy$ defined on $[0, \infty) \times E \times \mathscr{B}(E)$ is a (time homogeneous) transition function if:

1) $\mathscr{P}(t, x, \cdot)$ is a probability measure on $(E, \mathscr{B}(E))$.
2) $\mathscr{P}(0, x, \cdot) = \delta_x$ (Dirac measure).
3) $\mathscr{P}(\cdot, \cdot, \Gamma)$ is a Borel measurable function on $[0, \infty) \times E$.
4) The function satisfies the Chapman–Kolmogorov equation:

$$\mathscr{P}(t + s, x, \Gamma) = \int \mathscr{P}(s, y, \Gamma) \mathscr{P}(t, x, dy),$$

where $s, t \geq 0, x \in E$ and $\Gamma \in \mathscr{B}(E)$.

The Chapman–Kolmogorov equation is a property of all Markov processes basically saying that to go from x to some point in Γ in $t + s$ units of time, one has to go through some point at time t. Rewriting the expression using the transition function,

$$p(t+s,x,y) = \int p(s,z,y)p(t,x,z)dz,$$

where the integral in z is over all the state space points.

The connection between a function with such properties and the Markov process is the following. A transition function is the transition function for a time homogeneous Markov process X_t if and only if

$$\mathscr{P}(X_{t+s} \in \Gamma | \mathcal{F}_t^x) = \mathscr{P}(s, X_t, \Gamma),$$

which means it actually expresses the probability that the process goes into a set Γ at time $t+s$ given that it was at X_t at time t. Furthermore, all these properties may be expressed in terms of the density function $p(t,x,y)$

$$\mathscr{P}(t,x,\Gamma) = \int_\Gamma p(t,x,y)dy,$$

if the transition density function exists.

The next theorem stated without proof ensures that our discussion of transition functions is highly relevant.

Theorem 6.6.1 Any Markov process has a transition function. If the metric space (E, r) is complete and separable, then for any transition function there exists a unique Markov process with that transition function.

If the Markov process is defined on $\mathbb{R}$ (which is a separable space), then the theorem is valid and there is a one-to-one equivalence between Markov processes and their transition functions. However, $\mathbb{R}^2$ and generally $\mathbb{R}^n$ are not separable spaces. So, there may exist multiple Markov processes with the same transition function. This seemingly disappointing property is in fact very useful, and it is the basis on the Markov Chain Monte Carlo (MCMC) methods. Specifically, suppose we have a very complicated process and we need to create Monte Carlo paths but the transition distribution is complicated. So the idea is to map the process in $\mathbb{R}^2$ or higher dimensions and then find a simpler process with the same transition distribution. Then use the simpler process to generate the transition distribution of the more complicated one.

Definition 6.6.3 (Strong Markov Process). A Markov process $\{X_t\}_{t\geq 0}$ with respect to $\{\mathcal{G}_t\}_{t\geq 0}$ is called a strong Markov process at ζ if

$$\mathscr{P}(X_{t+\zeta} \in \Gamma | \mathcal{G}_\zeta) = \mathscr{P}(t, X_\zeta, \Gamma) \tag{6.30}$$

where ζ is a stopping time with respect to $\mathcal{G}_t$ and $\zeta < \infty$ almost surely (a.s.). A process is strong Markov with respect to $\mathcal{G} = \{\mathcal{G}\}_t$ if Eq. (6.30) holds for all ζ stopping times with respect to $\{\mathcal{G}_t\}$.

In other words what a strong Markov process has going for it is that the Markov property holds not only at any times but also at any stopping time. This is important since it allows us to use Markov processes in very interesting ways. Clearly any strong Markov process is Markov.

Remark 6.6.1 Any Markov process is a strong Markov at ζ if ζ is discrete valued.

Example 6.6.1 The 1-D Brownian motion is a homogeneous Markov process, with the transition density function

$$\mathscr{P}(t,x,\Gamma) = \int_\Gamma \frac{1}{\sqrt{2\pi t}} e^{-\frac{(y-x)^2}{2t}} dy.$$

The Brownian motion is one of the few examples where we can actually write the transition distribution explicitly. In general, it is not possible to calculate these transition probabilities exactly. The next section introduces a generalization that will allow us to tap into a much older and more developed theory.

6.7 Basic Elements of Operators and Semigroup Theory

To be able to better characterize Markov processes, we need to present some elements of semigroup theory. To this end, using the transition probability we define an operator

$$T(t)f(x) = \int f(y)\mathscr{P}(t,x,dy) = \int f(y)p(t,x,y)dy.$$

Note that $T(t)$ is a functional operator. Specifically, it takes as the argument a function $f(x)$ and outputs another function $T(t)f(x)$ (which is a function of x).

We say that $\{X_t\}$ a Markov process corresponds to a semigroup operator $T(t)$ if and only if

$$T(t)f(X_s) = \int f(y)\mathscr{P}(t,X_s,dy) = \int f(y)p(t,X_s,y)dy = E[f(X_{t+s})|X_s],$$

for any measurable function f. Recall the definition of transition probability: $p(t,x,y)$ is the density of the random variable $X_t|X_0$ which, by homogeneity of the process, is the same as the distribution of $X_{t+s}|X_s$. This operator $T(t)$ and the initial distribution of the process at time 0 completely characterize the Markov process.

Since $\mathscr{P}(t,x,dy)$ is a transition distribution, we can use the Chapman–Kolmogorov equation, and in this operator notation, we obtain

$$T(s+t)f(x) = \int f(y)p(t+s,x,y)dy = \int f(y) \int p(s,z,y)p(t,x,z)dzdy$$
$$= \int \int f(y)p(s,z,y)p(t,x,z)dydz$$
$$= \int \left(\int f(y)p(s,z,y)dy \right) p(t,x,z)dz$$
$$= \int (T(s)f(z))p(t,x,z)dz = T(t)(T(s)f(x))$$
$$= T(s) \circ T(t)f(x).$$

This particular expression tells us that $T(t)$ is a semigroup operator, specifically a contraction operator $||T(t)|| \leq 1$. We shall define exactly what this means next.

Let L be a Banach space of functions. A Banach space is a normed linear space that is a complete metric space with respect to the metric derived from its norm. To clarify some of these notions we will provide several definitions.

Definition 6.7.1 (Linear Space/Vector Space). A linear space (sometimes called vector space) L is a collection of elements endowed with two basic operations generally termed addition and multiplication with a scalar. A linear space L has the following properties:

1) Commutativity: $f + g = g + f$.
2) Associativity of addition: $(f + g) + h = f + (g + h)$.
3) Null element: there exists an element 0 such that $f + 0 = 0 + f = f$, for all $f \in L$.
4) Inverse with respect to addition: for any $f \in L$ there exists an element $-f \in L$ such that $f + (-f) = (-f) + f = 0$.
5) Associativity of scalar multiplication: $a(bf) = (ab)f$.
6) Distributivity: $(a+b)f = af + bf$ and $a(f+g) = af + ag$.
7) Identity scalar: there exists a scalar element denoted 1 such that $1f = f1 = f$.

Throughout the definition these properties hold for all $f, g, h \in L$ and a, b scalars. Scalars are typically one-dimensional real numbers.

Example 6.7.1 Most spaces of elements you know are linear spaces. For example, $L = \mathbb{R}^n$ with the scalar space $\mathbb{R}$ is the typical vector space and has all the properties discussed earlier. L, the space of squared matrices with dimension n and whose scalar space is $\mathbb{R}$, is another classical example. In fact this is sometimes called the general linear space and is the prime object of study for linear algebra. Here we shall use L as a space of functions. For example, $L = \mathscr{C}(0, \infty)$ the space of all continuous functions defined on the interval $(0, \infty)$. The notions that follow are common in functional analysis which studies spaces of functions and operations on spaces of functions. For example, the common derivation is just an operator (operation) on these spaces of functions.

A normed space is nothing more than a space that has a norm defined on it. The definition of a norm follows:

Definition 6.7.2 Given a linear space L a norm on this space is any function $\|\cdot\|: L \to [0, \infty)$ with the following properties:

1) Positive definite: if $\|f\| = 0$ then f is the 0 element of L.
2) Homogeneous: $\|af\| = |a| \|f\|$, for any scalar a and $f \in L$.
3) Subadditive (triangle inequality) $\|f + g\| \leq \|f\| + \|g\|$.

As a side note a seminorm is a function with properties 2 and 3 only.

Example 6.7.2 We can introduce many norms on the same space. Take $L = \mathbb{R}^n$ the space of vectors. The classical absolute value

$$\|f\| = |f| = \sqrt{f_1^2 + f_2^2 + \cdots + f_n^2},$$

where f_i's are the components of f, is a norm. Typically it is denoted by $\|f\|_2$. The generalization of this is the p-norm:

$$\|f\|_p = \sqrt[p]{\sum_{i=1}^n f_i^p} = \left(\sum_{i=1}^n f_i^p\right)^{\frac{1}{p}}.$$

The infinity norm is

$$\|f\|_\infty = \max_{1 \leq i \leq n} |f_i|.$$

These norms can be defined generally on any linear space. Other spaces, for example, when L is a space of squared matrices n dimensional have other classical norms. Say A is a matrix we can define

$$\|A\| = \max_{1 \leq j \leq n} \sum_{i=1}^n |a_{ij}|,$$

which is the maximum of sum of elements of each column in the matrix. The *Frobenius norm* is an equivalent to the absolute value for vectors:

$$\|A\|_F = \sqrt{\sum_{i=1}^n \sum_{j=1}^n a_{ij}^2}.$$

For a space of functions, for example, $L = C(0, \infty)$, we can define similar norms, for example,

$$\|f\|_0 = \max_{x \in (0,\infty)} f(x)$$

or norm 2:

$$\|f\|_2 = \left(\int_0^\infty f^2(x) dx \right)^{1/2}$$

Throughout this chapter and indeed throughout the constructions involving stochastic processes, we will be using L the space of continuous, bounded functions. On this space of functions we will work with operators which essentially are transformations of functions. As a simple example the differential of a function $\frac{\partial}{\partial x}$ is an operator that transform a function into its derivative (itself a function).

Definition 6.7.3 (Semigroup). A one parameter family $\{T(t)\}_{t \geq 0}$ of bounded linear operators is called a semigroup if

1) $T(0) = I$ (identity operator)
2) $T(s + t) = T(s) \circ T(t) \; \forall t, s \geq 0$

A semigroup is called **strongly continuous** if

$$\lim_{t \to 0} T(t)f = f \text{ for all } f \in L \quad \text{(equivalently } \|T(t)f - f\| \to 0 \quad \text{as} \quad t \to 0)$$
(6.31)

A semigroup is called a **contraction** if $\|T(t)\| \leq 1$ for all t.

The norm in the definition needs to be specified and the family is a semigroup with respect to that particular norm.

Example 6.7.3 Let $E = \mathbb{R}^n$ and let B a $n \times n$ matrix.
Define

$$e^{tB} = \sum_{k=0}^{\infty} \frac{1}{k!} t^k B^k,$$

where B^k are powers of the matrix which are well defined since the matrix is square and the convention $B^0 = I_n$ the identity matrix.
Then, the operator $T(t) = e^{tB}$ forms a strongly continuous semigroup and

$$\| e^{tB} \| \leq \sum_{k=0}^{\infty} \frac{1}{k!} t^k \| B^k \| \leq \sum \frac{1}{k!} t^k \| B \|^k \leq e^{t\|B\|},$$

where the norm is the supremum norm.

In general for a strongly continuous semigroup, we have $\| T(t) \| \leq Me^{mt}$, $t \geq 0$ for some constants $M \geq 1$ and $m \geq 0$.

In our specific case when the semigroup is defined from a Markov process X_t, i.e.

$$T(t)f(X_s) = \int f(y)p(t, X_s, y)dy = E[f(X_{t+s})|X_s],$$

we already shown that Chapman–Kolmogorov implies the second defining property of a semigroup. For the first one it is simply

$$T(0)f(X_s) = E[f(X_s)|X_s] = f(X_s),$$

since $f(X_s)$ is measurable with respect to X_s. For the contraction we have

$$\| T(t)f(x) \| = \left\| \int f(y)p(t, x, y)dy \right\| \leq \| f(y) \| \left\| \int p(t, x, y)dy \right\| = \| f(x) \|.$$

This will imply that $\| T(t) \| \leq 1$. The contraction property is true for any norm and is based on the fact that $\int p(t, x, y)dy = 1$. It is left as an exercise to prove this inequality for various specific norms.

6.7.1 Infinitesimal Operator of Semigroup

The notion of semigroup is important since it allows us to characterize long behavior $T(s + t)$ by looking at short behavior $T(s)$ and $T(t)$. Think about transition probability for a two-year period $T(2)$ as being characterized by transitions in only one year $T(1)$. Also note (very importantly) that this only works for homogeneous Markov processes. If the process is not homogeneous then we have $T(0, 2) = T(0, 1)T(1, 2)$ and even though the two transitions on the right are one-year long, they are different functions that need to be studied separately.

In the case when the process is homogeneous, we can extend the idea to go from a shot interval extended to obtain everything. But how short an interval may we characterize? In real analysis we look at the derivative of a function that tells us at an infinitesimal distance what the next value of the function will be. So the idea here would be: can we define a sort of derivative of an operator? If you remember from calculus the derivative is defined as a limit.

Definition 6.7.4 An *infinitesimal operator* of a semigroup $\{T(t)\}_{t \geq 0}$ on L is a linear operator defined as

$$Af = \lim_{t \to 0} \frac{1}{t}(T(t)f - f).$$

The domain of definition is those functions for which the operator exists (i.e. $D(A) = \{f | Af \text{ exists}\}$). The infinitesimal generator can be thought in some sense as the right derivative of the function $t \to T(t)f$.

If you remember that a strong continuous semigroup has $\lim_{t \to 0} T(t)f = f$, then the infinitesimal generator can be exactly interpreted as the derivative. The next theorem formalizes this case.

Theorem 6.7.1 If A is the infinitesimal generator of $T(t)$ and $T(t)$ is strongly continuous, then the integral $\int_0^t T(s)fds \in D(A)$ and we have:

1) If $f \in L$ and $t \geq 0$ then
$$T(t)f - f = A \int_0^t T(s)fds$$

2) If $f \in D(A)$, then
$$\frac{d}{dt} T(t)f = AT(t)f = T(t)Af$$
where $T(t)f \in D(A)$

3) $f \in D(A)$, then
$$T(t)f - f = \int_0^t AT(s)fds = \int_0^t T(s)Afds$$

6.7.2 Feller Semigroup

A Feller semigroup is a strongly continuous, positive contraction semigroup defined by
$$T(t)f(x) = \int f(y) \mathscr{P}(t, x, dy) \quad \forall \ t, x$$
with a corresponding infinitesimal generator A. A Feller semigroup's infinitesimal generator is defined for all continuous bounded functions.

This $T(t)$ corresponds to a special case of a homogeneous Markov process with transition $\mathscr{P}(t, x, dy)$ for which the infinitesimal generator exists. The Markov process with these properties is called a Feller process. Recall, if the Feller process is denoted X_t, we have
$$T(t)f(X_s) = E[f(X_{t+s})|F_s^X] = \int f(y) \mathscr{P}(t, X_s, dy) = \int f(y) p(t, X_s, y) dy.$$

The next theorem makes the connection between the infinitesimal generator of a Markov process and a corresponding martingale expressed in terms of the original process. The big deal about the Feller processes is that if X_t is Feller, the next theorem works for any function f. Often the domain $D(A)$ of the infinitesimal generator of a regular Markov process is hard to find and thus describe.

Theorem 6.7.2 Let X be a Markov process with generator A.

1) If $f \in D(A)$ then $M_t = f(X_t) - f(X_0) - \int_0^t Af(X_0)ds$ is a martingale.
2) If $f \in C_0(E)$ and there exists a function $g \in C_0(E)$ such that $f(X_t) - f(X_0) - \int_0^t g(X_s)ds$ is a martingale, then $f \in D(A)$ and $Af = g$.

Proof. In this proof for simplicity of notations we neglect $f(X_0)$. We have

$$E(M_{t+s}|F_t) = E[f(X_{t+s})|F_t] - E\left[\int_0^{t+s} Af(X_u)du \Big| F_t\right]$$

$$= T(s)f(X_t) - \int_0^t Af(X_u)du - \int_t^{t+s} E[Af(X_u)|F_t]du$$

$$= T(s)f(X_t) - \int_0^t Af(X_u)du - \int_0^s T(z)Af(X_t)dz.$$

But A is an infinitesimal generator and so

$$T(t)f - f = \int_0^t T(s)Afds,$$

which implies that

$$f = T(t)f - \int_0^t T(s)Afds$$

Substituting we obtain

$$E(M_{t+s}|F_t) = T(s)f(X_t) - \int_0^s T(z)Af(X_t)dz - \int_0^t Af(X_u)du = M_t.$$

where $f(X_t) = T(s)f(X_t) - \int_0^s T(z)Af(X_t)dz$. Note there is an $f(X_0)$ missing from the expression, which is the one we neglected.

We finally set

$$T(t)f - f = \int_0^t T(s)gds,$$

which is the property of the infinitesimal generator. □

6.8 General Diffusion Process

In this section, we briefly describe a general diffusion process and present some useful examples. We are finally in position to connect with the semigroup theory we have been describing. The martingale representation Theorem 6.7.2 is the connector with option pricing theory.

Let $a = (a_{ij})_{ij}$ be a continuous symmetric, nonnegative definite $d \times d$ matrix valued function on $\mathbb{R}^d$. Let $b : \mathbb{R}^d \to \mathbb{R}^d$ be a continuous function. Define

$$Af = \frac{1}{2}\sum_{i,j=1}^d a_{ij}\frac{\partial}{\partial x_i}\frac{\partial}{\partial x_j}f + \sum_{i=1}^d b_i\frac{\partial}{\partial x_i}f, \text{ for all } f \in C_c^\infty(\mathbb{R}^d).$$

6.8 General Diffusion Process

In general, for nonhomogeneous process we need to define

$$A_t f(x) = \frac{1}{2} \sum_{i,j=1}^{d} a_{ij}(t,x) \partial_{x_i} \partial_{x_j} f(x) + \sum_{i=1}^{d} b_i(t,x) \partial_{x_i} f(x).$$

A Markov process is a diffusion process with infinitesimal generator L if it has continuous paths and

$$E[f(X_t) - f(X_0)] = E\left[\int_0^t A(f(X_s))ds\right].$$

Conversely, if $\{X_t\}$ has continuous paths its generator is given by

$$Af = c(x)f + \sum_i b_i(x)\frac{\partial f}{\partial x_i} + \sum_{ij} a_{ij}(x)\frac{\partial^2 f}{\partial x_i \partial x_j}.$$

Given the connection mentioned earlier, we use Itô's lemma to give a general characterization for diffusion processes.

Definition 6.8.1 If X_t solves a d-dimensional SDE:

$$dX_t = b(X_t)dt + \sigma(X_t)dB_t$$

where b is a vector and σ is a $d \times d$ matrix $(\sigma_{ij})_{i,j}$, then X_t is a diffusion (Feller) process with infinitesimal generator:

$$A = \sum_{j=1}^{d} b_j(X_t)\frac{\partial}{\partial x_j} + \frac{1}{2} \sum_{i=1}^{d} \sum_{j=1}^{d} \sum_{k=1}^{d} \sigma_{ik}(X_t)\sigma_{kj}(X_t)\frac{\partial^2}{\partial x_i \partial x_j}.$$

Note that writing $a_{ij} = \sum_{k=1}^{d} \sigma_{ik}\sigma_{kj}$ puts the process in the classical form presented earlier.

Next, we present some useful examples of diffusion processes.

Example 6.8.1 (Pure Jump Markov Process). Let X_t be a compound Poisson process, with λ the jump intensity function and μ the probability distribution of jumps. That is in the interval $[0,t]$ the process jumps N times where N is a Poisson random variable with mean λt. Each times it jumps it does so with a magnitude Y with distribution μ. Each jump is independent. The process X_t cumulates (compounds) all these jumps. Mathematically,

$$X_t = \sum_{i=1}^{N} Y_i.$$

This process is a Markov process and its infinitesimal generator is

$$Af(x) = \lambda(x) \int (f(y) - f(x))\mu(x, dy)$$

Example 6.8.2 (Lévy Process). Levy processes will be discussed in details in Chapter 12 of this book. Their infinitesimal generator has the form

$$Af = \frac{1}{2} \sum_{ij} a_{ij}(t,x) \partial_i \partial_j f(x) + \sum_i b_i(t,x) \partial_i f(x)$$
$$+ \int_{\mathbb{R}^d} \left(f(x+y) - f(x) - \frac{y \cdot \nabla f(x)}{1+|y|^2} \right) \mu(t,x;dy)$$

In the case when jumps are not time dependent or dependent on the current state of the system (i.e. μ is independent of (t,x)), then the integral is

$$\int_{\mathbb{R}} \left(f(x+y) - f(x) - \frac{y \cdot \nabla f(x)}{1+|y|^2} \right) \mu(dy)$$

A more specific example is presented in the following.

Example 6.8.3 Suppose a stock process can be written as

$$dS_t = rS_t dt + \sigma S_t dW_t + S_t dJ_t,$$

where J_t is a compound Poisson process. This model is known as a jump diffusion process.

Let

$$X_t = \log S_t$$

then using Itô lemma, under the equivalent martingale measure

$$dX_t = \mu dt + \sigma dB_t + dJ_t,$$

where μ is $r - \frac{\sigma^2}{2} + \lambda(1 - E(e^Z))$ and Z is the random variable denoting the jump rate. The process is

$$J_t = \sum_i^{N_t} Z_i,$$

where $N_t \sim$ Poisson(λt) and Z_i are the jump rates. In this case we can write the infinitesimal generator:

$$Af(x) = \frac{1}{2}\sigma^2 \frac{\partial^2 f}{\partial x^2} + \mu \frac{\partial f}{\partial x} + \lambda \int_{\mathbb{R}} [f(x+z) - f(x)] p(z) dz,$$

where $p(z)$ is the distribution of jumps Z_i.

The following are some specific examples of the jump distributions $p(z)$.

Merton:

$$p(z) = \frac{1}{\sqrt{2\pi s^2}} e^{-\frac{(z-\mu)^2}{2s^2}}$$

and
$$E(e^z) = e^{\mu + \frac{\sigma^2}{2}}$$

Kou:
$$p(z) = py_1 e^{-y_1 z} 1_{\{z \geq 0\}} + (1-p)y_2 e^{y_2 z} 1_{\{z \leq 0\}} \quad \text{(double exponential)}$$

and
$$E(e^z) = 1 - (1-p)\frac{1}{y_2 + 1} + p\frac{1}{y_1 + 1}$$
$$= (1-p)\frac{y_2}{y_2 + 1} + p\frac{y_1}{y_1 - 1}$$

6.8.1 Example: Derivation of Option Pricing PDE

We recall from Theorem 6.7.2 that for all f, $f(X_t) - f(X_0) - \int_0^t Af(X_s)ds$ is a martingale, if the process X_t is Feller and A is its infinitesimal generator. Note that this martingale has expected value at time 0 equal to 0.

Suppose we want to price an option on X_t and further assume that its value is a function of the current value X_t only. This has been shown to happen for European type option when the final payoff is of the form $\psi(X_T) = F(e^{X_T})$. We know that in this case,

$$V(t, X_t) = E[e^{-r(T-t)}\psi(x_T)|\mathcal{F}_t] = e^{-r(T-t)}E[\psi(x_T)|\mathcal{F}_t] = e^{-r(T-t)}g(X_t)$$

is the option value at any time t, where the expectation is under the equivalent martingale measure. We used the function $g(\cdot)$ to denote the expectation in the case when X_t is a Markov process.

If X_t solves a homogeneous SDE, it will be a Feller process and applying the Theorem 6.7.2 for g we obtain $g(X_t) - g(X_0) - \int_0^t Ag(X_s)ds$ is a martingale. The initial value of this martingale is 0 and taking the derivative we get

$$\frac{\partial g}{\partial t} - Ag(X_t) = 0.$$

Now using that $g(x) = e^{r(T-t)}V(t, x)$ and substituting the derivatives and finally since the process is homogeneous in A involves only derivatives in x we finally get

$$\frac{\partial V}{\partial t} - AV - rV = 0.$$

This creates the PDE whose solution is the option price. Please note that this is the same PDE we derived in the previous chapters with a much different derivation, and furthermore that this derivation is much more general.

As an example, say X_t solves the one-dimensional SDE:

$$dX_t = b(X_t)dt + \sigma(X_t)dB_t$$

The infinitesimal generator is:
$$A = b(x)\frac{\partial}{\partial x} + \frac{1}{2}\sigma^2(x)\frac{\partial^2}{\partial x^2}.$$
Therefore the option price will solve
$$\frac{\partial V}{\partial t} - b(x)\frac{\partial V}{\partial x} - \frac{1}{2}\sigma^2(x)\frac{\partial^2 V}{\partial x^2} - rV = 0,$$
which in the case when the coefficients are constant reduces to the Black–Scholes–Merton PDE.

6.9 A General Diffusion Approximation Method

In this section, we discuss how to approximate general diffusion processes with discrete processes. This discussion forms the basics of all tree approximations. We follow the ideas presented in our original paper [71].

For our underlying continuous-time stochastic process model, we assume that the price process S_t and the volatility driving process Y_t solve the equations

$$\begin{cases} dS_t &= rS_t dt + \sigma(Y_t)S_t dW_t \\ dY_t &= \alpha(v - Y_t)dt + \psi(Y_t)dZ_t \end{cases} \qquad (6.32)$$

This model is written under the equivalent martingale measure, and it is a generalization of all the stochastic volatility models traditionally used in finance. For example, the Heston model is obtained by chosing $\sigma(x) = \sqrt{x}$ and $\psi(x) = \beta\sqrt{x}$. SABR is the notable exception since the diffusion term contains a S_t^β term.

For simplicity, we assume W_t and Z_t are two *independent* Brownian motions. The case when they are correlated is more complex, but it may be treated by constructing an approximation to a modified pair $X_t, \overline{Y}_t$ where $\overline{Y}_t$ is suitably modified to be independent of X_t. The ideas have never been written in the general model, but the transformation may be found in the tree approximation for the Heston model in [20].

For convenience, we work with the logarithm of the price $X_t = \log S_t$, and the Eq. (6.32) become

$$\begin{cases} dX_t &= \left(r - \frac{\sigma^2(Y_t)}{2}\right)dt + \sigma(Y_t)dW_t \\ dY_t &= \alpha(v - Y_t)dt + \psi(Y_t)dZ_t \end{cases} \qquad (6.33)$$

Here r is the short-term risk-free rate of interest. The goal is to obtain discrete time discrete space versions of the processes (X_t, Y_t) which would converge in distribution to the continuous processes in (6.33). Using the fact that the price of the European option can be written as a conditional expectation of a continuous function of the price, we can use the continuous mapping theorem to ensure

6.9 A General Diffusion Approximation Method

convergence in distribution of the option price calculated using the discrete approximation to the real price of the option.

We will present such a tree approximation in Section 6.11. In this section, we present a general Markov chain convergence theorem. The theorem is based on Section 11.3 in the book [198] (see also [62]).

We assume historical stock prices $S_{t_1}, S_{t_2}, \ldots, S_{t_K}$ (and also $X_{t_1} = \log S_{t_1}, X_{t_2} = \log S_{t_2}, \ldots, X_{t_n} = \log S_{t_K}$. We will use this history of prices to estimate the volatility process in the next Section 6.10. This method will produce an approximating process Y_t^n that converges in distribution, for each time $t = t_i$, $i = 1, 2, \ldots, K$, as $n \to \infty$, to the conditional law of the volatility process Y_t in (6.33) given $S_{t_1}, S_{t_2}, \ldots, S_{t_i}$.

We run this filter for all the past times, and at the final time t_K (the present time $t = 0$) we will price our option. For simplicity of notation we will drop the subscript: we use $Y^n = Y_0^n$ to denote the discrete process distribution at time $t_K = 0$, and we use $Y = Y_0$ for the distribution of the continuous process at time $t_K = 0$. The convergence result we prove in this section will be applied to provide a quadrinomial-tree approximation to the solution of the following equation:

$$dX_t = \left(r - \frac{\sigma^2(Y)}{2}\right)dt + \sigma(Y)dW_t. \tag{6.34}$$

In modeling terms, we shall construct two tree approximations. In one approximation the "static model" uses the distribution of the random variable $Y = Y_0$ for all future steps $\{Y_s : s \leq 0\}$ in (6.33). The distribution of Y is unchanged from time 0 (the present) into the future. This static assumption is in sharp contrast to the dynamic in (6.33); however it will construct a simpler and faster tree which will be shown to converge to the same number. We shall also present a second tree approximation, the "dynamic model," which is a straightforward approximation to the SV model in (6.33). For the latter model a simple Euler simulation, rather than a tree, will be used.

Let T be the maturity date of the option we are trying to price and N the number of steps in our tree. Let us denote the time increment by $h = \Delta t = \frac{T}{N}$. We start with a discrete Markov chain $(x(ih), \mathcal{F}_{ih})$ with transition probabilities denoted by $p_x^z = p(h, x, z)$ of jumping from the point x to the point z in h units of time. For a homogeneous Markov chain (as is our case), these transition probabilities only depend on h, x, and z. For each h let $\mathbf{P}_x^h$ be the probability measure on $\mathbb{R}$ characterized by

$$\begin{cases} \text{(i)} & \mathbf{P}_x^h\left(x(0) = x\right) = 1 \\ \text{(ii)} & \mathbf{P}_x^h\left(x(t) = \frac{(i+1)h-t}{h}x(ih) + \frac{t-ih}{h}x((i+1)h)\right. \\ & \left. , \; ih \leq t < (i+1)h\right) = 1, \quad \forall i \geq 0 \\ \text{(iii)} & \mathbf{P}_x^h\left(x((i+1)h) = z | \mathcal{F}_{ih}\right) = p_x^z, \quad \forall z \in \mathbb{R} \text{ and } \forall i \geq 0 \end{cases} \tag{6.35}$$

6 Tree Methods

Remark 6.9.1 The obscure equations earlier say the following:

1) Properties (i) and (iii) say that $(x(ih), \mathcal{F}_{ih})$, $i \geq 0$ is a time-homogeneous Markov chain starting at x with transition probability p_x^z under the probability measure $\mathbf{P}_x^h$.
2) Condition (ii) assures us that the process $x(t)$ is linear between $x(ih)$ and $x((i+1)h)$. In turn, this means that the process $x(t)$ we construct is a tree.
3) We will show in Section 6.11 precisely how to construct this Markov chain $x(ih)$.

Conditional on being at x and on the Y^n variable, we construct the following quantities as functions of $h > 0$:

$$b_h(x, Y^n) = \frac{1}{h} \sum_{z \text{ successor of } x} p_x^z (z - x) = \frac{1}{h} E^Y \left[\Delta x(ih) \right]$$

$$a_h(x, Y^n) = \frac{1}{h} \sum_{z \text{ successor of } x} p_x^z (z - x)^2 = \frac{1}{h} E^Y \left[\Delta^2 x(ih) \right],$$

where the notation $\Delta x(ih)$ is used for the increment over the interval $[ih, (i+1)h]$ and E^Y denotes conditional expectation with respect to the sigma algebra $\mathcal{F}_{t_K}^Y$ generated by the Y variable. Here the successor z is determined using both the predecessor x and the Y^n random variable. We will see exactly how z is defined in Section 6.11 when we construct our specific Markov chain. Similarly, we define the following quantities corresponding to the infinitesimal generator of the Eq. (6.33):

$$b(x, Y) = r - \frac{\sigma^2(Y)}{2},$$

$$a(x, Y) = \sigma^2(Y).$$

We make the following assumptions, where $\to^D$ denotes convergence in distribution:

$$\lim_{h \searrow 0} b_h(x, Y^n) \to^D b(x, Y), \text{ when } n \to \infty \tag{6.36}$$

$$\lim_{h \searrow 0} a_h(x, Y^n) \to^D a(x, Y), \text{ when } n \to \infty \tag{6.37}$$

$$\lim_{h \searrow 0} \max_{z \text{ successor of } x} |z - x| = 0. \tag{6.38}$$

Theorem 6.9.1 Assume that the martingale problem associated with the diffusion process X_t in (6.34) has a unique solution $\mathbf{P}_x$ starting from $x = \log S_K$ and that the functions $a(x,y)$ and $b(x,y)$ are continuous and bounded. Then conditions (6.36), (6.37), and (6.38) are sufficient to guarantee that $\mathbf{P}_x^h$ as defined in (6.35) converges to $\mathbf{P}_x$ as $h \searrow 0$ and $n \to \infty$. Equivalently, $x(ih)$ converges in distribution to X_t the unique solution of the equation (6.34)

Proof. The proof of the theorem mirrors that of the theorem 11.3.4 in [198]. In fact the theorem states the same result but in ours, it is a, more specific case.

The proof consists in showing the convergence of the infinitesimal generators formed using the discretized coefficients $b_h(.,.)$ and $a_h(.,.)$ to the infinitesimal generator of the continuous version.

It should be noted also that hypothesis (6.38), though implying condition (2.6) in the aforementioned book, is a much stronger assumption. However, since it is available to us, we shall use it. □

It should be noted that lost throughout this mathematical notation, the theorem discussed earlier states a very important fact. When approximating diffusion processes, one needs to calculate the expected instantaneous increment $b_h(x, Y^n)$ and the instantaneous second moment of the increment $a_h(x, Y^n)$ and make sure that when $h \to 0$ they converge to the drift coefficient and the squared diffusion coefficient of the continuous time stochastic process.

6.10 Particle Filter Construction

To construct a tree approximation we need the distribution of the Y process at time $t = 0$. One can use a theoretical distribution for Y if such distribution exists. For example, in the Heston model the Y distribution is given by the Cox–Ingersoll–Ross (CIR) [47] process and this process has a stationary noncentral Chi squared (χ^2) distribution. If the volatility is known analytically, we just need to generate random numbers from the distribution as detailed in the next section.

For more general processes the distribution of Y cannot be calculated analytically. In this section we present a general method to calculate an approximating distribution. We assume that the coefficients v, α and the functions $\sigma(y)$ and $\psi(y)$ are known or have already been estimated.

The particle filtering method is based on an algorithm due to Del Moral, Jacod, and Protter [53] adapted to our specific case. The method approximates for all $i = 1, \cdots, K$,

$$p_i(dy) = \mathbf{P}\left[Y_{t_i} \in dy | X_{t_1}, \cdots, X_{t_i}\right],$$

which is the filtered stochastic volatility process at time i given all discrete passed observations of the stock price. If $X_{t_1}, \cdots, X_{t_i}$ are observed, then p_i's depend explicitly on these observed values.

In [53], section 5, the authors provide the particle filtering algorithm. The algorithm produces n time-varying particles $\left\{Y_i^j : i = 1, \cdots, K; j = 1, \cdots, n\right\}$ and their corresponding probabilities $\{p_i^j : i = 1, \cdots, K; j = 1, \cdots, n\}$. These form an approximate probability distribution that converges for each i to the limiting probability defined by $p_i(dy)$. The algorithm is a two-step genetic-type algorithm with a mutation step followed by a selection step. We refer to the aforementioned article (Theorem 5.1) for the proof of convergence.

We present the algorithm in detail next. The data we work with is a sequence of returns: $\{x_0 = \log S_0, x_1 = \log S_1, \ldots, x_K = \log S_K\}$, observed from the market. We note that we are abusing notation here. Recall that the goal is the distribution at time $t = 0$ (now). However, all these observations $x_0, \ldots, x_K$ are in the past. It will become too confusing to use negative indices. Rather we use this notation and the goal is to approximate the distribution of the variable Y_K (now).

We need an initial distribution for the volatility process Y_t at time $t = 0$. In our implementation we use $\delta_{\{v\}}$ (all paths start from v the long term mean of the distribution). We can use any distribution here, for example, a uniform distribution centered on v. Here $\delta_{\{x\}}$ is a notation for the Dirac point mass. The only mathematical condition we need is that the functions $\sigma(x)$ and $\psi(x)$ be twice differentiable with bounded derivatives of all orders up to 2.

We need to define a weight function that will be used to measure how far the particles are from their target. The only requirement on this function (denoted using ϕ) is to have finite L^1 norm. In order to obtain good results, we need ϕ to be concentrated near 0.

In our application we use

$$\phi(x) = \begin{cases} 1 - |x| & \text{if } -1 < x < 1 \\ 0 & \text{otherwise.} \end{cases}$$

Another function that also produces good results is $\phi(x) = e^{-2|x|}$.

The algorithm generates n particles. For $n > 0$ we define the contraction corresponding to $\phi(x)$ as

$$\phi_n(x) = \sqrt[3]{n}\, \phi(x\sqrt[3]{n}) = \begin{cases} \sqrt[3]{n}\left(1 - |x\sqrt[3]{n}|\right) & \text{if } -\frac{1}{\sqrt[3]{n}} < x < \frac{1}{\sqrt[3]{n}} \\ 0 & \text{otherwise.} \end{cases} \quad (6.39)$$

We choose $m = m_n$ an integer.

<u>Step 1</u>: We start with $X_0 = x_0$ and $Y_0 = y_0 = v$.

Mutation step: This part calculates a random variable with approximately the same distribution as (X_1, Y_1) using the well-known Euler scheme for the Eq. (6.33). More precisely we set:

$$Y(m, y_0)_{i+1} := Y_{i+1} = Y_i + \frac{1}{m}\alpha(v - Y_i) + \frac{1}{\sqrt{m}}\psi(Y_i)U_i$$

$$X(m, x_0)_{i+1} := X_{i+1} = X_i + \frac{1}{m}(r - \frac{\sigma^2(Y_i)}{2}) + \frac{1}{\sqrt{m}}\sigma(Y_i)U_i'. \quad (6.40)$$

Here U_i and U'_i are iid Normal random variables with mean 0 and variance 1 so that $\frac{1}{\sqrt{m}}U_i$ is the distribution of the increment of the Brownian motion. At the end of this first evolution step we obtain

$$X_1 = X(m, x_0)_m,$$
$$Y_1 = Y(m, y_0)_m. \tag{6.41}$$

Selection step: We repeat the **mutation step** n times and obtain n pairs: $\{(X_1^j, Y_1^j)\}_{j=\overline{1,n}}$.

For each resulting Y_1^j we assign a discrete weight given by $\phi_n(X_1^j - x_1)$. Since ϕ_n has most weight at 0, the closer the resulting endpoint X_1^j is to actual observation x_1, the larger the weight. Since we can have multiple paths ending in the same value for Y_1, we accumulate these into a probability distribution. Mathematically

$$\Phi_1^n = \begin{cases} \frac{1}{C}\sum_{j=1}^n \phi_n(X_1^j - x_1)\delta_{\{Y_1^j\}} & \text{if } C > 0 \\ \delta_{\{0\}} & \text{otherwise.} \end{cases} \tag{6.42}$$

Here the constant C is chosen so that Φ_1^n is a probability distribution ($C = \sum_{j=1}^n \phi_n(X_1^j - x_1)$). The idea is to "select" only the values of Y_1 which correspond to values of X_1 not far away from the realization x_1. We end the first selection step by simulating n iid variables $\{Y_1'^j\}_{j=\overline{1,n}}$ from the distribution Φ_1^n we just obtained. These are the starting points of the next particle mutation step.

<u>Steps 2 to K:</u> For each subsequent step $i = 2, 3, \ldots, K$, we first apply the mutation step to each of the particles selected at the end of the previous step. Specifically, in the same Eq. (6.40), we start with $X_0 = x_{i-1}$ and $Y_0 = Y'^j_{i-1}$ for each $j = 1, 2, \ldots, n$. Again we obtain n mutated pairs $\{(X_i^j, Y_i^j)\}_{j=1,2,\ldots,n}$. Then we apply the selection step to these pairs. That is, we use them in the distribution (6.42) using $\{(X_1^j, Y_1^j)\}_{j=1,2,\ldots,n}$ for the pairs and using x_i instead of x_1 in the weight function.

At the end of each step i we obtain a discrete distribution Φ_i^n, and this is our estimate for the transition probability of the process Y_t at the respective time t_i. In our construction of the quadrinomial tree, we use only the latest estimated probability distribution, i.e. Φ_K^n. We will denote this distribution using the set of particles $\{\overline{Y}_1, \overline{Y}_2, \ldots, \overline{Y}_n\}$ together with their corresponding probabilities $\{\overline{p}_1, \overline{p}_2, \ldots, \overline{p}_n\}$.

Next we present an implementation of the filter. The implementation is in R and uses two input vectors. x is a vector containing logarithm of stock values and $DATE$ is a corresponding vector of dates. The model we are exemplifying is

$$dX_t = \left(r - \frac{e^{-|Y_t|}}{2}\right) dt + e^{-|Y_t|} dW_t$$
$$dY_t = \alpha \left(m - Y_t\right) dt + \beta Y_t dZ_t$$

The dynamics can of course be changed in the four functions at the beginning of the code. The parameter values have to be known and in the example are fixed $r = 0.05$, $\alpha = 0.1$, $m = \log(0.125)$, and $\beta = 0.1$. We assume we are working with daily data and in the code that follows we are using $\Delta t = \frac{1}{252}$ the conventional number of trading days in a year. The vector *logvect* contains logarithm of daily equity values.

```
#Coefficients functions:

miu.y <- function(x, alpha=.1, m=log(0.125)) alpha*(m-x)
sigma.y <- function(x, alpha=.1) alpha*x

sigma.x <- function(x) exp(x)
miu.x <- function(x,r=0.05) r-sigma.x(x)2/2

psi.select<-function(x) #this is the weight function
{ifelse((x>-1)&&(x<1),1-abs(x),0)}
# an alternative second function:
#psi.select<-function(x)
# {exp(-abs(x))/2}

psi.select.n<-function(x,n) # this is the contraction
{n(1/3)*psi.select(x*n(1/3))}

#the mutation step function
euler.mutation <-function(x0,y0,m)
{w<-rnorm(m); z<-rnorm(m); y<-y0; x<-x0;delta<-1/(252*m);
for(i in 1:m) }
    y<- y + miu.y(y)*delta +sigma.y(y)*w[i]*sqrt(delta);
    x <- x + miu.x(y)*delta + sigma.x(y)*z[i]*sqrt(delta)
        };
return(c(x,y))
}

#the selection step function
euler.selection <- function(a,x1,n)
{b<-sapply(a[1,]-x1,psi.select.n,n=n);
cc<-matrix(c(round(a[2,],2),b/sum(b)),2,length(b),
    byrow=T)
if(all(b==0)) {d<-matrix(c(0,1),2,1)}
    else{d<-density.estimation(cc[,cc[2,]!=0])};
return(d)
}
```

```
#this function combines the two steps and creates the
 final
#distribution
vol.distrib<-function(logvect,a2,n,m)
{
for (i in 1:(length(logvect)-1))
{y<-sample(a2[1,],n,replace=T,prob=a2[2,]);
a1<-sapply(y,euler.mutation,x0=logvect[i],m=m);
a2<-euler.selection(a1,logvect[i+1],n)
};
return(a2)
}
```

Please note the code earlier contains a function called "density estimation." At the end of the selection step, we obtain a large number of particles (n) with associated probabilities. We could use this distribution further but it is much more efficient if we create an actual histogram. That is what the function does. To be specific it constructs a grid for the range of Y (e.g. $[0, 0.01)$ and $[0.01, 0.02)$) and cumulates all probabilities for all particles in the respective range. It is just a histogram of the variable Y. You need to construct this function yourself.

In the last function the parameters n and m have to be given. n is the number of paths (particles) simulated at each selection step and m is the number of the intermediate steps when mutating particles. In the last function *logvect* is a vector containing logarithms of asset prices while *a2* is a matrix with two rows. The first row contains the values for the filtered distribution at the previous step $\{\overline{Y}_1, \overline{Y}_2, \cdots, \overline{Y}_n\}$, while the second row contains the respective probabilities, $\{\overline{p}_1, \overline{p}_2, \ldots, \overline{p}_n\}$.

6.11 Quadrinomial Tree Approximation

The goal is to price an option with maturity T written on the underlying process S_t. We note that if the option is about variability (a variance swap for example), the static model constructed here will not provide a good approximation. We refer to our subsequent work [214, 215] for an approach in this case.

The purpose of this section is to construct a discrete tree which will assist in calculating an estimate of the option's price. The data available is the value of the stock price today S and a history of earlier stock prices. As described in the previous section, we use the historical values to compute a set Y^n of particles $\{\overline{Y}_1, \overline{Y}_2, \cdots, \overline{Y}_n\}$ with weights $\{\overline{p}_1, \overline{p}_2, \ldots, \overline{p}_n\}$, whose empirical law approximates the volatility process Y_0 at time 0.

Remark 6.11.1 The market is incomplete. Thus, the option price is not unique.

It is easy to see that the remark is true as is the case with all the stochastic volatility models since the number of sources of randomness (2) is bigger than the number of tradable assets (1). Remember that the volatility process is not a tradable asset and cannot, in practice, be observed. This means that the price of a specific derivative will not be completely determined by just observing the dynamics of (X, Y) in Eq. (6.33) and by the arbitrage-free assumption. However, the requirement of no arbitrage will imply that the prices of various derivatives will have to satisfy certain internal consistency relationships, in order to avoid arbitrage possibilities on the derivative market.

To take advantage of this fact and to be able to use the classical pricing idea in incomplete markets, we make an assumption.

Assumption 6.11.1 There is a liquid market for every contingent claim. □

This assumption assures us that the derivatives are tradable assets. Thus, taking the price of one particular option (called the "benchmark" option) as given will allow us to find a unique price for all the other derivatives. Indeed, we would then have two sources of randomness and two tradable assets (the stock and the benchmark), and the price of any derivative would be uniquely determined.

Let us divide the interval $[0, T]$ into N subintervals each of length $\Delta t = \frac{T}{N} = h$. At each of the points $i\Delta t = ih$ the tree is branching. The nodes on the tree represent possible values for $X_t = \log S_t$.

6.11.1 Construction of the One-Period Model

Now, assume that we are at a point x in the tree. What are the possible successors of x?

We sample a volatility value from the discrete approximating distribution Y^n at each time period ih, $i \in \{1, 2, \ldots, N\}$. Denote the value drawn at step i corresponding to time ih by Y_i. Corresponding to this volatility value, Y_i we will construct the successors in the following way.

We consider a grid of points of the form $l\sigma(Y_i)\sqrt{\Delta t}$ with l taking integer values. The parent node x will fall at one such point or between two such grid points. We let j denote the integer that corresponds to the point directly above x. Mathematically, j is the point that attains

$$\min\left\{l \in \mathbf{N} \mid l\,\sigma(Y_i)\sqrt{\Delta t} \geq x\right\}.$$

We will have two possible cases: either the point $j\,\sigma(Y_i)\sqrt{\Delta t}$ on the grid corresponding to j is closer to x or the point $(j-1)\sigma(Y_i)\sqrt{\Delta t}$ corresponding to $j-1$ is closer. We will treat the two cases separately. This is needed so that that math works.

Figure 6.2 The basic successors for a given volatility value. Case 1.

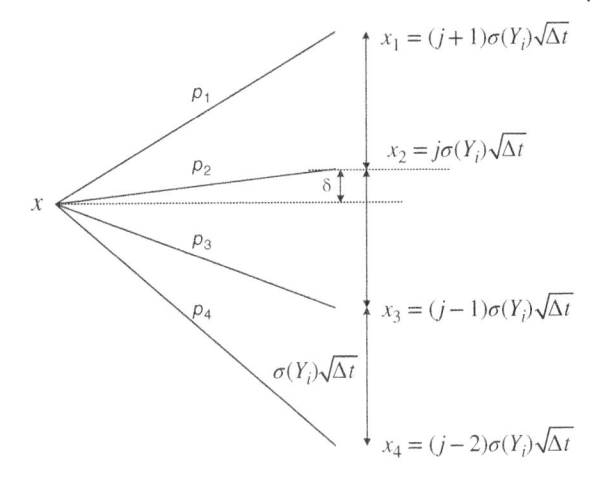

$$x_1 = (j+1)\sigma(Y_i)\sqrt{\Delta t}$$
$$x_2 = j\sigma(Y_i)\sqrt{\Delta t}$$
$$x_3 = (j-1)\sigma(Y_i)\sqrt{\Delta t}$$
$$x_4 = (j-2)\sigma(Y_i)\sqrt{\Delta t}$$

Case 1. $j\,\sigma(Y_i)\sqrt{\Delta t}$ is the point on the grid closest to x.
Figure 6.2 refers to this case.
Let us denote $\delta = x - j\,\sigma(Y_i)\sqrt{\Delta t}$.

Remark 6.11.2 In this case we have $\delta \in \left[-\frac{\sigma(Y_i)\sqrt{\Delta t}}{2}, 0\right]$ or $\frac{\delta}{\sigma(Y_i)\sqrt{\Delta t}} \in \left[-\frac{1}{2}, 0\right]$.

One of the assumptions we need to verify is (6.36), which asks the mean of the increment to converge to the drift of the X_t process in (6.34). In order to simplify this requirement, we add the drift quantity to each of the successors. This trick will simplify the conditions (6.36) to ask how the convergence of the mean increment to zero. This idea has been previously used by many authors including Leisen as well as Nelson and Ramaswamy.

Explicitly, we take the four successors to be

$$\begin{cases} x_1 = (j+1)\sigma(Y_i)\sqrt{\Delta t} + \left(r - \frac{\sigma^2(Y_i)}{2}\right)\Delta t \\ x_2 = j\sigma(Y_i)\sqrt{\Delta t} + \left(r - \frac{\sigma^2(Y_i)}{2}\right)\Delta t \\ x_3 = (j-1)\sigma(Y_i)\sqrt{\Delta t} + \left(r - \frac{\sigma^2(Y_i)}{2}\right)\Delta t \\ x_4 = (j-2)\sigma(Y_i)\sqrt{\Delta t} + \left(r - \frac{\sigma^2(Y_i)}{2}\right)\Delta t \end{cases} \quad (6.43)$$

First notice that condition (6.38) is trivially satisfied by this choice of successors. The plan is to set a system of equations verifying the variance condition (6.37) and the mean condition (6.36). We then solve the system to find the joint probabilities p_1, p_2, p_3, and p_4. Algebraically, we write $j\,\sigma(Y_i)\sqrt{\Delta t} = x - \delta$, and using this we infer that the increments over the period Δt are

$$\begin{cases} x_1 - x = \sigma(Y_i)\sqrt{\Delta t} - \delta + \left(r - \frac{\sigma^2(Y_i)}{2}\right)\Delta t \\ x_2 - x = -\delta + \left(r - \frac{\sigma^2(Y_i)}{2}\right)\Delta t \\ x_3 - x = -\sigma(Y_i)\sqrt{\Delta t} - \delta + \left(r - \frac{\sigma^2(Y_i)}{2}\right)\Delta t \\ x_4 - x = -2\sigma(Y_i)\sqrt{\Delta t} - \delta + \left(r - \frac{\sigma^2(Y_i)}{2}\right)\Delta t \end{cases} \quad (6.44)$$

Conditions (6.36) and (6.37) translate here as

$$\mathbb{E}[\Delta x | Y_i] = \left(r - \frac{\sigma^2(Y_i)}{2}\right)\Delta t$$

$$\mathbb{V}[\Delta x | Y_i] = \sigma^2(Y_i)\Delta t$$

where by Δx we denote the increment over the period Δt.

We will solve the following system of equations with respect to p_1, p_2, p_3, and p_4:

$$\begin{cases} \left(\sigma(Y_i)\sqrt{\Delta t} - \delta\right)p_1 + (-\delta)p_2 + \left(-\sigma(Y_i)\sqrt{\Delta t} - \delta\right)p_3 + \left(-2\sigma(Y_i)\sqrt{\Delta t} - \delta\right)p_4 = 0 \\ \left(\sigma(Y_i)\sqrt{\Delta t} - \delta\right)^2 p_1 + (-\delta)^2 p_2 + \left(-\sigma(Y_i)\sqrt{\Delta t} - \delta\right)^2 p_3 + \left(-2\sigma(Y_i)\sqrt{\Delta t} - \delta\right)^2 p_4 \\ \quad - \mathbb{E}[\Delta x | Y_i]^2 = \sigma^2(Y_i)\Delta t \\ p_1 + p_2 + p_3 + p_4 = 1 \end{cases}$$

$$(6.45)$$

Eliminating the terms in the first equation of the system, we get

$$\sigma(Y_i)\sqrt{\Delta t}\,(p_1 - p_3 - 2p_4) - \delta = 0$$

or

$$p_1 - p_3 - 2p_4 = \frac{\delta}{\sigma(Y_i)\sqrt{\Delta t}}. \quad (6.46)$$

Neglecting the terms of the form $\left(r - \frac{\sigma^2(Y_i)}{2}\right)\Delta t$ when using (6.46) in the second equation in (6.45), we obtain the following:

$$\sigma^2(Y_i)\Delta t = \sigma^2(Y_i)\Delta t\,(p_1 + p_3 + 4p_4) + 2\delta\sigma(Y_i)\sqrt{\Delta t}\,(p_3 - p_1 + 2p_4) + \delta^2 \\ - \left(\sigma(Y_i)\sqrt{\Delta t}\,(p_1 - p_3 - 2p_4) - \delta\right)^2.$$

After simplifications, we obtain the equation

$$(p_1 + p_3 + 4p_4) - (p_1 - p_3 - 2p_4)^2 = 1.$$

6.11 Quadrinomial Tree Approximation

So now the system of equations to be solved looks like as follows:

$$\begin{cases} p_1 + p_3 + 4p_4 = 1 + \dfrac{\delta^2}{\sigma^2(Y_i)\Delta t} \\ p_1 - p_3 - 2p_4 = \dfrac{\delta}{\sigma(Y_i)\sqrt{\Delta t}} \\ p_1 + p_2 + p_3 + p_4 = 1 \end{cases} \quad (6.47)$$

Note that this is a system with four unknowns and three equations. Thus, there exists an infinite number of solutions to the aforementioned system. Since we are interested in the solutions in the interval $[0, 1]$, we are able to reduce somewhat the range of the solutions. Let us denote by p the probability of the branch furthest away from x. In this case $p := p_4$. Also, let us denote $q := \delta / \left(\sigma(Y_i) \sqrt{\Delta t} \right)$ and, using Remark 6.11.2, we see that $q \in [-\tfrac{1}{2}, 0]$. Expressing the other probabilities in term of p and q, we obtain:

$$\begin{cases} p_1 = \tfrac{1}{2}(1 + q + q^2) - p \\ p_2 = 3p - q^2 \\ p_3 = \tfrac{1}{2}(1 - q + q^2) - 3p \end{cases} \quad (6.48)$$

Now using the condition that every probability needs to be between 0 and 1, we solve the following three inequalities:

$$\begin{cases} \tfrac{1}{2}(-1 + q + q^2) \leq p \leq \tfrac{1}{2}(1 + q + q^2) \\ \tfrac{q^2}{3} \leq p \leq \tfrac{1 + q^2}{3} \\ \tfrac{1}{6}(-1 - q + q^2) \leq p \leq \tfrac{1}{6}(1 - q + q^2) \end{cases} \quad (6.49)$$

It is not difficult to see that the solution of the inequalities (6.49) is $p \in [\tfrac{1}{12}, \tfrac{1}{6}]$. Thus we have the following result.

Lemma 6.11.1 *If we are in the conditions of* **Case 1** *with the successors given by (6.43), then the relations (6.48) together with $p_4 = p$ give an equivalent martingale Measure for every $p \in [\tfrac{1}{12}, \tfrac{1}{6}]$.*

It is clear earlier that we obtain an equivalent martingale measure for every $p \in [\tfrac{1}{12}, \tfrac{1}{6}]$, thanks to the first equation in (6.45).

Case 2. $(j - 1)\sigma(Y_i)\sqrt{\Delta t}$ is the point on the grid closest to x.
Figure 6.3 refers to this case.
Let us denote $\delta := x - (j - 1)\sigma(Y_i)\sqrt{\Delta t}$.

Remark 6.11.3 *In this case we have* $\delta \in \left[0, \dfrac{\sigma(Y_i)\sqrt{\Delta t}}{2} \right]$ *or* $\dfrac{\delta}{\sigma(Y_i)\sqrt{\Delta t}} \in \left[0, \tfrac{1}{2} \right]$.

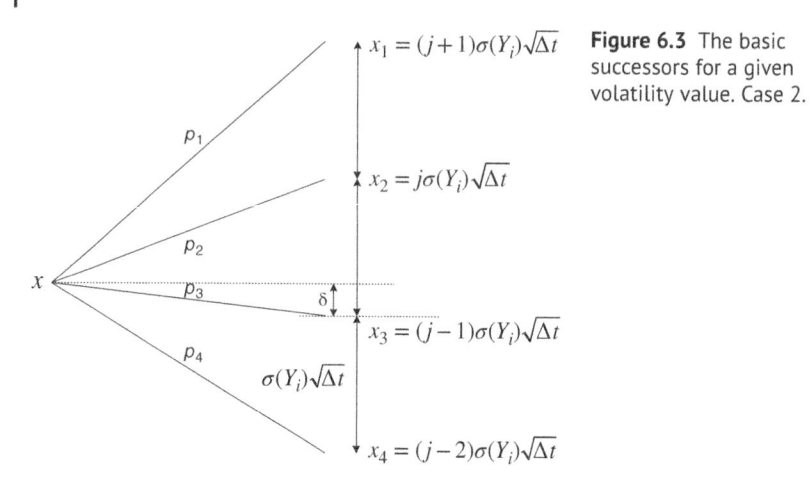

Figure 6.3 The basic successors for a given volatility value. Case 2.

$x_1 = (j+1)\sigma(Y_i)\sqrt{\Delta t}$
$x_2 = j\sigma(Y_i)\sqrt{\Delta t}$
$x_3 = (j-1)\sigma(Y_i)\sqrt{\Delta t}$
$x_4 = (j-2)\sigma(Y_i)\sqrt{\Delta t}$

The 4 successors are the same as in Case 1; the increments are going to be:

$$\begin{cases} x_1 - x = 2\sigma(Y_i)\sqrt{\Delta t} - \delta + \left(r - \frac{\sigma^2(Y_i)}{2}\right)\Delta t \\ x_2 - x = \sigma(Y_i)\sqrt{\Delta t} - \delta + \left(r - \frac{\sigma^2(Y_i)}{2}\right)\Delta t \\ x_3 - x = -\delta + \left(r - \frac{\sigma^2(Y_i)}{2}\right)\Delta t \\ x_4 - x = -\sigma(Y_i)\sqrt{\Delta t} - \delta + \left(r - \frac{\sigma^2(Y_i)}{2}\right)\Delta t \end{cases} \quad (6.50)$$

Remark 6.11.4 This second case is just the mirror image of the first case with respect to x.

Using the previous remark, we can see that the same conditions (6.36) and (6.37) will give the following system:

$$\begin{cases} 4p_1 + p_2 + p_4 = 1 + \frac{\delta^2}{\sigma^2(Y_i)\Delta t} \\ 2p_1 + p_2 - p_4 = \frac{\delta}{\sigma(Y_i)\sqrt{\Delta t}} \\ p_1 + p_2 + p_3 + p_4 = 1 \end{cases} \quad (6.51)$$

Notice that this is simply the system (6.47) with the roles of p_1 and p_4, and the roles of p_2 and p_3, reversed.

Again, denoting by p the probability of the successor furthest away, in this case p_1, and by $q := \delta / \left(\sigma(Y_i)\sqrt{\Delta t}\right)$ and using this time Remark 6.11.3 (i.e. $q \in [0, \frac{1}{2}]$) we obtain

6.11 Quadrinomial Tree Approximation

$$\begin{cases} p_2 = \frac{1}{2}\left(1+q+q^2\right) - 3p \\ p_3 = 3p - q^2 \\ p_4 = \frac{1}{2}\left(1-q+q^2\right) - p \end{cases} \qquad (6.52)$$

This is just the solution given in (6.48) with $p_1 \rightleftarrows p_4$ and $p_2 \rightleftarrows p_3$ taking into account the interval for δ. Thus, we will have the following result exactly like in Case 1.

Lemma 6.11.2 *If we are in the conditions of* **Case 2** *with the successors given by (6.43), then the relations (6.52) together with $p_1 = p$ give an equivalent martingale measure for every $p \in [\frac{1}{12}, \frac{1}{6}]$.*

Next we present a simple implementation in R of the one-step quadrinomial tree.

```
#this function determines which case is applicable case 1 or 2.
odd<-function(x,y)
{return(ifelse(trunc(x/y)/2-trunc(trunc(x/y)/2)==0,trunc(x/y)+1,
trunc(x/y)))}

#construction of the one step successors
successors.quatri<-function(x,y,r,delta,p)
{ss<-sigma.x(y)*sqrt(delta);s<-sigma.x(y);
if(trunc(x/ss)==round(x/ss))
    {    j<-trunc(x/ss)+1;
        x1<-(j+1)*ss+(r-s2/2)*delta;
        x2<-j*ss+(r-s2/2)*delta;
        x3<-(j-1)*ss+(r-s2/2)*delta;
        x4<-(j-2)*ss+(r-s2/2)*delta;
        dd<-(x-(j-1)*ss)/ss;
        p1<-p;
        p2<-(1+dd+dd2)/2-3*p;
        p3<-3*p-dd2;
        p4<-(1-dd+dd2)/2-p;;
    }
else
    {    j<-round(x/ss);
        x1<-(j+1)*ss+(r-s2/2)*delta;
        x2<-j*ss+(r-s2/2)*delta;
        x3<-(j-1)*ss+(r-s2/2)*delta;
        x4<-(j-2)*ss+(r-s2/2)*delta;
        dd<-(x-j*ss)/ss;
        p1<-(1+dd+dd2)/2-p;
        p2<-3*p-dd2;
        p3<-(1-dd+dd2)/2-3*p;
```

```
            p4<-p;
    };
    return(c(x1,x2,x3,x4,p1,p2,p3,p4))
}
```

The p value in the successors function is to be chosen in the interval $p \in [\frac{1}{12}, \frac{1}{6}]$. In practical examples we found that there isn't one particular value for p that guarantees a good valuation. We found that a lower range closer to $1/12$ is good for put options while a higher value close to $1/6$ is good for call options. The better range of values change depending on whether the option is out/in/at the money. In our practical experiments we picked a particular value fixed. We calibrated p using this chosen option value and then the p thus found served to generate the entire option chain.

We will continue these functions in the next section that talks about the multiple steps construction.

6.11.2 Construction of the Multiperiod Model: Option Valuation

Suppose now that we have to compute an option value. For illustrative purposes we will use a European type option, but the method should work with any kind of path-dependent option on S_t (e.g. American, Asian, barrier, etc).

Assume that the payoff function is $\Phi(X_T)$. The maturity date of the option is T, and the purpose is to compute the value of this option at time $t = 0$ (for simplicity) using our model (6.33). We divide the interval $[0, T]$ into N smaller ones of length $\Delta t := \frac{T}{N}$. At each of the points $i\Delta t$ with $i \in \{1, 2, \ldots, N\}$, we then construct the successors in our tree as in the previous section. This tree converges in distribution to the solution of the stochastic model (6.34). A proof of this may be found in [71].

In order to calculate an estimate for the option price, we will use the approximate discrete distribution for the initial volatility Y calculated in Section 6.10. Assume we know the initial Y distribution, i.e. we know the stochastic volatility particle filter values $\{\overline{Y}_1, \overline{Y}_2, \ldots, \overline{Y}_n\}$, each with probability $\{\overline{p}_1, \overline{p}_2, \ldots, \overline{p}_n\}$. To construct a tree with N steps, we sample N values from this distribution, and use them like the realization of volatility process Y along the future N steps of the tree. Call these sampled values $Y_1, \cdots, Y_N$; we start with the initial value x_0. We then compute the four successors of x_0 as in the previous section using for the volatility the first sampled value, Y_1. After this, for each one of the four successors we compute their respective successors using the second sampled volatility value Y_2, and so on.

The tree we construct this way allows us to compute one instance of the option price using the standard no-arbitrage pricing technique. That is, after creating the quadrinomial tree based on the sampled values, we compute the

6.11 Quadrinomial Tree Approximation

value of the payoff function Φ at the terminal nodes of the tree. Then, working backward in the path tree, we compute the value of the option at time $t = 0$ as the discounted expectation of the final node values. Because the tree is recombining by construction, the level of computation implied is manageable, typically of a polynomial order in N.

However, unlike the regular binomial tree, a single tree is not enough for the stochastic volatility models. We iterate this procedure by using repeated samples $\{Y_1, \cdots, Y_N\}$, and we take the average of all prices obtained for each tree generated using each separate sample. This volatility sampling Monte Carlo method converges, as the number of particles n increases, to the true option price for the quadrinomial tree in which the original distribution of the volatility is the true law of Y_0 given past observations of the stock price. The proof of this uses the following fact proved by Pierre del Moral, known as a *propagation of chaos* result: as n increases, for a fixed number N of particles $\{Y_1, \cdots, Y_N\}$ sampled from the distribution of particles $\{\overline{Y}_1, \overline{Y}_2, \ldots, \overline{Y}_n\}$ with probabilities $\{\overline{p}_1, \overline{p}_2, \ldots, \overline{p}_n\}$, the Y_i's are asymptotically independent and all identically distributed according to the law of Y_0 given all past stock price observations. Chapter 8, and in particular theorem 8.3.3 in [52], can be consulted for this fact. The convergence proof in the next section is also based on del Moral's propagation of chaos.

The next R code presents an implementation of the one-instance calculation using the quadrinomial tree. The code that follows contains two special functions: concatenation and deconcatenation. Due to the way the tree is constructed even though each node has four successors, many of these successors overlap. Specifically, suppose that we have k nodes at a particular level (step) in the tree. Initially, at the next step there are $4k$ generated successors. However, many of these are the same since the tree is recombining. The concatenate function creates the list of distinct successor nodes to be used for the next step in the tree construction. However, this maping needs to be used later when we go backward through the tree and calculate the option value. This is accomplished by the deconcatenate function.

```
payoff<-function(x,k) ifelse(x-k>0,x-k,0)

concatenation<-function(a)
{y<-a[1];
for (i in 2:length(a))
{
if (all(y!=a[i])) y<-c(y,a[i])};
return(sort(y))
}

deconcatenation<-function(a,d,b)
{dd<-a;
```

```
for(i in 1:length(d)) {dd[a==d[i]]<-b[i]};
return(dd)
}
```

The tree.eur.quatri.optim function constructs one instance of the multi-period quadrinomial tree and evaluates the option value.

```
tree.eur.quatri.optim<-function(x,y.val,y.prob,n,r,
Maturity,Strike,p)
{delta<-Maturity/n;
y<-sample(y.val,n,replace=T,prob=y.prob);prob<-rep(0,
length(y));
for(i in 1:n){prob[i]<-y.prob[y.val==y[i]]} ;
a<-list("");b<-list("");d<-list("");
a[[1]]<-matrix(successors.quatri(x,y[1],r,delta,p),8,1);
d[[1]]=a[[1]][1:4,];
for(i in 2:n)
    {
    a.succ<-c(a[[i-1]][1,],a[[i-1]][2,],a[[i-1]][3,],
a[[i-1]][4,]);
    d[[i]]<-concatenation(a.succ);
    a[[i]]<-sapply(d[[i]],successors.quatri,y=y[i],r=r,
delta=delta,p=p)
    }
b[[n]]<-matrix(payoff(c(exp(a[[n]][1,]),exp(a[[n]][2,]),
            exp(a[[n]][3,]), exp(a[[n]][4,])),Strike),4,
            length(a[[n]][1,]),byrow=T);
for(i in n:2)
    {
    counter<-b[[i]][1,]*a[[i]][5,]+b[[i]][2,]*a[[i]][6,]
            +b[[i]][3,]*a[[i]][7,]+b[[i]][4,]*a[[i]]
            [8,];
    b[[i-1]]<-deconcatenation(matrix(c
    (a[[i-1]][1,],a[[i-1]][2,],
            a[[i-1]][3,], a[[i-1]][4,]),4,
    length(a[[i-1]][1,]),
            byrow=T),d[[i]],counter);
    };
return(b[[1]][1,]*a[[1]][5,]+b[[1]][2,]*a[[1]][6,]+
            b[[1]][3,]*a[[1]][7,]+b[[1]][4,]*a[[1]]
            [8,])
}
```

In this function x is the current log stock value and y.val and y.prob are vectors of filtered volatility values and probabilities. In the function the a list contains the list of successors and probabilities at every step. The d list contains the concatenated node values. Finally, the list b contains the option value associated with every node in a. We use lists since we do not need to specify dimensions

Table 6.1 An illustration of the typical progression of number of elements in lists a, d, and b.

Step	a dim	d dim	b dim	binomial	Nonrecombining
1	8	4	4	4	8
2	32	7	16	6	32
3	56	9	28	8	128
4	72	13	36	10	512
5	104	17	52	12	2,048
6	136	19	68	14	8,192
7	152	21	76	16	32,768
8	168	27	84	18	131,072

For comparison we also included the number of elements (node values and probabilities) needed to be stored for a binomial tree and for a nonrecombining (exploding tree).

for elements in a list, even though it looks like there will be a lot of elements in these lists in practice that are not the case and the code is very fast. To illustrate this, Table 6.1 shows an example of the progression of the number of elements in these lists for a few steps.

These numbers were obtained for a specific example pricing IBM options. Please note that the dimension of d for the concatenated map is random and varies every time the tree is ran.

We refer to [71] for a full discussion of the convergence order of the tree and for empirical results obtained applying the tree methodology to pricing options.

We also note that the code mentioned earlier only calculates the option value using one tree.

6.12 Problems

1. For a Markov process define its infinitesimal operator:

$$T(t)f(X_s) = \int f(y)p(t, X_s, y)dy = E[f(X_{t+s})|X_s],$$

Show that this operator is a contraction with respect to:
(a) Norm 0: $\|f(x)\|_0 = \max_x f(x)$
(b) Norm 1: $\|f(x)\|_1 = |\int f(x)dx|$
(c) Norm 2: $\|f(x)\|_2 = (\int f(x)^2 dx)^{\frac{1}{2}}$
(d) Norm p: $\|f(x)\|_p = (\int f(x)^p dx)^{\frac{1}{p}}$, for all $p \geq 1$

174 | 6 Tree Methods

2. (The Martingale problem on L) Let (f,g) be a pair of functions in L. Find a process $\{X_t\}_t$ defined on E such that

$$M_t = f(X_t) - f(X_0) - \int_0^t g(X_0) ds$$

is a martingale.

3. Construct a binomial tree ($u = \frac{1}{d}$) with three quarters for an asset with present value $100. If $r = 0.1$ and $\sigma = 0.4$, using the tree compute the prices of:
   (a) A European Call option with strike 110
   (b) A European Put option with strike 110
   (c) A American Put option with strike 110
   (d) A European Call option with strike 110, assuming that the underlying asset pays out a dividend equivalent to $1/10$ of its value at the end of the second quarter
   (e) An American Call option with strike 110 for an underlying as in iv)
   (f) An *Asian option*, with strike equal to the average of the underlying asset values during the period
   (g) A *barrier option* (for different barriers)

4. In the binomial model obtain the values of u, d, and p given the volatility σ and the risk-free interest rate r for the following cases:
   (a) $p = \frac{1}{2}$.
   (b) $u = \frac{1}{d}$. **Hint:** $E\left(S_{t+\Delta t}^2\right) = S_t^2 e^{(2r+\sigma^2)\Delta t}$

5. Compute the value of an option with strike $100 expiring in four months on underlying asset with present value by $97, using the binomial model. The risk-free interest rate is 7% per year and the volatility is 20%. Assume that $u = \frac{1}{d}$.

6. In a binomial tree with n steps, let $f_j = f_{u...ud...d}$ (j times u and $n-j$ times d). For a European Call option expiring at $T = n\Delta t$ with strike K, show that

$$f_j = \max\{Su^j d^{n-j}, 0\}.$$

In particular, if $u = \frac{1}{d}$, we have that

$$f_j \geq 0 \Leftrightarrow j \geq \frac{1}{2}(n + \frac{\log(K/S)}{\log(u)}).$$

   (a) Is it true that the probability of positive payoff is the probability of $S_T \geq K$? Justify your answer.
   (b) Find a general formula for the present value of the option.

7. We know that the present value of a share is $40 and that after one month it will be $42 or $38. The risk-free interest rate is 8% per year continuously compounded.

(a) What is the value of a European Call option that expires in one month, with strike price $39?

(b) What is the value of a Put option with the same strike price?

8. Explain the difference of pricing a European option by using a binomial tree with one period and assuming no arbitrage and by using risk neutral valuation.

9. The price of a share is $100. During the following six months the price can go up or down in a 10% per month. If the risk-free interest rate is 8% per year, continuously compounded,

(a) what is the value of a European Call option expiring in one year with strike price $100?

(b) compare with the result obtained when the risk-free interest rate is monthly compounded.

10. The price of a share is $40, and it is incremented in 6% or it goes down in 5% every three months. If the risk-free interest rate is 8% per year, continuously compounded, compute:

(a) The price a European Put option expiring in six months with a strike price of $42

(b) The price of an American Put option expiring in six months with a strike price of $42

11. The price of a share is $25, and after two months it will be $23 or $27. The risk-free interest rate is 10% per year, continuously compounded. If S_T is the price of the share after two months, what is the value of an option with the same expiration date (i.e. two months) and payoff S_T^2?

12. Calculate the price for a forward contract, that is, there exists the obligation to buy at the expiration. Hint: construct a portfolio so that taking it back on time, it is risk free.

13. The price of a share is $40. If $\mu = 0.1$ and $\sigma^2 = 0.16$ per year, find a 95 % confidence interval for the price of the share after six months (i.e. an interval $I_{0.95} = (\underline{S}, \overline{S})$ so that $p(S \in I_{0.95}) = 0.95$). Hint: use that if Z is a stochastic variable with standard normal distribution, then $p(-1.96 \leq Z \leq 1.96) \simeq 0.95$).

14. Suppose that the price of a share verifies that $\mu = 16\%$ and the volatility is 30%. If the closing price of the share at a given day is $50, compute:

(a) The closing expected value of the share for the next day

(b) The standard deviation of the closing price of the share for the next day

(c) A 95 % confidence interval for the closing price of the share for the next day

15. American Down and Out Call. Given $K = 100$, $T = \frac{1}{30}$, $S = 100$, $\sigma = 0.6$, $r = 0.01$, and $B = 95$. Construct a tree for all nodes below the barrier value $= 0$.

16. Consider the parabolic equation:
$$\begin{cases} \frac{\partial u}{\partial t} + \alpha(x,t) \frac{\partial^2 u}{\partial x^2} + \beta(x,t) \frac{\partial u}{\partial x} = 0; \alpha > 0 \\ u(x,T) = \phi(x) \end{cases}$$

Use a convenient change of variable transform problem 16 into the one-dimensional heat equation.

7

Approximating PDEs

Finite Difference Methods

7.1 Introduction

In this chapter, we discuss a different approach for approximating the solutions of partial differential equations (PDEs), namely, finite difference methods and functional approximation methods. The finite difference methods approximate the differential operator by replacing the derivatives in the equation using differential quotients. The domain is partitioned in space and time, and approximations of the solution are computed at the space and time points. For the functional approximation method, we will study the numerical problems arising in the computation of the implied volatilities and the implied volatility surface.

In the previous chapter, the central idea in the construction of trees is to divide the time to maturity $[0, T]$ into n intervals of length $\Delta t = T/n$. The Finite Difference methods take this idea one step further dividing the space into intervals as well. Specifically it divides the space domain into equally spaced nodes at distance Δx apart, and the time domain into equally spaced nodes a distance Δt apart in the return interval of possible values.

We recall from Section 3.4.1 that if a European Call option on an asset does not pay dividends, then the Black–Scholes (B-S) model to evaluate is:

$$\frac{\partial V}{\partial t} + rS\frac{\partial V}{\partial S} + \frac{1}{2}\sigma^2 S^2 \frac{\partial^2 V}{\partial S^2} - rV = 0. \tag{7.1}$$

We also recall from Section 2.6 that the variable $x = \log S$ is known as the return process. Thus making a change of variable by using the return process, Eq. (7.1) becomes:

$$-\frac{\partial V}{\partial t} = v\frac{\partial V}{\partial x} + \frac{1}{2}\sigma^2 \frac{\partial^2 V}{\partial x^2} - rV, \tag{7.2}$$

where $v = r - \frac{1}{2}\sigma^2$.

Quantitative Finance, First Edition. Maria C. Mariani and Ionut Florescu.
© 2020 John Wiley & Sons, Inc. Published 2020 by John Wiley & Sons, Inc.

Before we begin discussing the finite difference techniques, we derive how Eq. (7.2) is obtained. Given the B-S PDE (7.1), let the return $x = \log S$ (i.e. $S = e^x$) and choose $V(S, t) = V(e^x, t) = U(x, t)$. Now we consider an example where

$$V(S, t) = \cos S + e^S \log t + 1. \tag{7.3}$$

Using the change of variable $S = e^x$, Eq. (7.3) becomes

$$V(e^x, t) = \cos e^x + e^{e^x} \log t + 1. \tag{7.4}$$

If we write the function $V(e^x, t)$ such that $V(e^x, t) = U(x, t)$ then Eq. (7.4) reduces to

$$U(x, t) = \cos e^x + e^{e^x} \log t + 1. \tag{7.5}$$

Using the chain rule, generally from (7.3) and (7.5) we have that:

$$\frac{\partial U}{\partial x} = \frac{\partial V}{\partial x} = \frac{\partial V}{\partial S} \cdot \frac{\partial S}{\partial x} = \frac{\partial V}{\partial S} \cdot e^x = \frac{\partial V}{\partial S} \cdot S$$

Similarly,

$$\frac{\partial^2 U}{\partial x^2} = \frac{\partial}{\partial x}\left(\frac{\partial U}{\partial x}\right)$$
$$= \frac{\partial}{\partial x}\left(\frac{\partial V}{\partial S} \cdot \frac{\partial S}{\partial x}\right)$$
$$= \frac{\partial^2 V}{\partial S^2} \cdot \frac{\partial S}{\partial x} \cdot \frac{\partial S}{\partial x} + \frac{\partial V}{\partial S} \cdot \frac{\partial^2 S}{\partial x^2}$$

From the return formula, $S = e^x$ and so $\frac{\partial S}{\partial x} = \frac{\partial^2 S}{\partial x^2} = e^x$. This implies that

$$\frac{\partial^2 U}{\partial x^2} = \frac{\partial^2 V}{\partial S^2}(e^x)^2 + \frac{\partial V}{\partial S} \cdot e^x$$

thus

$$\frac{\partial^2 U}{\partial x^2} = \frac{\partial^2 V}{\partial S^2} S^2 + \frac{\partial V}{\partial S} \cdot S.$$

Also, as discussed earlier, we know that

$$\frac{\partial U}{\partial t} = \frac{\partial V}{\partial t}.$$

Therefore the B-S PDE can be rewritten as,

$$\frac{\partial U}{\partial t} + r\frac{\partial U}{\partial x} + \frac{1}{2}\sigma^2\left(\frac{\partial^2 U}{\partial x^2} - \frac{\partial U}{\partial x}\right) - rU = 0$$

hence

$$\frac{\partial U}{\partial t} + \left(r - \frac{\sigma^2}{2}\right)\frac{\partial U}{\partial x} + \frac{1}{2}\sigma^2\frac{\partial^2 U}{\partial x^2} - rU = 0 \tag{7.6}$$

Equation (7.6) is a constant coefficient parabolic PDE which is easy to approximate.

The finite difference methods that will be used in this chapter to discretize Eq. (7.2) are the explicit finite difference method, the implicit finite difference method, and the Crank–Nicolson finite difference method. The differences between these methods come from the specific way the discretization is performed. The following notation will be used throughout this chapter: Assume that the time $[0, T]$ is discretized into n time steps of the form $i\Delta t$; furthermore assume that the possible values for the return are in some interval $[X_{min}, X_{max}]$ that we discretize into N equally spaced points at distance Δx, i.e. $x_j = X_{min} + j\Delta x$ with $\Delta x = (X_{max} - X_{min})/N$.

We will use the notation $V_{i,j}$ to denote the price of the option at time $i\Delta t$ when the return price is $X_{min} + j\Delta x$.

7.2 The Explicit Finite Difference Method

We will present the explicit finite difference method that will be used to discretize (7.2).

For the explicit method the Black–Scholes–Merton PDE (7.2) is discretized using the following formula:

Use a forward approximation for $\partial V/\partial t$, i.e.

$$\frac{\partial V}{\partial t} = \frac{V_{i+1,j} - V_{i,j}}{\Delta t}. \tag{7.7}$$

Use a central approximation for $\partial V/\partial x$, i.e.

$$\frac{\partial V}{\partial x} = \frac{V_{i+1,j+1} - V_{i+1,j-1}}{2\Delta x}. \tag{7.8}$$

Use a central approximation for $\partial^2 V/\partial x^2$, i.e.

$$\frac{\partial^2 V}{\partial x^2} = \frac{V_{i+1,j+1} + V_{i+1,j-1} - 2V_{i+1,j}}{\Delta x^2}. \tag{7.9}$$

$V_{i,j}$ denotes the value of option price at the (i, j) point and the indices i and j represent nodes on the pricing grid. Substituting these approximations into (7.2) gives

$$-\frac{V_{i+1,j} - V_{i,j}}{\Delta t} = \nu \frac{V_{i+1,j+1} - V_{i+1,j-1}}{2\Delta x} + \frac{1}{2}\sigma^2 \frac{V_{i+1,j+1} - 2V_{i+1,j} + V_{i+1,j-1}}{\Delta x^2} - rV_{i+1,j}. \tag{7.10}$$

Equation (7.10) can be rewritten as formula (3.18) in [38] as:

$$V_{i,j} = F(V_{i+1,j-1}, V_{i+1,j}, V_{i+1,j+1}). \tag{7.11}$$

In other words if one knows what happens one step in the future (at $i + 1$), then one can calculate what happened at i. Starting with values at time T, Eq. (7.11) allows us to work back in time until time $t = 0$.

7.2.1 Stability and Convergence

In this subsection, we discuss about the stability and convergence of the explicit finite difference method. The condition that guarantees the convergent of the explicit finite difference method is

$$\Delta x \geq \sigma\sqrt{3\Delta t}. \tag{7.12}$$

In fact the best choice for Δx is

$$\Delta x = \sigma\sqrt{3\Delta t}. \tag{7.13}$$

This condition is the same as the trinomial tree case discussed in Section 6.5 of this book. Using this condition enables one to determine for a specific volatility σ the minimum number of steps in the grid. See [38] for more details.

The error made in the calculation due to the fact that we have used a difference approximation rather than the PDE is called the order of the algorithm. The order of the explicit algorithm discussed earlier is $O(\Delta x^2 + \Delta t)$ as in the trinomial tree case. Obviously, we would expect that a higher order algorithm (that is, one where the degrees of the increments are higher) would perform better than a lower order one, given that both are stable.

7.3 The Implicit Finite Difference Method

Next, we discuss the implicit finite difference method used to discretize (7.2). In order to implement the implicit method for the Black–Scholes–Merton PDE, we approximate (7.2) using the central differences for the derivative in x at time $i\Delta t$ rather than at $(i+1)\Delta t$. We discretize (7.2) using the following approximations:

$$\frac{\partial V}{\partial t}(i,j) \simeq \frac{V_{i+1,j} - V_{i,j}}{\Delta t}$$

$$\frac{\partial V}{\partial x}(i,j) \simeq \frac{V_{i,j+1} - V_{i,j-1}}{2\Delta x}$$

$$\frac{\partial^2 V}{\partial x^2}(i,j) \simeq \frac{V_{i,j+1} - 2V_{i,j} + V_{i,j-1}}{\Delta x^2}.$$

and the last term in (7.2) is discretized using $V = V_{i,j}$. Where $V_{i,j}$ denote the value of option price at the (i,j) point and the indices i and j represent nodes on the pricing grid. Substituting the aforementioned approximations into (7.2) we obtain:

$$AV_{i,j+1} + BV_{i,j} + CV_{i,j-1} = V_{i+1,j} \tag{7.14}$$

where

$$A = -\frac{1}{2}\Delta t \left(\frac{\sigma^2}{\Delta x^2} + \frac{v}{\Delta x} \right),$$

$$B = 1 + \Delta t \frac{\sigma^2}{\Delta x^2} + r\Delta t, \text{ and}$$

$$C = -\frac{1}{2}\Delta t \left(\frac{\sigma^2}{\Delta x^2} - \frac{v}{\Delta x} \right)$$

We note that A, B, and C are not probabilities any more and at each (i,j), they are the same. Since we have one equation for each node from $N_j - 1$ to $-N_j + 1$, we have $2N_j - 1$ equations. However we have $2N_j + 1$ unknowns. Therefore we need two more equations and these will depend on the boundary conditions. We will add extra boundary conditions determined by the specific type of option being considered. We discuss the technique for obtaining the additional equations by providing the following example.

Consider a Put option which has the following two conditions:

- When $S \uparrow \infty$: $\frac{\partial V}{\partial S} = 0$
- When $S \downarrow 0$: $\frac{\partial V}{\partial S} = -1$

Here $\uparrow$ denotes increasing pointwise convergence and $\downarrow$ denotes decreasing pointwise convergence. Since $S \uparrow \infty$ is equivalent to $j = N_j$, we can also state that $S \downarrow 0$ is equivalent to $j = -N_j$. This translates into boundary conditions for a Put option as shown in the following. The first condition translates into the boundary conditions for a Put option as follows:

$$\frac{V_{i,N_j} - V_{i,N_j-1}}{e^{N_j \Delta x + x_0} - e^{(N_j-1)\Delta x + x_0}} = 0. \tag{7.15}$$

Equation (7.15) implies that

$$V_{i,N_j} - V_{i,N_j-1} = 0. \tag{7.16}$$

Similarly from the second condition,

$$\frac{V_{i,-N_j+1} - V_{i,-N_j}}{e^{(-N_j+1)\Delta x + x_0} - e^{-N_j \Delta x + x_0}} = -1. \tag{7.17}$$

This implies that

$$V_{i,-N_j+1} - V_{i,-N_j} = e^{-N_j \Delta x + x_0}\left(1 - e^{\Delta x}\right).$$

Either way, the boundary equations will be of the form:

$$V_{i,N_j} - V_{i,N_j-1} = \lambda_V \tag{7.18}$$

and

$$V_{i,-N_j+1} - V_{i,-N_j} = \lambda_L. \tag{7.19}$$

So putting Eqs. (7.14), (7.18), and (7.19) together at all steps, we will need to solve a linear system $MX = N$, which is of the form:

$$\begin{bmatrix} 1 & -1 & 0 & 0 & 0 & \cdots & 0 \\ A & B & C & 0 & 0 & \cdots & 0 \\ 0 & A & B & C & 0 & \cdots & 0 \\ \vdots & \ddots & \ddots & \ddots & \ddots & \ddots & \vdots \\ 0 & 0 & \ddots & \ddots & B & C & 0 \\ 0 & 0 & 0 & \ddots & A & B & C \\ 0 & 0 & 0 & \cdots & 0 & 1 & -1 \end{bmatrix} \begin{bmatrix} V_{i,N_j} \\ V_{i,N_j-1} \\ V_{i,N_j-2} \\ \vdots \\ \vdots \\ V_{i,-N_j+1} \\ V_{i,-N_j} \end{bmatrix} = \begin{bmatrix} \lambda_V \\ V_{i+1,N_j-1} \\ V_{i+1,N_j-2} \\ \vdots \\ \vdots \\ V_{i+1,-N_j+1} \\ \lambda_L \end{bmatrix} \quad (7.20)$$

The system (7.20) has $2N_j + 1$ equations with $2N_j + 1$ unknowns. Solving the system (7.20) may be challenging when considering its dimension. For instance, when $N_j = 100$ points in the finite difference grid, it implies $2N_j + 1 = 201$. Inverting a 201×201 matrix is not easy and it gets more complex if N_j is larger. Fortunately this matrix has a special form (i.e. tridiagonal). In Section (7.6), we will discuss a technique to solve a tridiagonal system. For more details of this method; see Section 3.6 in [38].

7.3.1 Stability and Convergence

In this subsection, we discuss the stability of the implicit finite difference method. An important question that arises is as follows: when is the method stable? And if it is stable, how fast does it converge? An iterative algorithm that is unstable will lead to the calculation of ever increasing numbers that will at some point diverge (approach infinity). On the other hand, a stable algorithm will converge to a finite solution. Typically the faster that a finite solution is reached, the better the algorithm.

While everything was straightforward for the explicit finite difference scheme, it is not so for the implicit finite difference scheme. This is due to the fact that the stability condition does not exist any more as discussed in Section 7.2.1. The implicit method is more stable and it does not require condition (7.12). In other words it is faster than the explicit method since it will not require the same number of points in the grid. However, the order of convergence of option values for the implicit finite difference scheme (i.e. $O(\Delta x^2 + \Delta t)$) is the same as the explicit finite difference scheme. The main advantage of the implicit finite difference method over the explicit finite difference method is the fact that the iterative algorithm is always numerically stable. In a numerically stable algorithm, errors in the input lessen in significance as the algorithm executes, having little effect on the final output. In fact the implicit methods are unconditionally stable and convergent whereas the explicit method fails if the time step Δt is too large. For more details on the stability discussion, please consult [38].

7.4 The Crank–Nicolson Finite Difference Method

The Crank–Nicolson finite difference method represents an average of the implicit method and the explicit method. This method uses a better approximation of the derivatives in (7.2) than either the explicit or implicit method. It averages the space derivatives at i and $i+1$. We discretized using the following approximations:

Use a central approximation for $\frac{\partial V}{\partial t}$,

$$\frac{\partial V}{\partial t} = \frac{V_{i+1,j} - V_{i,j}}{\Delta t} \qquad (7.21)$$

Use a central approximation for $\frac{\partial V}{\partial x}$,

$$\frac{\partial V}{\partial x} = \frac{\frac{V_{i,j+1} + V_{i+1,j+1}}{2} - \frac{V_{i,j-1} + V_{i+1,j-1}}{2}}{2\Delta x} \qquad (7.22)$$

Use a standard approximation for $\frac{\partial^2 V}{\partial x^2}$,

$$\frac{\partial^2 V}{\partial x^2} = \frac{\left(\frac{V_{i,j+1} + V_{i+1,j+1}}{2}\right) - 2\left(\frac{V_{i,j} + U_{i+1,j}}{2}\right) + \left(\frac{V_{i,j-1} + V_{i+1,j-1}}{2}\right)}{\Delta x^2} \qquad (7.23)$$

and for the last term in (7.2), we discretize using $\frac{V_{i+1,j} + V_{i,j}}{2}$, where the indices i and j represent nodes on the pricing grid. Substituting these approximations into (7.2) and collecting equal terms gives a system which is of the form:

$$AV_{i,j+1} + BV_{i,j} + CV_{i,j-1} = A'V_{i+1,j+1} + B'V_{i+1,j} + C'V_{i+1,j-1} = y_j \qquad (7.24)$$

The terms after the equal to sign, i.e. $y_j = A'V_{i+1,j+1} + B'V_{i+1,j} + C'V_{i+1,j-1}$, are known. The solution of (7.24) is identical to the finite difference implicit scheme. The implementation is very similar with the implicit finite difference method. It uses the same boundary conditions as the implicit finite difference method discussed in the previous subsection.

7.4.1 Stability and Convergence

Both implicit and explicit are unconditionally convergent, that is, the convergent condition is not based on the number of grid points. Thus there is no need for a convergence condition. On the other hand, the order of convergence of the

Crank–Nicolson difference algorithm is directly related to the truncation error introduced when approximating the partial derivatives. It is given as:

$$O\left(\Delta x^2 + \left(\frac{\Delta t}{2}\right)^2\right). \tag{7.25}$$

Hence the Crank–Nicolson method converges at the rates of $O(\Delta x^2)$ and $O(\Delta t^2)$. This is a faster rate of convergence than the explicit method and the implicit method. The explicit and implicit finite difference methods are known to converge at the rates of $O(\Delta t)$ and $O(\Delta x^2)$. The Crank–Nicolson difference scheme is always stable and convergent since it is an average of the explicit and the implicit finite difference method; thus it uses a better approximation of the derivatives in (7.2). Please see [122] for more details.

7.5 A Discussion About the Necessary Number of Nodes in the Schemes

One question that normally arises is how many nodes should we have in the grid. The time is in the interval $[0, T]$ and so we would like to determine the number n of time intervals. Furthermore, the return x can take values in increments of Δx and there are N upper values and N lower values from 0. Thus the range for x is from $-N\Delta x$ to $N\Delta x$. For all of the finite difference methods discussed so far, there is a relationship between N and n.

Now, suppose we need the order of the error of our scheme to be ε. How large should n and N be? The answer depends on the particular scheme we are looking at. The tool to use is the theoretical order of convergence. The theoretical order of convergence for the schemes is directly related to the truncation error introduced when approximating the partial derivatives. More details on the number of nodes needed to achieve a higher accuracy is found in [38].

We now present some discussions on the number of nodes for each of the three (3) difference schemes.

7.5.1 Explicit Finite Difference Method

The explicit finite difference method unlike the others has a requirement for convergence as $\Delta x \geq \sigma\sqrt{3\Delta t}$. Furthermore, its order of convergence is $O(\Delta x^2 + \Delta t)$. We recall in Section 7.2.1 that due to the constraint, the best convergence is obtained when $\Delta x = \sigma\sqrt{3\Delta t}$.

Therefore finding the number n of time intervals is very simple; we just set the order of error equal to ε, i.e.:

$$\sigma^2 3\Delta t + \Delta t = \varepsilon \quad \text{or} \quad \Delta t = \frac{\varepsilon}{1 + 3\sigma^2}.$$

Since $\Delta t = T/n$, we can easily solve for n. Furthermore, due to the way the explicit scheme works, one just has to construct a number $N = n$ of nodes.

For example, suppose we are interested in the minimum number of spatial steps required for the explicit finite difference solution for a European Put option with the exact B-S formula given by (7.2) to be numerically stable. Given the volatility $\sigma = 0.3$ and time step $\Delta t = \frac{1}{50}$ and then by using condition (7.13), we have that $\Delta x = \sigma\sqrt{3\Delta t}$. Therefore plugging the aforementioned values into this equation, the scheme will be numerically stable if Δx is 0.0735.

7.5.2 Implicit Finite Difference Method

In the case of the implicit finite difference method, we use a similar reasoning as in the case of the explicit finite difference method. We recall in Section 7.3.1 that the order of convergence for the implicit finite difference method is $O(\Delta x^2 + \Delta t)$, which is the same as the explicit finite difference scheme.

Thus finding the number n of time intervals is very simple; we choose n and Δx to have $\Delta x^2 + \Delta t = \varepsilon$. For example we could have each equal to $\varepsilon/2$. Then we need enough values for x so N can be large. How large is up to the reader since there is no condition. However, it is recommended that one chooses an N so that it has the same order of magnitude as n (see [38]). The order of magnitude is an exponential change of plus or minus 1 in the value of a quantity. The term is generally used in conjunction with power of 10. In particular, if the process follows the usual geometric Brownian motion, then one needs to cover at least the range $[-3\sigma\sqrt{T}, 3\sigma\sqrt{T}]$ which is the range containing 99.7% of the possible values for the return. Then one can fix N so that the interval $[-N\Delta x, N\Delta x]$ contains at least this range.

For example if we are interested in the order of convergence of the implicit finite difference method scheme, given the spatial step, $\Delta x = \frac{1}{n}$ and the time step, $\Delta t = \frac{1}{k}$, and then using the relation $\Delta x^2 + \Delta t = \varepsilon$, the implicit finite difference method scheme is convergent if and only if

$$\frac{1}{n^2} + \frac{1}{k} = \varepsilon. \tag{7.26}$$

The relation (7.26) was obtained by substituting $\Delta x = \frac{1}{n}$ and $\Delta t = \frac{1}{k}$ into $\Delta x^2 + \Delta t = \varepsilon$.

7.5.3 Crank-Nicolson Finite Difference Method

Finally for the Crank–Nicolson method, we specifically choose Δt and Δx to have

$$\Delta x^2 + \left(\frac{\Delta t}{2}\right)^2 = \varepsilon. \tag{7.27}$$

The determination of N is identical to the one for the implicit scheme. The Crank–Nicolson method converges much faster than the implicit and explicit finite difference methods; see [38, 122] for details.

Suppose we are interested in the order of convergence of the Crank–Nicolson finite difference method scheme. Using the same parameters as in the previous example with $\Delta x = \frac{1}{n}$ and $\Delta t = \frac{1}{k}$, then from (7.27), we have that

$$\frac{1}{n^2} + \frac{1}{4k^2} = \varepsilon. \tag{7.28}$$

where (7.28) was obtained by substituting $\Delta x = \frac{1}{n}$ and $\Delta t = \frac{1}{k}$ into (7.27). This implies that when we specifically choose Δx and Δt such that (7.28) is satisfied, then our method converges faster.

We recall that the solution of the implicit finite difference method and the Crank–Nicolson method requires the inversion of a matrix which has a very special form known as a tridiagonal form. In the following section, we study how one can solve a tridiagonal system. These systems will be used later in this chapter when discussing the obstacle problem in Section 7.8 of this chapter.

7.6 Solution of a Tridiagonal System

We will begin the discussion with how to invert a tridiagonal matrix and finally present an algorithm to solve a tridiagonal system.

7.6.1 Inverting the Tridiagonal Matrix

A matrix is said to be in a tridiagonal form if it is a band matrix (i.e. a sparse matrix in which the nonzero elements are located in a band about the main diagonal) that has nonzero elements and only on the main diagonal – the first diagonal below this and the first diagonal above the main diagonal.

Specifically, when stepping back in the grid at every step, one has to solve a system of the following type:

$$Ac_n = b_{n+1},$$

where A is a fixed square matrix, c_n is a vector at time t_n that needs to be determined, and b_{n+1} is another vector that has already been calculated at the previous step. The matrix A in both cases has a very nice form where the only nonzero elements are situated on the longest three diagonals.

The solution for such a system is:

$$c_n = A^{-1} b_{n+1},$$

and the main difficulty is finding the inverse matrix A^{-1}. If the matrix A is of no particular form, then one uses a LU decomposition or the Gauss–Jordan elimination algorithm to find the inverse. The LU decomposition of a matrix is the factorization of a given square matrix into two triangular matrices, one upper triangular matrix and one lower triangular matrix, such that the product of these two matrices gives the original matrix. Please refer to [11, 122] for the discussions of the LU decomposition and the Gauss–Jordan algorithm. The LU decomposition and the Gauss–Jordan elimination algorithm for finding the inverse of a matrix is not efficient for a large dimension matrix since it is a time-consuming process. In [38], the authors provided an algorithm that takes advantage of the special form of the matrix A and performs the inversion efficiently.

In the next subsection, we briefly discuss the algorithm for obtaining the solution of a tridiagonal system.

7.6.2 Algorithm for Solving a Tridiagonal System

Consider the system $AX = Y$ where

$$A = \begin{bmatrix} a_{11} & a_{12} & 0 & 0 & 0 & \cdots & 0 \\ a_{21} & a_{22} & a_{23} & 0 & 0 & \cdots & 0 \\ 0 & a_{32} & a_{33} & a_{34} & 0 & \cdots & 0 \\ \vdots & \ddots & \ddots & \ddots & \ddots & \ddots & \vdots \\ 0 & 0 & \ddots & \ddots & \ddots & \ddots & 0 \\ 0 & 0 & 0 & \ddots & \ddots & \ddots & a_{n-1n} \\ 0 & 0 & 0 & \cdots & 0 & a_{nn-1} & a_{nn} \end{bmatrix}, X = \begin{bmatrix} x_1 \\ x_2 \\ x_3 \\ \vdots \\ \vdots \\ x_{n-1} \\ x_n \end{bmatrix}, Y = \begin{bmatrix} y_1 \\ y_2 \\ y_3 \\ \vdots \\ \vdots \\ y_{n-1} \\ y_n \end{bmatrix}$$

The algorithm for solving the tridiagonal system $AX = Y$ is presented as follows:

1) Initialize vectors C and D

$$C_1 = \frac{1}{a_{11}} y_1; \quad D_1 = -\frac{a_{12}}{a_{11}}$$

2) For (i in 2 to n – 1) calculate the vectors C_i and D_i using the following relations:

$$C_i = \frac{y_i - a_{i,i-1} C_{i-1}}{a_{i,i-1} D_{i-1} + a_{ii}}; \quad D_i = -\frac{a_{i,i+1}}{a_{i,i-1} D_{i-1} + a_{ii}}$$

3) Next calculate the solution x_n using the formula:

$$x_n = \frac{y_n - a_{n,n-1} C_{n-1}}{a_{n,n-1} D_{n-1} + a_{nn}}$$

4) For (i in n – 1 to 1), calculate the solution using the formula:

$$x_i = C_i + D_i x_{i+1}$$

5) Finally return the solution vector as follows:

$$x = \begin{bmatrix} x_1 \\ x_2 \\ x_3 \\ \vdots \\ \vdots \\ x_{n-1} \\ x_n \end{bmatrix}$$

In the next section, we discuss how we can approximate the Heston PDE by using a finite difference scheme.

7.7 Heston PDE

We recall in Section 3.5 that the Heston PDE is given as:

$$\frac{\partial U}{\partial t} = \frac{1}{2}vS^2\frac{\partial^2 U}{\partial S^2} + \rho\sigma vS\frac{\partial^2 U}{\partial v \partial S} + \frac{1}{2}\sigma^2 v\frac{\partial^2 U}{\partial v^2} - rU + (r-2)S\frac{\partial U}{\partial S} + \kappa(\theta - v)\frac{\partial U}{\partial v} \quad (7.29)$$

The Heston PDE (7.29) can be rewritten as a system in the form:

$$\frac{\partial U}{\partial t} = LU \quad (7.30)$$

where

$$L = \frac{1}{2}vS^2\frac{\partial^2}{\partial S^2} + \frac{1}{2}\sigma^2 v\frac{\partial^2}{\partial v^2} + \rho\sigma vS\frac{\partial}{\partial v \partial S} + \frac{1}{2}\sigma^2 v\frac{\partial^2}{\partial v^2} - r$$
$$+ (r-2)S\frac{\partial}{\partial S} + \kappa(\theta - v)\frac{\partial}{\partial v}$$

Here time is $t = T - \tau$, which is the time to maturity. When $t = 0$, we get a boundary since t is the time to maturity. Therefore we have an initial value problem.

We now consider a grid in three dimensions where $t \in [0, \tau]$. Suppose we have ΔS, Δv, and Δt; then for a uniform grid we have:

$$S_i = i\Delta S, \quad i = 0, \ldots, N_S$$
$$v_j = j\Delta v, \quad j = 0, \ldots, N_v$$
$$t_n = n\Delta t, \quad n = 0, \ldots, N_T$$

where ΔS, Δv, and Δt are defined as follows $\Delta S = \frac{S_{max} - S_{min}}{N_S}$, $\Delta v = \frac{v_{max} - v_{min}}{N_v}$ and $\Delta t = \frac{\tau}{N_T}$. Then we use $U_{ij}^n = U(S_i, v_j, t_n) = U(i\Delta S, j\Delta v, n\Delta t)$.

Unlike the B-S PDE the coefficients are not constant. So the solution at every point in the grid will be specific to the point. There are two ways to approach this issue:

1) Take a fixed grid. In this case, the coefficients will be different at every point (i, j, k).
2) Take a nonuniform grid in such a way that the coefficients will be constant.

We now present an example of a nonuniform grid for the Heston PDE:

Example 7.7.1 Consider the equation:

$$S_i = K + C \sinh(\zeta_i), i = 0, 1, \ldots, N_S$$

where K is the strike price, $C = \frac{K}{5}$, $\zeta_i = \sinh^{-1}(-\frac{K}{C}) + i\Delta\zeta$, and $\Delta\zeta = \frac{1}{N_s}\left[\sinh^{-1}\left(\frac{S_{max}-K}{C}\right) - \sinh^{-1}(-\frac{K}{C})\right]$. For volatility, $v_j = d \sinh(j\Delta\eta), j = 0, 1, \ldots, N_v$, where $d = \frac{V_{max}}{500}$ and $\Delta\eta = \frac{1}{N_v}\sinh^{-1}(\frac{V_{max}}{d})$

7.7.1 Boundary Conditions

We recall that for the Heston PDE, the time to maturity is $t = T - \tau$ where τ is the initial time. At $t = 0$, the terminal condition, $\tau = T$ is an initial condition. We will exemplify with a Call option. At maturity (i.e. $\tau = T, t = 0$) the value is the intrinsic value of the option (the payoff) and is defined as:

$$U(S_i, v_j, 0) = (S_i - K)_+$$

The boundary condition for $S = S_{min}$ and $S = S_{max}$ are defined as follows: When $S = S_{min} = 0$ the Call is worthless and so $U(0, v_j, t_n) = 0 = U_{0,j}^n$. When $S = S_{max}$ the delta for the option becomes 1 and so $\frac{\partial U}{\partial S}(S_{max}, v_j, t_n) = 1$ and so $\frac{U_{N_S,j}^n - U_{N_S-1,j}^n}{S_{N_S} - S_{N_S-1}} = 1$.

Next, we define the boundary condition at v_{min} and v_{max}. When v is large the value of the option is essentially $S - K$ and so $\frac{\partial U}{\partial S}(S, v_{max}, t_n) = 1$ thus $\frac{U_{i,N_v}^n - U_{i-1,N_v}^n}{S_i - S_{i-1}} = 1$. On the other hand, when $v = v_{min} = 0$ the condition is complex. When we set $v = 0$ in the Heston PDE, all the terms involving v vanishes and (7.29) reduces to:

$$\frac{\partial U}{\partial t} = -rU + (r-2)S\frac{\partial U}{\partial S} + \kappa\theta\frac{\partial U}{\partial v}. \tag{7.31}$$

We discretize Eq. (7.31) using the following approximations:

$$\frac{\partial U}{\partial t}(S_i, 0, t_n) = \frac{U_{i,0}^{n+1} - U_{i,0}^n}{\Delta t} \tag{7.32}$$

and

$$\frac{\partial U}{\partial S}(S_i, 0, t_n) = \frac{U_{i+1,0}^n - U_{i-1,0}^n}{S_{i+1} - S_{i-1}} \tag{7.33}$$

and finally,

$$\frac{\partial U}{\partial v}(S_i, 0, t_n) = \frac{U_{i,1}^n - U_{i,0}^n}{v_1 - 0} \tag{7.34}$$

Substituting the aforementioned approximations into (7.31) gives an equation in $U_{i,0}^{n+1}$ as a function of the quantities at $t = t_n$.

In the next subsection, we will discuss the derivative approximation for a nonuniform grid.

7.7.2 Derivative Approximation for Nonuniform Grid

In the Heston PDE, it is much simpler working with a uniform grid. We set $S_{i+1} - S_i = \Delta S$, $v_{i+1} - v_i = \Delta v$ and $t_{n+1} - t_n = \Delta t$.

We use the following approximations to discretize Eq. (7.29):

$$\frac{\partial^2 U}{\partial S^2}(S_i, v_j) = \frac{\frac{U_{i+1,j}^n - U_{i,j}^n}{S_{i+1} - S_i} - \frac{U_{i,j}^n - U_{i-1,j}^n}{S_i - S_{i-1}}}{S_{i+1} - S_i}$$

$$\frac{\partial^2 U}{\partial v^2}(S_i, v_j) = \frac{\frac{U_{i,j+1}^n - U_{i,j}^n}{v_{j+1} - v_j} - \frac{U_{i,j}^n - U_{i,j-1}^n}{v_j - v_{j-1}}}{v_{j+1} - v_j}$$

$$\frac{\partial^2 U}{\partial S \partial v}(S_i, v_j) = \frac{U_{i+1,j+1}^n - U_{i-1,j+1}^n - U_{i+1,j-1}^n + U_{i-1,j-1}^n}{(S_{i+1} - S_{i-1})(v_{j+1} - v_{j-1})}$$

$$\frac{\partial U}{\partial S}(S_i, v_j) = \frac{U_{i+1,j}^n - U_{i-1,j}^n}{S_{i+1} - S_{i-1}}$$

$$\frac{\partial U}{\partial v}(S_i, v_j) = \frac{U_{i,j+1}^n - U_{i,j-1}^n}{v_{j+1} - v_{j-1}}$$

$$\frac{\partial U}{\partial t}(S_i, v_j) = \frac{U_{i,j}^{n+1} - U_{i,j}^n}{t_{n+1} - t_n}$$

We note that in these derivatives, the only term at $n+1$ is $U_{i,j}^{n+1}$. The explicit finite difference is essentially an expression of the form:

$$U^{n+1} = L(U^n) \tag{7.35}$$

where $L(U)$ is a function and U^n contains close values to each of the terms $U_{i,j}^{n+1}$. Specifically using the derivatives notation and just the finite difference for $\frac{\partial U}{\partial t}$, we obtain:

$$U_{ij}^{n+1} = U_{ij}^n + dt \left[\frac{1}{2} v_j S_i^2 \frac{\partial^2 U}{\partial S^2} + \frac{1}{2} \sigma^2 v_j \frac{\partial^2 U}{\partial v^2} \right.$$
$$\left. + \rho \sigma v_j S_i \frac{\partial^2 U}{\partial v \partial S} + (r-2) S_i \frac{\partial U}{\partial S} + \kappa(\theta - v_j) \frac{\partial U}{\partial v} - rU \right]$$

We now substitute the finite difference expressions of all derivatives at time $t = t_n$. For simplicity we consider the uniform grid case. After some few steps, we obtain:

$$U_{ij}^{n+1} = \left[1 - \Delta t \left(i^2 j \Delta v + \frac{\sigma^2 j}{\Delta v} + r \right) \right] U_{i,j}^n$$
$$+ \left[\frac{i \Delta t}{2} (ij \Delta v - r + 2) \right] U_{i-1,j}^n$$
$$+ \left[\frac{i \Delta t}{2} (ij \Delta v + r - 2) \right] U_{i+1,j}^n$$
$$+ \left[\frac{\Delta t}{2 \Delta v} (\sigma^2 j - \kappa(\theta - j \Delta v)) \right] U_{i,j-1}^n$$
$$+ \left[\frac{\Delta t}{2 \Delta v} (\sigma^2 j + \kappa(\theta - j \Delta v)) \right] U_{i,j+1}^n$$
$$+ \frac{ij \Delta t \sigma}{4} \left(U_{i+1,j+1}^n + U_{i-1,j-1}^n - U_{i-1,j+1}^n - U_{i+1,j-1}^n \right)$$

where $S_i = i \Delta S$, $v_j = j \Delta v$, $t_n = n \Delta t$, $\Delta t = t_n - t_{n-1}$, $\Delta S = S_i - S_{i-1}$ and $\Delta v = v_j - v_{j-1}$.

For a nonuniform grid the expression is more complicated but all are numerical expressions. Only the trinomial values $U_{i,j}^{N_T}$ are needed. These are the option values when the stock is S_i and the volatility is v_j. Generally for implementation it is easier using a three-dimensional array. However due to the large memory requirements of this implementation in practice, it is best to implement a two-dimensional array $U_{i,j}$ in stock and volatility and overwrite this array at every step n.

7.8 Methods for Free Boundary Problems

Using finite difference methods for pricing European options is relatively easy compared to American options since there is no possibility of early exercise. The likelihood of early exercise leads to variability or free boundaries. From the numerical analysis point of view, these free boundaries are difficult to solve since they are not fixed; therefore imposing the boundary conditions is difficult. There are different strategies to overcome this problem. One is to track down the free boundary as a part of the time stepping process. In the context of American option's valuation, this is not a very good method as the free boundary conditions are both implicit. That means that, they do not give a direct expression for the free boundary or its time derivatives. Another strategy is to find a possible transformation that will reduce the problem to a fixed boundary value problem. It turns out that there are many ways to get such transformation. For example, the common ones in practice are *linear complementarity problems* (LCP) and *variational inequalities* (see [89]). The LCP will be discussed later in this section. The LCP method will be introduced with reference to the simplest free boundary problem. This is known as the obstacle problem [77].

7.8.1 American Option Valuations

We recall that an American option is one in which the holder has the right to exercise the option on or before the expiration date. Consider a simple Call option $(S_T - K)_+$ or a Put option $(K - S_T)_+$ that may be exercised at any time τ and receive $(S_\tau - K)_+$ or $(K - S_\tau)_+$.

In order to proceed with the American option valuations, we state the following theorem.

Theorem 7.8.1 If the underlying asset does not pay dividends, the price of an American Call option with strike K and maturity T is identical to the price of a European Call with the same strike K and maturity T.

Proof. Suppose we have an American option; let $\mathscr{P}_A(S, t)$ be the price at time t of the American. Let $\mathscr{P}_E(S_t, t)$ be the price of the corresponding European. Then we must have: $\mathscr{P}_A(S, t) \geq \mathscr{P}_E(S, t) = (K - S_t)_+$. If $\mathscr{P}_A(S, t) < (K - S_t)_+$; we just buy the option and exercise immediately and receive $(K - S_t)_+ - \mathscr{P}_A(S, t) > 0$. So we must have at any t : $\mathscr{P}_A(S, t) \geq (K - S_t)_+$. Hence the proof. □

7.8.2 Free Boundary Problem

At any time t, there must exist a price for which exercising the option is optimal. We call this price S_f. Suppose we consider an American Put option with payoff = $(K - S_\tau)_+$ and we can exercise at t to get $(K - S_t)$. Clearly we observe that $S_f(t) < K$ (if not we exercise and get 0). Assume that the option should be exercised if $S_t < S_f(t)$ and held if $S_t > S_f$. In order to proceed, we state the following lemma.

Lemma 7.8.1 Delta at the free boundary is -1.

Proof. There are three possibilities for $\frac{\partial \mathscr{P}_A}{\partial S}(S_f(t))$, namely:

1) $\frac{\partial \mathscr{P}_A}{\partial S} < -1$
2) $\frac{\partial \mathscr{P}_A}{\partial S} > -1$
3) $\frac{\partial \mathscr{P}_A}{\partial S} = -1$

For the first possibility, if $\frac{\partial \mathscr{P}_A}{\partial S} < -1$ as $S > S_f$ then $\frac{\mathscr{P}_A(S) - \mathscr{P}_A(S_f)}{S - S_f} < -1$. Thus $\mathscr{P}_A(S) < \mathscr{P}_A(S_f) = (K - S_f)_+$. This leads to a contradiction since at any time $t: \mathscr{P}_A(S, t) \geq (K - S_f)_+$.

Next, if $\frac{\partial \mathscr{P}_A}{\partial S} > -1$, we get $\mathscr{P}_A(S) > \mathscr{P}_A(S_f)$ for $S > S_f$ and so we get a better value by taking S_f small enough. Therefore minimizing S_f, we eventually get $S_f = 0$ and that contradicts the existence of S_f. Therefore $\frac{\partial \mathscr{P}_A}{\partial S} = -1$. □

We can set up a no arbitrage arguments to eventually obtain the B-S equation:

$$\frac{\partial \mathscr{P}}{\partial t} + \frac{1}{2}\sigma^2 S^2 \frac{\partial^2 \mathscr{P}}{\partial S^2} + rS\frac{\partial \mathscr{P}}{\partial S} - r\mathscr{P} \leq 0.$$

However, when $S > S_f$ we do not exercise and this example must be held with equality. If $0 \leq S \leq S_f$ we exercise and so $\mathscr{P}_A(S, t) = (K - S)$; hence

$$\frac{\partial \mathscr{P}}{\partial t} + \frac{1}{2}\sigma^2 S^2 \frac{\partial^2 \mathscr{P}}{\partial S^2} + rS\frac{\partial \mathscr{P}}{\partial S} - r\mathscr{P} < 0.$$

If $S_f \leq S < \infty$ we do not exercise and $\mathscr{P}_A(S, t) > (K - S)_+$; thus

$$\frac{\partial \mathscr{P}}{\partial t} + \frac{1}{2}\sigma^2 S^2 \frac{\partial^2 \mathscr{P}}{\partial S^2} + rS\frac{\partial \mathscr{P}}{\partial S} - r\mathscr{P} = 0.$$

For free boundary conditions, $\frac{\partial \mathscr{P}_A}{\partial S} = -1$ and $\mathscr{P}(S_f(t), t) = (K - S_f(t))_+$. This generally works for any American type options.

For example consider an American option with discrete constant dividend $V(S, t)$

$$\frac{\partial V}{\partial t} + \frac{1}{2}\sigma^2 S^2 \frac{\partial^2 V}{\partial S^2} + (r - D_0)S\frac{\partial V}{\partial S} - rV \leq 0. \tag{7.36}$$

So when exercise is optimal, we have $V(S, t) = A(S)$ and $\frac{\partial V}{\partial t} + \frac{1}{2}\sigma^2 S^2 \frac{\partial^2 V}{\partial S^2} + (r - D_0)S\frac{\partial V}{\partial S} - rV < 0$. However, when the exercise is not optimal, we have $V(S, t) > A(S)$ and $\frac{\partial V}{\partial t} + \frac{1}{2}\sigma^2 S^2 \frac{\partial^2 V}{\partial S^2} + (r - D_0)S\frac{\partial V}{\partial S} - rV = 0$. At the price $S_f(t)$, both $V(S, t)$ and $\frac{\partial V}{\partial S}$ must be continuous.

We now formally describe a strategy that can be used to find a possible transformation that will reduce a free boundary problem to a fixed boundary value problem.

7.8.3 Linear Complementarity Problem (LCP)

For any time t, consider the following equation:

$$\left(\frac{\partial \mathscr{P}}{\partial t} + \frac{1}{2}\sigma^2 S^2 \frac{\partial^2 \mathscr{P}}{\partial S^2} + rS\frac{\partial \mathscr{P}}{\partial S} - r\mathscr{P}\right)\left(\mathscr{P}_A(S, t) - (K + S)_+\right) = 0 \tag{7.37}$$

with $\frac{\partial \mathscr{P}}{\partial t} + \frac{1}{2}\sigma^2 S^2 \frac{\partial^2 \mathscr{P}}{\partial S^2} + rS\frac{\partial \mathscr{P}}{\partial S} - r\mathscr{P} \geq 0$ and $\mathscr{P}_A(S, t) \geq (K - S)_+$ for all (S, t).

The following free boundary conditions are defined for the problem.

At S_f we have $\mathscr{P}(S_f, t) = (K - S_f)_+$ and $\frac{\partial \mathscr{P}}{\partial S}(S_f, t) = -1$.

For the terminal condition we have: $\mathscr{P}(S, T) = (K - S)_+$ and finally for the boundary condition we have: $\mathscr{P}(S_{\min}, t) = (K - S_{\min})_+$ and $\mathscr{P}(S_{\max,t}) = 0$.

Next we discuss the reduction problem. This reduction problem is the same as the B-S equation. We begin by first setting $S = Ke^x$; $t = T - \frac{\tau}{1/2\sigma^2}$; $\mathscr{P}(S, t) = Kv(x, \tau)$ or $x = \log(\frac{S}{t})$; $\tau = \frac{1}{2}\sigma^2(T - t)$; $v(x, \tau) = \frac{1}{K}\mathscr{P}(S, t)$. For changes to initial time we have $v(x, 0) = \frac{1}{K}\mathscr{P}(S, T - \frac{0}{1/2\sigma^2}) = \frac{1}{K}(K - Ke^x)_+$; $v(x, 0) = (1 - e^x)_+$ and $K = \frac{r}{1/2\sigma^2}$. Then we take $U(x, \tau) = e^{+\frac{1}{2}(K-1)x + \frac{1}{4}(K+1)^2\tau}v(x, \tau)$ and then the equation, for example, is:

$$\frac{\partial U}{\partial \tau} = \frac{\partial^2 U}{\partial x^2} \quad x \in \mathbb{R}, \tau > 0 \tag{7.38}$$

and

$$U(x, 0) = e^{\frac{1}{2}(K-1)x + \frac{1}{4}(K+1)^2\tau}(1 - e^x)_+$$
$$= e^{\frac{1}{4}(K+1)^2\tau}\left(e^{\frac{1}{2}(K-1)x} - e^{\frac{1}{2}(K+1)x}\right)_+$$

Therefore using the following substitutions, $S = Ke^x$, $t = T - \frac{\tau}{1/2\sigma^2}$, $K = \frac{r}{1/2\sigma^2}$ we get

$$U(x, \tau) = e^{\frac{1}{2}(K-1)x + \frac{1}{4}(K+1)^2\tau}\frac{\mathscr{P}(S, t)}{K}$$

and

$$e^{\frac{1}{4}(K+1)^2\tau}\left(e^{\frac{1}{2}(K-1)x} - e^{\frac{1}{2}(K+1)x}\right)_+ = g(x, \tau)$$

with

$$U(x, 0) = g(x, 0)$$

Thus the LCP reduces to:

$$\left(\frac{\partial U}{\partial \tau} - \frac{\partial^2 U}{\partial x^2}\right)(U(x, \tau) - g(x, \tau)) = 0 \tag{7.39}$$

$$\frac{\partial U}{\partial t} - \frac{\partial^2 U}{\partial x^2} \geq 0 \tag{7.40}$$

$$U(x, \tau) - g(x, \tau) \geq 0$$

When the free boundary condition at $U = g$, we exercise the option. When $U > g$ we do not exercise the option. The boundary conditions are:

$$U(x, 0) = g(x, 0)(\text{terminal condition})$$
$$U(x_{\min}, \tau) = g(x_{\min}, \tau)$$
$$U(x_{\max}, \tau) = 0$$

7.8 Methods for Free Boundary Problems

We now solve the LCP (7.39) using the finite difference method. We construct a grid similar to the regular finite difference method. Given the parameters x_{min}, x_{max}, Δx, Δt, and $2N_{j+1}$ points for x and $2N_{j+1}$ points for t, we discretize (7.39) using the following approximations:

$$\frac{\partial U}{\partial \tau}\left(x, \tau + \frac{\Delta \tau}{2}\right) = \frac{U_{i+1,j} - U_{i,j}}{\Delta t}$$

$$\frac{\partial^2 U}{\partial x^2}\left(x, \tau + \frac{\Delta \tau}{2}\right) = \frac{1}{2}\left(\frac{U_{i+1,j+1} - 2U_{i+1,j} + U_{i+1,j-1}}{\Delta x^2}\right)$$

Substituting the above equation into (7.40), we obtain:

$$U_{i+1,j} - \frac{1}{2}\frac{\Delta \tau}{\Delta x^2}\left(U_{i+1,j+1} - 2U_{i+1,j} + U_{i+1,j-1}\right)$$

$$\geq U_{i,j} + \frac{1}{2}\alpha\left(U_{i,j+1} - 2U_{i,j} + U_{i,j-1}\right)$$

where $\alpha = \frac{\Delta \tau}{(\Delta x)^2}$. We denote $g_{i,j} = g(i\Delta x, j\Delta \tau)$ then we have

$$U_{i,j} \geq g_{i,j}$$
$$U_{i,-N\Delta x} = g_{i,-N\Delta x}$$
$$U_{i,N\Delta x} = g_{i,N\Delta x}$$
$$U_{0,j} = g_{0,j}$$

Let $Z_{i,j} = (1-\alpha)U_{i,j} + \frac{1}{2}\alpha(U_{i,j+1} + U_{i,j-1})$ be known then $(1+\alpha)U_{i+1,j} - \frac{1}{2}\alpha U_{i+1,j+1} - \frac{1}{2}\alpha U_{i+1,j-1} \geq Z_{i,j}$. Thus the complementarity condition is:

$$\left((1+\alpha)U_{i+1,j} - \frac{1}{2}\alpha U_{i+1,j-1} - \frac{1}{2}\alpha U_{i+1,j-1} - Z_{i,j}\right)\left(U_{i+1,j} - g_{i+1,j}\right) = 0$$

(7.41)

If we write (7.41) in matrix form we will have

$$U_{i+1} = \begin{bmatrix} U_{i+1,N_j} \\ \vdots \\ U_{i+1,-N_j} \end{bmatrix},$$

$$b_i = \begin{bmatrix} \frac{1}{2}\alpha g_{i+1,-N} \\ Z_{i,-N+1} \\ \vdots \\ Z_{i,N-1} \\ \frac{1}{2}\alpha g_N \end{bmatrix}$$

and

$$A = \begin{bmatrix} 1 & -1 & 0 & 0 & \cdots & 0 \\ -\frac{1}{2}\alpha & 1+\alpha & -\frac{1}{2}\alpha & 0 & \cdots & 0 \\ 0 & 0 & \ddots & \ddots & \ddots & \vdots \\ \vdots & \vdots & & \ddots & \ddots & \vdots \\ \vdots & \vdots & & \ddots & 1+\alpha & 0 \\ 0 & 0 & \cdots & 0 & 1 & -1 \end{bmatrix},$$

is a tridiagonal matrix. Thus we obtain the tridiagonal system $AU_{i+1} = b_i$. This system is complicated to solve. Specifically

$$AU_{i+1} \geq b_i, U_{i+1} \geq g_{i+1} \tag{7.42}$$

and so

$$(U_{i+1} - g_{i+1})(AU_{i+1} - b_i) = 0 \tag{7.43}$$

We now discuss the simplest example of a free boundary problem. This is known as the obstacle problem.

7.8.4 The Obstacle Problem

The obstacle problem is a classic example in the mathematical study of free boundary problems. The problem is to find the equilibrium position of an elastic string whose boundary is held fixed. We consider a string stretched over an obstacle. If $u(x)$ is the displacement of the string and $f(x)$ is the height of the obstacle, then the problem may be written as

$$u''(u-f) = 0, -u'' \geq 0, \quad (u-f) \geq 0, \tag{7.44}$$

subject to the conditions that

$$u(-1) = u(1) = 0 \text{ and } u, u' \text{ are continuous.} \tag{7.45}$$

We approximate the second derivative $u''(x)$ in (7.44) by the two point centered finite differences, i.e.:

$$u'' = \frac{u_{n+1} - 2u_n + u_{n-1}}{(\Delta x)^2}, \tag{7.46}$$

where $u_n = u(n\Delta x)$ and thus obtain a finite difference approximation to the problem. Using the notation $f_n = f(n\Delta x)$ and with v_n as finite difference approximation of u_n, (7.44) becomes

$$(v_{n+1} - 2v_n + v_{n+1})(v_n - f_n) = 0, \tag{7.47}$$

for $-N < n < N$ and $v_{-N} = v_N = 0$. Equation (7.47) can be compactly written as the matrix equation in the form:

$$\mathbf{Av} \cdot (\mathbf{v} - \mathbf{f}) = \mathbf{0}, \quad \mathbf{Av} \geq 0, \quad (\mathbf{v} - \mathbf{f}) \geq 0$$

where the notation $\mathbf{v} - \mathbf{f} \geq 0$ implies each component of $\mathbf{v}$ is greater than or equal to the corresponding component of $\mathbf{f}$. Rewriting (7.47) in matrix form, we obtain:

$$\mathbf{A} = \begin{pmatrix} 2 & -1 & 0 & \cdots & 0 \\ -1 & 2 & -1 & \cdots & 0 \\ 0 & -1 & 2 & \cdots & 0 \\ \vdots & \vdots & \ddots & \ddots & -1 \\ 0 & \cdots & 0 & -1 & 2 \end{pmatrix}, \quad \mathbf{v} = \begin{pmatrix} v^i_{N-1} \\ \vdots \\ v^i_{-N+1} \end{pmatrix}, \quad \mathbf{f} = \begin{pmatrix} f^i_{N-1} \\ \vdots \\ f^i_{-N+1} \end{pmatrix},$$

where $\mathbf{A}$ is a tridiagonal matrix and $\mathbf{v}$ and $\mathbf{f}$ are vectors.

At each point in time we need to solve the matrix equation in order to calculate the $v_{i,j}$ values. One approach is to solve the matrix equation via an iterative method.

In subsequent subsections, we will discuss four iterative methods namely, Jacobi, Gauss–Seidel, successive over relaxation (SOR) and Projected SOR. The SOR method can be easily adapted to value the American options.

Given the linear system of equations,

$$U_{i+1,j} = \alpha b_{i,j} + \beta U_{i+1,j-1} + \gamma U_{i+1,j+1}. \tag{7.48}$$

we solve the system (7.48) using the four iterative schemes mentioned earlier. We begin our discussion with the Jacobi method.

7.8.4.1 Jacobi Method

The Jacobi method is a method of solving a matrix equation on a matrix that has no zeros along its main diagonal. Each diagonal element is solved for, and an approximate value plugged in. By repeated iterations, a sequence of approximations are formed that often converges to the actual solution.

In practice, for the Jacobi method, we pick some initial guess $U_{i+1,j}$ (for example, values from the previous step) and substitute into the right-hand side. We obtain new estimates using the relation:

$$U^k_{i+1,j} = \alpha b_{i,j} + \beta U^{k-1}_{i+1,j-1} + \gamma U^{k-1}_{i+1,j+1}$$

so $U^0_{i+1} \to U^1_{i+1} \to \cdots \to U^k_{i+1}$. We repeat this until

$$\|U^k_{i+1} - U^{k-1}_{i+1}\| = \sum_{j=-N}^{N} (U^k_{i+1,j} - U^{k-1}_{i+1,j})^2$$

is smaller than a tolerance ϵ.

We now discuss the Gauss–Seidel method iterative scheme.

7.8.4.2 Gauss–Seidel Method

The Gauss–Seidel method is a technique for solving the n equations of the linear system of equations $Ax = b$ one at a time in sequence and uses previously computed results as soon as they are available.

In the Gauss–Seidel method, updates cannot be done simultaneously as in the Jacobi method. New estimates are obtained using the relations:

$$U_{i+1,j}^k = \alpha b_{i,j} + \beta U_{i+1,j-1}^k + \gamma U_{i+1,j+1}^{k-1}$$

since $\beta U_{i+1,j-1}^k$ is already calculated. The Gauss–Seidel method converges faster than the Jacobi method since in the Gauss–Seidel method, the updated variables are used in the computations as soon as they are updated.

The method of SOR which is a variant of the Gauss–Seidel method is discussed next.

7.8.4.3 The Successive Over Relaxation Method

The SOR technique observes that

$$U_{i+1,j}^k = U_{i+1,j}^{k-1} + (U_{i+1,j}^k + U_{i+1,j}^{k-1})$$

Therefore, we let

$$U_{i+1,j}^k = U_{i+1,j}^{k-1} + \omega(y_{i+1,j}^k - U_{i+1,j}^{k-1})$$

where

$$y_{i+1,j}^k = \alpha b_{i,j} + \beta U_{i+1,j+1}^k + \gamma U_{i+1,j-1}^{k-1}$$

where ω is the over-relaxation parameter and $\omega \in (0, 2)$. When $\omega \in (0, 1)$ the term under relaxation is used. When $\omega \in (1, 2)$ the term over relaxation is used. We observe that if $\omega = 1$ then the SOR technique reduces to the Gauss–Seidel method. In the SOR method, the idea is to choose a value for ω that will accelerate the rate of convergence of the iterates to the solution. In practice, the best choice of ω is when $\omega \in (1, 2)$ since that choice significantly accelerate convergence; however, it is difficult to obtain. The best approach is to change ω at each time step until we obtain convergence. In practice the iteration is stopped when a significant accuracy is obtained, i.e until $||U_{i+1}^k - U_{i+1}^{k-1}|| = \sum_{j=-N}^{N} (U_{i+1,j}^k - U_{i+1,j}^{k-1})^2$ is smaller than a tolerance ϵ.

The SOR method is very useful for pricing American options.

7.8.4.4 The Projected SOR Method

For the American option, we need a modification. We recall that

$$AU_{i+1} \geq b_i, U_{i+1} \geq g_{i+1}$$
$$(U_{i+1} - g_{i+1})(AU_{i+1} - b_i) = 0$$

where each example in $AU_{i+1} = b_i$ is of the form:

$$U_{i+1,j} = \beta b_{i,j} + \gamma U_{i+1,j-1} + \delta U_{i+1,j+1}$$

The SOR will solve the equation $AU_{i+1} = b_i$. However we need $U_{i+1} \geq g_{i+1}$. Thus we enforce this by using the following relations:

$$U_{i+1,j}^k = \max\left(U_{i+1,j}^{k-1} + \omega\left(y_{i+1,j}^k - U_{i+1,j}^{k-1}\right), g_{i+1,j}\right)$$

where

$$y_{i+1,j}^k = \beta b_{i,j} + \gamma U_{i+1,j-1}^k + \delta U_{i+1,j+1}^{k-1}$$

We iterate until difference is negligible. After solving it the efficient frontier is found by going backward finding S_f for which we have $U_{i+1,j} = g_{i+1,j}$ at every step $i + 1$.

Now we present in more details a method to find the value of an American option.

7.9 Methods for Pricing American Options

In this section we will discuss a method to find the value of an American option. From (7.38), the model for an American Put option can be written as:

$$\frac{\partial u}{\partial \tau} = \frac{\partial^2 u}{\partial x^2},$$

$$u(x, 0) = \max\left(e^{\frac{1}{2}(k-1)x} - e^{\frac{1}{2}(k+1)x}, 0\right),$$

$$u(x, \tau) = e^{\frac{1}{2}(k+1)^2 \tau} \max\left(e^{\frac{1}{2}(k-1)x} - e^{\frac{1}{2}(k+1)x}, 0\right),$$

$$\lim_{x \to \infty} u(x, \tau) = 0.$$

Suppose we define a Put option to be the function $g(x, \tau)$, i.e.:

$$g(x, \tau) = e^{\frac{1}{2}(k+1)^2 \tau} \max\left(e^{\frac{1}{2}(k-1)x} - e^{\frac{1}{2}(k+1)x}, 0\right), \tag{7.49}$$

and for a Call option we define a function

$$g(x, \tau) = e^{\frac{1}{2}(k+1)^2 \tau} \max\left(e^{\frac{1}{2}(k+1)x} - e^{\frac{1}{2}(k-1)x}, 0\right), \tag{7.50}$$

and then we observe that Eqs. (7.49) and (7.50) look similar but are very different. The difference is as a result of the functions that we are maximizing. For the Put option, we are maximizing the function $(e^{\frac{1}{2}(k-1)x} - e^{\frac{1}{2}(k+1)x}, 0)$, and for the Call option we are maximizing the function $(e^{\frac{1}{2}(k-1)x} - e^{\frac{1}{2}(k-1)x}, 0)$. Using Eq. (7.44), we can rewrite this option valuation problem in a more compact form as:

$$\left(\frac{\partial u}{\partial \tau} - \frac{\partial^2 u}{\partial x^2}\right)(u(x, \tau) - g(x, \tau)) = 0 \tag{7.51}$$

where

$$\left(\frac{\partial u}{\partial \tau} - \frac{\partial^2 u}{\partial x^2}\right) \geq 0 \quad \text{and} \quad (u(x,\tau) - g(x,\tau)) \geq 0,$$

for any $x \in \mathbb{R}, 0 < \tau < T$ and $u(x,\tau)$ and $\frac{\partial u}{\partial x}$ are continuous. The initial condition and boundary condition become:

$$u(x,0) = g(x,0),$$

and

$$\lim_{x \to \pm\infty} u(x,\tau) = \lim_{x \to \pm\infty} g(x,\tau).$$

The advantage of the aforementioned formulation is the absence of explicit free boundary term. If we can solve this problem, then we find the free boundary $X(\tau)$ by the condition that defines it. The condition for Put option is

$$u(X(\tau),\tau) = g(X(\tau),\tau),$$

with $u(x,\tau) > g(x,\tau)$ for $x > X(\tau)$, and for Call option the condition is

$$u(X(\tau),\tau) = g(X(\tau),\tau),$$

with $u(x,\tau) > g(x,\tau)$ for $x < X(\tau)$. Note that this formulation is valid even when there are several free boundaries or no free boundary at all.

Next we formulate a finite difference analysis of Eq. (7.51). We approximate $\frac{\partial u}{\partial \tau} - \frac{\partial^2 u}{\partial x^2}$ by the finite difference formula on a regular mesh with step size $\Delta \tau$ and Δx and by truncation so that $x \in [-x^-, x^+]$, with

$$-x^- = -N^- \Delta x \leq x = n\Delta x \leq N^+ \Delta x = x^+,$$

where N^- and N^+ are appropriate large numbers. We approximate $\frac{\partial u}{\partial \tau}$ by the central difference formula, i.e.

$$\frac{\partial u}{\partial \tau} = \frac{u_n^{m+1} - u_n^m}{\Delta \tau}, \tag{7.52}$$

and $\frac{\partial^2 u}{\partial x^2}$ by the two point central difference formula,

$$\frac{\partial^2 u}{\partial x^2} = \theta \frac{u_{n+1}^{m+1} - 2u_n^{m+1} + u_{n-1}^{m+1}}{(\Delta x)^2} + (1-\theta) \frac{u_{n+1}^m - 2u_n^m + u_{n-1}^m}{(\Delta x)^2}, \tag{7.53}$$

where $\theta \in [0,1]$ and $u_n^m = u(n\Delta x, m\Delta \tau)$. We write

$$g_n^m = g(n\Delta x, m\Delta \tau)$$

for the discretized payoff function.

Let v_n^m denote the finite difference approximate solution of u_n^m. Then using the fact that $u(x,\tau) \geq g(x,\tau)$, we obtain $v_n^m \geq g_n^m$, for $m \geq 1$. Also, the initial condition and boundary condition will give:

$$v_{-N^-}^m = g_{-N^-}^m, \quad v_{N^+}^m = g_{N^+}^m, \quad v_n^0 = g_n^0.$$

Substituting (7.52) and (7.53) into

$$\left(\frac{\partial u}{\partial \tau} - \frac{\partial^2 u}{\partial x^2}\right) \geq 0$$

we obtain:

$$v_n^{m+1} - \alpha\theta\left(v_{n+1}^{m+1} - 2v_n^{m+1} + v_{n-1}^{m+1}\right) \geq v_n^m - \alpha(1-\theta)\left(v_{n+1}^m - 2v_n^m + v_{n-1}^m\right),$$

where $\alpha = \frac{\Delta\tau}{(\Delta x)^2}$. Next we define

$$b_n^m = v_n^m = (v_{n+1}^m - 2v_n^m + v_{n-1}^m).$$

Thus

$$v_n^{m+1} - \alpha\theta\left(v_{n+1}^{m+1} - 2v_n^{m+1} + v_{n-1}^{m+1}\right) \geq b_n^m.$$

At the time step $(m+1)\Delta\tau$ we can find b_n^m explicitly. Finally (7.51) is approximated by the formula:

$$\left(v_n^{m+1} - \alpha\theta\left(v_{n+1}^{m+1} - 2v_n^{m+1} + v_{n-1}^{m+1}\right) - b_n^m\right)\left(v_n^{m+1} - g_n^{m+1}\right) = 0.$$

7.10 Problems

1. Suppose we want to find the explicit finite difference solution for a European Put with the exact B-S formula (7.2), with $\sigma = 0.2$, $\Delta x = \frac{1}{20}$, and $\Delta t = \frac{1}{50}$. Now suppose that we increase our spatial resolution by refining $\Delta x = \frac{1}{30}$. What must the time step be in order to maintain the same stability condition?

2. Using the implicit finite difference method, how large should Δx be chosen given that $\varepsilon = 0.001$ and $\Delta t = \frac{1}{10000}$?

3. Using the Crank–Nicolson finite difference method, how large should Δx be chosen given that $\varepsilon = 0.0001$ and $\Delta t = \frac{1}{1000}$?

4. Use the implicit finite difference method to solve the heat conduction problem on the unit square:

$$\frac{\partial^2 u}{\partial x^2} = \frac{\partial u}{\partial t}$$
$$u(x, 0) = xe^x$$
$$u(0, t) = u(1, t) = 0$$

5. Find the finite difference solution of the problem

$$\frac{\partial^2 u}{\partial x^2} = \frac{\partial u}{\partial t} \quad 0 < x < 1, 0 < t < \infty$$
$$u(x, 0) = \sin(\pi x) \quad 0 \leq x \leq 1$$
$$u_x(0, t) = 0$$
$$u(1, t) = 0$$

by the explicit method. Plot the solution using $h = 0.1$ and $k = 0.005$ for the first three time levels ($j = 0, 1, 2$).

6. Prove that the order of convergence of the Crank–Nicolson finite difference method is

$$O\left(\Delta x^2 + \left(\frac{\Delta t}{2}\right)^2\right).$$

7. Consider a linear system $\mathbf{Av} = \mathbf{f}$, where

$$\mathbf{A} = \begin{pmatrix} 2 & -1 & 0 \\ -1 & 2 & -1 \\ 0 & -1 & 2 \end{pmatrix}, \quad \mathbf{f} = \begin{pmatrix} 1 \\ 4 \\ -4 \end{pmatrix}$$

(a) Compute the first two iterations using the Gauss–Seidel method starting with the initial guess $x^{(0)} = (0, 0, 0)^T$.
(b) Verify that the SOR method with relaxation parameter $w = 1.5$ can be used to solve this system.
(c) Compute the first two iterations using the SOR method starting with the initial guess $x^{(0)} = (0, 0, 0)^T$.
(d) Compare the results obtained by using the Gauss–Seidel and SOR method.

8

Approximating Stochastic Processes

Monte Carlo Simulations and Generating and Generating Random Variables

8.1 Introduction

This chapter is divided into two parts. The first part discusses the Monte Carlo simulation methods for pricing options, and the second part deals with methods used for simulating and generating random variables. The Monte Carlo simulation method is another technique that can also be used to price options. It has been extensively studied by many authors which includes [32, 38, 81, 101, 140] and several others.

It is very important to know how to simulate a random experiment to find out what expectations one may have about the results of the phenomenon. We shall see later that the central limit theorem (CLT) and the Monte Carlo method allow us to draw conclusions about the expectations even if the distributions involved are very complex. We assume that given a Uniform (0,1) random number generator, any software can produce a uniform random variable. The typical name for such a uniform random number is RAND. For more current references and a very efficient way to generate exponential and normal random variables without going to uniform, we refer the reader to [174].

We begin the first part of this chapter with a discussion of the Plain vanilla Monte Carlo method.

8.2 Plain Vanilla Monte Carlo Method

Pricing options using the Monte Carlo simulation method is tractable since it departs from the regular geometric Brownian motion assumption for the stock price. The main idea is to simulate many paths of the underlying asset price. For every simulated path CLT for example, path i, one can observe what the simulated price of the option would be. For instance, at the path i, depending on the evolution of the underlying asset, one can calculate the value of the option at the time of maturity, T. We call the value of the option corresponding to path

Quantitative Finance, First Edition. Maria C. Mariani and Ionut Florescu.
© 2020 John Wiley & Sons, Inc. Published 2020 by John Wiley & Sons, Inc.

i as $C_{T,i}$. Then we proceed to calculate the value of the option at the present time $t = 0$ by discounting back as: $C_i = \exp(-rT)C_{T,i}$.

The Monte Carlo option price will be just the average of all the values calculated for each path:

$$C = \frac{1}{M} \sum_{i=1}^{M} C_i,$$

where M is the number of paths one chooses to simulate.

In order to proceed with the discussion of the Monte Carlo simulation method, we briefly recall the CLT which is discussed in Section 1.7. Assume that we are given $X_1, X_2, \ldots, X_M$ values from some distribution with mean μ and standard deviation σ. Then the CLT states that the distribution of the average of these values denoted, $\overline{X}$ is normally approximated with mean μ and standard deviation $\sigma/\sqrt{M}$. The CLT helps us understand what the approximate distribution of the sample average looks like. This is the result that makes the Monte Carlo method possible. Thus to approximate μ we use the average of the values, $\overline{X}$ i.e.:

$$\hat{\mu} = \frac{\sum_{i=1}^{M} X_i}{M}$$

and to approximate the standard deviation σ we use:

$$\hat{\sigma} = \sqrt{\frac{\sum_{i=1}^{M} \left(X_i - \overline{X}\right)^2}{M-1}}$$

To estimate the standard deviation of $\overline{X}$, we calculate the standard error as: $\hat{\sigma}/\sqrt{M}$, where $\hat{\sigma}$ is the standard deviation estimate and M the sample size. In summary, the precision of the estimate C is of the order $\hat{\sigma}/\sqrt{M}$. Even more precisely a 95% confidence interval for the true option price when M is large is given as:

$$\left[C - 1.96\frac{\hat{\sigma}}{\sqrt{M}}, C + 1.96\frac{\hat{\sigma}}{\sqrt{M}}\right] \tag{8.1}$$

This is all from the CLT.

The standard error in (8.1) is a measure of our estimation precision. Ideally we would require a small standard error and consequently a precise confidence interval. However, this is not likely to happen with the plain Monte Carlo method. This is because computing the average of the simulated values using the plain Monte Carlo method results in large standard error because the plain Monte Carlo algorithm samples points randomly.

8.3 Approximation of Integrals Using the Monte Carlo Method

The Monte Carlo method can be used to numerically approximate the value of an integral. For a function of one variable the steps are:

1) Pick M randomly distributed points $X_1, X_2, \ldots, X_n$ in the interval $[a, b]$
2) Determine the average value of the function:

$$\bar{f} = \frac{1}{M} \sum_{i=1}^{M} f(X_i).$$

3) Compute the approximation to the integral:

$$\int_a^b f(x)dx \approx \bar{f}(b-a).$$

4) Compute an estimate for the error as:

$$\text{Error} \approx (b-a)\sqrt{\frac{\bar{f^2} - (\bar{f})^2}{M}}$$

where $\bar{f^2} = \frac{1}{M} \sum_{i=1}^{M} f^2(X_i)$.

Every time a Monte Carlo simulation is made using the same sample size, it will come up with a slightly different value. Larger values of M will produce more accurate approximations.

8.4 Variance Reduction

We have seen earlier in Section 8.2 that one of the problems with the Monte Carlo simulations is the big variance (standard error) when computing the average of the simulated values. The variance can be decreased by increasing the number of simulated values; however this will also increase the run time of the whole simulation.

An alternative approach is to use variance reduction techniques, which will be discussed in the next two subsections. These methods are variance reduction technique used in Monte Carlo methods.

8.4.1 Antithetic Variates

The antithetic variates method reduces the variance of the simulation results. The main idea behind this technique is to examine the asset equation that we are trying to simulate:

$$dS_t = rS_t dt + \sigma S_t dz_t \tag{8.2}$$

and recognize that since z_t is a standard Brownian motion we can replace z_t with $-z_t$ and that will have the same distribution as (8.2). This implies that

$$dS_t = rS_t dt - \sigma S_t dz_t$$

will generate sample paths of the same asset as (8.2).

When we want to generate the increments for a Brownian motion in the simulation of sample paths, one will use the increments $\varepsilon_1, \varepsilon_2, \ldots, \varepsilon_n$ to generate one path and also use the negative of these increments, i.e. $-\varepsilon_1, -\varepsilon_2, \ldots, -\varepsilon_n$ to generate an antithetic path that is given a path $\{\varepsilon_1, \ldots, \varepsilon_M\}$ to also take $\{-\varepsilon_1, \ldots, -\varepsilon_M\}$. Both of these paths will simulate the same asset; thus one can use them to find two instances of the Call option. Afterwards, we use both to calculate the final Monte Carlo value and that reduces the variance of the estimate.

8.4.2 Control Variates

The control variates method exploits the information about the errors in estimates of known quantities to reduce the error of an estimate of an unknown quantity. We begin the discussion of control variates with delta hedging. Delta hedging is an option strategy that reduces the risk associated with price movements in the underlying asset by canceling long and short positions.

For example, suppose that at time $t = 0$ we receive C_0 the price of an option that pays C_T at time T. The price of this option at any time t is a function $C(t, S)$. Then, if we hold at any moment in time $\frac{\partial C}{\partial S}(t, S) = \frac{\partial C_t}{\partial S}$ units of stock, we will be able to replicate the payout of this option C_T at time T. This scenario is very theoretical since in practice we cannot trade continuously. So in practice we perform a partial hedge where we only recalibrate at some discrete moments in time say $t_1, t_2, \ldots, t_N$. For simplicity of writing the formulas, we take $t_i = i\Delta t$ equidistant in time and $t_N = N\Delta t = T$. Table 8.1 presents the adjustments we make at various times. The notations $\frac{\partial C_i}{\partial S} = \frac{\partial C}{\partial S}(t, S)$ and $S_i = S_{t_i} = S_{i\Delta t}$ will be used.

Table 8.1 Calibrating the delta hedging.

At time	Receive from before	Pay to hold $\frac{\partial C}{\partial S}(t, S)$ units of stock
$t_0 = 0$	C_0	$\frac{\partial C_0}{\partial S} S_0$
$t_1 = \Delta t$	$\frac{\partial C_0}{\partial S} S_1$	$\frac{\partial C_1}{\partial S} S_1$
$t_2 = 2\Delta t$	$\frac{\partial C_1}{\partial S} S_2$	$\frac{\partial C_2}{\partial S} S_2$
$\vdots$	$\vdots$	$\vdots$
$t_{N-1} = (N_1)\Delta t$	$\frac{\partial C_{N-2}}{\partial S} S_{N-1}$	$\frac{\partial C_{N-1}}{\partial S} S_{N-1}$
$t_N = T$	$\frac{\partial C_{N-1}}{\partial S} S_N$	C_T

Based on Table 8.1, if we bring all the money in their value at time T by multiplying each row with the corresponding factor $e^{(T-i\Delta t)r}$ and we take the received money with plus and the money we payout with minus, we should approximately obtain a zero sum. Thus we should have:

$$C_0 e^{rT} - \frac{\partial C_0}{\partial S} S_0 e^{rT} + \frac{\partial C_0}{\partial S} S_1 e^{r(T-\Delta t)} - \frac{\partial C_1}{\partial S} S_1 e^{r(T-\Delta t)} + \cdots + \frac{\partial C_{N-1}}{\partial S} S_N - C_T = \eta \quad (8.3)$$

where η is the pricing error. By factorizing similar derivatives, (8.3) can be rewritten as:

$$C_0 e^{rT} + \sum_{i=0}^{N-1} \frac{\partial C_i}{\partial S} \left(S_{i+1} - S_i e^{r\Delta t} \right) e^{r(T-(i+1)\Delta t)} = C_T + \eta \quad (8.4)$$

Dividing through (8.4) by e^{rT} we obtain:

$$C_0 = C_T e^{-rT} - \left(\sum_{i=0}^{N-1} \frac{\partial C_i}{\partial S} \left(S_{i+1} - S_i e^{r\Delta t} \right) e^{r(T-(i+1)\Delta t)} \right) e^{-rT} + \eta e^{-rT} \quad (8.5)$$

The sum in parenthesis in Eq. (8.5) represents the control variate. The pricing error η is further multiplied with a small number that will be averaged and so it is neglected.

In order to use this formula, we need an equation for the dynamics of S_t. Suppose we have such equation and we use the Euler's method, for example, to generate a sample path of prices $S_0, S_1, \ldots, S_N$, the payoff value C_T can be determined based on the sequence of prices.

However, in order to use this formula, we need to approximate $\frac{\partial C_i}{\partial S}$ for any time t_i. For complex price dynamics, approximating $\frac{\partial C_i}{\partial S}$ for any time t_i is much harder than finding the price of the option itself. But for the Black–Scholes equation, this can be accomplished because we have a formula for the price of the option at any moment in time. Next we present an example.

Example 8.4.1 (European Call option). We recall in Section 3.3 that the Black–Scholes formula for European Call option when the stock price is modeled as a geometric Brownian motion is given as:

$$C(S,t) = S N(d_1) - K e^{-r(T-t)} N(d_2), \quad (8.6)$$

where

$$d_1 = \frac{\ln\left(\frac{S}{K}\right) + \left(r + \frac{1}{2}\sigma^2\right)(T-t)}{\sigma\sqrt{T-t}}$$

$$d_2 = d_1 - \sigma\sqrt{T-t} = \frac{\ln\left(\frac{S}{K}\right) + \left(r - \frac{1}{2}\sigma^2\right)(T-t)}{\sigma\sqrt{T-t}}.$$

Note that d_1 and d_2 are themselves functions of S, thus we need to apply the chain rule when calculating the derivative $\frac{\partial C}{\partial S}(t, S)$. The partial derivatives of d_1 and d_2 with respect to S are given as:

$$\frac{\partial d_1}{\partial S} = \frac{1}{\sigma\sqrt{T-t}}\frac{1}{S}$$

$$\frac{\partial d_2}{\partial S} = \frac{\partial d_1}{\partial S} = \frac{1}{S\sigma\sqrt{T-t}}$$

respectively.

Thus;

$$\frac{\partial C}{\partial S}(t, S) = N(d_1) + SN'(d_1)\frac{\partial d_1}{\partial S} - Ke^{-r(T-t)}N'(d_2)\frac{\partial d_2}{\partial S}$$

$$= N(d_1) + \left(SN'(d_1) - Ke^{-r(T-t)}N'(d_2)\right)\frac{1}{S\sigma\sqrt{T-t}} \quad (8.7)$$

In the formula (8.7), $N(x)$ is the cumulative density function (CDF) of a standard normal distribution calculated at x, and $N'(x)$ is the probability density function (pdf) of a standard normal distribution calculated at x, i.e.:

$$N'(x) = \frac{1}{\sqrt{2\pi}}e^{-\frac{x^2}{2}}.$$

Thus, at time $i\Delta t$ if we plug $t = i\Delta t$ and $S = S_{i\Delta t}$ into (8.7), the stock price is generated according to the current path.

We remark that the control variates method is much faster than any other Monte Carlo method since it can be used to speed up stochastic gradient Monte Carlo algorithms by enabling smaller minibatches to be used.

In the next section, we will study the pricing of path dependent options. In particular, we will study the pricing of an American option using the Monte Carlo simulation technique.

8.5 American Option Pricing with Monte Carlo Simulation

This section introduces methods to price American options with the Monte Carlo simulation. Recent progress on application of Monte Carlo simulation for pricing American options includes the martingale optimization and the least squares Monte Carlo (LSM). The martingale optimization formulation for the American option price is used to obtain an upper bound on the price, which is complementary to the trivial lower bound in [32]. The LSM on the other hand provides a direct method for pricing American options. In the subsequent subsections, we present a brief introduction to the aforementioned methods.

8.5.1 Introduction

The Monte Carlo simulation technique for valuation of American options is a method based on the probability distribution of complete history of the underlying option process. The Monte Carlo method lends itself naturally to the evaluation of security prices represented as expectations.

We recall from Section 2.3 the definition of an American option: the type of option that may be exercised at any time prior to expiry. Thus the option holder has the right to sell an underlying security for a specified price K at any time between the initiation of the agreement ($t = 0$) and the expiration date ($t = T$).

Consider an equity price process $S(t)$ that follows the geometric Brownian motion process according to the following stochastic differential equation:

$$dS = \mu S dt + \sigma S dW \tag{8.8}$$

where μ and σ are the growth rate and volatility, respectively, and $W = W(t)$ is the standard Brownian motion. Suppose the option payout function is given as $u(S, t)$, and then a path dependent option is one for which $u(S, t)$ depends on the entire path $(S(t') : 0 < t' < t)$ whereas a non-path dependent option has $u(S, t) = u(S(t), t)$. For an American option, exercising an option may be at any time before the expiry date T so that the payout is $f(\tau) = u(S(\tau), \tau)$ where τ is an optimally chosen stopping time. τ is selected to be the stopping time since the decision to exercise at time t can only depend on the values of S up to and including t. We infer from Section 3.3 that for an American option with value P, the Black–Scholes partial differential inequality is:

$$\frac{\partial P}{\partial t} + \frac{1}{2}\sigma^2 S^2 \frac{\partial^2 P}{\partial S^2} + rS\frac{\partial P}{\partial S} - rP \leq 0 \tag{8.9}$$

where r is the risk-free rate of return. The risk-neutral valuation formula which is applicable to path dependent options and other derivatives for which the PDE is intractable is given as:

$$P(S, t) = \max_{t < \tau < T} E'[e^{-r(\tau-t)} u(S(\tau), \tau) | S(t) = S] \tag{8.10}$$

where E' is the risk-neutral expectation, for which the growth rate μ in (8.8) is replaced by the risk-free interest rate r. Equation (8.10) is the formula to which the Monte Carlo simulation can be applied. The risk-neutral valuation approach provides a stochastic characterization of the early exercise boundary. For example, consider the exercise decision at a point (S, t); then the value of the early exercise is just the payoff $u(S, t)$. The expected value of deferred exercise $\tilde{P}$ is given by:

$$\tilde{P}(S, t) = \max_{t < \tau < T} E'[e^{-r(\tau-t)} u(S(\tau), \tau) | S(t) = S]. \tag{8.11}$$

The holder of the option will choose to exercise if $u \geq \tilde{P}$ so that

$$P = \max(u(S, t), \tilde{P}) \tag{8.12}$$

and $\tilde{P} = u(S, t)$ on the early exercise boundary. A lower bound on the American option price follows from (8.12). Let τ' be any stopping time and let

$$P' = E'[e^{-r(\tau-t)}u(S(\tau), \tau)|S(t) = S]$$

be the price using this stopping time; then from the aforementioned relations we have that

$$P \geq P'. \tag{8.13}$$

8.5.2 Martingale Optimization

Rogers in [169] derived the formula:

$$P(0) = \min_M E'[\max_{t<t'<T} \left(e^{-rt}u(t') - M(t')\right)] \tag{8.14}$$

for the American option price that is dual to the formula in (8.10) in which the minimum is taken over all martingales for which $M(0) = 0$.

By inserting a nonoptimal martingale M into (8.14), one obtains an upper bound on P. This method has been carried out for various choices of M in [169].

For the martingale optimization method, it is difficult to determine the accuracy of the upper bound since it is hard to quantify the degree of nonoptimality for the martingale that is used in (8.14) in order to generate the upper bound. The following characterization of the optimal martingale may be a starting point for determining the degree of optimality. The optimal martingale $M^*(t)$ is the characterization coming from the American option itself, for which the martingale decomposition of $F(t)$ is expressed as:

$$M^*(t) = e^{-rt}P(t) - P(0) + B(t) \tag{8.15}$$

where $B(t)$ is a nondecreasing process with $B(0) = 0$. Rogers [169] proved that

$$P(0) \leq E[\sup_{0<t<T} \left(e^{-rt}u(t) - M^*(t)\right)]$$

$$= P(0) + E[\sup_{0<t<T} \left(e^{-rt}(u(t) - P(t)) - B(t)\right)]. \tag{8.16}$$

Since both $-u(t) + P(t)$ and $B(t)$ are nonnegative, they must both equal 0, that is,

$$\sup_{0<t<T} \left(e^{-rt}u(t) - M^*(t)\right) = P(0). \tag{8.17}$$

Therefore Eq. (8.17) uniquely characterizes the correct price $P(0)$. Please refer to [169] for more details.

8.5.3 Least Squares Monte Carlo (LSM)

The Least LSM [140] is another method used to price American options by using the Monte Carlo simulation. The implementation is done by replacing

8.5 American Option Pricing with Monte Carlo Simulation

the future expectation by a least squares interpolation. At each exercise time point, option holders compare the payoff for immediate exercise with the expected payoff for continuation. If the payoff for the immediate exercise is higher, then they exercise the options. Otherwise, they do not exercise the options. The expected payoff for continuation is conditional on the information available at that point in time. Since the conditional expectation can be estimated from the cross-sectional information in the simulation by using least squares, this makes the LSM approach readily applicable in path dependent and multifactor situations, where other methods such as the finite difference techniques cannot be used.

The method starts with generating N random sample paths (S_n^k, t_n) for $1 \leq k \leq N$ and $t_n = ndt$. The valuation is performed by rolling back on these sample paths. Suppose that $P_{n+1}^k = P(S_{n+1}^k, t_{n+1})$ is known. For the points (S_n^k, t_n), we set the current equity value, $X = S_n^k$, and the value of deferred exercise, $Y^k = e^{rdt}P(S_{n+1}^k, t_{n+1})$. Next, we perform regression of Y as a function of the polynomials $X, X^2, \ldots, X^m$ for some small value of m, i.e.: approximate Y by a least squares fit of polynomials $X, X^2, \ldots, X^m$. Finally, the regressed value is used as an approximation to $\tilde{P}$ in (8.11) and we apply it in deciding whether to exercise the option early.

In the LSM, the least squares fit provides coupling between the prices on different Monte Carlo paths. This coupling replaces the Monte Carlo-on-Monte Carlo feature of direct Monte Carlo evaluation of American options, without the computational intractability of the direct method. The efficiency of the LSM method depends on only using a small number m of polynomials in the least squares fit, and the strong accuracy that is attained with such small degree for the polynomial fit is remarkable. Please refer to [195] for the convergence proof of the LSM method.

We present a numerical example of using the LSM method to value an American option.

Example 8.5.1 Consider an American Put option on a share of non-dividend-paying stock. The Put option is exercisable at a strike price of 1.10 at times $t = 1, 2,$ and 3, where time $t = 3$ is the expiration date of the option. The risk-less rate is $r = 6\%$.

For simplicity, the algorithm is illustrated using only eight sample paths for the price of the stock. The first step is to simulate eight sample paths generated under the risk-neutral measure. These sample paths are shown in Table 8.2.

We need to use this information to determine the continuation value at each point in time for each path. To do this we will construct a "cash flow matrix" at each point in time.

8 Approximating Stochastic Processes

Table 8.2 Stock price paths.

Path	t = 0	t = 1	t = 2	t = 3
1	1.00	1.09	1.08	1.34
2	1.00	1.16	1.26	1.54
3	1.00	1.22	1.07	1.03
4	1.00	0.93	0.97	0.92
5	1.00	1.11	1.56	1.52
6	1.00	0.76	0.77	0.90
7	1.00	0.92	0.84	1.01
8	1.00	0.88	1.22	1.34

Table 8.3 Cash flow matrix at time $t = 3$.

Path	t = 1	t = 2	t = 3
1	—	—	0.00
2	—	—	0.00
3	—	—	0.07
4	—	—	0.18
5	—	—	0.00
6	—	—	0.20
7	—	—	0.09
8	—	—	0.00

Table 8.3 denotes the cash flows at $t = 3$ assuming that we held the option that far:

The next step is to find a function that describes the continuation value at time $t = 2$ as a function of the value of S at time $t = 2$.

To do this we use a regression technique, which takes the values at time $t = 2$ as the "X" values and the discounted payoff at time $t = 3$ as the "Y" values.

The regression is only carried out on paths that are on the money at time $t = 2$, i.e. Y regressed upon X and X^2.

The resulting conditional expectation function is $E[Y|X] = -1.070 + 2.983X - 1.813X^2$. With this conditional expectation function, we now compare the value of immediate exercise at time $t = 2$ with the value from continuation from Table 8.4.

8.5 American Option Pricing with Monte Carlo Simulation

Table 8.4 Regression at $t = 2$.

Path	Y	X
1	0.00 × 0.94176	1.08
2	—	—
3	0.07 × 0.94176	1.07
4	0.18 × 0.94176	0.97
5	—	—
6	0.20 × 0.94176	0.77
7	0.09 × 0.94176	0.84
8	—	—

Table 8.5 Optimal early exercise decision at time $t = 2$.

Path	Y	X
1	0.02	0.0369
2	—	—
3	0.03	0.0461
4	0.13	0.1176
5	—	—
6	0.33	0.1520
7	0.26	0.1565
8	—	—

This then allows you to decide at which points in time you would exercise (see Table 8.5) and thus determine the cash flows at $t = 2$ as observed in Table 8.6. Notice that for each path, if we exercise at $t = 2$ then we do not also exercise at $t = 3$.

We can apply the same process to $t = 1$, for each of the paths that are on the money (in the money means that the stock option is worth money and we can turn around and sell or exercise it), we regress the discounted future cash flows (Y) on the current value of the underlying asset (X), where X and Y are as given in Table 8.7.

The regression equation here is $Y = 2.038 - 3.335X + 1.356X^2$ and again we use this to estimate the continuation value and decide on an early exercise strategy. Substituting the values of X into this regression gives the estimated conditional expectation function. These estimated continuation values and immediate exercise values at time $t = 1$ are given in the first and second columns of Table 8.8. Comparing the two columns shows that exercise at time $t = 1$ is optimal for the fourth, sixth, seventh, and eighth paths.

Table 8.6 Cash flow matrix at time $t = 2$.

Path	$t = 1$	$t = 2$	$t = 3$
1	—	0.00	0.00
2	—	0.00	0.00
3	—	0.00	0.07
4	—	0.13	0.00
5	—	0.00	0.00
6	—	0.33	0.00
7	—	0.26	0.00
8	—	0.00	0.00

Table 8.7 Regression at time $t = 1$.

Path	Y	X
1	0.00 × 0.94176	1.09
2	—	—
3	—	—
4	0.13 × 0.94176	0.93
5	—	—
6	0.33 × 0.94176	0.76
7	0.26 × 0.94176	0.92
8	0.00 × 0.94176	0.88

Having identified the exercise strategy at times $t = 1, 2$, and 3, the stopping rule can now be represented by Table 8.9, where the ones denote exercise dates at which the option is exercised.

With this specification of the stopping rule, it is now straightforward to determine the cash flows realized by following this stopping rule. This is done by simply exercising the option at the exercise dates where there is a 1 in Table 8.9. This leads to the optimal cash flow observed in Table 8.10.

Having identified the cash flows generated by the American Put at each date along each path, the option can now be valued by discounting each cash flow in the option cash flow matrix back to time zero and averaging over all paths. So in this case, applying this procedure results in a value of:

Table 8.8 Optimal early exercise decision at time $t = 1$.

Path	Exercise	C
1	0.01	0.0139
2	—	—
3	—	—
4	0.17	0.1092
5	—	—
6	0.34	0.2866
7	0.18	0.1175
8	0.22	0.1533

Table 8.9 Stopping rule.

Path	$t = 1$	$t = 2$	$t = 3$
1	0	0	0
2	0	0	0
3	0	0	1
4	1	0	0
5	0	0	0
6	1	0	0
7	1	0	0
8	1	0	0

$$P_0 = \frac{1}{8}(0.07e^{-3r} + 0.17e^{-r} + 0.34e^{-r} + 0.18e^{-r} + 0.22e^{-r}) = 0.1144$$

for the American Put.

Example 8.5.1 illustrates how least squares can use the cross-sectional information in the simulated paths to estimate the conditional expectation function. In turn, the conditional expectation function is used to identify the exercise decision that maximizes the value of the option at each date along each path. As shown by this example, the LSM approach is easily implemented since it only involves simple regression.

In the next section, we will discuss nonstandard Monte Carlo methods. Many real-world problems involve elements of non-Gaussianity, nonlinearity,

Table 8.10 Optimal cash flow.

Path	$t = 1$	$t = 2$	$t = 3$
1	0	0	0
2	0	0	0
3	0	0	0.07
4	0.17	0	0
5	0	0	0
6	0.34	0	0
7	0.18	0	0
8	0.22	0	0

and nonstationarity. Nonstandard Monte Carlo method is a type of Monte Carlo method which avoids making linearity or normality assumptions.

8.6 Nonstandard Monte Carlo Methods

Nonstandard Monte Carlo methods arise as a result of the departure from Gaussianity. This departure is due to the fact that there has been a constant rise in very sophisticated models that deal with high-dimensional and complex patterns. Hence there is the need to characterize these models by considering another family of Monte Carlo algorithms. In this section, we briefly present some examples of nonstandard Monte Carlo methods.

8.6.1 Sequential Monte Carlo (SMC) Method

Sequential Monte Carlo (SMC) methods are a set of simulation-based methods which provide a convenient and attractive approach for computing to posterior distributions. Suppose one is uncertain about something, then the uncertainty is described by a probability distribution called a prior distribution. If on the other hand you obtain some data relevant to that thing, the data changes your uncertainty, which is then described by a new probability distribution called posterior distribution. Posterior distribution reflects the information both in the prior distribution and the data. SMC methods are used to calculate approximations to posterior distributions.

Unlike grid-based methods, SMC methods are very flexible and applicable in very generic settings. For example, when one has a large number of samples drawn from a posterior distribution, it is easy to approximate the intractable integrals appearing in the posterior distributions. Please refer to [56] for more details of the SMC methods.

The SMC methods can be used to solve complex nonlinear non-Gaussian problems. They are also very popular in physics where they are used to calculate eigenvalues of positive operators, the solution of PDEs/integral equations or simulate polymers.

8.6.2 Markov Chain Monte Carlo (MCMC) Method

When computing posterior distributions, many situations arise in which the posterior distribution is difficult to either calculate or simulate. This is as a result of the necessity to calculate integrals of complex and commonly high-dimensional expressions.

Markov Chain Monte Carlo (MCMC) is a general purpose technique for generating fair samples from a probability in high-dimensional spaces, using random numbers drawn from uniform probability in certain range. It does this by constructing a Markov Chain which converges after a certain number of steps to the desired probability distribution. As already discussed in Chapters 1 and 6 of this book, a Markov Chain is a process in which the next step or iteration of the process only depends upon the current step and not upon any previous steps in the process. If a Markov Chain has certain properties, then the process will evolve in a random fashion until it reaches a certain state in which it remains thereafter, called equilibrium.

The algorithm of the MCMC constructs a Markov Chain in which the random evolution of the chain is probabilistic. Thus, each step in the chain constructs an empirical distribution which is a Monte Carlo approximation of the posterior, and the chain converges to an equilibrium distribution which is the posterior distribution. It is this fact regarding the guaranteed convergence to the desired equilibrium distribution which enables the empirical distribution generated to serve as a Monte Carlo simulation of the posterior distribution. The simplest MCMC algorithms are random walks where each step in the algorithm takes small, random steps around the target equilibrium distribution. An example is the Metropolis–Hastings algorithm. The Metropolis–Hastings method generates a random walk using a proposal density and a method for rejecting some of the proposed moves.

MCMC methods are primarily used for calculating numerical approximations of multidimensional integrals, for example, in Bayesian statistics, computational physics, computational biology, and computational linguistics. Please refer to [9] for more details.

We now begin the second part of this chapter.

8.7 Generating One-Dimensional Random Variables by Inverting the cdf

Let X be a one-dimensional random variable defined on any probability space $(\Omega, \mathscr{F}, \mathbf{P})$ with distribution function $F(x) = \mathbf{P}(X \leq x)$. All the distribution generation methods discussed in this section are based on the following lemma:

Lemma 8.7.1 The random variable $U = F(X)$ is distributed as a $U(0, 1)$ random variable. If we let $F^{-1}(u)$ denote the inverse function, that is,

$$F^{-1}(u) = \{x \in \mathbb{R} \mid F(x) = u\},$$

then the variable $F^{-1}(U)$ has the same distribution as X.

Here, $F^{-1}(u)$ as defined in the lemma is a set.

Proof. If F is a bijective function. Note that in this case we have

$$\mathbf{P}(U \leq u) = \mathbf{P}(F(X) \leq u)$$

But recall that F is a probability itself, so the result of the aforementioned is zero if $u < 0$ and 1 if $u \geq 1$. If $0 < u < 1$, since F is an increasing function, we can write

$$\mathbf{P}(U \leq u) = \mathbf{P}(X \leq F^{-1}(u)) = F(F^{-1}(u)) = u$$

and this is the distribution of a $U(0, 1)$ random variable.

If F is not a bijective function, the proof still holds but we have to work with sets. The relevant case is, once again, $0 < u < 1$. Recall that F is increasing. Because of this fact and using the definition of F^{-1} as shown earlier, we have

$$F^{-1}((-\infty, u]) = \{x \in \mathbb{R} \mid F(x) \leq u\}.$$

If the set $F^{-1}(u)$ has only one element (F is bijective at u), there is no problem and the set mentioned earlier is just $(-\infty, F^{-1}(u)]$. Therefore the same derivation as shown earlier works. If $F^{-1}(u) = \{x \in \mathbb{R} \mid F(x) = u\}$ has more than one element, then let $x_{\max}$ be the maximum element in the set. This element exists and it is in the set since $u < 1$. We may write

$$F^{-1}((-\infty, u]) = (-\infty, x_{\max}].$$

Thus we have

$$\mathbf{P}(U \leq u) = \mathbf{P}(F(X) \leq u) = \mathbf{P}\left(X \in F^{-1}((-\infty, u])\right)$$
$$= \mathbf{P}\left(X \in (-\infty, x_{\max}]\right) = \mathbf{P} \circ X^{-1}((-\infty, x_{\max}])$$
$$= F(x_{\max})$$

by the definition of distribution function. Now, recall that $x_{\max} \in F^{-1}(u)$ and so multiplying through by F produces $F(x_{\max}) = u$. Once again, we reach the distribution of a uniform random variable. □

This lemma is very useful for generating random variables with prescribed distribution. We only need to figure out the inverse cumulative distribution function (cdf) F^{-1} to generate random variables with any distribution starting with the uniform density. This approach works best when the distribution function F has an analytical formula and, furthermore, when F is a bijective function. A classical example is the exponential distribution.

Example 8.7.1 (Generating an exponential random variable). Suppose we want to generate an Exponential(λ) random variable, that is, a variable with density:

$$f(x) = \lambda e^{-\lambda x} \mathbf{1}_{\{x>0\}}$$

Note that the expectation of this random variable is $1/\lambda$. This distribution can also be parameterized using $\lambda = 1/\theta$, in which case the expectation will be θ. The two formulations are equivalent.

We may calculate the distribution function in this case as:

$$F(x) = (1 - e^{-\lambda x}) \mathbf{1}_{\{x>0\}}$$

We need to restrict this function to $F : (0, \infty) \to (0, 1)$ to have a bijection. In this case, for any $y \in (0, 1)$, the inverse is calculated as

$$F(x) = 1 - e^{-\lambda x} = y \Rightarrow x = -\frac{1}{\lambda} \log(1 - y)$$

$$\Rightarrow F^{-1}(y) = -\frac{1}{\lambda} \log(1 - y)$$

Therefore to generate an Exponential(λ) random variable, first generate U a Uniform(0,1) random variable and then calculate

$$-\frac{1}{\lambda} \log(1 - U);$$

this will have the desired distribution.

We remark that a further simplification may be made since $1 - U$ has the same distribution as U, and we obtain the same exponential distribution by taking

$$-\frac{1}{\lambda} \log U.$$

Note that one should use one of the forms $-\frac{1}{\lambda} \log U$ or $-\frac{1}{\lambda} \log(1 - U)$, but not both, since the two variables are related and therefore not independent.

For all discrete random variables, the distribution function is a step function. In this case, the F is not bijective, so we need to restrict it somehow to obtain the desired distribution. The main issue is that every element of the codomain is not mapped to by at least one element of the domain. Therefore, we need to know what to do when a uniform is generated.

Example 8.7.2 (Generating any discrete random variable with finite number). Suppose we need to generate the outcomes of the discrete random variable Y which takes n values $a_1, \ldots, a_n$ each with probability $p_1, \ldots, p_n$, respectively, so that $\sum_{j=1}^{n} p_j = 1$.

To generate such outcomes, we first generate a random variable U as a Uniform(0,1) random variable. Then we find the index j such that

$$p_1 + \cdots + p_{j-1} \leq U < p_1 + \cdots + p_{j-1} + p_j.$$

The generated value of the Y variable is the outcome a_j.

Note that, theoretically, the case when the generated uniform values of U are exactly equal to p_j does not matter since the distribution is continuous and the probability of this event happening is zero. However, in practice such events do matter since the cycle used to generate the random variable is finite, and thus the probability of the event is not zero – it is extremely small but not zero. This is dealt with by ignoring 1 if it is generated and keeping the rest of the algorithm as mentioned earlier.

Remark 8.7.1 The previous example explains the need for generating random variables – the tossing of a coin or generating a Bernoulli(p) random variable. Specifically, generate U a Uniform(0,1). If $U < p$, output 1, else output 0.

8.8 Generating One-Dimensional Normal Random Variables

Generating normal (Gaussian) random variables is important because this distribution is the most widely encountered distribution in practice. In the Monte Carlo methods, one needs to generate millions of normally distributed random numbers. That means that the precision with which these numbers are generated is quite important. Imagine that one in 1000 numbers is not generated properly. This translates on average into about 1000 numbers being bad in 1 million numbers simulated. Depending on the simulation complexity, a single Monte Carlo path may need 1 million generated numbers, and therefore the simulated path has 1000 places where the trajectory is not what it ought to be.

8.8 Generating One-Dimensional Normal Random Variables

Let us first remark that, if we know how to generate X a standard normal variable with mean 0 and variance 1, that is, with pdf:

$$f(x) = \frac{1}{\sqrt{2\pi}} e^{-\frac{x^2}{2}},$$

then we know how to generate any normal variable Y with mean μ and standard deviation σ. This is accomplished by simply taking

$$Y = \mu + \sigma X.$$

Thus, to generate any normal it is enough to learn how to generate $N(0, 1)$ random variables. The inversion methodology presented in the previous section cannot be applied directly since the normal cdf does not have an explicit functional form, and therefore inverting it directly is impossible.

One of the fastest and best methods uses a variant of the inversion method. Specifically, the algorithm developed by [208] calculates quantiles (i.e. dividing the range of a probability distribution into contiguous intervals with equal probabilities) corresponding to the generated probability values. The algorithm has two subroutines to deal with the hard-to-estimate quantiles from the tails of the Gaussian distribution. First, the algorithm generates p using a Uniform(0,1) distribution. Then it calculates the corresponding normally distributed value z_p by inverting the distribution function

$$p = \int_{-\infty}^{z_p} \frac{1}{\sqrt{2\pi}} e^{-x^2/2} dx = \Phi(z_p),$$

where $z_p = \Phi^{-1}(p)$. The respective Fortran subroutines PPND7 or PPND16 are chosen depending on the generated uniform value p, more specifically if $|p - 0.5| \leq 0.425$, respectively, greater than 0.425. These routines are polynomial approximations of the inverse function $\Phi^{-1}(\cdot)$. The algorithm has excellent precision (of the order 10^{-16}) and it runs relatively fast (only requires a logarithmic operation besides the polynomial operations). Please refer to [208] for more details.

More classical methods of generating normally distributed numbers are presented in the following. All of the methods take advantage of transformations of random variables.

8.8.1 The Box–Muller Method

This method of generating normally distributed random numbers is named after George Edward Pelham Box and Mervin Edgar Muller, who developed the algorithm in 1958. The algorithm uses two independent Uniform(0,1) random numbers U and V. Then two random variables X and Y are calculated using

$$X = \sqrt{-2\ln U} \cos 2\pi V$$
$$Y = \sqrt{-2\ln U} \sin 2\pi V$$

Then the two random variables X and Y have the standard normal distribution and are independent.

This result is easy to derive since, for a bivariate normal random vector (XY), the variable $X^2 + Y^2$ is distributed as a chi-squared random variable with two degrees of freedom. A chi-squared with two degrees of freedom is in fact the same as the exponential distribution with parameter $1/2$; note that, in fact, the generated quantity $-2\ln U$ has this distribution. Furthermore, the projection of this quantity on the two axes is determined by the angle that is made by the point and the origin, and this angle is chosen by the random variable V. The angle is uniform between $[0, \pi]$.

An advantage of the Box–Muller method unlike the previous method is the fact that the Box–Muller method uses two independent uniforms and produces two independent normal random numbers which may both be used in the algorithms. However the method requires a logarithm operation, a square root operation, and two trigonometric function calculations, and these may be slower when repeated many times.

8.8.2 The Polar Rejection Method

This method is due to Marsaglia and is a simple modification of the Box–Muller algorithm. Recall that one of the things slowing down the Box–Muller method was the calculation of the trigonometric functions sin and cos. The polar rejection method avoids the trigonometric functions calculations, replacing them with the rejection sampling method presented in the next section.

In this method, two random numbers U and V are drawn from the Uniform$(-1, 1)$ distribution. Note the difference from the previous uniform numbers. Then, the quantity

$$S = U^2 + V^2$$

is calculated. If S is greater than or equal to 1, then the method starts over by regenerating two uniforms until it generates two uniforms with the desired property. Once those are obtained, the variables

$$X = U\sqrt{-2\frac{\ln S}{S}}$$
$$Y = V\sqrt{-2\frac{\ln S}{S}}$$

are calculated. These X and Y are independent, standard, normal random numbers. Intuitively, we need to generate coordinates of the points inside the unit circle, which is why we perform the accept/reject step.

Marsaglia's polar rejection method is both faster and very accurate (because it does not require the approximation of three functions) compared to the Box–Muller method. The drawback of the method is that, unlike the Muller method, it may require multiple sets of uniform numbers until it reaches values that are not rejected. In fact, for the one-dimensional case presented, we can easily calculate the probability that the generated pair is accepted. This is the probability that a two-dimensional uniform vector falls inside the unit circle and it is the ratio of the area of this circle over the square with center at origin and sides 2 (that is from −1 to 1). Mathematically,

$$P(S < 1) = P(U^2 + V^2 < 1) = \frac{\pi}{2^2} = \pi/4 \approx 0.79.$$

Therefore, about 79% of the generated pairs fall inside the circle and are used in the algorithm. Note that the number of times one needs to generate these pairs to obtain a usable one is easily seen as having a *Geometric* (0.79) distribution.

Next we take advantage of the CLT to generate a number n of Uniform(0,1) random numbers. Then calculate their sum $Y = \sum_{i=1}^{n} U_i$. The exact distribution of Y is the so-called *Irwin–Hall* distribution, named after Joseph Oscar Irwin and Philip Hall, which has the probability density

$$f_Y(y) = \frac{1}{2(n-1)!} \sum_{k=0}^{n} (-1)^k \binom{n}{k} (y-k)^{n-1} \text{sign}(y-k),$$

where sign(x) is the sign function. The sign function is defined on $(-\infty, 0) \cup (0, \infty)$ as

$$\text{sign}(x) = \frac{|x|}{x} = \begin{cases} 1 & \text{if } x > 0 \\ -1 & \text{if } x < 0 \end{cases}$$

The Irwin–Hall distribution has mean $n/2$ and variance $n/12$. However, the CLT (Section 1.7) guarantees that, as $n \to \infty$, the distribution of Y approaches the normal distribution. So a simple algorithm will generate, say, 12 uniforms, and then it would calculate the standardized random variable

$$Y = \frac{\sum_{i=1}^{12} U_i - 12/2}{\sqrt{12/12}}$$

which can further be simplified as

$$Y = \sum_{i=1}^{12} U_i - 6$$

This new random variable is therefore approximately distributed as a N(0,1) random variable. Since we use the 12 uniforms, the range of the generated values is [−6, 6], which is different from the real normal random numbers which are distributed on $\mathbb{R}$. Of course, taking a larger n will produce normals with better precision. However, recall that we need n uniform variables to create *one* normally distributed number, so the algorithm slows down considerably as n gets larger. Please consult [67] for more details.

One of the most popular methods of generating normals today is the Ziggurat algorithm (due to [148]). This is the default algorithm in MATLAB for generating normals. We will describe this algorithm after we present the rejection sampling method in the next section. The Ziggurat algorithm is based on this method.

8.9 Generating Random Variables: Rejection Sampling Method

The polar method is a particular case of the rejection sampling method. In rejection sampling (also named the *accept–reject method*), the objective is to generate a random variable X having the *known* density function $f(x)$. The idea of this method is to use a different but easy-to-generate-from distribution $g(x)$. The method is very simple and was originally presented by John Von Neumann. You can see the idea of this algorithm in the Buffon needle problem (throwing the needle and accepting or rejecting depending whether the needle touches the lines or not).

First determine a constant M such that

$$\frac{f(x)}{g(x)} < M, \quad \forall x$$

Once such M is determined, the algorithm is as follows:

Step 1. Generate a random variable Y from the distribution $g(x)$.
Step 2. Accept $X = Y$ with probability $f(Y)/Mg(Y)$. If reject, go back to step 1.

The accept–reject step can be easily accomplished by a Bernoulli random variable. Specifically, step 2 is: generate $U \sim \text{Uniform}(0, 1)$ and accept if

$$U < \frac{f(Y)}{Mg(Y)};$$

Go back to step 1 if reject.

Proposition 8.9.1 *The random variable X created by the rejection sampling algorithm shown earlier has the desired density $f(x)$.*

Proof. Let N be the number of necessary iterations to obtain the final variable X. Let us calculate the distribution of X. Since each trial is independent, we have

8.9 Generating Random Variables: Rejection Sampling Method

$$P\{X \leq x\} = P\left\{Y \leq x \bigg| U \leq \frac{f(Y)}{Mg(Y)}\right\} = \frac{P\left(\{Y \leq x\} \cap \left\{U \leq \frac{f(Y)}{Mg(Y)}\right\}\right)}{P\left\{U \leq \frac{f(Y)}{Mg(Y)}\right\}}$$

Now, the numerator is

$$P\left(\left\{U \leq \frac{f(Y)}{Mg(Y)}\right\} \mid \{Y \leq x\}\right) P(\{Y \leq x\})$$

$$= \int_{-\infty}^{x} P\left(\left\{U \leq \frac{f(y)}{Mg(y)}\right\} \mid \{Y = y\}\right) g(y) dy$$

$$= \int_{-\infty}^{x} \frac{f(y)}{Mg(y)} g(y) dy = \frac{1}{M} \int_{-\infty}^{x} f(y) dy$$

Similarly, the denominator is

$$P\left\{U \leq \frac{f(Y)}{Mg(Y)}\right\} = \int_{-\infty}^{\infty} P\left\{U \leq \frac{f(y)}{Mg(y)} \mid \{Y = y\}\right\} g(y) dy$$

$$= \int_{-\infty}^{\infty} \frac{f(y)}{Mg(y)} g(y) dy = \frac{1}{M} \int_{-\infty}^{\infty} f(y) dy = \frac{1}{M}$$

Now taking the ratio, i.e. dividing the denominator with the numerator, shows that X has the desired distribution. □

Note that calculating the denominator in the proof mentioned earlier shows that the probability of accepting the generated number is always $1/M$. So, if the constant M is close to 1, then the method works very efficiently. However, this is dependent on the shape of the densities f and g. If the density g is close in shape to f, then the method works very well. Otherwise, a large number of generated random variables are needed to obtain one random number with density f.

Corollary 8.9.1 (Slight generalization). Suppose that we need to generate random variables from the density $f(x) = Cf_1(x)$, where we know the functional form f_1 and C is a normalizing constant, potentially unknown. Then suppose we can find a density $g(x)$ which can easily be generated from and a constant M such that

$$\frac{f(x)}{g(x)} < M, \quad \forall x.$$

Then the rejection sampling procedure described earlier will create random numbers with density f. □

The corollary can be proved in exactly the same way as the main proposition. Sometimes, the constant M is difficult to calculate and this is why the corollary is useful in practice.

8.9.1 Marsaglia's Ziggurat Method

The rejection sampling is a straightforward method. One needs to find a distribution $g(x)$ such that

1) It is easy to generate random numbers.
2) It dominates the target density $f(x)$ – eventually using a constant M.

So, the first basic idea when trying to apply it for the normal random variables (when $f(x)$ is the $N(0, 1)$ pdf) is to try taking $g(x)$ a uniform density. This density is easy to generate, and indeed Figure 8.1 illustrates an example where we use a scaled uniform (dashed lines) which dominates a normal density.

However, there are two problems with this approach:

1) Recall that the probability with which we accept the generated number is equal to the ratio between the two pdfs at the respective generated number. Looking at Figure 8.1, it is easy to see that this ratio gets drastically close to 0 if the generated number is in the normal tails. Thus, the algorithm would get stuck trying to generate such numbers.
2) Another problem is the tails of the normal. The uniform generates on a fixed interval while the normal takes values all the way to $\pm\infty$.

The Ziggurat method deals with both these problems separately. It also uses several tricks to simplify as well as speed up the generation of numbers.

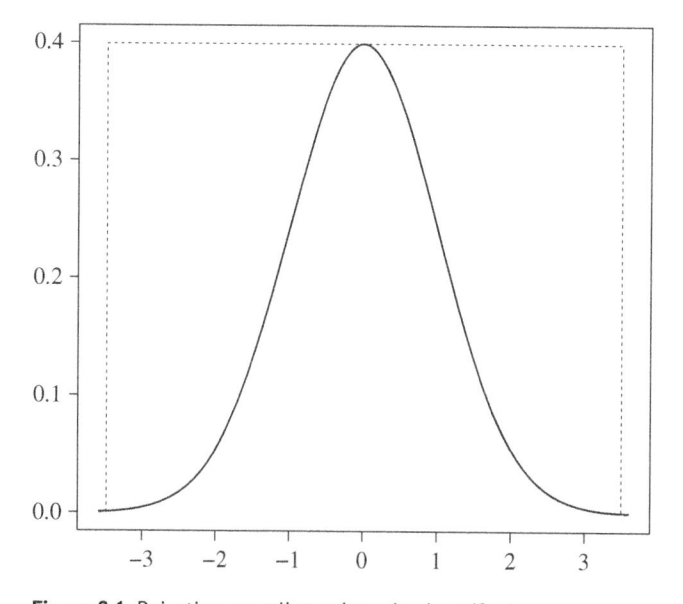

Figure 8.1 Rejection sampling using a basic uniform.

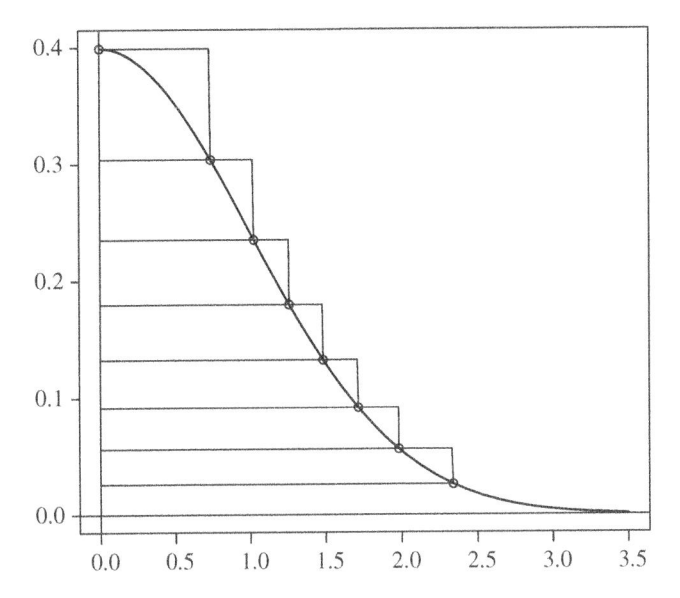

Figure 8.2 The Ziggurat distribution for $n = 8$.

The Ziggurat method is actually general and can be applied to any density function.

In the case of the normal distribution, since the distribution is symmetric, a first simplification is made by restricting to the positive axis. If we generate a positive normal number, then we just generate its sign by generating a separate +1 or −1 each with probability 1/2.

The method first proceeds by constructing a dominating distribution $g(x)$ as shown in Figure 8.2. The plot contains a distribution g constructed using $n = 8$ areas. We shall describe the algorithm to give a better understanding of the method.

First, the points determining the rectangles plotted are chosen according to the uniform distribution on the g area. The rectangles themselves are constructed in such a way that it is easy to choose the point and to use the rejection sampling method. We illustrate the method for $n = 8$ (refer to Figure 8.2).

First, we choose $x_1, x_2, \ldots, x_7$, such that the areas of the rectangles in the figure are all equal and equal to the bottom area containing the infinite tail as well. Next, we assume that the points are known, namely, $0 = x_0 < x_1 < x_2 < \ldots < x_7$ and denote the rectangles in Figure 8.2 with $R_1, R_2, \ldots R_7$ and with R_8 for the bottom one which is a rectangle plus the tail. This last one needs to be dealt with differently. So the rectangle R_i is determined by points x_{i-1} and x_i.

The point with distribution g is generated in the following way:

1) We first choose one rectangle from $R_1, R_2, \ldots R_7$, and R_8 at random (we can do this since all have the same area).
2) Suppose we generated numbers such that $i \leq 7$. Next we need to generate a point in the rectangle R_i. Note that the x coordinate of such a point is between 0 and x_i and uniformly distributed. To generate the x coordinate of this point, just generate

$$X = Ux_i,$$

where U is $Uniform(0, 1)$. We need to see if the generated point is rejected or not. We have two situations possible:
   a) If the generated $X < x_{i-1}$, then clearly its corresponding y coordinate (thus the g density) is less than f and therefore the point is accepted.
   b) If the generated X is between x_{i-1} and x_i, then we need to calculate $f(x)$ and decide if we reject it or not. If we reject, we go back to the beginning (choosing an i). In practice, if n is large, then the interval x_{i-1} to x_i is very small, so this step happens very rarely.
3) If the chosen i is 8, thus corresponding to the bottom area R_8, the previous step is modified a bit. Denote the last point x_7 with r, and denote with v the area of all rectangles (which is the same for all). We generate the x coordinate of the point by taking

$$X = \frac{v}{f(v)} U,$$

where f is the normal pdf and once again U is uniform on $(0, 1)$. Again, we have two situations:
   a) If $X < r$, then similarly we accept the generated X.
   b) If $X > r$, we return an X from the tail. To do so, we may use Marsaglia's procedure (polar rejection again – but only in the tail). Since we are in the tail, the curve there is less than an exponential, so we can use that distribution to generate from the tail. Specifically, keep generating U_1, U_2 uniform between $(0, 1)$ and calculate

$$X = -\frac{\log U_1}{r}$$
$$Y = -\log U_2$$

until $2Y > X^2$; then return $r + X$ as the generated number.

Then, given the sequence $0 = x_0 < x_1 < \ldots < x_{n-1}$, the pseudo-code for the Ziggurat algorithm is as follows:

1) Generate i a discrete uniform from $\{1, \ldots, n\}$. Generate $U \sim Unif(0, 1)$. Set $X = Ux_i$.
2) For all i: If $X < x_{i-1}$, deliver X; jump to step 4.

3) a) If $i \in \{1, \ldots, n-1\}$, generate $V \sim Unif(0,1)$.
   i) If $\frac{f(X)-f(x_i)}{f(x_{i-1})-f(x_i)} > V$ deliver the generated X and jump to step 4.
   ii) Else we reject and go back to step 1.
   b) If $i = n$ (bottom rectangle), return an X from the tail (see previous text).
4) Generate $W \sim Unif(0,1)$. If $W < 0.5$, set $X = -X$.

The last step ensures that we generate positive and negative values. The $f(x_i)$ values can be precalculated, so the cost of the algorithm is just the generation of uniforms and the calculation of $f(X)$.

8.9.1.1 The Rectangle Construction

Choosing the points x_i so that the rectangles all have the same area is nontrivial. The determination of these points are presented as follows.

Given n, find points $0 = x_0 < x_1 < \cdots < x_{n-1}$ such that all rectangles have the same area. Specifically, denote v this common area (which we do not know at the moment). For this v, the points must satisfy

$$v = x_1(f(0) - f(x_1))$$
$$= x_2(f(x_1) - f(x_2))$$
$$\vdots$$
$$= x_{n-1}(f(x_{n-2}) - f(x_{n-1}))$$
$$= x_{n-1}f(x_{n-1}) + \int_{x_{n-1}}^{\infty} f(x)dx \qquad (8.18)$$

Here we have n equations with n unknowns $(x_1, \ldots, x_{n-1}$ and $v)$ so the system should be solvable. The problem is that the equations are highly nonlinear. Therefore an approximation method needs to be used.

Marsaglia proposes the following method. Denote the last point x_{n-1} with r. Simplify everything in terms of r and make the system as close to exact as possible.

For any r, define a function $z(r)$ by doing the following. First calculate

$$v = rf(r) + \int_r^{\infty} f(x)dx$$

the common area size as a function of r. Then by setting each of the expressions in (8.18) equal to this v, we obtain the rest of the points $x_1, \ldots, x_{n-2}$ as:

$$x_{n-2} = f^{-1}\left(\frac{v}{r} + f(r)\right)$$
$$\vdots ,$$
$$x_1 = f^{-1}\left(\frac{v}{x_2} + f(x_2)\right)$$

Note that we did not use the first equation. If the r is the right one, then this first equation would be verified as well; therefore we need to find the r which makes this equation as close to 0 as possible. Specifically, output the value of the function $z(r)$ as

$$z(r) = v - x_1(f(0) - f(x_1)),$$

where v and x_1 have just been calculated in terms of r.

To calculate the grid points, one needs to search for the r which makes the last expression as close to 0 as possible. In practice, this is very difficult. The function f^{-1} is the inverse of the normal pdf and it is given as

$$f^{-1}(y) = \sqrt{-(\log 2\pi + 2 \log y)}.$$

When using this function to compute the points and the final probability, everything is fine if r is actually larger than the optimal r (in this case the $z(r)$ is negative). However, if r is less than the optimal, the expression under the square root becomes negative and the value of $z(r)$ becomes a complex number. This is why finding the grid for any n is very difficult in practice. For reference, when implementing the grid in R, with 16 bit accuracy, we obtained for $n = 8$ the optimal $r = 2.3383716982$ and the corresponding $z(r) = 6.35934 \times 10^{-11}$. Using $n = 256$ we obtained $r = 3.654152885361009$ and $z(r) = 5.50774 \times 10^{-16}$.

We will conclude this section with an R code to calculate the function $z(r)$ and give the points that will be used to make the plots in the section. Any line which starts with the # character is a comment line.

```
#The function below calculates for given $r$ and $n$ the value
#of the function z(r)

ziggurat.z=function(r,n)
{v=r*dnorm(r)+pnorm(-r);points=r;
for(i in 1:(n-2))
    {points=c(points,invdnorm(v/points[i]+dnorm (points[i])))};
return(c(points,v-points[n-1]\*(dnorm(0)-dnorm (points[n-1]))))
}

#Since there is a problem if r is less than the optimal one,
#the procedure is the following:
# set a fix n
#Find the r such that the last point in the output is as
#close to 0 as possible (bisection method by hand);

#Finally, here is the function to create the ziggurat plot;
#Last lines drawn are for the bottom rectangle and the axes

plot.ziggurat=function(passedgrid)
{x=seq(0,3.5,by=.001);
plot(x,dnorm(x),type="l'',col="blue'',lty=1,xlab="'', ylab="'');
points(passedgrid,dnorm(passedgrid))
forplot=c(0,passedgrid)
```

```
for(i in 1:(length(forplot)-1))
    {lines(c(seq(0,forplot[i+1],by=0.01),forplot[i+1]),
    c(rep(dnorm(forplot[i]), length(seq(0,forplot[i+1],
    by=0.01))),
    dnorm(forplot[i+1])))}

lines(seq(0,forplot[length(forplot)],by=0.01),
    rep(dnorm(forplot[length(forplot)]),
    length(seq(0,forplot[length(forplot)],by=0.01))))

abline(0,0);abline(v=0)
}

#We need to pass the grid points (x_1,...,x_{n-1})
#in increasing order so here is for example how to call:

$n=8$
pict.8=(ziggurat.z(2.3383716982,n))[n:1]
plot.ziggurat(pict.8)

$n=256$
pict.256=(ziggurat.z(3.6541528852885361009,n))[n:1]
plot.ziggurat(pict.256)
```

8.9.1.2 Examples for Rejection Sampling

Example 8.9.1 Let us exemplify Corollary 8.9.1. Suppose we want to generate from the density

$$f(x) = Cx^2 (\sin x)^{\cos x} |\log x|, \quad x \in \left(\frac{\pi}{6}, \frac{\pi}{2}\right)$$

The constant C is chosen to make the density f integrate to 1. Note that actually calculating C is impossible.

We wish to apply the rejection sampling to generate from the distribution f. To this end, we will use the uniform distribution in the interval $\left(\frac{\pi}{6}, \frac{\pi}{2}\right)$, and we shall calculate the constant M so that the resulting function is majoring the distribution. To do so, we calculate the maximum of the function

$$m = \max_{x \in \left(\frac{\pi}{6}, \frac{\pi}{2}\right)} x^2 (\sin x)^{\cos x} |\log x| = 1.113645$$

and we take

$$M = Cm \left(\frac{\pi}{2} - \frac{\pi}{6}\right).$$

With this constant M we are guaranteed that $f(x) < Mg(x)$ for every $x \in \left(\frac{\pi}{6}, \frac{\pi}{2}\right)$. To see this, recall that the probability density of the uniform distribution on the desired interval $x \in \left(\frac{\pi}{6}, \frac{\pi}{2}\right)$ is a constant $g(x) = \left(\frac{\pi}{2} - \frac{\pi}{6}\right)^{-1}$. Furthermore, the ratio that needs to be calculated is

$$\frac{f(x)}{Mg(x)} \qquad (8.19)$$

Substituting the values of $f(x)$, M and $g(x)$ into the formula (8.19), we obtain

$$\frac{f(x)}{Mg(x)} = \frac{x^2(\sin x)^{\cos x}|\log x|}{m}.$$

Obviously, this ratio is very good (approaches 1) when x is close to $\pi/2$, and it is close to 0 (as it should) when x is close to 1.

The following code is written in R and implements the rejection sampling example presented earlier. Any line which starts with the # character is a comment line.

```
##Example: Rejection Sampling R code
#We calculate the constant m used later

m=max(x2*(sin(x)cos(x))*abs(log(x)))

# Next defines the function used to calculate the ratio
# f(x)/M*g(x)

ratio.calc=function(x)
{return(x2*(sin(x)cos(x))*abs(log(x))/m)}

#The next function returns n generated values from the
#distribution f(x)

random.f=function(n)
{GeneratedRN=NULL;
    for(i in 1:n)
        {OK=0;
        while(OK!=1){
                Candidate=runif(1,pi/6,pi/2);
                U=runif(1);
        if(U<ratio.calc(Candidate))
                {OK=1;GeneratedRN=c(GeneratedRN,Candidate)}
        }
        }
return(GeneratedRN)
}

#Now we call the function we just created to generate
#10,000 numbers

estimated.dist=random.f(10000)
```

```
#Finally, to check here is the histogram of these numbers
hist(estimated.dist,nclass=75)
```

Next we present how to generate random variables by mixing two or more classical distributions. The classical distributions used are the beta distribution, gamma distribution, and Uniform distribution. Such random variables are much easier to generate (and much faster) as the next section details.

8.9.1.3 Generating from a Mixture of Distributions

Suppose that the density we need to generate from is a mixture of elemental distributions. Specifically,

$$f(x) = \sum_{i=1}^{n} w_i f_i(x|\theta_i)$$

where the weights w_i sum to 1, and the densities f_i may all be different and dependent on the vectors of parameters θ_i. It is much easier to generate from such distributions provided that we have implemented already generators for each of the distributions f_i. The idea is that the weights determine which distribution generates the respective random number.

Specifically, we first generate a random number U as a Uniform(0,1) random variable. Then we find the weight index j such that

$$w_1 + \cdots + w_{j-1} \leq U < w_1 + \cdots + w_{j-1} + w_j.$$

Next, the desired random number is generated from the distribution f_j. Let us exemplify this generating strategy by looking at the following example.

Example 8.9.2 Consider a mixture of beta distributions

$$f(x) = 0.5\ \beta(x; 10, 3) + 0.25\ \beta(x; 3, 15) + 0.25\ \beta(x; 7, 10), \quad x \in (0, 1),$$

where the $\beta(x, a, b)$ denotes the beta distribution pdf with shape parameters a and b,

$$\beta(x, a, b) = \frac{\Gamma(a+b)}{\Gamma(a)\Gamma(b)} x^{a-1}(1-x)^{b-1},$$

and $\Gamma(\cdot)$ is the gamma function,

$$\Gamma(x) = \int_0^\infty t^{x-1} e^{-t}\ dt.$$

The objective is to generate random numbers from the mixture beta distribution

$$f(x) = 0.5\ \beta(x; 10, 3) + 0.25\ \beta(x; 3, 15) + 0.25\ \beta(x; 7, 10), \quad x \in (0, 1)$$

We will implement the code in R once again and we will take this opportunity to display some advanced R programming features.

```
## Method of generation for mixture distributions.
random.mixturebeta.v2=function(n)
{GeneratedRN=NULL;
for(i in 1:n)
{U=runif(1);
RandomVal=ifelse(U<0.5,rbeta(1,10,3),ifelse(U<0.75,
                            rbeta(1,3,15),rbeta(1,7,10)))
GeneratedRN=c(GeneratedRN,RandomVal)
}
return(GeneratedRN)}

#And calling the function
random.mixturebeta.v2(100000)
```

The function "ifelse(CONDITION,VALUEIFYES,VALUEIFNO)" is R specific but the code earlier does not take advantage of this built-in function in R, which is working with vectors and large objects. The next function accomplishes the same thing but it is much faster as we shall see.

```
## Method of generation for mixture distributions (opti-
mized
#   code).
random.mixturebeta.v2.optimal=function(n)
{U=runif(n);   GeneratedRN=rep(0,n)
beta1=(U<0.5);beta2=(U>=0.5)&(U<0.75);beta3=(U>=0.75);
n1=GeneratedRN[beta1]; n2=GeneratedRN[beta2];
   3=GeneratedRN[beta3];
GeneratedRN[beta1]=rbeta(n1,10,3);
GeneratedRN[beta2]=rbeta(n2,3,15)
GeneratedRN[beta3]=rbeta(n3,7,10)
return(GeneratedRN)}
```

In the code earlier, the *beta*1, *beta*2, and *beta*3 are vectors containing values TRUE and FALSE depending on whether the respective condition is satisfied. When such vectors (containing TRUE and FALSE values) are applied as indices as in *GeneratedRN[beta*1], they select from the vector *GeneratedRN* only those values which correspond to the TRUE indices. This allows us to operate inside vectors very fast without going through the vectors one by one. Furthermore, the code takes advantage of the internal R method of generating beta distributions which is one of the best and fastest available in any statistical software.

8.9 Generating Random Variables: Rejection Sampling Method

Table 8.11 Average running time in seconds for 30 runs of the three methods.

	Rejection sampling	Mixture gen	Mixture gen (optimal)
Average time (s)	20.122	20.282	0.034
Standard deviation	0.409	0.457	0.008

Table 8.11 provides the running times for the two methods as well as the optimized algorithm mentioned earlier. Each run generates a vector of 100 000 random numbers.

For each number having the desired distribution, the rejection sampling procedure generates a minimum of two uniform random numbers, and since it may reject the numbers produced, the number of uniforms generated may actually be larger. The mixture generating algorithm, on the other hand, always generates one uniform and one beta distributed number. Both algorithms produce numbers one after another until the entire set of 100 000 values are produced. The fact that the times are so close to each other tells us that generating uniforms only (even more than two) may be comparable in speed with generating from a more complex distribution.

Compare the numbers in the second column with the numbers in the third column and recall that it actually is the same algorithm with the only difference being that the optimized version works with vectors. The 100,000 numbers are generated in about one-third of a second, which basically means that simulating a path takes nothing at all if done in this way, and thus long simulations may be made significantly quicker by using a vectorized code.

Rejection sampling works even if one only knows the approximate shape of the target density. As we have seen from the examples, any distribution $g(\cdot)$ may be used, but in all the examples, we have used the uniform distribution for it and this in general is done in practice. There are two reasons for choosing this. One is the fact that generating random uniforms is the most common random number generator as well as the fastest, and secondly, the constant M must be chosen so that $f(x) < Mg(x)$ for all x in support of f. This is generally hard to assess unless one uses the uniform distribution, and the constant M becomes related to the maximum value of $f(\cdot)$ over its support as we have already seen. However, this translates into rejecting a number of generated values, which is *proportional* to the difference in the area under the constant function M and the area under $f(\cdot)$.

As we have seen, this is not so bad for one-dimensional random variables. However, it gets really bad quickly as the dimension increases. The importance sampling method tries to deal with this by sampling more from certain parts of the $f(\cdot)$ density.

8.10 Generating Random Variables: Importance Sampling

Importance sampling is a general technique for estimating properties of a particular distribution while only having samples generated from a different distribution than the distribution of interest.

It is important to realize that, unlike the methods presented thus far, the importance sampling method does not generate random numbers with specific density $f(\cdot)$. Instead, the purpose of the importance sampling method is to estimate expectations. Specifically, suppose that X is a random variable (or vector) with a known density function $f(x)$ and h is some other known function on the domain of the random variable X; then the importance sampling method will help us to estimate

$E[h(X)]$.

If we recall that the probability of some set A may be expressed as an expectation

$P(X \in A) = E[\mathbf{1}_A(X)]$,

then we see that the importance sampling method may be used to calculate any probabilities related to the random variable X as well. For example, the probability of the random variable X tails is

$P(|X| > M) = E[\mathbf{1}_{(-\infty,-M)}(X)] + E[\mathbf{1}_{(M,\infty)}(X)]$,

for some suitable M, and both may be estimated using importance sampling.

We now consider how one can estimate expectations using samples.

The laws of large numbers (either weak or strong) say that if $X_1, \ldots, X_n$ are i.i.d. random variables drawn from the distribution $f(\cdot)$ with a finite mean $E[X_i]$, then the sample mean converges to the theoretical mean $E[X]$ in probability or a.s. Thus if we have a way to draw samples from the distribution $f(\cdot)$, be these $x_1, \ldots, x_n$, then for each function $h(\cdot)$ defined on the codomain of X, we must have:

$$\frac{1}{n} \sum_{i=1}^{n} h(x_i) \to \int h(x)f(x)\,dx = E[h(X)].$$

Therefore, the idea of estimating expectations is to use generated numbers from the distribution $f(\cdot)$. However, in many cases we may not draw from the density f and, instead, use any simple density g. The method is modified using the following observation:

$$E_f[h(X)] = \int h(x)f(x)\,dx = \int h(x)\frac{f(x)}{g(x)}g(x)\,dx,$$

which can be rewritten as

$$E_f[h(X)] = E_g\left[h(X)\frac{f(X)}{g(X)}\right],$$

where we used the notations E_f and E_g to denote expectations with respect to density f and g, respectively. The expression earlier is correct only if the support of g includes the support of f; otherwise, we can have points where $f(x) \neq 0$ and $g(x) = 0$, and thus the ratio $f(x)/g(x)$ becomes undefined.

Combining the approximating idea with the expression earlier, we describe the importance sampling algorithm to estimate $E_f[h(X)]$ s follows:

1) Find a distribution g which is easy to sample from and its support includes the support of f (i.e. if $f(x) = 0$ for some x will necessarily imply $g(x) = 0$).
2) Draw n sampled numbers from the distribution g: $x_1, \ldots, x_n$.
3) Calculate and output the estimate:

$$\frac{1}{n}\sum_{i=1}^{n} h(x_i)\frac{f(x_i)}{g(x_i)} = \sum_{i=1}^{n} h(x_i)\frac{f(x_i)}{ng(x_i)}.$$

The reason why this method is called importance sampling is due to the so-called importance weight $\frac{f(x_i)}{ng(x_i)}$ given to x_i. The ratio $\frac{f(x_i)}{g(x_i)}$ may be interpreted as the number modifying the original weight $1/n$ given to each observation x_i. Specifically, if the two densities are close to each other at x_i, then the ratio $\frac{f(x_i)}{g(x_i)}$ is close to 1 and the overall weight given to x_i is close to the weight $1/n$ (the weight of x_i if we would be able to draw directly from f). Suppose that x_i is in a region of f which is very unlikely. Then the ratio $\frac{f(x_i)}{g(x_i)}$ is going to be close to 0 and thus the weight given to this observation is very low. On the other hand, if x_i is from a region where f is very likely, then the ratio $\frac{f(x_i)}{g(x_i)}$ is going to be large and thus the weight $1/n$ is much increased.

We note the following three observations of the importance sampling methodology.

Firstly, the weights $\frac{f(x_i)}{ng(x_i)}$ may not sum to 1. However, their expected value is 1:

$$E_g\left[\frac{f(X)}{g(X)}\right] = \int \frac{f(x)}{g(x)} g(x)\, dx = \int f(x)\, dx = 1.$$

Thus the sum $\sum_{i=1}^{n} \frac{f(x_i)}{ng(x_i)}$ tends to be close to 1.

Secondly, the estimator

$$\hat{\mu} = \sum_{i=1}^{n} h(X_i)\frac{f(X_i)}{ng(X_i)}.$$

is unbiased and we can calculate its variance. If the average estimate of several random samples is equal to the population parameter then the estimate is said to be unbiased. Thus we have that,

$$E[\hat{\mu}] = E_f[h(X)]$$
$$\text{Var}(\hat{\mu}) = \frac{1}{n}\text{Var}_g\left(h(X)\frac{f(X)}{g(X)}\right). \tag{8.20}$$

Finally, the variance of the estimator obviously depends on the choice of the distribution g. However, we may actually determine the best choice for this distribution. Minimizing the variance of the estimator with respect to the distribution g means minimizing

$$\text{Var}_g\left(h(X)\frac{f(X)}{g(X)}\right) = E_g\left[h^2(X)\left(\frac{f(X)}{g(X)}\right)^2\right] - E_f^2[h(X)].$$

The second term does not depend on g, while using the Jensen inequality [216] in the first term provides

$$E_g\left[\left(h(X)\frac{f(X)}{g(X)}\right)^2\right] \geq \left(E_g\left[|h(X)|\frac{f(X)}{g(X)}\right]\right)^2 = \left(\int |h(x)|f(x)\,dx\right)^2.$$

However, the right side is not a distribution, but it does provide the *optimal importance sampling distribution*:

$$g^*(x) = \frac{|h(x)|f(x)}{\int |h(x)|f(x)\,dx}.$$

This is not really useful from a practical perspective since typically sampling from $f(x)h(x)$ is harder than sampling from $f(x)$. However, it does tell us that the best results are obtained when we sample from $f(x)$ in regions where $|h(x)|f(x)$ is relatively large. As a consequence of this, using the importance sampling is better at calculating $E[h(X)]$ than using a straight Monte Carlo approximation, that is, sampling directly from f and taking a simple average of the $h(x_i)$ values.

In practice, it is important that the estimator has finite variance (otherwise it never improves with n). To see this, we recall the formula for variance in (8.20), i.e.

$$\text{Var}(\hat{\mu}) = \frac{1}{n}\text{Var}_g\left(h(X)\frac{f(X)}{g(X)}\right).$$

Below are the sufficient conditions for the finite variance of the estimator $\hat{\mu}$:

- There exists some M such that $f(x) < Mg(x)$ for all x and $\text{Var}_f(h(X)) < \infty$.
- The support of f is compact, f is bounded above, and g is bounded below on the support of f.

Remark 8.10.1 Choosing the distribution g is crucial. For example, if f has support on $\mathbb{R}$ and has heavier tails than g, the weights $w(X_i) = f(X_i)/g(X_i)$ will have infinite variance and the estimator will fail.

Next we present an application of importance sampling methodology.

Example 8.10.1 In this example, we will showcase the importance of the choice of distribution g. Suppose we want to estimate $E[|X|]$ where X is distributed as a Student-t random variable with three degrees of freedom. We select $h(x) = |x|$ and $f(x)$ is the t-density function given by:

$$\frac{\Gamma\left(\frac{v+1}{2}\right)}{\sqrt{v\pi}\,\Gamma\left(\frac{v}{2}\right)}\left(1+\frac{x^2}{v}\right)^{-\frac{v+1}{2}},$$

with degrees of freedom $v = 3$ where $\Gamma(x)$ denotes the gamma function used earlier in this chapter.

First, we note that the target density does not have compact support; thus the use of a uniform density for g is not possible. To exemplify the practical aspects of the importance sampling algorithm, we shall use two candidate densities: $g_1(x)$ the density of a t distribution with one degree of freedom and $g_2(x)$ the standard normal density ($N(0,1)$). We shall also use a straight Monte Carlo where we generate directly from the distribution f. The plot of these densities may be observed in Figure 8.3.

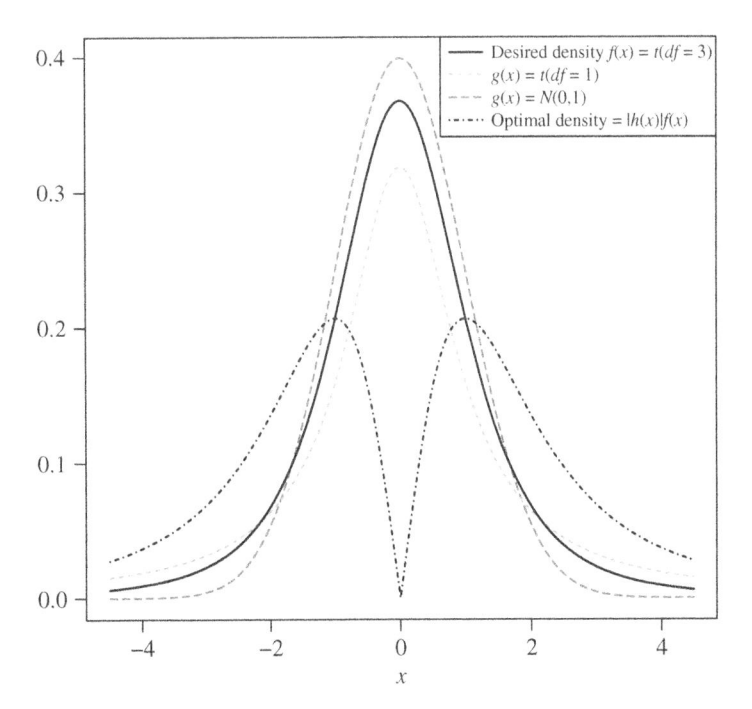

Figure 8.3 Candidate densities for the importance sampling procedure as well as the target density.

We know that the optimal choice is $|h(x)|f(x)$ (plotted in blue in the figure); however, generating from this density is very complex. We may also observe that while the t density with one degree of freedom (green) dominates the tails of f, the normal density (red) does not, so we expect the estimator produced using the normal density to be not very good (the weights $f(X_i)/g(X_i)$ have infinite variance).

Next we present the R-code used for the importance sampling technique.

```
#Straight Monte Carlo:
n=10:1500
nsim=100

straightMC=NULL;
for(i in n)
{mu=NULL;
for(j in 1:nsim)
{a=rt(i,3);mu=c(mu,mean(abs(a)))}
straightMC=cbind(straightMC,c(i,mean(mu),sd(mu)))
}

#Importance Sampling using first candidate:

usingt1=NULL;
for(i in n)
{mu=NULL;
for(j in 1:nsim)
{a=rt(i,1);mu=c(mu,mean(abs(a)*dt(a,3)/dt(a,1)))}
usingt1=cbind(usingt1,c(i,mean(mu),sd(mu)))
}

#Importance Sampling using second candidate:

usingnorm=NULL;
for(i in n)
{mu=NULL;
for(j in 1:nsim)
{a=rnorm(i);mu=c(mu,mean(abs(a)*dt(a,3)/dnorm(a)))}
usingnorm=cbind(usingnorm,c(i,mean(mu),sd(mu)))
}
```

8.10.1 Sampling Importance Resampling

As we mentioned, the importance sampling technique is primarily used to calculate expectations. However, it is possible to adapt the technique to obtain samples from the distribution f. To do this, recall that

$$P(X \in A) = E_f[\mathbf{1}_A(X)].$$

Thus, following the procedure already detailed, one approximates the probability with

$$P(X \in A) = \frac{1}{N} \sum_{x_i} 1_A(x_i) \frac{f(x_i)}{g(x_i)}$$

Thus, only the generated random numbers that fall into the set A are actually counted.

One may take this argument further and obtain an approximate discrete distribution for f:

$$\hat{p}(x) = \sum_{i=1}^{N} \frac{f(x_i)}{g(x_i)} 1_{\{x_i\}}(x) = \sum_{i=1}^{N} w(x_i) 1_{\{x_i\}}(x)$$

where the x_i values are generated from the density g and obviously the corresponding weights $w(x_i) = \frac{f(x_i)}{g(x_i)}$ may not sum to 1. To generate M independent random variables from the distribution f, one normalizes this discrete distribution $\hat{p}$ and just generates from it M values where M is much larger than N. Once the M new values are obtained, denoted by say $\{y_1, \ldots, y_M\}$, the new estimated f distribution is given by:

$$\tilde{p}(x) = \frac{1}{M} \sum_{i=1}^{M} 1_{\{y_i\}}(x).$$

Note that, since the values y_i are generated from the x_i values, the new sample $\{y_1, y_2, \ldots y_M\}$ contains repeated observations. This method is known as *sampling importance resampling* and is as a result of the work by [173].

8.10.2 Adaptive Importance Sampling

As we have shown in Example 8.10.1, the proper choice of g can lead to efficient estimation algorithms. Specifically, using a g distribution which is close to $|h(x)|f(x)$ is much more efficient than using a straight Monte Carlo method (i.e. with $g(x) = f(x)$). However, as the dimension of x increases (x becomes a random vector with many dimensions), it becomes more complicated to obtain a suitable $g(x)$ from which to draw the samples. A strategy to deal with this problem is the *adaptive importance sampling* technique [30].

The method considers a parameterized distribution $g(x|\theta)$, where θ is a parameter vector which is adaptable depending on the sampling results. The idea of the method is to minimize the variance of the estimator $\hat{\mu}$. Specifically, consider a parameterized distribution $g(x|\theta)$. We want to minimize

$$E_g[f^2(X)w^2(X)] - E_f^2[h(X)] \tag{8.21}$$

with respect to g, where $w(x|\theta) = \frac{f(x)}{g(x|\theta)}$. Note that the second term does not depend on g and minimizing (8.21) involves calculating the derivative of the first term with respect to θ. Since derivative and expectation commute on a probability space if the expectation exists (i.e. we can interchange the differentiation and expectation operators) the problem reduces to finding the roots of the derivative

$$D(\theta) = 2E_g \left[f^2(x) w(x|\theta) \frac{\partial w}{\partial \theta}(x|\theta) \right].$$

where $D(\theta)$ was obtained by differentiating (8.21) with respect to θ. A Newton–Raphson iteration procedure would find the minimum of the original expression by using

$$\theta_{n+1} = \theta_n - \left(\nabla D(\theta_n) \right)^{-1} D(\theta_n).$$

However, the expectation $D(\theta_n)$ and the inverse of the gradient of D (or the Hessian) of the original expression to be minimized is very complex to calculate. Therefore, the algorithm simply replaces the expression $(\nabla D(\theta_n))^{-1}$ with a learning constant and the $D(\theta_n)$ with its sample value.

A pseudo-code is presented in the following. The technique starts with a general distribution $g(x|\theta)$ capable of many shapes as the θ parameter varies in some parameter space Θ. Then, the method *adapts* the parameter θ to fit the problem at hand. This is done as follows:

- Start with an initial parameter value θ_0 (which produces some indifferent shape of the distribution g), and let $n = 0$.
- Do the following until the difference $|\theta_{n+1} - \theta_n| < \varepsilon$, where ε is a pre-specified tolerance level.
  - Generate N values $x_1, \ldots, x_N$ from the distribution $g(x|\theta_n)$.
  - Update the θ value using

$$\theta_{n+1} = \theta_n - \alpha \frac{1}{N} \sum_{i=1}^{N} f^2(x_i) w(x|\theta_n) \frac{\partial w}{\partial \theta_n}(x_i|\theta_n).$$

- Check the condition $|\theta_{n+1} - \theta_n| < \varepsilon$. If not satisfied, let $n = n + 1$ and repeat the loop.

8.11 Problems

1. Given the Irwin–Hall distribution with pdf,

$$f_Y(y) = \frac{1}{2(n-1)!} \sum_{k=0}^{n} (-1)^k \binom{n}{k} (y-k)^{n-1} \operatorname{sign}(y-k),$$

where the sign function $\operatorname{sign}(x)$ is defined on $(-\infty, 0) \cup (0, \infty)$ as

$$\operatorname{sign}(x) = \frac{|x|}{x} = \begin{cases} 1 & \text{if } x > 0 \\ -1 & \text{if } x < 0 \end{cases}$$

verify that the Irwin–Hall distribution has mean $n/2$ and variance $n/12$.

2. A call option pays an amount $V(S) = \frac{1}{1+\exp(S(T)K)}$ at time T for some predetermined price K. Discuss what you would use for a control variate and generate a simulation to determine how it performs, assuming geometric Brownian motion for the stock price, interest rate of 6%, annual volatility 25%, and various initial stock prices and values of K and T.

3. Suppose we want to value a Bermudan Put option where exercise is possible now and at three future dates, i.e at times $t = 1, 2$, and 3, where time $t = 3$ is the expiration date of the option. $S_0 = 1, X = 1.2, r = 0.09$. Using the same approach used in Section 8.5.1, apply the least squares Monte Carlo to compute the prices of the Bermudan Put option.

4. Compare and contrast the sequential Monte Carlo method and the Markov Chain Monte Carlo method. Give applications of each.

5. Let $f(x) = \frac{4}{1+x^2}$. Use the Monte Carlo method to calculate approximations to the integral $\int_0^1 \frac{4}{1+x^2} dx$.

6. Let $f(x) = \sqrt{x + \sqrt{x}}$. Use the Monte Carlo method to calculate approximations to the integral $\int_0^1 \sqrt{x + \sqrt{x}} dx$.

7. The following Matlab script calculates the value of π using the Monte Carlo simulation:

```
n = 2000000;              # number of Monte Carlo samples
x = rand(n, 1);           # sample the input random variable x
y = rand(n, 1);           # sample the input random variable y
R = (x.2 + y.2 < 1);      # is the point inside a unit cir-
cle?
percentage = sum(R)/n;    # compute statistics: the inside
                            percentage
pi_estimate = percentage * 4
```

Run the previous script for the following cases:
(a) $n = 200$
(b) $n = 20000$
(c) $n = 2000000$
How accurate is the estimated π for each case?

8. Using the Box–Muller method and the two resulting variables X and Y. Calculate the joint and marginal distributions of these variables and show that they are independent.

9. Show that the two variables given by the polar rejection algorithm are independent.

10. Consider the following normal mixture density:

$$f(x) = 0.7 \frac{1}{\sqrt{2\pi 9}} e^{-\frac{(x-2)^2}{18}} + 0.3 \frac{1}{\sqrt{2\pi 4}} e^{-\frac{(x+1)^2}{8}}$$

(a) Calculate the expected value of a random variable with this distribution.
(b) Write code and implement it to generate random variables with this distribution.
(c) Use the previous part to generate 100 independent random numbers with this density. Calculate the sample mean (average of generated numbers). Compare with answer in part (a).
(d) Repeat the previous part, but this time generate 1000 numbers. Comment.

9

Stochastic Differential Equations

9.1 Introduction

In Section 1.8, we briefly defined a stochastic differential equation (SDE) and presented some examples. In this chapter, we will discuss a more intuitive way to arise to an SDE. We will begin by discussing the construction of a stochastic integral and an SDE and we will present some properties. Methods for approximating stochastic differential equation (SDEs) will also be discussed. This chapter is very crucial for many applications of stochastic processes to finance. SDEs have been briefly discussed in Chapter 1 of this textbook. However in order to make this chapter self contained, we will repeat some of the concepts earlier introduced and then discuss other concepts that will be used throughout this book.

We recall in Chapter 1 of this book that an SDE is a deterministic differential equation which is perturbed by random noise. Suppose we define an equation of the form:

$$\frac{dx(t)}{dt} = b(t, x(t)) + \sigma(t, x(t)) \times \text{``noise.''}$$

The term "noise" needs to be properly defined. Assuming we define a stochastic process w_t such that "noise"= w_t. Then we have:

$$dx(t) = b(t, x(t))dt + \sigma(t, x(t))w_t dt. \tag{9.1}$$

This process was named the "white noise" process, and from practical applications the following properties were postulated as being necessary for this process:

1) For any times $t_1 \neq t_2$, the random variables w_{t_1} and w_{t_2} are independent.
2) w_t has a stationary distribution.
3) $E[w_t] = 0$.
4) w_t has continuous paths.

Quantitative Finance, First Edition. Maria C. Mariani and Ionut Florescu.
© 2020 John Wiley & Sons, Inc. Published 2020 by John Wiley & Sons, Inc.

Norbert Wiener first introduced a form of this noise term for the case when $\sigma(t, x) = \sigma(t)$ was just a function of time (1933). Later, in 1944, Kiyoshi Itô handled the case when $\sigma(t, x)$ was a general function, and this led to what we now call the *stochastic integral* or the *Itô integral* and Itô's lemma previously mentioned in Chapter 2. This introduction was revolutionary; it led to another way of thinking about stochastic processes, and in fact it changed completely one of the fundamental calculus formulas, namely, the integration by parts formula. This led to an entirely different type of calculus which is now called *stochastic calculus*.

The main idea is to look at the associated integral form of the ODE. Thus, instead of trying to construct a white noise process, we shall try to construct the integral with respect to this white noise process.

9.2 The Construction of the Stochastic Integral

We will start this section by continuing the motivating example shown earlier. We consider a discrete version of Eq. (9.1) for times t_i in a partition $\pi_n = (0 = t_0 < t_1 < \cdots < t_n = t)$. We have for $i \in \{1, 2, \ldots, n\}$

$$X_{t_{i+1}} - X_{t_i} = b(t_i, X_{t_i})\Delta t_i + \sigma(t_i, X_{t_i}) w_{t_i} \Delta t_i, \tag{9.2}$$

where $\Delta t_i = t_i - t_{i-1}$.

The main idea is to replace the term $w_{t_i} \Delta t_i$ with the increment of a stochastic process $\Delta B_i = \Delta B_{t_i} - \Delta B_{t_{i-1}}$. Again, the requirements on the increments of the process are the same as before: independent, stationary, and has zero mean and continuous paths for the process. But we already know a stochastic process whose paths have these properties. That is the Brownian motion we introduced in Chapter 1. In fact, there are several processes with these properties. Within this chapter, we only consider the Brownian motion. If we take B_t to be a standard Brownian motion and sum the increments for all i in (9.2), we obtain

$$X_{t_n} - X_0 = X_t - X_0 = \sum_{i=1}^{n} b(t_i, X_{t_i})\Delta t_i + \sum_{i=1}^{n} \sigma(t_i, X_{t_i})\Delta B_i \tag{9.3}$$

As we make $n \to \infty$ and consequently the partition norm $\|\pi_n\| = \max_i (t_i - t_{i-1}) \to 0$, the first sum in (9.3) is a usual Riemann sum and it goes to a regular integral almost surely (in fact always for every path ω):

$$\int_0^t b(t, X_t)dt.$$

However, for the second one we do not know whether the limit exists or even how to define it due to the increment of the stochastic process ΔB_i. This is why we need to construct and define the concept of the stochastic integral.

9.2 The Construction of the Stochastic Integral

Consider a function of the type:

$$f(t,\omega) = \sum_{i=1}^{n} e_i(\omega) \mathbf{1}_{(t_{i-1}, t_i]}(t).$$

where the e_i's are random variables. Before we present an example, the goal is to define:

$$\int_s^t f(t,\omega) dB_t(\omega) = e_i(\omega)(B_{t_m}(\omega) - B_s(\omega)) + \sum_{i=m+1}^{n} e_i(\omega)(B_{t_i}(\omega) - B_{t_{i-1}}(\omega)),$$

for some m such that $t_{m-1} < s < t_m < \cdots < t_n = t$.

Note that we will work on the interval $[0, t]$ and will define the integral for this interval. We could work on more general intervals $[s, t]$, but the construction is identical, with changes related to s.

Remark 9.2.1 The way the e_j's are defined is very important as the next example shows.

Example 9.2.1 In the Riemann integral, the e_j's have to be such that they are evaluated at some point in the interval $[t_{j-1}, t_j]$, that is, $e_j(\omega) = e(\xi_j, \omega)$, with ξ_j anywhere in the interval $[t_{j-1}, t_j]$. The equivalent concept in the theory of stochastic processes is measurability. In the stochastic integral definition, the sigma algebra is crucial. We proceed by defining a simple function. A simple function is a finite linear combination of indicator functions of measurable sets.

Consider two simple functions as before:

$$f_1(t,\omega) = \sum_{i=0}^{n} B_{t_{i-1}}(\omega) \mathbf{1}_{(t_{i-1}, t_i]}(t),$$

and

$$f_2(t,\omega) = \sum_{i=0}^{n} B_{t_i}(\omega) \mathbf{1}_{(t_{i-1}, t_i]}(t)$$

where $\mathbf{1}_{(t_{i-1}, t_i]}(t)$ is the indicator function.

Note that the only thing that has changed is the time chosen in each interval for the Brownian term. In particular, for f_1, the term e_i is measurable with respect to $\mathcal{F}_{t_{i-1}}$, that is, the left point on the interval. In f_2, the term e_i is measurable with respect to $\mathcal{F}_{t_i}$. This makes a huge difference. The integrals as defined earlier are (we drop the omega from the notation)

$$\int_0^t f_1(t) dB_t = \sum_{i=1}^{n} B_{t_{i-1}} (B_{t_i} - B_{t_{i-1}}),$$

and

$$\int_0^t f_2(t) dB_t = \sum_{i=1}^{n} B_{t_i} (B_{t_i} - B_{t_{i-1}}).$$

For the limit to be the same, their expected values must be identical. However, we have:

$$E\left[\int_0^t f_1(t)dB_t\right] = \sum_{i=1}^n E[B_{t_{i-1}}(B_{t_i} - B_{t_{i-1}})]$$

$$= \sum_{i=1}^n E[B_{t_{i-1}}]E[(B_{t_i} - B_{t_{i-1}})] = 0$$

since $B_0 = 0$, $B_{t_{i-1}} = B_{t_{i-1}} - B_0$, and $B_{t_i} - B_{t_{i-1}}$ are increments which are independent. On the other hand

$$E\left[\int_0^t f_2(t)dB_t\right] = \sum_{i=1}^n E[B_{t_i}(B_{t_i} - B_{t_{i-1}})]$$

$$= \sum_{i=1}^n E[(B_{t_{i-1}} + B_{t_i} - B_{t_{i-1}})(B_{t_i} - B_{t_{i-1}})]$$

$$= \sum_{i=1}^n E[B_{t_{i-1}}(B_{t_i} - B_{t_{i-1}})] + \sum_{i=1}^n E[(B_{t_i} - B_{t_{i-1}})^2]$$

$$= 0 + \sum_{i=1}^n (t_i - t_{i-1}) = t$$

So, clearly the nature of the function to be integrated makes a big difference.

In general, if one chooses a $t_i^* \in [t_{i-1}, t_i]$, the following are the two resulting integrals:

- For $t_i^* = t_{i-1}$, we obtain the Itô integral (covered in this chapter).
- For $t_i^* = \frac{t_{i-1}+t_i}{2}$, we obtain the Fisk–Stratonovich integral. Regular calculus rules apply to this integral. For details about this, integral see [159].

Given any probability space, let $\mathscr{F}_t$ denote the filtration generated by the Brownian motion. Recall that a stochastic process X_t is adapted with respect to a filtration $\mathscr{F}_t$ if and only if X_t is $\mathscr{F}_t$ measurable for all t.

The integrand of a stochastic integral $\int_0^t X_t dB_t$ for any stochastic process X_t satisfies the following conditions:

- The function $(t, \omega) \to X_t(\omega)$ is measurable with respect to the combined space sigma algebra $\mathscr{B}([0, \infty)) \times \mathscr{F}$
- The process X_t is $\mathscr{F}_t$ adapted
- $E[\int_0^T X_t(\omega)^2 dt] < \infty$, for $T > 0$.

Note that the first property is normal and is in fact satisfied by pretty much everything. The second is needed to properly define the integral as mentioned earlier. The third property simplifies the mathematical derivations by allowing us to concentrate on limits in L^2 (space of square integrable functions).

9.2.1 Itô Integral Construction

We will denote with v_T the class of integrands for which we can define a stochastic integral. Suppose that $f(t) \in v_T$; specifically

$$f(t, \omega) = \sum_{i=1}^{n} e_{t_{i-1}}(\omega) \mathbf{1}_{(t_{i-1}, t_i]}(t), \tag{9.4}$$

where we used the index to signify that $e_{t_{i-1}}$ is $F_{t_{i-1}}$ measurable and $\pi_n = (0 = t_0 < t_1 < \cdots < t_n = T)$. For this type of simple functions, we define

$$\int_0^T f(t) dB_t = \sum_{i=1}^{n} e_{t_{i-1}} (B_{t_i} - B_{t_{i-1}}).$$

Lemma 9.2.1 (Itô isometry for simple functions). If f is a simple function in v_T and bounded, then

$$E\left[\left(\int_0^T f(t) dB_t\right)^2\right] = E\left[\int_0^T f(t)^2 dt\right] = \int_0^T E[f(t)^2] dt.$$

Proof. Let $\Delta B_i = B_{t_i} - B_{t_{i-1}}$. Then

$$E[e_{t_{i-1}} e_{t_{j-1}} \Delta B_i \Delta B_j] = \begin{cases} 0, & \text{if } i \neq j \\ E[e_{t_{i-1}}^2](t_i - t_{i-1}), & \text{if } i = j \end{cases}$$

The expression is easy to prove since $e_{t_{i-1}}$ is $\mathcal{F}_{t_{i-1}}$ measurable and ΔB_i is independent of $\mathcal{F}_{t_{i-1}}$. Using this result, we have

$$E\left[\left(\int_0^T f(t) dB_t\right)^2\right] = E\left[\left(\sum_i e_{t_{i-1}} \Delta B_i\right)\left(\sum_j e_{t_{j-1}} \Delta B_j\right)\right]$$

$$= \sum_{i,j} E\left(e_{t_{i-1}} e_{t_{j-1}} \Delta B_i \Delta B_j\right)$$

$$= \sum_i E[e_{t_{i-1}}^2](t_i - t_{i-1})$$

$$= \int_0^T E[f(t)^2] dt \qquad \square$$

For the rest of the construction, we will state the steps. For the full proof of the steps, please see [158] for more details.

Step 1. Let $f \in v_T$ and be bounded such that all the paths are continuous ($f(., \omega)$ is continuous for all ω). Then there exist simple functions f_n of the form (9.4) such that

$$E\left[\int_0^T (f - f_n)^2 dt\right] \to 0$$

as $n \to \infty$. Step 1 says that any bounded, adapted process with continuous paths may be approximated with these simple functions.

Step 2. Let $g \in v_T$ and bounded. Then there exist bounded functions in v_T denoted g_n with continuous paths and

$$E\left[\int_0^T (g-g_n)^2 dt\right] \to 0$$

as $n \to \infty$. Step 2 is a slight generalization. It says that the bounded (not necessarily continuous) integrands may be approximated with bounded continuous functions and thus with simple functions from Step 1. Again, the approximation is in L^2.

Step 3. Let $h \in v_T$. Then there exist bounded functions in v_T denoted by h_n such that

$$E\left[\int_0^T (h-h_n)^2 dt\right] \to 0$$

as $n \to \infty$.

Any integrand in v_T can be approximated with simple functions by steps 1, 2, and 3. Because of these three steps, we can state the following definition:

Definition 9.2.1 (Itô integral). Let $f \in v_T$. Then the stochastic integral or Itô integral of f with respect to the Brownian motion B_t is defined as

$$\int_0^T f(t,\omega) dB_t(\omega) = \lim_{n\to\infty} \int_0^T f_n(t,\omega) dB_t(\omega),$$

where f_n's form a sequence of simple functions as in (9.4) such that

$$E\left[\int_0^T (f-f_n)^2 dt\right] \to 0, \quad \text{as } n \to \infty.$$

By steps 1, 2 and 3 shown earlier, such a sequence exists. It therefore remains to show that the limit is unique regardless of the approximating sequence, but for this, one needs to go to the proof of the steps. We refer to [158] once again.

Remark 9.2.2 The approximation sequence is not relevant when we need to approximate an integral. For example, given

$$\int_0^T \cos(B_t^2) X_t dB_t,$$

consider a partition of $[0, t]$: $\pi_n = (0 = t_0 < t_1 < t_2 < \cdots < t_n = t)$ then we have that

$$\sum_{i=1}^n \cos\left(B_{t_{i-1}}^2\right) X_{t_{i-1}} (B_{t_i} - B_{t_{i-1}}).$$

The only requirement is that the integrand is $\mathcal{F}_{t_{i-1}}$ measurable and is approximate with the given function in L^2. Note that the expression is the stochastic integral corresponding to the integrand

$$f(t) = \sum_{i=1}^{n} \cos\left(B_{t_{i-1}}^2\right) X_{t_{i-1}} \mathbf{1}_{(t_{i-1}, t_i]}(t),$$

which is in v_T and of the form (9.4).

Corollary 9.2.1 (Itô isometry). Let f be adapted and measurable. Then

$$E\left[\left(\int_0^T f(t)dB_t\right)^2\right] = \int_0^T E[f(t)^2]dt$$

$\square$

This is easily proven by using Lemma 9.2.1 and taking L^2 limits.

9.2.2 An Illustrative Example

We have seen before that the stochastic integral really depends on the construction. In this example, we will point out that the stochastic integral is really different from a regular integral.

Suppose we have a deterministic function; let us call it $f(t)$ and suppose we want to calculate the integral

$$\int_0^T f(t)df(t).$$

For a partition π_n, we have

$$\lim_{||\pi|| \to 0} \sum_{i=1}^{n} f(\xi_i)(f(t_i) - f(t_{i-1})) = \int_0^T f(t)df(t)$$

for any ξ_i in the interval $[t_{i-1}, t_i]$. For more details, we refer the reader to [172, 207].

Since we can write the integral as the limit using any point in the interval, we can calculate

$$\sum_{i=1}^{n} f(t_i)(f(t_i)) - f(t_{i-1}) = \sum_{i=1}^{n} \left(f(t_i)^2 - f(t_i)f(t_{i-1}) - f(t_{i-1})^2 + f(t_{i-1})^2\right)$$

$$= \sum_{i=1}^{n} \left(f(t_i)^2 - f(t_{i-1})^2\right) - \sum_{i=1}^{n} f(t_{i-1})(f(t_i) - f(t_{i-1}))$$

$$= f(T)^2 - f(0)^2 - \sum_{i=1}^{n} f(t_{i-1})(f(t_i) - f(t_{i-1})) \quad (9.5)$$

Since both end sums go to the same integral by taking $||\pi_n|| \to 0$, we obtain

$$\int_0^T f(t)df(t) = f(T)^2 - f(0)^2 - \int_0^T f(t)df(t)$$

which implies that

$$2\int_0^T f(t)df(t) = f(T)^2 - f(0)^2 \quad (9.6)$$

Therefore, in the case when $f(t)$ is deterministic, from Eq. (9.6) we have:

$$\int_0^T f(t)df(t) = \frac{f(T)^2 - f(0)^2}{2}. \quad (9.7)$$

If $f(0) = 0$ in (9.6), we will have:

$$\int_0^T f(t)df(t) = \frac{f(T)^2}{2}.$$

Now consider the case of the stochastic integral $\int_0^T B_t dB_t$, where B_t is a Brownian motion. Remembering the definition, we cannot use the argument mentioned earlier since the sum we considered in (9.5) will not converge to the Itô integral. Instead we use,

$$\sum_{i=1}^n B_{t_{i-1}}(B_{t_i} - B_{t_{i-1}}) = \sum_{i=1}^n \left(B_{t_{i-1}} B_{t_i} - B_{t_{i-1}}^2 + \frac{1}{2}B_{t_i}^2 - \frac{1}{2}B_{t_i}^2 \right)$$

which can be rewritten as

$$\sum_{i=1}^n B_{t_{i-1}}(B_{t_i} - B_{t_{i-1}}) = \sum_{i=1}^n \left(B_{t_{i-1}} B_{t_i} - \frac{1}{2}B_{t_i}^2 - \frac{1}{2}B_{t_{i-1}}^2 \right) + \sum_{i=1}^n \left(\frac{1}{2}B_{t_i}^2 - \frac{1}{2}B_{t_{i-1}}^2 \right)$$

$$= -\sum_{i=1}^n \frac{1}{2}\left(B_{t_i} - B_{t_{i-1}} \right)^2 + \frac{1}{2}\sum_{i=1}^n \left(B_{t_i}^2 - B_{t_{i-1}}^2 \right)$$

$$= \frac{1}{2}\sum_{i=1}^n \left(B_{t_i} - B_{t_{i-1}} \right)^2 - \frac{1}{2}B_T^2$$

Notice now that the remaining sum in the expression converges to the quadratic variation of the Brownian motion which is equal to T. Therefore, taking the limit in L^2 as $||\pi_n|| \to 0$, we obtain

$$\int_0^T B_t dB_t = \frac{1}{2}(B_T^2 - T) \quad (9.8)$$

since $\sum_{i=1}^n \left(B_{t_i} - B_{t_{i-1}} \right)^2 \to T$ in L^2 as $t_i - t_{i-1} \to 0$. The extra term $-\frac{1}{2}T$ shows that the Itô stochastic integral does not behave like ordinary integrals. As we can easily see, the stochastic integral is really a different notion from the regular integral.

9.3 Properties of the Stochastic Integral

First, we would like to recall the Wiener integral which was developed by Norbert Wiener and then preceded the Itô integral and Itô calculus. The Wiener integral is obtained when the integrand of the stochastic integral is a deterministic function, that is,

$$\int_0^T f(t) dB_t$$

where f is deterministic. For this integral, all the regular calculus rules apply. For example, the integration by parts formula is

$$\int_0^T f(t) dB_t = f(T) B_T - \int_0^T B_t df(t),$$

where the latter is a Riemann–Stieltjes integral. Furthermore, this is a random variable which is normally distributed with mean 0 and variance $\int_0^T f^2(t) dt$.

Theorem 9.3.1 (Properties of the Itô integral). Let X_t, Y_t be stochastic processes in v_T, and let $\mathcal{F}_t$ denote the standard filtration generated by B_t. We have:

1) $\int_0^T X_t dB_t = \int_0^S X_t dB_t + \int_S^T X_t dB_t$, for all $0 \le S \le T$;
2) $\int_0^T (aX_t + bY_t) dB_t = a \int_0^T X_t dB_t + b \int_0^T Y_t dB_t$, for all $a, b \in \mathbb{R}$.
3) $\int_0^t X_t dB_t$ is $\mathcal{F}_T$ measurable.
4) $E\left[\int_0^T X_t dB_t\right] = 0$.
5) $E\left[\left(\int_0^T X_t dB_t\right)^2\right] = \int_0^T E\left[X_t^2\right] dt$, (Itô isometry).

Please refer to [158] for the proof of the theorem. All these properties are proven first for simple functions and then going to the L^2 limit in the general stochastic integrands in v_T.

Lemma 9.3.1 (Quadratic variation of the stochastic integral). If $X_t \in v_T$ and we denote

$$I_t = \int_0^t X_s dB_s$$

the Itô integral of X_t, then the quadratic variation of the process I_t is

$$[I, I]_t = \int_0^T X_t^2 dt.$$

Theorem 9.3.2 (Continuity of the stochastic integral). For $X_t \in v_T$, we denote

$$I_t = \int_0^t X_s dB_s$$

Then the Itô integral mentioned earlier has a continuous version J_t such that

$$\mathbf{P}\left(J_t = \int_0^t X_s dB_s\right) = 1, \quad \text{for all } t, \text{with } 0 \le t \le T.$$

For the proof of these two results, we refer to [[158], chapter 3]. The last theorem is especially important. It says that there always exists a version of the stochastic integral which is continuous a.s. regardless of whether the integrand has jumps (discontinuous) or is not well behaved. From now on, we will always assume that the Itô integral means a continuous version of the integral.

Proposition 9.3.1 The stochastic integral $\int_0^t X_t dB_t$ is an $\mathscr{F}_T$-martingale. Furthermore, we have

$$\mathbf{P}\left(\sup_{0 \le s \le T}\left|\int_0^t X_s dB_s\right| \ge \lambda\right) \le \frac{1}{\lambda^2} E\left[\int_0^T X_t^2 dt\right].$$

Proof. The proof is very simple. One needs to use the previous theorem, Doob's martingale inequality [158], and Itô isometry. □

9.4 Itô Lemma

In this section, we briefly revisit the Itô lemma which was presented in Section 2.6.1 of this book. Here, we will discuss several of its applications and generalizations. The Itô lemma is an important tool of stochastic calculus. However, before we reintroduce the lemma for the completeness of this chapter, let us introduce the type of processes that will be the object of the lemma.

Definition 9.4.1 Let B_t be a standard one-dimensional Brownian motion on $(\Omega, \mathscr{F}, \mathbf{P})$. Then the process

$$X_t(\omega) = X_0(\omega) + \int_0^t \mu(s, \omega) ds + \int_0^t \sigma(s, \omega) dB_s(\omega) \tag{9.9}$$

is called an Itô process. The processes μ and σ need to be adapted to the filtration generated by the Brownian motion with the condition that

$$\mathbf{P}\left(\int_0^t |\mu(s, \omega)| ds < \infty \text{ for all } t > 0\right) = 1.$$

Note that the functions $\mu, \sigma \in v(0, \infty)$ are random in general. Sometimes in order to simplify notation, the equation above is written as:

$$dX_t = \mu dt + \sigma dB_t.$$

This latter equation has no meaning other than providing a symbolic notation for Equation (9.9). We shall use this latter notation at all times, meaning an equation of the type (9.9).

Lemma 9.4.1 (The one-dimensional Itô formula). Let X be an Itô process as in Equation (9.9). Suppose that $f(t,x)$ is a function defined on $[0,\infty) \times \mathbb{R}$, twice differentiable in x, and one time differentiable in t. Symbolically, $f \in \mathscr{C}^{1,2}([0,\infty) \times \mathbb{R})$.

Then, the process $Y_t = f(t, X_t)$ is also an Itô process and

$$df(t, X_t) = \frac{\partial f}{\partial t}(t, X_t)dt + \frac{\partial f}{\partial x}(t, X_t)dX_t + \frac{1}{2}\frac{\partial^2 f}{\partial x^2}(t, X_t)d[X, X]_t,$$

where $[X, X]_t$ is the quadratic variation of X_t. To simplify the calculation, we rewrite the formula as

$$dY_t = \frac{\partial f}{\partial t}dt + \frac{\partial f}{\partial x}dX_t + \frac{1}{2}\frac{\partial^2 f}{\partial x^2}(dX_t)^2.$$

The proof of Lemma 9.4.1 was presented in Section 2.6.1 of this book. Next we concentrate on the interpretation and reaching SDEs.

Example 9.4.1 Recall the derivation in (9.8). Here is the same result using a much shorter derivation. Suppose $X_t = B_t$ (i.e. $dX_t = dB_t$) which is an Itô process. Let the function $f(t,x) = \frac{1}{2}x^2$. Then $Y_t = \frac{1}{2}B_t^2$. We can calculate easily $\frac{\partial f}{\partial t} = 0$, $\frac{\partial f}{\partial x} = x$, and $\frac{\partial^2 f}{\partial x^2} = 1$. We apply Itô formula to this function:

$$dY_t = \frac{\partial f}{\partial t}dt + \frac{\partial f}{\partial x}dX_t + \frac{1}{2}\frac{\partial^2 f}{\partial x^2}(dX_t)^2$$

$$d\left(\frac{1}{2}B_t^2\right) = B_t dB_t + \frac{1}{2}(dB_t)^2 = B_t dB_t + \frac{1}{2}dt,$$

or in the proper notation:

$$\frac{1}{2}B_t^2 - \frac{1}{2}B_0^2 = \int_0^t B_s dB_s + \frac{1}{2}t$$

Rewriting this expression and using that $B_0 = 0$, we obtain the integral of the Brownian motion with respect to itself:

$$\int_0^t B_s dB_s = \frac{1}{2}\left(B_t^2 - t\right)$$

Lemma 9.4.2 (The general Itô formula). Suppose $\mathbf{B}_t = (B_1(t), \ldots, B_d(t))$ is a d-dimensional Brownian motion. Recall that each component is a standard Brownian motion. Let $\mathbf{X} = (X_1(t), X_2(t), \ldots, X_d(t))$ be an n-dimensional Itô process, that is,

$$dX_i(t) = \mu_i dt + \sigma_{i1} dB_1(t) + \sigma_{i2} dB_2(t) + \cdots + \sigma_{id} dB_d(t)$$

9 Stochastic Differential Equations

for all i from 1 to n. This expression can be represented in matrix form as

$$d\mathbf{X}_t = Udt + \sum d\mathbf{B}_t,$$

where

$$U = \begin{pmatrix} \mu_1 \\ \mu_2 \\ \vdots \\ \mu_n \end{pmatrix} \text{ and } \sum = \begin{pmatrix} \sigma_{11} & \sigma_{12} & \cdots & \sigma_{1d} \\ \sigma_{21} & \sigma_{22} & \cdots & \sigma_{2d} \\ \vdots \\ \sigma_{n1} & \sigma_{n2} & \cdots & \sigma_{nd} \end{pmatrix}$$

Suppose that $f(t, x) = (f_1(t, \mathbf{x}), f_2(t, \mathbf{x}), \ldots, f_m(t, \mathbf{x}))$ is a function defined on $[0, \infty) \times \mathbb{R}^n$ with values in $\mathbb{R}^m$ with $f \in \mathscr{C}^{1,2}([0, \infty) \times \mathbb{R}^n)$.

Then, the process $\mathbf{Y}_t = f(t, \mathbf{X}_t)$ is also an Itô process and its component k is given by

$$dY_k(t) = \frac{\partial f_k}{\partial t} dt + \sum_{i=1}^n \frac{\partial f_k}{\partial x_i} dX_i(t) + \frac{1}{2} \sum_{i,j=1}^n \frac{\partial^2 f_k}{\partial x_i \partial x_j} (dX_i(t))(dX_j(t)),$$

for all k from 1 to m. The last term is calculated using the following rules:

$$dtdt = dB_i(t)dt = dtdB_i(t) = dB_i(t)dB_j(t) = 0, \quad \forall i \neq j$$
$$dB_i(t)dB_i(t) = dt.$$

Example 9.4.2 Suppose the process $\mathbf{X}_t$ is

$$\mathbf{X}_t = \begin{pmatrix} B_1(t) + B_2(t) + 3t \\ B_1(t)B_2(t)B_3(t) \end{pmatrix}$$

where $\mathbf{B}_t = (B_1(t), B_2(t), B_3(t))$ is a three-dimensional Brownian motion. We want to show whether this process is a general Itô process.

Let

$$f(t, \mathbf{x}) = \begin{pmatrix} x_1 + x_2 + 3t \\ x_1 x_2 x_3 \end{pmatrix}.$$

We apply the general Itô's lemma to the process $\mathbf{X}_t = f(t, \mathbf{B}_t)$. We should have

$$dX_1(t) = 3dt + dB_1(t) + dB_2(t), \tag{9.10}$$

and

$$dX_2(t) = B_2(t)B_3(t)dB_1(t) + B_1(t)B_3(t)dB_2(t) + B_1(t)B_2(t)dB_3(t) \tag{9.11}$$
$$+ B_3(t)dB_1(t)dB_2(t) + B_2(t)dB_1(t)dB_3(t) + B_1(t)dB_2(t)dB_3(t) \tag{9.12}$$
$$= B_2(t)B_3(t)dB_1(t) + B_1(t)B_3(t)dB_2(t) + B_1(t)B_2(t)dB_3(t)$$

Therefore, $\mathbf{X}_t$ is a general Itô process since it's a stochastic process that solves the SDEs (9.10) and (9.10).

9.5 Stochastic Differential Equations (SDEs)

If the functions μ and σ in (9.9) depend on ω through the process X_t itself, then (9.9) defines an SDE. Specifically, assume that $X_t \in v(0, \infty)$. Assume that the functions $\mu = \mu(t, x)$ and $\sigma = \sigma(t, x)$ are twice differentiable with continuous second derivative in both variables. Then

$$X_t = X_0 + \int_0^t \mu(s, X_s)ds + \int_0^t \sigma(s, X_s)dB_s \tag{9.13}$$

defines an SDE.

Example 9.5.1 (Brownian motion on the unit circle). Recall that the point $(\cos x, \sin x)$ belongs to the unit circle. Thus, naturally if we take the process

$$Y_t = (\cos B_t, \sin B_t),$$

this will belong to the unit circle. We wish to obtain an SDE such that its solution will be the process Y_t. To obtain such an SDE, we use the Itô lemma. Let $f(t, x) = (\cos x, \sin x)$. We note that $Y_t = (Y_1(t), Y_2(t)) = f(t, B_t)$, and applying the general Itô lemma we obtain

$$\begin{cases} dY_1(t) = -\sin B_t dB_t - \frac{1}{2}\cos B_t dt \\ dY_2(t) = \cos B_t dB_t - \frac{1}{2}\sin B_t dt \end{cases}$$

Replacing $Y_1(t) = \cos x$ and $Y_2(t) = \sin x$ in the Eq. (9.14), the Brownian motion on the unit circle is a solution of

$$\begin{cases} dY_1(t) = -\frac{1}{2}Y_1(t)dt - Y_2(t)dB_t, \\ dY_2(t) = -\frac{1}{2}Y_2(t)dt + Y_1(t)dB_t \end{cases}$$

From the example earlier, we see that the Brownian motion on the unit circle must be a solution of the system presented. But is it the only solution? The next theorem gives conditions on the SDE coefficients μ and σ under which a solution exists and it is unique.

Theorem 9.5.1 (The existence and uniqueness of the solution to an SDE). Given $T > 0$, let $\mu : [0, T] \times \mathbb{R}^n \to \mathbb{R}^n$, $\sigma : [0, T] \times \mathbb{R}^n \to \mathbb{R}^{n,d}$ be two measurable functions. Let $\mathbf{X}_t$ and $\mathbf{B}_t$ denote a d-dimensional Brownian motion and consider the SDE

$$d\mathbf{X}_t = \mu(t, \mathbf{X}_t)dt + \sigma(t, \mathbf{X}_t)d\mathbf{B}_t, \tag{9.14}$$

with initial condition $\mathbf{X}_0 = \mathbf{Z}$ for some random vector with $E[|\mathbf{Z}|^2] < \infty$.

Suppose the coefficients μ and σ have the following properties:

1) **Linear growth condition:**

$$|\mu(t, \mathbf{x})| + |\sigma(t, \mathbf{x})| \leq C(1 + |\mathbf{x}|),$$

for some constant C, for all $\mathbf{x} \in \mathbb{R}^n$ and $t \in [0, T]$.

2) **Lipschitz condition:**

$$|\mu(t, \mathbf{x}) - \mu(t, \mathbf{y})| + |\sigma(t, \mathbf{x}) - \sigma(t, \mathbf{y})| \leq D|\mathbf{x} - \mathbf{y}|, \qquad (9.15)$$

for some constant D, for all $\mathbf{x}, \mathbf{y} \in \mathbb{R}^n$ and $t \in [0, T]$.

Then the SDE (9.14) has a unique solution $\mathbf{X}_t$ adapted to the filtration generated by Z and B_t and

$$E\left[\int_0^T |\mathbf{X}_t|\right] < \infty.$$

For the proof of the theorem, please consult [158].

Remark 9.5.1 We chose to give the more general form when the SDE is n dimensional. For the one-dimensional case, the conditions are identical.

The absolute value in the theorem is the modulus in R^n for μ, that is,

$$|\mathbf{x}|^2 = \sum_{i=1}^n x_i^2,$$

and for the matrix *sigma* we have

$$|\sigma|^2 = \sum_{i=1}^n \sum_{j=1}^d \sigma_{ij}^2.$$

The linear growth condition does not allow the coefficients to grow faster than x. The Lipschitz condition is a regularity condition. In fact this condition is stronger than requiring continuity but weaker than requiring that μ and σ be derivable on their domain. These conditions are the usual ones in the ODE theory. In practice, the Lipschitz condition is rarely verified. Indeed, if the functions μ and σ are differentiable in the x variable with continuous derivative, then the Lipschitz condition is satisfied.

Example 9.5.2 (Violating the linear growth condition). Consider the following SDE:

$$dX_t = X_t^2 dt, \quad X_0 = 1$$

Note that $\mu(t, x) = x^2$, and it does not satisfy the linear growth condition. Since there is no stochastic integral, i.e. we do not have a dB_t term, this SDE can be solved like a regular ODE as follows:

$$\frac{dX_t}{X_t} = dt \quad \Rightarrow \quad -\frac{1}{X_t} = t + k \quad \Rightarrow \quad X_t = -\frac{1}{t+k}$$
$$\text{Since } X_0 = 1 \quad \Rightarrow \quad k = -1 \quad \Rightarrow \quad X_t = \frac{1}{1-t}$$

Note that $\frac{1}{1-t}$ is the unique solution; however, the solution does not exist at $t = 1$, and therefore there is no global solution defined on the entire $[0, \infty)$.

Example 9.5.3 (Violating the Lipschitz condition). Consider the following SDE:

$$dX_t = 3\sqrt[3]{X_t^2} dt, \quad X_0 = 0$$

Note that $\mu(t,x) = 3\sqrt[3]{x^2}$ does not satisfy the Lipschitz condition at $x = 0$. This SDE has multiple solutions:

$$X_t = \begin{cases} 0, & \text{if } t \leq a \\ (t-a)^3, & \text{if } t > a \end{cases}$$

is a solution for any value of a.

9.5.1 Solution Methods for SDEs

Definition 9.5.1 (Strong and weak solutions). If the Brownian motion B_t is given and we can find a solution X_t which is adapted with respect to $\mathcal{F}_t$, the filtration generated by the Brownian motion, then X_t is called a **strong solution**.

If only the functions μ and σ are given and we need to find both X_t and B_t that solve the SDE, then the solution is a **weak solution**.

A strong solution is a weak solution. However, the converse is not necessarily true. The most famous example is the Tanaka equation

$$dX_t = \text{sign}(X_t) dB_t, \quad X_0 = 0,$$

where

$$\text{sign}(x) = \begin{cases} +1, & \text{if } x \geq 0 \\ -1, & \text{if } x < 0 \end{cases}$$

Note that the sign function does not satisfy the Lipschitz condition (9.15) at 0. Thus this equation has no strong solution, but it does have a weak solution. The idea is to use $-B_t$ which is also a standard Brownian motion. For more details, refer to [[158], exercise 4.10 and chapter 5].

Definition 9.5.2 (Strong and weak uniqueness). For any SDE, we say that the solution is **strongly unique** if for any two solutions $X_1(t)$ and $X_2(t)$, we have

$$X_1(t) = X_2(t), \text{ a.s. for all } t, T.$$

The solution is called **weakly unique** if for any two solutions $X_1(t)$ and $X_2(t)$, they have the same finite dimensional distribution.

Lemma 9.5.1 If μ and σ satisfy the conditions in Theorem 9.5.1, then a solution is weakly unique.

9.6 Examples of Stochastic Differential Equations

In this section, we present some examples of SDEs.

Example 9.6.1 Show that the solution to the SDE

$$dX_t = \frac{1}{2} X_t dt + X_t dB_t, \quad X_0 = 1$$

is e^{B_t}.

Note that the coefficients are $\mu(t, x) = \frac{1}{2}x$ and $\sigma(t, x) = x$, which satisfy the linear growth condition and the Lipschitz condition. Therefore, the solution exists and it is weakly unique by Section 9.5.1.

This is an example we have seen before of the geometric Brownian motion. We again take $f(t, x) = \log x$ and $Y_t = \log X_t$. Applying the Itô's formula, we obtain

$$dY_t = \frac{1}{X_t} dX_t + \frac{1}{2} \left(-\frac{1}{X_t^2} \right)(dX_t)^2$$

$$= \frac{1}{X_t} \left(\frac{1}{2} X_t dt + X_t dB_t \right) - \frac{1}{2X_t^2} \left(\frac{1}{2} X_t dt + X_t dB_t \right)^2$$

$$= \frac{1}{2} dt + dB_t - \frac{1}{2} dt$$

$$= dB_t$$

Hence,

$$dY_t = dB_t, \tag{9.16}$$

The integral form of (9.16) is given as:

$$Y_t - Y_0 = B_t. \tag{9.17}$$

9.6 Examples of Stochastic Differential Equations

We recall that $Y_t = \log X_t$ and therefore substituting Y_t into (9.17) and solving for X_t, we obtain the explicit solution $X_t = e^{B_t}$.

Example 9.6.2 Let $X_t = \sin B_t$, with $B_0 = \alpha \in (-\frac{\pi}{2}, \frac{\pi}{2})$. Show that X_t is the unique solution of

$$dX_t = -\frac{1}{2}X_t dt + \sqrt{1 - X_t^2}\, dB_t,$$

for $t < \inf\{s > 0 : B_s \notin (-\frac{\pi}{2}, \frac{\pi}{2})\}$.

We take $f(t, x) = \sin x$ and applying the Itô's formula to the process $X_t = f(t, B_t)$, we obtain:

$$dX_t = \cos B_t\, dB_t + \frac{1}{2}(-\sin B_t)dt$$

which can be rewritten as

$$dX_t = \sqrt{1 - X_t^2}\, dB_t - \frac{1}{2}X_t dt$$

by using the transformation $X_t = \sin B_t$. Thus the process X_t is a solution of the equation. To show that it is the unique solution, we need to show that its coefficients are Lipschitz. The coefficient $\mu(t, x) = -\frac{1}{2}x$ clearly satisfies the Lipschitz condition. For $\sigma(t, x) = \sqrt{1 - x^2}$, we need to restrict the time domain to the interval 0 until the first time the argument leaves the interval $(-\frac{\pi}{2}, \frac{\pi}{2})$. Otherwise, we will have multiple solutions. If B_t is in that interval, the corresponding x, y are in $(-1, 1)$ and we have

$$|\sqrt{1 - x^2} - \sqrt{1 - y^2}| = \frac{|\sqrt{1 - x^2} - \sqrt{1 - y^2}||\sqrt{1 - x^2} + \sqrt{1 - y^2}|}{|\sqrt{1 - x^2} + \sqrt{1 - y^2}|}$$

$$= \frac{|1 - x^2 - 1 + y^2|}{\sqrt{1 - x^2} + \sqrt{1 - y^2}} = \frac{|x - y||x + y|}{\sqrt{1 - x^2} + \sqrt{1 - y^2}}$$

$$\leq \frac{2}{\sqrt{1 - x^2} + \sqrt{1 - y^2}}|x - y|.$$

Since x and y are bounded away from -1 and 1, there exists a constant c such that

$$\min(\sqrt{1 - x^2}, \sqrt{1 - y^2}) > c,$$

and therefore

$$|\sqrt{1 - x^2} - \sqrt{1 - y^2}| \leq \frac{1}{c}|x - y|$$

for all $x, y \in (-1, 1)$, so the function $\sigma(t, x)$ is Lipschitz on the interval of definition. This shows that the SDE has the unique solution stated for $t < \inf\{s > 0 : B_s \notin (-\frac{\pi}{2}, \frac{\pi}{2})\}$.

Example 9.6.3 (Ørnstein–Uhlenbeck (O-U) SDE). In this example, we start the presentation of a technique which applies to linear SDEs. Specifically, we present the idea of using an integrating factor to simplify the equation. The O-U SDE written in the differential form is

$$dX_t = \mu X_t dt + \sigma dB_t$$

with $X_0 = x \in \mathbb{R}$.

In order to find the solution of the O-U SDE, we group the terms

$$dX_t - \mu X_t dt = \sigma dB_t \qquad (9.18)$$

which is similar to regular ODEs, and so the term on the left looks like what we obtain once we multiply with the integrating factor and take derivatives. Specifically, it looks like the derivative of

$$d(X_t e^{-\mu t})$$

However, these being stochastic processes, we need to verify that this is indeed the case. This is verified as follows:

$$d(X_t e^{-\mu t}) = e^{-\mu t} dX_t - \mu X_t e^{-\mu t} dt - \mu e^{-\mu t} dt dX_t = e^{-\mu t} dX_t - \mu X_t e^{-\mu t} dt$$

where the last expression was obtained by applying the product rule on two Itô processes and noticing that no matter what the stochastic process X_t is, the last term only contains terms of the type $dtdt$ and $dtdB_t$.

We observe that the left side of Eq. (9.18) is the derivative if we consider the integrating term. So multiplying both sides of (9.18) with the integrating factor $e^{-\mu t}$ we obtain:

$$e^{-\mu t} dX_t - \mu e^{-\mu t} X_t dt = \sigma e^{-\mu t} dB_t$$
$$d(X_t e^{-\mu t}) = \sigma e^{-\mu t} dB_t$$

Writing both sides in the actual integral form, we have

$$X_t = x e^{\mu t} + \int_0^t \sigma e^{\mu(t-s)} dB_s.$$

This guarantees a unique solution since the coefficients of the equations satisfy the existence and uniqueness conditions of Theorem 9.5.1.

We can also calculate the moments of the process X_t. To do so, it is important to remember that the stochastic integral which is a martingale started from 0 and therefore its expectation is always 0. Thus, we have

$$E[X_t] = x e^{\mu t}.$$

Thus the expectation of this process goes to 0 or ∞ in the long run depending on the sign of the parameter μ.

For the variance, we use Itô's isometry which was presented in Section 1.9 as follows:

$$V(X_t) = E\left[(X_t - E[X_t])^2\right] = E\left[\left(e^{\mu t}\int_0^t \sigma e^{-\mu s}dB_s\right)^2\right]$$

$$= e^{2\mu t}\int_0^t \sigma^2 e^{-2\mu s}ds = \frac{\sigma^2}{2\mu}e^{2\mu t}\left(1 - e^{-2\mu t}\right)$$

$$= \frac{\sigma^2}{2\mu}\left(e^{2\mu t} - 1\right)$$

So, again we reach the same conclusion that the variance becomes either 0 or ∞.

9.6.1 An Analysis of Cox–Ingersoll–Ross (CIR)-Type Models

We conclude the examples section with a depth analysis of CIR-type models. These types of models are heavily used in finance, especially in fixed-income security markets.

The CIR model was first formulated in [46]. Let r_t denote the stochastic process modeled as a CIR-type process:

$$dr_t = \alpha(\mu - r_t)dt + \sigma\sqrt{r_t}dB_t, \tag{9.19}$$

with α, μ, and σ as real constants and B_t a standard Brownian motion adapted to $\mathcal{F}_t$. μ, sometimes denoted with $\bar{r}$, as we shall see, is the long-term mean of the process. α is called the rate of return (or speed of return), a name inherited from the simpler mean-reverting O-U process but the meaning is similar here as well. Finally, σ is a constant that controls the variability of the process.

In general, this process does not have an explicit (stochastic) solution. However, its moments may be found by using a standard stochastic technique. Before we analyze the actual process, let us present some other models that are related to the CIR, namely, the O-U process and the *mean-reverting O-U process*. The O-U process is

$$dr_t = \alpha r_t dt + \sigma dB_t. \tag{9.20}$$

As mentioned in the examples section, it requires an integrating factor to obtain an explicit solution.

The *mean-reverting O-U process* is

$$dr_t = \alpha(\mu - r_t)dt + \sigma dB_t. \tag{9.21}$$

Note that this process is slightly more general than the regular O-U process. This process is called mean-reverting because of the dt term. To understand the name, consider the behavior of the increment for a small time step Δt. We can write

$$r_t - r_{t-\Delta t} = \alpha(\mu - r_{t-\Delta t})\Delta t + \sigma\Delta B_t.$$

Now look at the deterministic term. If $r_{t-\Delta t} < \mu$, then $\mu - r_{t-\Delta t}$ and the whole drift term will be positive. Thus the increment will tend to be positive, which means that r_t will tend to be closer to μ. If $r_{t-\Delta t} > \mu$, then $\mu - r_{t-\Delta t}$ is negative. Thus the increment will tend to be negative and once again r_t tends to get closer to the mean μ. The magnitude of the parameter α can amplify or decrease this mean reversion trend, and thus it is called the speed of reversion. The parameter σ governs the size of random fluctuations.

The *mean-reverting O-U process* model can be solved in exactly the same way as the regular O-U process. In finance, the *mean-reverting O-U process* model is known as the *Vasiček model* after the author of the paper that introduced the model [202].

The *Longstaff and Schwartz model*. This model writes the main process as a sum:

$$r_t = \alpha X_t + \beta Y_t,$$

where each of the stochastic processes X_t, Y_t are written as a CIR model:

$$dX_t = (\gamma - \delta X_t)dt + \sqrt{X_t}\, dW_t^1, \qquad (9.22)$$
$$dY_t = (\eta - \theta Y_t)dt + \sqrt{Y_t}\, dW_t^2,$$

Note that the form is a bit different, but, of course, each of the equations may be written as a CIR by simply factoring the coefficient of the process in the drift (i.e. $\gamma - \delta X_t = \delta\left(\frac{\gamma}{\delta} - X_t\right)$) and taking the parameter σ in CIR equal to 1.

The *Fong and Vasiček model* once again uses a couple of CIR models but this time embedded into each other:

$$dr_t = \alpha(\mu - r_t)dt + \sqrt{v_t}dB_1(t)$$
$$dv_t = \gamma(\eta - v_t)dt + \xi\sqrt{v_t}dB_2(t),$$

where the Brownian motions are correlated, that is, $dB_1(t)dB_2(t) = \rho dt$. This model is an example of what is called a *stochastic volatility model*. Note that the coefficient of the Brownian motion driving the main process $B_1(t)$ is stochastic, and the randomness is coming from the outside process v_t. Any model that has this property is called a stochastic volatility model.

The *Black–Karasinski model* is very popular in the financial markets. This process is in fact a generalization of the simple mean-reverting O-U model mentioned earlier. Specifically, if we denote $X_t = \log r_t$, the logarithm of the short rate, then the Black–Karasinski model is

$$dX_t = (\theta(t) - \phi(t)X_t)dt + \sigma(t)dB_t. \qquad (9.23)$$

We note that

$$dX_t = \phi(t)\left(\frac{\theta(t)}{\phi(t)} - X_t\right)dt + \sigma(t)dB_t,$$

and thus it is a mean-reverting O-U model where the coefficients are deterministic functions of time. The reason for the time-varying coefficients and their wide use in finance is the inability of the constant coefficient model to fit the yield curve produced by observed bond prices. This problem is common in finance for any of the models with constant coefficients. Having time-dependent coefficients means that one cannot solve the SDE analytically anymore. The reason for the Black–Karasinski model being popular is the fact that one can construct a tree that approximates its dynamics.

9.6.2 Moments Calculation for the CIR Model

In this subsection we will talk about calculating moments for stochastic processes described by SDEs. This is a very useful methodology which is applicable even in the case where the SDE cannot be solved explicitly.

The CIR process in (9.19) can be written in its integral form as

$$r_t - r_0 = \int_0^t \alpha(\mu - r_s)ds + \int_0^t \sigma\sqrt{r_s}\,dW_s.$$

Next we apply expectations on both sides and use Fubini's theorem in the first integral. The Fubini's theorem is a result which gives conditions under which it is possible to compute a double integral using iterated integrals. As a consequence it allows the order of integration to be changed in iterated integrals.

$$E[r_t] - r_0 = E\left[\int_0^t \alpha(\mu - r_s)ds\right] + E\left[\int_0^t \sigma\sqrt{r_s}\,dW_s\right]$$
$$= \int_0^t E[\alpha(\mu - r_s)]ds = \int_0^t \alpha(\mu - E[r_s])ds,$$

where we assume that at time 0 the process is a constant r_0. The second integral disappears since any stochastic integral is a martingale and any martingale has the property that $E[M_t] = E[M_0]$. But since at 0 the stochastic integral is equal to zero, then any stochastic integral has an expected value 0.

Next, let us denote $u(t) = E[r_t]$, which is just a deterministic function. Substituting in the aforementioned and going back to the differential notation, we have

$$du(t) = \alpha(\mu - u(t))dt.$$

But this is a first order linear differential equation. We can solve it easily by multiplying with the integrating factor.

$$du(t) + \alpha u(t)dt = \alpha\mu dt$$
$$e^{\alpha t}du(t) + \alpha e^{\alpha t}u(t)dt = \alpha\mu e^{\alpha t}dt$$
$$d\left(e^{\alpha t}u(t)\right) = \alpha\mu e^{\alpha t}dt$$

and finally, integration gives:

$$e^{at}u(t) - u(0) = \mu(e^{at} - 1)$$

Now using that $u(0) = E[r_0] = r_0$ and after rearranging terms, we finally obtain:

$$E[r_t] = u(t) = \mu + (r_0 - \mu)e^{-at} \qquad (9.24)$$

We now calculate the variance as follows:

$$Var(r_t) = E\left[(r_t - E[r_t])^2\right] = E[r_t^2] - (E[r_t])^2$$

So we will need to calculate the second moment $E[r_t^2]$. The second term is just the square of the expectation calculated earlier. We cannot just square the dynamics of r_t; we need to calculate the proper dynamics of r_t^2 by using Itô's lemma and apply expectations like we did earlier. Take the function $f(x) = x^2$ and apply Itô's lemma to this function and r_t to obtain the dynamics of $r_t^2 = f(r_t)$. We get

$$r_t^2 = 2r_t dr_t + \frac{1}{2}2(dr_t)^2 = 2r_t(\alpha(\mu - r_t)dt + \sigma\sqrt{r_t}dW_t) + \sigma^2 r_t dt$$
$$= (2\alpha\mu + \sigma^2)r_t dt - 2\alpha r_t^2 dt + 2\sigma r_t \sqrt{r_t}dW_t$$

Next, we write the equation in the integral form, apply expectation on both sides, use Fubini, and use the fact that the expectation of a stochastic integral is zero.

We denote $y_t = E[r_t^2]$, and we note that $y_0 = r_0^2$ to obtain

$$dy_t + 2\alpha y_t dt = (2\alpha\mu + \sigma^2)E[r_t]dt. \qquad (9.25)$$

We then multiply through (9.25) with the integrating factor, which in this case is $e^{2\alpha t}$, to get

$$d\left(e^{2\alpha t}y_t\right) = e^{2\alpha t}(2\alpha\mu + \sigma^2)E[r_t]dt.$$

Integrating both sides, we get

$$e^{2\alpha t}y_t - y_0 = \int_0^t e^{2\alpha s}(2\alpha\mu + \sigma^2)E[r_s]ds. \qquad (9.26)$$

Next replacing $E[r_s]$ with $(\mu + (r_0 - \mu)e^{-\alpha s})$ on the right-hand side of the above equation, we obtain:

$$\int_0^t e^{2\alpha s}(2\alpha\mu + \sigma^2)E[r_s]ds = \int_0^t e^{2\alpha s}(2\alpha\mu + \sigma^2)(\mu + (r_0 - \mu)e^{-\alpha s})ds.$$

Hence the integral on the right-hand side of (9.26) is simplified to:

$$(2\alpha\mu + \sigma^2)\left(\frac{\mu}{2}(e^{2\alpha t} - 1) + (r_0 - \mu)(e^{\alpha t} - 1)\right)$$

Now going back, we finally obtain the second moment as

$$E[r_t^2] = r_0^2 e^{-2\alpha t} + (2\alpha\mu + \sigma^2)\left(\frac{\mu}{2}(1 - e^{-2\alpha t}) + (r_0 - \mu)(e^{-\alpha t} - e^{-2\alpha t})\right) \qquad (9.27)$$

In order to obtain the variance, we substitute Equation (9.27) and the square of the mean of Eq. (9.24) into the following equation:

$$Var(r_t) = E[r_t^2] - (E[r_t])^2.$$

Therefore,

$$Var(r_t) = \frac{\sigma^2 \mu}{2\alpha} + \frac{\sigma^2}{\alpha}(r_0 - \mu)e^{-\alpha t} + \frac{4\alpha^2 r_0 + \sigma^2(\mu - 2r_0)}{2\alpha}e^{-2\alpha t}. \quad (9.28)$$

9.6.3 Interpretation of the Formulas for Moments

If we look at the expectation formula in (9.24), we observe that the mean of r_t converges exponentially to μ. If the process starts from $r_0 = \mu$, then in fact the mean at t is constant and equal to μ, and the process is stationary in mean.

Furthermore, looking at the variance formula in (9.28), one sees that the variance converges to $\frac{\sigma^2 \mu}{2\alpha}$ and again the convergence is faster with the choice $r_0 = \mu$. However, the variance is not stationary unless either $r_0 = \mu = 0$ (this gives the trivial 0 process) or the following relationship between the parameters holds:

$$r_0 = \mu, \quad 4\alpha^2 - \sigma^2 = 0.$$

9.6.4 Parameter Estimation for the CIR Model

This section is presented only for the sake of financial applications. Suppose we have a history of the forward rate r_t. Let us denote these observations $r_0, r_1, \ldots, r_n$ equidistant at times Δt apart, and we calculate the increments of the process as $\Delta r_i = r_i - r_{i-1}$ (n total increments). The process r_t in the CIR specification is not Gaussian, nor are its increments. Note that the noise term is the Brownian motion increment but multiplied with the square root of the actual process. The estimation, however, proceeds as if the increments would be Gaussian. It is worth noting that this is in fact an approximate estimation. An essential step in the derivation is to realize that the process r_t is a Markov process since its future probabilities are determined by its most recent values.

Let us denote the mean in (9.24) with

$$m(r_0, t) = E[r_t] = \mu + (r_0 - \mu)e^{-\alpha t} \quad (9.29)$$

and the variance in (9.28) with

$$v(r_0, t) = \frac{\sigma^2 \mu}{2\alpha} + \frac{\sigma^2}{\alpha}(r_0 - \mu)e^{-\alpha t} + \frac{4\alpha^2 r_0 + \sigma^2(\mu - 2r_0)}{2\alpha}e^{-2\alpha t} \quad (9.30)$$

Since the process is Markov, we can calculate the mean and variance of r_t given the value of the process at time s, namely, r_s, by just using

$$m(r_s, t-s), \text{ and } v(r_s, t-s).$$

Thus the joint density function may be written as

$$f(\Delta r_n, \ldots, \Delta r_1, \theta) = f(\Delta r_n \mid \Delta r_{n-1} \ldots, \Delta r_1, \theta) \cdots f(\Delta r_2 \mid \Delta r_1, \theta)$$

$$= \prod_{i=1}^{n-1} f(\Delta r_{i+1} \mid \Delta r_i, \theta),$$

where $\theta = (\mu, \alpha, \sigma)$ is the vector of parameters. Therefore, the log-likelihood function (defined to be the natural logarithm of the likelihood function) is

$$L_n(\theta) = \sum_{i=1}^{n-1} \log f(\Delta r_{i+1} \mid \Delta r_i, \theta), \tag{9.31}$$

and this function needs to be maximized with respect to the vector of parameters $\theta = (\mu, \alpha, \sigma)$.

Next we briefly present how the approximation with the Gaussian distribution is performed. Let ϕ be the density of the normal distribution with mean μ and variance Δt, that is,

$$\phi(x, \mu, \Delta t) = \frac{1}{\sqrt{2\pi \Delta t}} e^{-\frac{(x-\mu)^2}{2\Delta t}}$$

This function will be substituted for the f in expression (9.31). The logarithm of this function has a simpler expression:

$$\log \phi(x, \mu, \Delta t) = \log \frac{1}{\sqrt{2\pi \Delta t}} e^{-\frac{(x-\mu)^2}{2\Delta t}}$$

$$= -\frac{1}{2} \log \Delta t - \frac{(x-\mu)^2}{2\Delta t},$$

where we neglected the constant $-\frac{1}{2} \log 2\pi$ since it is irrelevant for the maximization. With the notation shown earlier, the approximate log-likelihood function is

$$L_n(\theta) = \sum_{i=1}^{n-1} \log \phi(\Delta r_{i+1}, m(r_i, \Delta t), v(\Delta r_i, \Delta t)), \tag{9.32}$$

where Δt is the time between consecutive observations. Note that the expressions $m(r_i, \Delta t)$ and $v(\Delta r_i, \Delta t)$ contain the parameters $\theta = (\mu, \alpha, \sigma)$, and the optimization procedure is with respect to these variables.

9.7 Linear Systems of SDEs

In practice, we often need to solve systems of SDEs. Consider the following stochastic system:

$$dX_1(t) = 2t\,dt + 4dB_1(t)$$
$$dX_2(t) = X_1(t)dB_2(t)$$

9.7 Linear Systems of SDEs

We can write this system in a matrix form as

$$dX_t = Cdt + DdB_t$$

where the coefficients are the matrices

$$C = \begin{pmatrix} 2t \\ 0 \end{pmatrix}, \quad D = \begin{pmatrix} 4 & 0 \\ 0 & X_1(t) \end{pmatrix}.$$

This particular system can be easily solved by finding the solution for X_1 first and then using that solution to find X_2, but we are looking for a more general theory.

Assume that we have the following system of SDEs:

$$dX_t = Cdt + DdB_t,$$

for an n-dimensional process X_t and further assume that the vector C can be written as:

$$C = AX_t,$$

for an $n \times n$ matrix A. Also assume that the matrix D is only a function of t (does not contain the process X_t). In this case, the system can be written as

$$dX_t = AX_t dt + DdB_t, \tag{9.33}$$

and since the matrix D is independent of the variable X_t, we use an idea similar to the integrating factor where we are going to solve the system. But this idea is very similar to the integrating factor idea in the first order differential equations case. However, here we consider exponential matrices. If the matrix A is given, we define the exponential matrix to be of the form

$$e^A = \sum_{n=0}^{\infty} \frac{1}{n!} A^n,$$

where $A^0 = I$ is the identity matrix. Multiplying through Eq. (9.33) by the exponential matrix and rearranging terms, we obtain:

$$e^{-At} dX_t - e^{-At} AX_t dt = e^{-At} DdB_t.$$

Since the matrix A does not depend on X_t, the regular calculus rules apply and we will obtain

$$d(e^{-At} X_t) = e^{-At} DdB_t.$$

Thus we can obtain the solution

$$X_t = e^{At} X_0 + e^{At} \int_0^t e^{-As} DdB_s$$

$$= e^{At} X_0 + \int_0^t e^{A(t-s)} DdB_s$$

The issue when calculating the solution is to calculate the exponential. The idea is to write the matrix A as $A = UDU^{-1}$, where D is a diagonal matrix; thus the powers will be

$$A^n = UD^n U^{-1}.$$

If we have this, we can easily verify that

$$e^A = Ue^D U^{-1}.$$

We now present an example of a linear system SDE.

Example 9.7.1 Consider the system

$$dX_1(t) = X_2 dt + \alpha dB_1(t)$$
$$dX_2(t) = -X_1(t)dt + \beta dB_2(t)$$

with $\mathbf{X}_0 = (0,0)^t$.

This is a system of the type (9.33):

$$d\mathbf{X}_t = A\mathbf{X}_t dt + D d\mathbf{B}_t,$$

where

$$A = \begin{pmatrix} 0 & 1 \\ -1 & 0 \end{pmatrix}, \quad D = \begin{pmatrix} \alpha & 0 \\ 0 & \beta \end{pmatrix}$$

Therefore, the solution is

$$\mathbf{X}_t = \int_0^t e^{A(t-s)} D d\mathbf{B}_s$$

and we just need to calculate e^{At}. To do so, let us use Cholesky decomposition. Given a symmetric positive definite matrix A, the Cholesky decomposition is an upper triangular matrix U with strictly positive diagonal entries such that $A = U^T U$. We need to first calculate the eigenvalues and eigenvectors of the matrix A. We have

$$\det(I - \lambda A) = 1 + \lambda^2;$$

thus the eigenvalues are i and $-i$. We can calculate the corresponding eigenvectors as

$$u_1 = \begin{pmatrix} i \\ -1 \end{pmatrix}, \quad u_2 = \begin{pmatrix} i \\ 1 \end{pmatrix}.$$

We then need to calculate U^{-1}. We skip details and just show the result:

$$U = \begin{pmatrix} i & -1 \\ i & 1 \end{pmatrix}, \quad U^{-1} = \frac{1}{2i}\begin{pmatrix} 1 & -i \\ 1 & i \end{pmatrix}, \quad \text{and } D = \begin{pmatrix} i & 0 \\ 0 & -i \end{pmatrix}.$$

Therefore,

$$e^{At} = \sum_{i=0}^{\infty} \frac{1}{n!} U(Dt)^n U^{-1} = U \begin{pmatrix} e^{it} & 0 \\ 0 & e^{-it} \end{pmatrix} U^{-1}.$$

Thus we finally obtain:

$$e^{At} = \begin{pmatrix} \cos t & \sin t \\ -\sin t & \cos t \end{pmatrix}.$$

Substituting $e^{A(t-s)}$, we can finally obtain the solution

$$\mathbf{X}_t = \begin{pmatrix} \int_0^t \alpha \cos(t-s) dB_1(s) + \int_0^t \beta \sin(t-s) dB_2(s) \\ -\int_0^t \alpha \sin(t-s) dB_1(s) + \int_0^t \beta \cos(t-s) dB_2(s) \end{pmatrix}.$$

9.8 Some Relationship Between SDEs and Partial Differential Equations (PDEs)

In this section, we will discuss about some relationship that will allow us to find solutions for SDEs. There is a deeper relationship which involves a famous result: the Feynman–Kaç theorem. We recall that Feynman–Kaç theorem was discussed in Chapter 6 of this book. Please consult [117] for very general versions of this theorem.

Instead, let us consider the simple SDE in (9.13):

$$X_t = x_0 + \int_0^t \mu(s, X_s) ds + \int_0^t \sigma(s, X_s) dB_s. \tag{9.34}$$

A natural candidate solution would be a function of time and B_t, that is, we may look for a solution of the type

$$X_t = f(t, B_t).$$

If we can determine such a function, then using the existence and uniqueness of the solution (Theorem 9.5.1), we should be able to conclude that this is the unique solution. But how do we find the function $f(t,x)$? The only tool in this stochastic calculus is the Itô lemma. Applying it to the function f, we must have:

$$X_t = f(t, B_t) = f(0,0) + \int_0^t \left(\frac{\partial f}{\partial s} + \frac{1}{2} \frac{\partial^2 f}{\partial x^2} \right) ds + \int_0^t \frac{\partial f}{\partial x} dB_s \tag{9.35}$$

This indicates that, if we choose the function f so that it satisfies the following system of PDEs that follows, then the solution is going to provide us with a possible solution of the SDE. We still need to verify that the process solves the equation.

The system is obtained by equating the terms in (9.34) and (9.35) as follows:

$$\frac{\partial f}{\partial t}(t,x) + \frac{1}{2}\frac{\partial^2 f}{\partial x^2}(t,x) = \mu(t,f(t,x))$$
$$\frac{\partial f}{\partial x}(t,x) = \sigma(t,f(t,x))$$
$$f(0,0) = x_0 \qquad (9.36)$$

Very often this method is not helpful since the resulting PDE (9.36) is harder to solve than the original SDE (9.34).

Example 9.8.1 Consider the O-U SDE (Eq. (9.20))

$$dX_t = \alpha X_t dt + \sigma dB_t, \quad X_0 = x.$$

If we follow the technique in this section, we obtain the following PDE system:

$$\frac{\partial f}{\partial t}(t,x) + \frac{1}{2}\frac{\partial^2 f}{\partial x^2}(t,x) = -\alpha f(t,x)$$
$$\frac{\partial f}{\partial x}(t,x) = \sigma$$
$$f(0,0) = x_0$$

The second equation gives $f(t,x) = \sigma x + c(t)$. Substituting into the first equation we obtain:

$$c'(t) = -\alpha(\sigma x + c(t))$$

or

$$c'(t) + \alpha c(t) = -\alpha\sigma x.$$

This equation has to hold for all x and t. However, the left-hand side is a function of t only, while the right-hand side is a function of x only. Therefore, the only way this can happen is when both sides are equal to a constant, say λ. This will imply that $-\alpha\sigma x = \lambda$, and that happens only if x is a constant, which is absurd.

Thus the system cannot be solved, and we can see that even in this very simple case, the PDE method doesn't solve the problem.

There are several methods used to generate the sample path of SDEs. However, in the next section, we will describe the Euler method for approximating SDEs. Please refer to [146] and references there in for other techniques used to generate sample paths of SDEs.

9.9 Euler Method for Approximating SDEs

In this section, we present a simple pseudo-code to generate sample paths of a process defined by an SDE. There are many methods to generate the sample paths but we will focus on the simplest one which is the Euler's method.

Consider the following SDE:

$$dX_t = \alpha(t, X_t)dt + \beta(t, X_t)dW_t, \quad X_0 = x_0, \quad (9.37)$$

with W_t a one-dimensional standard Brownian motion.

The solution of (9.37) is given as:

$$X_t = x_0 + \int_0^t \alpha(s, X_s)ds + \int_0^t \beta(s, X_s)dW_s. \quad (9.38)$$

Therefore, approximating the path of X_t equates with approximating the integral. There are many ways to approximate the first integral; the second however has to be approximated using Euler method. The Euler method uses a simple rectangular rule. Assume that the interval $[0, t]$ is divided into n subintervals. For simplicity assume the subintervals to be equal in length but this is not a necessary condition. This implies that the increment is $\Delta t = t/n$ and that the points are $t_0 = 0, t_1 = \Delta t, \ldots, t_i = i\Delta t, \ldots, t_n = n\Delta t = t$. Thus using X_i to denote X_{t_i} we have:

$$\begin{cases} X_0 = x_0 \\ X_i = X_{i-1} + \alpha(t_{i-1}, X_{i-1})\Delta t + \beta(t_{i-1}, X_{i-1})\Delta W_i, \quad \forall i \in \{1, 2, \ldots, n\}, \end{cases}$$

where ΔW_i is the increment of a standard Brownian motion over the interval $[t_{i-1}, t_i]$. We recall from Section 1.5.2 that the Brownian motion has independent and stationary increments, so it follows that each such increment is independent of all others and is distributed as a normal (Gaussian) random variable with mean 0 and variance the length of the time subinterval (i.e. Δt). Therefore, the standard deviation of the increment is $\sqrt{\Delta t}$.

Next we present an example of generating sample paths for solutions of SDEs using Euler's method.

Example 9.9.1 Suppose X_0 is a random variable uniformly distributed between $[-15, 15]$. Generate a sample path for a one-year process that has the estimated functions to be $\alpha(t, x) = 15x$ and $\beta(t, x) = 12\log|x|$.

The solution is as follows: First we need to decide on the number of points of our simulated path. Since we are interested in generating sample path for a one year process, we will simulate daily values, i.e. choose $t = 1$ and $n = 365$ (assuming that there are price movements during weekends and holiday). Thus, $\Delta t = 1/365$. The functions *Rand()* and *Randn()* in Matlab are used to generate the random numbers.

- *Rand()* generates a uniform distribution between 0 and 1.
- *Randn()* generates a normal distribution with mean 0 and variance 1.

The algorithm for generating the sample paths of Example 9.9.1 using Euler's method is as follows:

Algorithm 1 Generating a sample path using Euler's method.
---
initialize $n, t, \Delta t$
X[0] = 30 * Rand − 15 {Generate the starting value}
for $i = 1$ to n **do**
$\quad \Delta W = \sqrt{\Delta t}$ * Randn {Generate the increment for BM}
$\quad$ X[i] = X[i − 1] + 15 * X[i − 1] * Δt + 12 * log |X[i − 1]| * ΔW
$\quad$ Store X[i]
end for
Plot the whole path.

This algorithm can be extended to general α and β functions by creating separate functions.

The Euler method gives a first order approximation for the stochastic integral. The Euler–Milstein method, detailed next, provides an improvement by including second order terms. The idea in this scheme is to consider expansions on the coefficients μ and σ. We refer to [81, 123] for an in-depth analysis of the order of convergence. This scheme is only one of the schemes based on Milstein's work, and it is applied when the coefficients of the process are functions of only the main process (do not depend on time). Specifically, the scheme is designed to work with SDEs of the type

$$dX_t = \alpha(X_t)dt + \beta(X_t)dB_t, \quad X_0 = x_0.$$

We consider expansions on the coefficients $\alpha(X_t)$ and $\beta(X_t)$ using Itô's lemma. We then have

$$d\alpha(X_t) = \alpha'(X_t)dX_t + \frac{1}{2}\alpha''(X_t)(dX_t)^2$$

$$d\alpha(X_t) = \left(\alpha'(X_t)\alpha(X_t) + \frac{1}{2}\alpha''(X_t)\beta^2(X_t)\right)dt + \alpha'(X_t)\beta(X_t)dB_t.$$

Proceeding in a similar way for $\beta(X_t)$ and writing in the integral form from t to u for any $u \in (t, t + \Delta t]$, we obtain

$$\alpha_u = \alpha_t + \int_t^u \left(\alpha'_s \alpha_s + \frac{1}{2}\alpha''_s \beta_s^2\right) ds + \int_t^u \alpha'_s \beta_s dB_s$$

$$\beta_u = \beta_t + \int_t^u \left(\beta'_s \alpha_s + \frac{1}{2}\beta''_s \beta_s^2\right) ds + \int_t^u \beta'_s \beta_s dB_s,$$

9.9 Euler Method for Approximating SDEs

where we used the notation $\alpha_u = \alpha(X_u)$. Substituting these expressions in the original SDE, we obtain

$$X_{t+\Delta t} = X_t + \int_t^{t+\Delta t} \left(\alpha_t + \int_t^u \left(\alpha_s' \alpha_s + \frac{1}{2}\alpha_s'' \beta_s^2 \right) ds + \int_t^u \alpha_s' \beta_s dB_s \right) du$$
$$+ \int_t^{t+\Delta t} \left(\beta_t + \int_t^u \left(\beta_s' \alpha_s + \frac{1}{2}\beta_s'' \beta_s^2 \right) ds + \int_t^u \beta_s' \beta_s dB_s \right) dB_u$$

In this expression, we eliminate all terms which will produce, after integration, higher orders than Δt. That means eliminating terms of the type $dsdu = O(\Delta_t^2)$ and $dudB_s = O(\Delta t^{\frac{3}{2}})$. The only terms remaining other than simply du and ds are the ones involving $dB_u dB_s$ since they are of the right order. Thus, after eliminating the terms, we obtain:

$$X_{t+\Delta t} = X_t + \alpha_t \int_t^{t+\Delta t} du + \beta_t \int_t^{t+\Delta t} dB_u + \int_t^{t+\Delta t} \int_t^u \beta_s' \beta_s dB_s dB_u \quad (9.39)$$

For the last term, we apply Euler discretization in the inner integral:

$$\int_t^{t+\Delta t} \left(\int_t^u \beta_s' \beta_s dB_s \right) dB_u \approx \int_t^{t+\Delta t} \beta_t' \beta_t (B_u - B_t) dB_u$$
$$= \beta_t' \beta_t \left(\int_t^{t+\Delta t} B_u dB_u - B_t \int_t^{t+\Delta t} dB_u \right)$$
$$= \beta_t' \beta_t \left(\int_t^{t+\Delta t} B_u dB_u - B_t B_{t+\Delta t} + B_t^2 \right). \quad (9.40)$$

For the integral inside, recall that we showed in Section 9.4 that:

$$\int_0^v B_u dB_u = \frac{1}{2}(B_t^2 - t).$$

Therefore, applying for t and $t + \Delta t$ and taking the difference, we obtain

$$\int_t^{t+\Delta t} B_u dB_u = \frac{1}{2}(B_{t+\Delta t}^2 - t - \Delta t) - \frac{1}{2}(B_t^2 - t).$$

Therefore, substituting back in (9.40), we have

$$\int_t^{t+\Delta t} \left(\int_t^u \beta_s' \beta_s dB_s \right) dB_u \approx \beta_t' \beta_t \left(\frac{1}{2}(B_{t+\Delta t}^2 - B_t^2 - \Delta t) - B_t B_{t+\Delta t} + B_t^2 \right)$$
$$= \beta_t' \beta_t \left(\frac{1}{2}B_{t+\Delta t}^2 + \frac{1}{2}B_t^2 - B_t B_{t+\Delta t} - \Delta t \right)$$
$$= \beta_t' \beta_t \left(\frac{1}{2}(B_{t+\Delta t} - B_t)^2 - \Delta t \right).$$

We recall that $B_{t+\Delta t} - B_t$ is the increment of the Brownian motion which we know is $N(0, \Delta t)$ or $\sqrt{\Delta t} Z$, where $Z \sim N(0, 1)$. This actually concludes the derivation of the Euler–Milstein scheme. To summarize, for the SDE

$$dX_t = \alpha(X_t)dt + \beta(X_t)dB_t, \quad X_0 = x_0,$$

the Euler–Milstein scheme starts with $X_0 = x_0$ and for each successive point, first generates $Z \sim N(0, 1)$, and then calculates the next point as

$$X_{t+\Delta t} = X_t + \alpha(X_t)\Delta t + \beta(X_t)\sqrt{\Delta t}Z + \frac{1}{2}\beta'(X_t)\beta(X_t)\Delta t(Z^2 - 1) \quad (9.41)$$

Example 9.9.2 (Euler–Milstein scheme for the Black–Scholes model). In the Black–Scholes model, the stock follows

$$dS_t = rS_t dt + \sigma S_t dB_t,$$

where S_t is the stock price at time t, r is the risk-free rate, and σ is the volatility of the stock. Therefore, the coefficients depend only on S_t, and thus we can apply the Euler–Milstein scheme. Specifically

$$\alpha(x) = rx, \quad \beta(x) = \sigma x$$

Therefore, the scheme is

$$S_{t+\Delta t} = S_t + rS_t\Delta t + \sigma S_t\sqrt{\Delta t}Z + \frac{1}{2}\sigma^2 S_t\Delta t(Z^2 - 1).$$

Since the Black–Scholes scheme is none other than the geometric Brownian motion and we have already seen in Chapter 2 of this book how to solve it by taking $X_t = \log S_t$ and applying the Itô's lemma, the equation for X_t is

$$dX_t = \left(r - \frac{\sigma^2}{2}\right) dt + \sigma dB_t.$$

Therefore, in this case

$$X_{t+\Delta t} = X_t + \left(r - \frac{\sigma^2}{2}\right)\Delta t + \sigma\sqrt{\Delta t}Z,$$

since the last term contains the derivative of $\beta(x) = \sigma$, which is a constant function. The new term in the Milstein scheme vanishes when the the SDE has additive noise since $\frac{\partial \beta}{\partial x} \equiv 0$. For such SDEs the stochastic Milstein and Euler are identical, thus we get no improvement by applying the more complex scheme.

Example 9.9.3 (A stochastic volatility example). We present an example of approximating using the Euler–Milstein scheme to a more complex model involving two related processes. These processes appear in finance and we shall use the notation which is common there. We will focus on the Heston model [95] since it is very popular with practitioners. This model is formulated in differential form as

$$dS_t = rS_t dt + \sqrt{Y_t}S_t dB_1(t)$$
$$dY_t = \kappa(\overline{Y} - Y_t)dt + \sigma\sqrt{Y_t}dB_2(t) \quad (9.42)$$

where the Brownian motions are correlated with the correlation coefficient ρ.

The Euler scheme for (9.42) is straightforward:

$$dS_{t+\Delta t} = S_t + rS_t\Delta t + \sqrt{Y_t}S_t\sqrt{\Delta t}Z_1$$
$$dY_{t+\Delta t} = Y_t + \kappa(\overline{Y} - Y_t)\Delta t + \sigma\sqrt{Y_t}\sqrt{\Delta t}Z_2$$

where we start from known values S_0 and Y_0 and the correlated increments are created by first generating two independent $N(0, 1)$ variables (call them W_1 and W_2) and taking

$$Z_1 = W_1$$
$$Z_2 = \rho W_1 + \sqrt{1 - \rho^2}W_2.$$

The Milstein scheme requires more work. For the Y_t process (which we note is just a CIR-type process), the coefficients are $\alpha(x) = \kappa(\overline{Y} - x)$ and $\beta(x) = \sigma\sqrt{x}$, and therefore a straightforward application of (9.41) produces

$$dY_{t+\Delta t} = Y_t + \kappa(\overline{Y} - Y_t)\Delta t + \sigma\sqrt{Y_t}\sqrt{\Delta t}Z_2 + \frac{1}{4}\sigma^2\Delta t(Z_2^2 - 1),$$

where Z_1 and Z_2 are generated in the same way as in the Euler scheme. For the S_t process, the coefficients are $\alpha(x) = rx$ and $\beta(x) = \sqrt{Y_t}x$. Therefore, the scheme is

$$dS_{t+\Delta t} = S_t + rS_t\Delta t + \sqrt{Y_t}S_t\sqrt{\Delta t}Z_1 + \frac{1}{2}Y_tS_t\Delta t(Z_2^2 - 1).$$

For the process $X_t = \log S_t$, we can apply Itô's lemma to the two-dimensional function $\log(x, y)$ to obtain the dynamics as

$$dX_t = \left(r - \frac{Y_t}{2}\right)dt + \sqrt{Y_t}dZ_1(t).$$

So the Milstein scheme for this process will be

$$X_{t+\Delta t} = X_t + \left(r - \frac{Y_t}{2}\right)\Delta t + \sqrt{Y_t}\sqrt{\Delta t}Z_1,$$

which, in fact, is the same as the Euler discretization in this case because the last term involves the derivative of a constant function.

9.10 Random Vectors: Moments and Distributions

As we saw in Chapter 8 a random vector $\mathbf{X}$ is any measurable function defined on the probability space $(\Omega, \mathscr{F}, \mathbf{P})$ with values in $\mathbb{R}^n$. Any random vector has a distribution function, defined similarly with the one-dimensional case. Specifically, if the random vector $\mathbf{X}$ has components $\mathbf{X} = (X_1, \ldots, X_n)$, its distribution function is

$$F_{\mathbf{X}}(\mathbf{x}) = \mathbf{P}(\mathbf{X} \leq \mathbf{x}) = \mathbf{P}(X_1 \leq x_1, \ldots X_n \leq x_n).$$

If the n-dimensional function F is differentiable, then the random vector will have a density (the joint density) f such that

$$F(\mathbf{x}) = F(x_1, \ldots, x_n) = \int_{-\infty}^{x_1} \cdots \int_{-\infty}^{x_n} f(t_1, \ldots, t_n) dt_n \ldots dt_1.$$

Using these notions, we can of course define the moments of the distribution. In fact, suppose that $g : \mathbb{R}^n \to \mathbb{R}$ is any function, then we can calculate the expected value of random variable $g(X_1, \ldots, X_n)$ when the joint density exists as:

$$E[g(X_1, \ldots, X_n)] = \int_{-\infty}^{\infty} \cdots \int_{-\infty}^{\infty} g(x_1, \ldots, x_n) f(x_1, \ldots, x_n) dx_1 \ldots dx_n$$

Now we can define the moments of the random vector. The first moment is a vector

$$E[\mathbf{X}] = \mu_{\mathbf{X}} = \begin{pmatrix} E[X_1] \\ \vdots \\ E[X_n] \end{pmatrix}.$$

We recall that the expectation applies to each component in the random vector. The second moment requires calculating all the combination of the components. The result can be presented in a matrix form. The second central moment can be presented as the covariance matrix.

$$\mathrm{Cov}(\mathbf{X}) = E[(\mathbf{X} - \mu_{\mathbf{X}})(\mathbf{X} - \mu_{\mathbf{X}})^t]$$

$$= \begin{pmatrix} \mathrm{Var}(X_1) & \mathrm{Cov}(X_1, X_2) & \cdots & \mathrm{Cov}(X_1, X_n) \\ \mathrm{Cov}(X_2, X_1) & \mathrm{Var}(X_2) & \cdots & \mathrm{Cov}(X_2, X_n) \\ \vdots & \vdots & \ddots & \vdots \\ \mathrm{Cov}(X_n, X_1) & \mathrm{Cov}(X_n, X_2) & \cdots & \mathrm{Var}(X_n) \end{pmatrix}, \quad (9.43)$$

where we used the transpose matrix notation, and since the $\mathrm{Cov}(X_i, X_j) = \mathrm{Cov}(X_j, X_i)$, the matrix is symmetric. The transpose matrix notation of a matrix A is A^t.

The matrix is also positive semidefinite (nonnegative definite) as we shall see. First we define a positive definite matrix.

Definition 9.10.1 (Positive definite and semidefinite matrix). A square $n \times n$ matrix A is called positive definite if, for any vector $u \in \mathbb{R}^n$ nonidentically zero, we have

$$u^t A u > 0$$

A matrix A is called positive semidefinite (or nonnegative definite) if, for any vector $u \in \mathbb{R}^n$ nonidentically zero, we have

$$u^t A u \geq 0$$

9.10 Random Vectors: Moments and Distributions

Now we explain why the covariance matrix has to be semidefinite. Take any vector $u \in \mathbb{R}^n$. Then the product $u^t \mathbf{X} = \sum u_i X_i$ is a random variable (one dimensional) and its variance must be nonnegative. But let us calculate its variance. Its expectation is $E[u^t \mathbf{X}] = u^t \mu_\mathbf{X}$. Then we can write (since for any number a, $a^2 = aa^t$)

$$\text{Var}(u^t \mathbf{X}) = E\left[\left(u^t \mathbf{X} - u^t \mu_\mathbf{X}\right)^2\right] = E\left[\left(u^t \mathbf{X} - u^t \mu_\mathbf{X}\right)\left(u^t \mathbf{X} - u^t \mu_\mathbf{X}\right)^t\right]$$

$$= E\left[u^t \left(\mathbf{X} - \mu_\mathbf{X}\right)\left(\mathbf{X} - \mu_\mathbf{X}\right)^t (u^t)^t\right] = u^t \text{Cov}(\mathbf{X})\, u$$

Since the variance is always nonnegative, the covariance matrix must be nonnegative definite (or positive semi-definite). A square symmetric matrix $A \in \mathbb{R}^{n \times n}$ is positive semidefinite if $v^t A v \geq 0$, $\forall v \in \mathbb{R}^n$. This difference is in fact important in the context of random variables since you may be able to construct a linear combination $u^t \mathbf{X}$ which is not always constant but whose variance is equal to zero.

We now present examples of multivariate distributions.

9.10.1 The Dirichlet Distribution

$Dir(\boldsymbol{\alpha})$, named after Johann Peter Gustav Lejeune Dirichlet (1805–1859), is a multivariate distribution parameterized by a vector $\boldsymbol{\alpha}$ of positive parameters $(\alpha_1, \ldots, \alpha_n)$.

Specifically, the joint density of an n-dimensional random vector $\mathbf{X} \sim Dir(\boldsymbol{\alpha})$ is

$$f(x_1, \ldots, x_n) = \frac{1}{\mathbf{B}(\boldsymbol{\alpha})} \left(\prod_{i=1}^n x_i^{\alpha_i - 1} \mathbf{1}_{\{x_i > 0\}}\right) \mathbf{1}_{\{x_1 + \cdots + x_n = 1\}}.$$

The components of the random vector $\mathbf{X}$ thus are always positive and have the property $X_1 + \cdots + X_n = 1$. The normalizing constant $\mathbf{B}(\boldsymbol{\alpha})$ is the multinomial beta function, which is defined as:

$$\mathbf{B}(\boldsymbol{\alpha}) = \frac{\prod_{i=1}^n \Gamma(\alpha_i)}{\Gamma\left(\sum_{i=1}^n \alpha_i\right)} = \frac{\prod_{i=1}^n \Gamma(\alpha_i)}{\Gamma(\alpha_0)},$$

where we used the notation $\alpha_0 = \sum_{i=1}^n \alpha_i$ and $\Gamma(x) = \int_0^\infty t^{x-1} e^{-t} dt$ for the gamma function.

Because the Dirichlet distribution creates n positive numbers that always sum to 1, it is extremely useful to create candidates for probabilities of n possible outcomes. This distribution is very popular and related to the multinomial distribution which needs n numbers summing to 1 to model the probabilities in the distribution.

With the notation mentioned earlier and α_0 as the sum of all parameters, we can calculate the moments of the distribution. The first moment vector has coordinates:

$$E[X_i] = \frac{\alpha_i}{\alpha_0}$$

The covariance matrix has elements:

$$\text{Var}(X_i) = \frac{\alpha_i(\alpha_0 - \alpha_i)}{\alpha_0^2(\alpha_0 + 1)},$$

and when $i \neq j$

$$\text{Cov}(X_i, X_j) = \frac{-\alpha_i \alpha_j}{\alpha_0^2(\alpha_0 + 1)}.$$

The covariance matrix is singular (its determinant is zero).

Finally, the univariate marginal distributions are all beta with parameters $X_i \sim \text{Beta}(\alpha_i, \alpha_0 - \alpha_i)$.

9.10.2 Multivariate Normal Distribution

A vector $\mathbf{X}$ is said to have a k-dimensional multivariate normal distribution with mean vector $\mu = (\mu_1, \ldots, \mu_k)$ and covariance matrix $\Sigma = (\sigma_{ij})_{ij \in \{1,\ldots,k\}}$ (denoted $MVN_k(\mu, \Sigma)$) if its density can be written as

$$f(\mathbf{x}) = \frac{1}{(2\pi)^{k/2} \det(\Sigma)^{1/2}} e^{-\frac{1}{2}(\mathbf{x}-\mu)^T \Sigma^{-1}(\mathbf{x}-\mu)},$$

where we used the usual notations for the determinant, transpose, and inverse of a matrix. The vector of means μ may have any elements in $\mathbb{R}$, but, just as in the one-dimensional case, the standard deviation has to be positive. In the multivariate case, the covariance matrix Σ has to be symmetric and positive definite.

The multivariate normal defined thus has many nice properties. The basic one is that the one-dimensional distributions are all normal, that is, $X_i \sim N(\mu_i, \sigma_{ii})$ and $\text{Cov}(X_i, X_j) = \sigma_{ij}$. This is also true for any marginal. For example, if $(X_r, \ldots, X_k)$ are the last coordinates, then

$$\begin{pmatrix} X_r \\ X_{r+1} \\ \vdots \\ X_k \end{pmatrix} \sim MVN_{k-r+1} \left(\begin{pmatrix} \mu_r \\ \mu_{r+1} \\ \vdots \\ \mu_k \end{pmatrix}, \begin{pmatrix} \sigma_{r,r} & \sigma_{r,r+1} & \cdots & \sigma_{r,k} \\ \sigma_{r+1,r} & \sigma_{r+1,r+1} & \cdots & \sigma_{r+1,k} \\ \vdots & \vdots & \ddots & \vdots \\ \sigma_{k,r} & \sigma_{k,r+1} & \cdots & \sigma_{k,k} \end{pmatrix} \right)$$

So any particular vector of components is normal.

Conditional distribution of a multivariate normal is also a multivariate normal. Given that $\mathbf{X}$ is a $MVN_k(\mu, \Sigma)$ and using the vector notations discussed earlier and assuming that $\mathbf{X}_1 = (X_1, \ldots, X_r)$ and $\mathbf{X}_2 = (X_{r+1}, \ldots, X_k)$, then we can write the vector μ and matrix Σ as

$$\mu = \begin{pmatrix} \mu_1 \\ \mu_2 \end{pmatrix} \text{ and } \Sigma = \begin{pmatrix} \Sigma_{11} & \Sigma_{12} \\ \Sigma_{21} & \Sigma_{22} \end{pmatrix},$$

where the dimensions are accordingly chosen to match the two vectors (r and $k - r$). Thus the conditional distribution of $\mathbf{X}_1$ given $\mathbf{X}_2 = \mathbf{a}$, for some vector $\mathbf{a}$, is

$$\mathbf{X}_1 | \mathbf{X}_2 = \mathbf{a} \sim MVN_r \left(\mu_1 - \Sigma_{12} \Sigma_{22}^{-1} (\mu_2 - \mathbf{a}), \Sigma_{11} - \Sigma_{12} \Sigma_{22}^{-1} \Sigma_{21} \right).$$

Furthermore, the vectors $\mathbf{X}_2$ and $X_1 - \Sigma_{21} \Sigma_{22}^{-1} X_2$ are independent. Finally, any affine transformation $AX + b$, where A is a $k \times k$ matrix and b is a k-dimensional constant, is also a multivariate normal with mean vector $A\mu + b$ and covariance matrix $A \Sigma A^T$. This can be easily verified by performing a joint change of variables.

9.11 Generating Multivariate (Gaussian) Distributions with Prescribed Covariance Structure

In this section, we will present a technique used to generate multivariate normal vectors with prescribed mean and covariance matrix. This technique can be applied to other distributions as well, but the covariance matrix needs to be calculated in terms of the parameters of the distribution.

9.11.1 Generating Gaussian Vectors

We want to generate a multivariate normal vector with a given mean vector

$$\mu = \begin{pmatrix} \mu_1 \\ \vdots \\ \mu_n \end{pmatrix}$$

and covariance matrix

$$\Sigma = \begin{pmatrix} \sigma_{11} & \sigma_{12} & \cdots & \sigma_{1n} \\ \sigma_{21} & \sigma_{22} & \cdots & \sigma_{2n} \\ \vdots & \vdots & \vdots & \vdots \\ \sigma_{n1} & \sigma_{n2} & \cdots & \sigma_{nn} \end{pmatrix}.$$

Suppose that we can find an n-dimensional square matrix R such that $R^T R = \Sigma$.

Then the algorithm is as follows: Generate a vector

$$\mathbf{X} = \begin{pmatrix} X_1 \\ \vdots \\ X_n \end{pmatrix}$$

where the X_i's are n independent standard normal random variables $N(0, 1)$.

Then the vector

$$\mathbf{Y} = R^T \mathbf{X} + \mu$$

is an n-dimensional vector with the multivariate normal distribution required.

The proof is direct, we just calculate the mean vector and covariance matrix of $R^T \mathbf{X} + \mu$ and using the fact that the covariance matrix of the vector $\mathbf{X}$ is the identity matrix.

In order to generate a multivariate normal vector with covariance matrix, Σ we need to find a matrix R such that $R^T R = \Sigma$. There are in fact many ways to find the matrix R, but here we mention the most popular way, the Cholesky decomposition, which uses the eigenvalues and eigenvectors of the matrix Σ.

Lemma 9.11.1 (Cholesky decomposition). Given a symmetric, positive, semidefinite matrix Σ, there exists a U upper triangular matrix and D a diagonal matrix such that

$$\Sigma = U^T D U.$$

The decomposition is unique when Σ is positive definite.

Note that an upper triangular matrix transposed is lower triangular, and in some other books the lower triangular form is used. As mentioned in previous sections of this chapter, a matrix Σ is positive semidefinite if, for every u vector with real elements, we have

$$u^T \sum u \geq 0.$$

A matrix Σ is positive definite if, for every u vector nonidentically 0 with real elements, we have

$$u^T \sum u > 0.$$

This condition fits the random vectors very well. Note that for any $\mathbf{X}$ vector and any u n-dimensional vector of constants we have, $u^T \mathbf{X}$ is a one-dimensional random variable. Therefore, its variance must be nonnegative. But its variance is exactly

$$\text{Var}(u^T \mathbf{X}) = E[(u^T \mathbf{X} - u^T \mu)^2] = u^T \sum u \geq 0.$$

In fact, if u is not identically 0, then the variance of the resulting random variable is positive.

Therefore, the input matrix should be symmetric and positive definite. It can be easily verified that symmetric positive definite matrix has positive eigenvalues. To check for the positive definiteness of a matrix using the R statistical software, we run the function *eigen(matrixname)* and inspect the resulting eigenvalues. If they are all positive, then the matrix is positive definite.

There is a drawback to this check. The eigenvalues of a matrix Σ are the roots of the characteristic polynomial $f(\lambda) = \det(\Sigma - \lambda I_n)$, where I_n denotes the identity matrix. The theory says that all the roots of the characteristic polynomial are real and positive if the matrix is symmetric and positive definite. However, in practice the computer does not calculate these roots exactly; instead, it approximates them. So, there may be issues especially if one of the roots is close to 0. There is another way to check by looking at the coefficients of the polynomial $f(\lambda)$ directly. Descartes' Rule of Signs [7] states that the number of positive roots of a polynomial equals the number of sign changes in the coefficients. Since the characteristic polynomial of a positive definite matrix has n positive roots, the coefficients must have exactly n changes of sign. Finally, there is another way if computing the polynomial is hard but calculating determinants of constant matrices of lower dimensions is straightforward. A matrix Σ is positive definite if and only if the diagonal elements are all positive and the determinants of all the upper left corners are positive.

9.12 Problems

1. State and prove the two-dimensional Itô formula.
2. Prove that if X_t, Y_t are two Itô processes of the form (9.9), then

   $$d(X_t Y_t) = X_t dY_t + Y_t dX_t + dX_t dY_t,$$

   where in the last term the same rules apply as in the Itô formula.
3. Calculate

   $$\int_0^t s \, dB_s.$$

   Hint: Apply Itô lemma to $X_t = B_t$ and $f(t, x) = tx$.
4. Calculate the solution to the following mean-reverting Ørnstein–Uhlenbeck SDE:

   $$dX_t = \mu X_t dt + \sigma dB_t$$

   with $X_0 = x$.
5. Calculate the solution to the following SDE:

   $$dX_t = \alpha(m - X_t)dt + \sigma dB_t$$

   with $X_0 = x$. The process satisfying this equation is called the mean-reverting Ørnstein–Uhlenbeck process.

6. Using a software of your choice, generate points $(Y_1(i\Delta t), Y_2(i\Delta t))$ solving

$$\begin{cases} dY_1(t) = -\frac{1}{2}Y_1(t)dt - Y_2(t)dB_t \\ dY_2(t) = -\frac{1}{2}Y_2(t)dt + Y_1(t)dB_t \end{cases}$$

using an Euler approximation. Use $Y_1(0) = 1$, $Y_2(0) = 0$, $\Delta t = 1/1000$, and $i \in \{1, 2, \ldots, 1000\}$. Plot the resulting pairs of points in a two-dimensional space where the first coordinate is the Y_1 value and the second coordinate is the Y_2 value.

7. Let B_t be a standard Brownian motion started at 0. Use that for any function f we have:

$$E[f(B_t)] = \frac{1}{\sqrt{2\pi t}} \int_{-\infty}^{\infty} f(x) e^{-\frac{x^2}{2t}} dx,$$

to calculate:

$$E[B_t^{2k}]$$

for some k, an integer. As a hint, you may want to use integration by parts and induction to come up with a formula for $E[B_t^{2k}]$.

8. Using your choice of software, simulate B_t a standard Brownian motion on the interval $t \in [0, 2]$. Using the simulated paths and the Central Limit Theorem, estimate

$$E[B_2^4], E[B_2^8], E[B_2^{20}].$$

Then simulate each of the processes B_t^4, B_t^8, and B_t^{20} separately and obtain the previous expectations at $t = 2$.
Compare all the numbers obtained and also compare with the values in the previous problem. Use a minimum of one million simulated paths, and for each path use a time increment of $\Delta t = 0.01$.

9. Let X_t, $t \geq 0$, be defined as

$$X_t = \{B_t \mid B_t \geq 0\}, \quad \forall t > 0,$$

that is, the process has the paths of the Brownian motion conditioned by the current value being positive.
   (a) Show that the pdf of X_t is

   $$f_{X_t}(x) = 2f_{B_t}(x), \forall x \geq 0.$$

   (b) Calculate $E[X_t]$ and $V(X_t)$.
   (c) Is X_t a Gaussian process?
   (d) Is X_t stationary?
   (e) Are X_t and $|B_t|$ identically distributed?

10. If $X_t \sim N(0, t)$, calculate the distribution of $|X_t|$. Calculate $E|X_t|$ and $V(|X_t|)$.

11. If $X_t \sim N(0, \sigma^2 t)$ and $Y_t = e^{X_t}$, calculate the pdf of Y_t. Calculate $\mathbf{E}[Y_t]$ and $V(Y_t)$. Calculate the transition probability

$$\mathbf{P}(Y_t = y \mid Y_{t_0} = y_0),$$

and give the density of this transition probability.

12. Prove by induction that

$$\int_0^T B_t^k dB_t = \frac{B_T^{k+1}}{k+1} - \frac{k}{2} \int_0^T B_t^{k-1} dt.$$

13. Solve the following SDEs using the general integrating factor method with $X_0 = 0$:
    (a) $dX_t = \frac{X_t}{t} dt + \sigma t X_t dB_t$.
    (b) $dX_t = X_t^\alpha + \sigma X_t dB_t$. For what values of α the solution of the equation explodes (is equal to ∞ or $-\infty$ for a finite time t).

14. Show that the stochastic process

$$e^{\int_0^t c(s)dB_s - \frac{1}{2}\int_0^t c^2(s)ds}$$

is a martingale for any deterministic function $c(t)$. Does the result change if $c(t, \omega)$ is a stochastic process such that the stochastic integral is well defined?

15. Give an explicit solution for the mean-reverting Ørnstein–Uhlenbeck SDE

$$dX_t = \alpha(\mu - X_t)dt + \sigma dB_t,$$

with $X_0 = x$.

16. Suppose $\theta(t, \omega) = (\theta_1(t, \omega), \theta_2(t, \omega), \ldots, \theta_n(t, \omega))$ is a stochastic process in $\mathbb{R}^n$ such that $\theta_i(t, \omega) \in v_T$ for all i. Define the process

$$Z_t = e^{\int_0^t \theta(s,\omega)dB_s - \frac{1}{2}\int_0^t \theta^2(s,\omega)ds},$$

where $\mathbf{B}_t$ is an n-dimensional Brownian motion and all terms in the integrals are calculated as scalar products, for example,

$$\theta(s, \omega)d\mathbf{B}_s = \sum_{i=1}^n \theta_i(s, \omega)dB_i(s)$$

and

$$\theta^2(s, \omega)ds = \sum_{i=1}^n \theta_i^2(s, \omega).$$

Note that the process Z_t is one dimensional. Use Itô formula to show that

$$dZ_t = Z_t \theta(t, \omega)d\mathbf{B}_t$$

and derive that Z_t is a martingale if and only if $Z_t \theta_i(t, \omega) \in v_T$ for all i.

17. Generate and plot five Monte Carlo simulations of the process

$$dX_t = 10(20 - X_t)dt + 0.7dB_t,$$

with $X_0 = 20$. Use the Euler scheme for this problem.

18. Give the Euler–Milstein approximation scheme for the following SDE:

$$dS_t = \mu_S S_t dt + \sigma S_t^\beta dB_t,$$

where $\beta \in (0, 1]$. Generate five paths and plot them for the following parameter values:

$$\mu_S = 0.1, \quad \sigma = 0.7, \quad \beta = \frac{1}{3}, \quad S_0 = 100$$

19. Consider the [102] model

$$dS_t = \mu_S S_t dt + \sqrt{Y_t} S_t dB_t$$
$$dY_t = \mu_Y Y_t dt + \xi Y_t dW_t.$$

The volatility process is $\sqrt{Y_t}$. Show that the volatility process has moments

$$E[\sqrt{Y_t}] = \sqrt{Y_0}\, e^{\frac{1}{2}\mu_Y t - \frac{1}{8}\xi^2 t}$$

$$V[\sqrt{Y_t}] = \sqrt{Y_0}^2\, e^{\mu_Y t}\left(1 - e^{-\frac{1}{4}\xi^2 t}\right)$$

and study what happens with these moments when $t \to \infty$ depending on the values of the parameters in the model.

Part III

Advanced Models for Underlying Assets

10

Stochastic Volatility Models

10.1 Introduction

In Chapter 3 of this book, we discussed the implied volatility and we observed that it is hardly constant for any given stock. This is a violation of the assumption of constant volatility in the model. Further evidence may be provided by calculating histograms of the returns and testing for normality. Most of such tests fail especially when the time interval for calculating returns is less than a day. Thus, the entire class of High Frequency models needs to be more complex than a geometric Brownian motion, hence the need for stochastic volatility (SV) model.

10.2 Stochastic Volatility

We begin this section with the following definition.

Definition 10.2.1 We call any model of the following type:

$$dS_t = \mu(t, S_t, Y_t)dt + \sigma(t, S_t, Y_t)dW_t \qquad (10.1)$$

a continuous time SV model. The functions $\mu(\cdot, \cdot, \cdot)$ and $\sigma(\cdot, \cdot, \cdot)$ are twice differentiable in all variables, and with continuous second partial derivatives, W_t is a standard Brownian motion and Y_t is a stochastic process.

The stochastic factor Y_t is typically not directly observable and it is the major innovation behind continuous time SV models. The conditions on the functions μ and σ are needed so that the solution to the stochastic Eq. (10.2) in S_t exists and is unique.

Using some change of variables and under the risk-neutral probability measure, Eq. (10.1) has an equivalent formulation as follows:

$$dS_t = rS_t dt + \sigma(t, S_t, Y_t)dW_t \qquad (10.2)$$

Quantitative Finance, First Edition. Maria C. Mariani and Ionut Florescu.
© 2020 John Wiley & Sons, Inc. Published 2020 by John Wiley & Sons, Inc.

This is the general formulation that will be used throughout this chapter.

The volatility is typically approximated using some sort of quadratic variation of the process. For example, the simple rolling standard deviation accomplishes this. Let $r_1, \ldots, r_N$ be a long sequence of returns calculated using some time interval Δt; the estimated rolling standard deviation σ_n at time n is:

$$\sigma_n^2 = \frac{1}{m-1} \sum_{i=0}^{m-1} (r_{n-i} - \bar{r})^2,$$

using a rolling window of m observations. Realized volatility uses a simpler equation; nevertheless we remark that if the coefficient of W_t is $\sigma(t, S_t, Y_t) = \sigma S_t$ with σ a constant as in the Black–Scholes model, then necessarily the estimate mentioned earlier will be a constant equal to $\sigma \sqrt{\Delta t}$.

If we take $\sigma(t, S_t, Y_t) = \sigma(t) S_t$, where $\sigma(t)$ is not a constant but a deterministic function of t, then the estimate mentioned earlier will converge to $\frac{1}{N} \int_0^{N\Delta t} \sigma(t) dt$.

In practice using real data, these estimates are never constant and in fact never deterministic (cannot be predicted in advance). It seems reasonable to assume that any model constructed for the asset price process would have a random standard deviation. The SV model is one way to accomplishing this modeling feature.

This is important since derivative prices are very sensitive to *future* (both implied and instantaneous) volatility levels.

A question that normally arises is which specific volatility to use. In order to answer this, we need to identify which features of the real data we want to capture and then we specify the model appropriately. However in practice, it is often difficult to understand how features of SV models are translated in resulting distributions. For example, consider the use of an SV model for the management of options on swaps versus the use of the same model in the plain vanilla option market. Obviously the two markets have different characteristics; thus an SV model that can be perfectly adequate to capture the risk in one of the aforementioned categories may completely miss the exposures in other products.

10.3 Types of Continuous Time SV Models

The ability of SV models to produce data similar with observed real data has been understood for a long time. Due to the wide range of possible applications, there is not a unique standard model that fits every occasion. As mentioned in the previous section, an SV model that can be perfectly adequate to capture the risk in some instances may completely miss the exposures in other products. This is the reason for which there are so many variants of SV models.

We begin our discussion by presenting the case where volatility is a function of stock and time only.

Consider the model

$$dS_t = rS_t dt + \sigma(t, S_t) dW_t \tag{10.3}$$

where S_t is the stock price, r is the risk-free interest rate, $\sigma(t, x)$ is a deterministic function, and W_t is a Brownian motion.

Note that the geometric Brownian motion in the Black-Scholes model [24] is a particular case of this model (when $\sigma(t, S_t) = \sigma S_t$). Also recall this is the only case where the corresponding log process ($X_t = \log S_t$) has independent identically distributed (i.i.d.) increments; the returns calculated over Δt intervals $r_t = \log S_t - \log S_{t-\Delta t}$ are i.i.d. as well.

The other well studied model of this type is the Constant Elasticity of Variance (CEV) model.

10.3.1 Constant Elasticity of Variance (CEV) Models

The general model was introduced in [43] and recently in [44]. The first official version of this model was presented in [45].

The CEV model is written as:

$$dS_t = rS_t dt + \sigma S_t^{\beta/2} dW_t, \quad \beta < 2, \tag{10.4}$$

where W_t is a Brownian motion and σ is a positive constant.

Applying Itô's lemma to the log process $X_t = \log S_t$ gives the equation:

$$dX_t = \left(r - \frac{\sigma^2 S_t^{\beta-2}}{2}\right) dt + \sigma S_t^{\frac{\beta-2}{2}} dW_t, \quad \beta < 2. \tag{10.5}$$

Therefore the variance of the return process is:

$$v(t, s) = \sigma^2 s^{\beta-2} \tag{10.6}$$

Elasticity (i.e. the ability to change and adapt) of the variance when stock is changing is defined as the change in variance corresponding to change in stock price. Differentiating (10.6) by s, we obtain:

$$\frac{dv(t, s)}{ds} = \sigma^2(\beta - 2) s^{\beta-3} \tag{10.7}$$

and (10.8) is obtained by dividing (10.6) by s as follows:

$$\frac{v(t, s)}{s} = \sigma^2 s^{\beta-3}. \tag{10.8}$$

Therefore the ratio of (10.7) and (10.8), i.e.

$$\frac{dv(t, s)/ds}{v(t, s)/s} = \beta - 2.$$

Rewriting the above equation gives:

$$\frac{dv(t,S_t)}{v(t,S_t)} = \frac{dv}{v}(t,S_t) = (\beta - 2)\frac{dS_t}{S_t}$$

which suggests the name constant elasticity of variance (which equals $\beta - 2$).

If $\beta = 2$ the elasticity is 0 and the variance does not react to stock prices. In fact, the stock process follows the usual geometric Brownian motion [24]. If $\beta = 1$ one obtains the original model proposed by [45] which has elasticity -1.

For the case where $\beta = 1$, we obtain the following option pricing formula for a European Call with strike price K and time to maturity $\tau = T - t$ as follows:

$$S_t e^{-r\tau} \sum_{n=0}^{\infty} \frac{e^{-x} x^n G(n + 1 + \frac{1}{2-\beta}, \kappa K^{2-\beta})}{\Gamma(n+1)} - Ke^{-r\tau} \sum_{n=0}^{\infty} \frac{e^{-x} x^{n + \frac{1}{2-\beta}} G(n+1, \kappa K^{2-\beta})}{\Gamma(n+1+\frac{1}{2-\beta})},$$

where $\kappa = \frac{2r}{\sigma^2(2-\beta)[e^{r(2-\beta)\tau}-1]}$ and $G(x,m)$ is the probability of the upper tail of a Gamma distribution with shape parameter m and scale parameter 1, i.e.

$$G(x,m) = \frac{1}{\Gamma(m)} \int_x^{\infty} u^{m-1} e^{-u} du.$$

For option pricing methodology for generalized CEV models, we refer the reader to [99].

Next we consider the case when the instantaneous volatility is modeled as a diffusion.

In this case the SV model has a form given by Eq. (10.2) where the process Y_t is specified as a separate diffusion process. Specifically, Y_t is the solution of a stochastic differential equation of the form:

$$dY_t = \alpha(Y_t)dt + \beta(Y_t)dZ_t,$$

and the Brownian motion Z_t may or may not be correlated with the original Brownian motion W_t.

We like to call these models "true" SV models since the randomness comes from an exogenous specified process. An exogenous change is one that comes from outside the model and is unexplained by the model. We present examples of these models.

10.3.2 Hull–White Model

The first SV model was introduced by Hull and White (see [102]) who took the correlation between the two Brownian motions ($\rho = 0$). This model was generalized by Wiggins (see [209]) to the more general case when the driving Brownian motions are correlated ($\rho \neq 0$).

The equation under the risk neutral measure is written as:

$$\begin{cases} dS_t = rS_t dt + \sqrt{Y_t} S_t dW_t \\ dY_t = \mu_Y Y_t dt + \xi Y_t dZ_t \end{cases} \tag{10.9}$$

The parameters μ_Y and ξ are the volatility drift and the volatility of volatility respectively. Note that the volatility process is a simple geometric Brownian motion. Hull, White and then Wiggins calculated the moments of the volatility (the process $\sqrt{Y_t}$) as:

$$E[\sqrt{Y_t}] = \sqrt{Y_0}\, e^{\frac{1}{2}\mu_Y t - \frac{1}{8}\xi^2 t}$$
$$V[\sqrt{Y_t}] = \sqrt{Y_0}^2 e^{\mu_Y t}\left(1 - e^{-\frac{1}{4}\xi^2 t}\right)$$

Note that the calculation is straightforward and is dependent only on the process Y_t; thus in fact these are the mean and variance of a geometric Brownian motion. In the forthcoming section, we shall present a more general calculation of moments for a diffusion process.

Analyzing these formulas we can see that if $\mu_Y < \frac{1}{4}\xi^2$, the expectation of the volatility converges to zero. Similarly, if $\mu_Y > \frac{1}{4}\xi^2$ the expectation approaches infinity.

Furthermore, if we analyze the variance, in the case when $\mu_Y > 0$, the variance grows unbounded. If $\mu_Y < 0$ the variance converges to zero. Neither of these properties is desirable. For the case when $\mu_Y > 0$, variance starts to oscillate, and when $\mu_Y < 0$, the variability dies out and essentially becomes deterministic. In the special case when $\mu_Y = 0$, the variance of the volatility converges to the initial value $\sqrt{Y_0}^2$; however the expectation of the volatility goes to zero.

10.3.3 The Stochastic Alpha Beta Rho (SABR) Model

This the most popular SV model used in practice. We chose to present it here rather than wait until its proper chronological order because the dynamic of the volatility is in fact identical with that in the Hull–White model.

The model was introduced by [86]. The original form introduced by the authors is:

$$\begin{cases} dS_t = rS_t dt + \sigma_t S_t^\beta dW_t \\ d\sigma_t = \alpha \sigma_t dZ_t \end{cases}, \quad \beta \in [0,1], \alpha \geq 0, \tag{10.10}$$

and ρ is the correlation of the geometric Brownian motions. Note that in practice the drift term $rS_t dt$ is typically missing. One may see the similarities in this formulation with the CEV model, and in fact this was the original motivation of the model that makes the coefficient stochastic.

In this model let us make a change of variables: $Y_t = \sigma_t^2$. Then apply Itô's lemma to obtain the dynamics of Y_t and finally rewrite the SABR model in terms of this Y_t. This gives the following formulation:

$$\begin{cases} dS_t = rS_t dt + \sqrt{Y_t} S_t^\beta dW_t \\ dY_t = \alpha^2 Y_t dt + 2\alpha Y_t dZ_t \end{cases} \tag{10.11}$$

Note that it is very clear now that the volatility follows the same process as it does in the Hull–White model equation [102] with the parameter values: $\mu_y = \alpha^2$ and $\xi = 2\alpha$. With this choice of parameters *the expectation of the volatility stays constant*; however *the variance of the volatility grows unbounded*. Thus in the SABR model even if the expectation is constant, the unbounded variance means that the actual volatility value will approach infinity.

This model is desirable in practice since it has an option pricing formula. Thus, there is also a formula for the implied volatility in this model. So it can be matched very easily (and fast) with the local volatility approach pioneered by [59]. The speed issue in particular is extremely important since any option chain (which includes all options written on a particular equity) needs to be able to update itself on the order of seconds (today milliseconds) to keep up with changes in the underlying equity price. Thus analytical formulas are extremely desirable.

We can find many examples in [6] which analyzed this and many other popular models and their resulting explosive behavior of volatility.

10.3.4 Scott Model

Scott [181] and later [36] are historically relevant models since these articles introduce for the first time mean reversion in the volatility dynamics. In [181], Scott assumed that two Brownian motions are uncorrelated ($\rho = 0$). The model is given as:

$$\begin{cases} dS_t = rS_t dt + e^{Y_t} S_t dW_t \\ dY_t = \alpha(\overline{Y} - Y_t)dt + \sigma_Y dZ_t \end{cases} \tag{10.12}$$

where α, $\overline{Y}$ and σ_Y are parameters, which are assumed to be positive. The dynamics for the volatility driving process Y_t are given by a a classical mean-reverting Ørnstein–Uhlenbeck (O-U) process.

In this model the volatility cannot converge to zero due to the exponential in the volatility term.

This model implies a strong negative correlation between spot price and volatility. The spot price is the current market price at which an asset is bought or sold for immediate payment and delivery. This is a very good feature of this model. A strong negative correlation between spot returns and instantaneous volatility is well documented in literature. In fact this relationship is named as *the leverage effect*. The leverage effect refers to the observed tendency of an asset's volatility to be negatively correlated with the asset's returns. We refer to [69] for an in-depth study of models that are capable of producing this effect.

The larger the value of the α parameter (which is in fact called the rate or speed of mean reversion), the faster the volatility distribution goes back to its asymptotic constant value ($\overline{Y}$). However, if the mean reversion speed is large, correlation between volatility and spot returns is not sufficient to generate a significant skew in the implied volatility plot (the plot of the implied volatility with respect to the strike price).

10.3.5 Stein and Stein Model

In [194], a variant of the Scott model is introduced. Specifically, the model used is:

$$\begin{cases} dS_t = rS_t dt + Y_t S_t dW_t \\ dY_t = \alpha(\overline{Y} - Y_t)dt + \sigma_Y dZ_t \end{cases} \quad (10.13)$$

The volatility driving process Y_t is modeled as a mean reverting O-U process.

The main feature of this model is that the sign of the correlation between spot returns and volatility can switch from negative to positive and vice versa.

The advantage of these two models is that the stochastic differential equation for Y_t can be solved, and accordingly the distribution of Y_t can be easily calculated.

Remarks 10.3.1 An important misconception about volatility is the belief that the volatility is always positive. The volatility is estimated as the square root of the quadratic variation process. So we choose to always use the positive value (just like we do for the standard deviation). So for identifiability purposes we always chose the process that generates positive values.

10.3.6 Heston Model

This is perhaps the best known SV model [95]. This model is also heavily used in practice. As in the case of the SABR model discussed earlier, the reason is that the Heston model provides analytical formulas for the European Call (and thus Put) options. The analytical formulas here involve calculating Fourier transforms which in practice is impossible thus typically an approximation using the discrete Fourier transform is used. Nevertheless the implementation is extremely fast.

The Heston model is:

$$\begin{cases} dS_t = rS_t dt + \sqrt{Y_t} S_t dW_t \\ dY_t = \alpha(\overline{Y} - Y_t)dt + \sigma_Y \sqrt{Y_t} dZ_t \end{cases} \quad (10.14)$$

We recall from Section 3.6 our discussion of the CIR model used in forward rate modeling that the volatility process Y_t cannot go below zero. Thus the square root appearing in the volatility of S_t is not a problem in theory. However, in practice (when we simulate the price process by Euler approximation), one needs to keep the time interval Δt very small to avoid these problems.

Unlike the previous models (the Scott and the Stein models), introducing the square root changes the properties of the volatility. This addition makes working directly with the dynamics of Y_t complex. However, having this specification makes the resulting return dynamics affine, and this is the reason behind why pricing formulas exist for European Call and Put options.

Since option formulas exist one may use the observed prices to estimate parameters. Again, as it is in the case of the SABR model, the existence of the formulas insure a fast matching methodology.

In the next section, we formally discuss the mean-reverting O-U process.

10.4 Derivation of Formulae Used: Mean-Reverting Processes

In this section we provide some details on simple stochastic calculus calculations. Let Y_t denote a mean-reverting O-U process formulated as:

$$dY_t = \alpha(m - Y_t)dt + \beta dZ_t, \tag{10.15}$$

with α, m, and β as real constants and Z_t a standard Brownian motion.

Applying the Ito's lemma to the function $f(t, y) = y e^{\alpha t}$ we can obtain the explicit solution:

$$Y_t = e^{-\alpha t} Y_0 + m\left(1 - e^{-\alpha t}\right) + e^{-\alpha t} \int_0^t \beta e^{\alpha s} dZ_s. \tag{10.16}$$

Now we obtain formulas for the moments as follows: The expectation is immediate by just applying expectation in both terms mentioned earlier. Recall that the stochastic integral is a martingale; therefore its expected value is always the same as the value at 0, so

$$E[Y_t] = E[Y_0] e^{-\alpha t} + m(1 - e^{-\alpha t}) \tag{10.17}$$

The next formulae are hard to find in literature, and so we state them as separate results.

Lemma 10.4.1 With the process Y_t a mean-reverting O-U process specified as in (10.15), we have the covariance function of the process given by:

$$\text{Cov}(Y_t, Y_s) = e^{-\alpha(s+t)} \left[V(Y_0) + \frac{\beta^2}{2\alpha} \left(e^{2\alpha\, t \wedge s} - 1\right) \right], \tag{10.18}$$

where $V(Y_0)$ denotes the variance of the initial variable Y_0 and $t \wedge s$ denotes the minimum of the two numbers t and s.

Proof. Using Eqs. (10.16) and (10.17), we can write:

$$Y_t - E[Y_t] = \left(Y_0 - E[Y_0]\right) e^{-\alpha t} + e^{-\alpha t} \int_0^t \beta e^{\alpha u} dZ_u,$$

10.4 Derivation of Formulae Used: Mean-Reverting Processes

and we can calculate the covariance function of the process as:

$$\text{Cov}(Y_t, Y_s) = \mathbf{E}\left[(Y_t - \mathbf{E}[Y_t])(Y_s - \mathbf{E}[Y_s])\right]$$

$$= e^{-\alpha(t+s)}\mathbf{E}\left[\left(Y_0 - \mathbf{E}[Y_0] + \int_0^t \beta e^{\alpha u} dZ_u\right)\left(Y_0 - \mathbf{E}[Y_0] + \int_0^s \beta e^{\alpha v} dZ_v\right)\right]$$

$$= e^{-\alpha(t+s)}\left(V(Y_0) + \mathbf{E}\left[(Y_0 - \mathbf{E}[Y_0])\int_0^s \beta e^{\alpha v} dZ_v\right]\right.$$

$$+ \mathbf{E}\left[(Y_0 - \mathbf{E}[Y_0])\int_0^t \beta e^{\alpha u} dZ_u\right] + \beta^2 \mathbf{E}\left[\int_0^t e^{\alpha u} dZ_u \int_0^s e^{\alpha v} dZ_v\right]\right)$$

$$= e^{-\alpha(t+s)}\left(V(Y_0) + \beta^2 \int_0^{t \wedge s} e^{2\alpha u} du\right).$$

To obtain the last equality, we use the Itô isometry (see Section 1.9) in the last integral and for the two middle integrals the fact that a stochastic integral is a martingale with zero expectation. Finally, computing the sole remaining integral, we obtain the result stated in the lemma. □

We observe that by taking $t = s$ in the previous lemma, one obtains the variance of the process.

Next we present more results about this process. The way to prove the results and the calculations involved are very important. We believe they provide great exercises in stochastic calculus.

First, we calculate a stochastic differential equation for Y_t^2. We apply Itô's lemma to the function $f(t, y) = y^2 e^{2\alpha t}$ and the process Y_t.

$$d(Y_t^2 e^{2\alpha t}) = 2\alpha e^{2\alpha t} Y_t^2 dt + 2Y_t e^{2\alpha t} dY_t + \frac{1}{2}(2e^{2\alpha t}) d<Y, Y>_t$$

$$= e^{2\alpha t}\left(2\alpha m Y_t + \beta^2\right) dt + 2\beta Y_t e^{2\alpha t} dZ_t.$$

Therefore, we can express the equation in integral form as:

$$Y_t^2 = Y_0^2 e^{-2\alpha t} + 2\alpha m e^{-2\alpha t}\int_0^t e^{2\alpha u} Y_u du + \frac{\beta^2}{2\alpha}\left(1 - e^{-2\alpha t}\right)$$

$$+ 2\beta e^{-2\alpha t}\int_0^t e^{2\alpha u} Y_u dZ_u \tag{10.19}$$

Lemma 10.4.2 *The second moment of a mean-reverting O-U process Y_t specified as in (10.15) is given by:*

$$\mathbf{E}[Y_t^2] = e^{-2\alpha t}\left[\mathbf{E}[Y_0^2] + \frac{\beta^2}{2\alpha}\left(e^{2\alpha t} - 1\right) + 2m\mathbf{E}[Y_0]\left(e^{\alpha t} - 1\right) + m^2\left(e^{\alpha t} - 1\right)^2\right].$$

$$\tag{10.20}$$

Furthermore, the integrated second moment is given by:

$$\int_0^t \mathbf{E}[Y_s^2]dt = t\left(m^2 + \frac{\beta^2}{2\alpha}\right) + \left(\mathbf{E}[(Y_0 - m)^2] - \frac{\beta^2}{2\alpha}\right)\frac{1 - e^{-2\alpha t}}{2\alpha}$$
$$+ 2m\mathbf{E}[Y_0 - m]\frac{1 - e^{-\alpha t}}{\alpha}. \qquad (10.21)$$

Proof. The proof is left as an exercise (See problem 5). □

Example 10.4.1 If Y_t is an O-U process as in (10.15), give an expression for:
$$\mathbf{E}\left[Y_t^2 Y_s\right] - \mathbf{E}\left[Y_t^2\right]\mathbf{E}\left[Y_s\right]$$

Note that some expressions discussed earlier will help with this calculations. To step in the write direction we write

$$\mathbf{E}\left[Y_t^2 Y_s\right] = e^{-2\alpha t}e^{-\alpha s}\mathbf{E}\Bigg[\left(Y_0^2 + 2\alpha m\int_0^t e^{2\alpha u}Y_u du + \frac{\beta^2}{2\alpha}\left(e^{2\alpha t} - 1\right)\right.$$
$$\left. + 2\beta\int_0^t e^{2\alpha u}Y_u dZ_u\right)\left(Y_0 + m\left(e^{\alpha s} - 1\right) + \int_0^s \beta e^{\alpha v}dZ_v\right)\Bigg]$$
$$= e^{-\alpha(2t+s)}\left(I + II + III + IV\right),$$

where

$$I = \mathbf{E}\Bigg[\left(Y_0^2 + 2\alpha m\int_0^t e^{2\alpha u}Y_u du + \frac{\beta^2}{2\alpha}\left(e^{2\alpha t} - 1\right)\right)\left(Y_0 + m\left(e^{\alpha s} - 1\right)\right)\Bigg]$$
$$= \mathbf{E}(Y_0^3) + 2\alpha m\int_0^t e^{2\alpha u}\mathbf{E}[Y_u Y_0]du + \frac{\beta^2}{2\alpha}\left(e^{2\alpha t} - 1\right)\mathbf{E}(Y_0) + m\left(e^{\alpha s} - 1\right)\mathbf{E}(Y_0^2)$$
$$+ 2\alpha m^2\left(e^{\alpha s} - 1\right)\int_0^t e^{2\alpha u}\mathbf{E}[Y_u]du + \frac{\beta^2 m}{2\alpha}\left(e^{2\alpha t} - 1\right)\left(e^{\alpha s} - 1\right)$$

II and *III* are expectations of the cross-product terms which are both zero, and

$$IV = 2\beta^2 \mathbf{E}\left[\int_0^t e^{2\alpha u}Y_u dZ_u \int_0^s e^{\alpha v}dZ_v\right]$$
$$= 2\beta^2 \int_0^{t\wedge s} e^{3\alpha u}\mathbf{E}(Y_u)du.$$

From this expression we need to subtract $\mathbf{E}\left[Y_t^2\right]\mathbf{E}\left[Y_s\right]$ which is calculated using (10.17) and (10.19) as:

$$\mathbf{E}\left[Y_t^2\right]\mathbf{E}\left[Y_s\right] = e^{-\alpha(2t+s)}\left(\mathbf{E}(Y_0^2) + 2\alpha m\int_0^t e^{2\alpha u}\mathbf{E}(Y_u)du + \frac{\beta^2}{2\alpha}\left(e^{2\alpha t} - 1\right)\right)$$
$$\left(\mathbf{E}[Y_0] + m(e^{\alpha s} - 1)\right)$$
$$= e^{-\alpha(2t+s)}\Bigg(\mathbf{E}(Y_0^2)\mathbf{E}[Y_0] + 2\alpha m\int_0^t e^{2\alpha u}\mathbf{E}(Y_u)\mathbf{E}[Y_0]du + \frac{\beta^2}{2\alpha}\left(e^{2\alpha t} - 1\right)\mathbf{E}[Y_0]$$
$$+ m(e^{\alpha s} - 1)\mathbf{E}(Y_0^2) + 2\alpha m^2(e^{\alpha s} - 1)\int_0^t e^{2\alpha u}\mathbf{E}(Y_u)du + \frac{\beta^2 m}{2\alpha}\left(e^{2\alpha t} - 1\right)(e^{\alpha s} - 1)\Bigg)$$

10.4 Derivation of Formulae Used: Mean-Reverting Processes

Finally, to calculate the $\mathbf{E}\left[Y_t^2 Y_s\right] - \mathbf{E}\left[Y_t^2\right]\mathbf{E}\left[Y_s\right]$ we take the difference of the expressions calculated earlier. After simplifications we obtain:

$$e^{-\alpha(2t+s)} \left(\mathbf{E}(Y_0^3) - \mathbf{E}(Y_0^2)\mathbf{E}[Y_0] + 2\alpha m \int_0^t e^{2\alpha u}(\mathbf{E}[Y_u Y_0] - \mathbf{E}(Y_u)\mathbf{E}[Y_0])du \right.$$

$$\left. + 2\beta^2 \int_0^{t \wedge s} e^{3\alpha u} \mathbf{E}(Y_u) du \right)$$

Noting that in the first integral we have the term $\mathrm{Cov}(Y_u, Y_0)$ which we know from (10.18) is equal to $e^{-\alpha u} V(Y_0)$ and that in the second integral we have the expectation of the mean-reverting O-U process which we calculated in (10.17), we can substitute these terms and integrate them to finally obtain the following expression:

$$\mathbf{E}\left[Y_t^2 Y_s\right] - \mathbf{E}\left[Y_t^2\right]\mathbf{E}\left[Y_s\right] = e^{-\alpha(2t+s)} \left(\mathbf{E}(Y_0^3) - \mathbf{E}(Y_0^2)\mathbf{E}[Y_0]\right.$$

$$\left. + 2mV(Y_0)\left(e^{\alpha t} - 1\right) + \frac{\beta^2}{\alpha}\mathbf{E}(Y_0)\left(e^{2\alpha t \wedge s} - 1\right) + \frac{2m\beta^2}{\alpha}\left(\frac{e^{3\alpha t \wedge s}}{3} - \frac{e^{2\alpha t \wedge s}}{2}\right)\right)$$

(10.22)

We note that this particular derivation may also be seen in the appendix of [69].

We conclude the discussion of the various types of SV models with the moment analysis for CIR type processes.

10.4.1 Moment Analysis for CIR Type Processes

The SV specification in the Heston model follows a CIR type dynamics. Thus the square root term appearing in the model makes it impossible to solve exactly as we have done in the simpler O-U model (10.15). However, we may still be able to calculate its moments by using a very classical technique. To illustrate this technique first we write the process in its integral form:

$$Y_t - Y_0 = \int_0^t \alpha(\overline{Y} - Y_s)ds + \int_0^t \sigma\sqrt{Y_s}\, dZ_s.$$

Next, we apply expectation to both sides as shown in the following:

$$\mathbf{E}[Y_t] - Y_0 = \mathbf{E}\left[\int_0^t \alpha(\overline{Y} - Y_s)ds\right] + \mathbf{E}\left[\int_0^t \sigma\sqrt{Y_s}\, dW_s\right]$$

and then we use Fubini's theorem to change the order of integration in the first integral as follows:

$$\mathbf{E}[Y_t] - Y_0 = \int_0^t \mathbf{E}[\alpha(\overline{Y} - Y_s)]ds = \int_0^t \alpha(\overline{Y} - \mathbf{E}[Y_s])ds$$

where we take at given the value of the process Y_0 at time $t = 0$. The second integral disappears for a fundamental reason. Any stochastic integral is a martingale and any martingale has the property that $\mathbf{E}[M_t] = \mathbf{E}[M_0]$. But since at time $t = 0$ the stochastic integral is equal to zero then any stochastic integral has expected value 0.

Next, let us denote $u(t) = E[Y_t]$ which is just a deterministic function. Substituting in the aforementioned and going back to the differential notation we have:

$$du(t) = \alpha(\overline{Y} - u(t))dt$$

But this is a first order linear differential equation. We can solve it easily by multiplying with the integrating factor.

$$du(t) + \alpha u(t)dt = \alpha \overline{Y} dt$$
$$e^{\alpha t} du(t) + \alpha e^{\alpha t} u(t)dt = \alpha \overline{Y} e^{\alpha t} dt$$
$$d\left(e^{\alpha t} u(t)\right) = \alpha \overline{Y} e^{\alpha t} dt$$

and integrating gives:

$$e^{\alpha t} u(t) - u(0) = \overline{Y}(e^{\alpha t} - 1)$$

Now using the fact that $u(0) = E[Y_0] = Y_0$ after rearranging terms we finally obtain:

$$E[Y_t] = u(t) = \overline{Y} + (Y_0 - \overline{Y})e^{-\alpha t} \qquad (10.23)$$

We now calculate the variance as follows:

$$\text{Var}(Y_t) = E\left[(Y_t - \exp[Y_t])^2\right] = E[Y_t^2] - \left(E[Y_t]\right)^2$$

So, we need to calculate the second moment $E[Y_t^2]$. Note that the second term is just the square of the first term.

To perform the calculation we cannot simply square the dynamics of Y_t. Instead, we need to calculate the proper dynamics of Y_t^2 and apply expectation. Take the function $f(x) = x^2$ and apply Itô's lemma to this function to obtain the dynamics of $Y_t^2 = f(Y_t)$. We get:

$$Y_t^2 = 2Y_t dY_t + \frac{1}{2}2(dY_t)^2 = 2Y_t(\alpha(\mu - Y_t)dt + \sigma\sqrt{Y_t}dW_t) + \sigma^2 Y_t dt$$
$$= (2\alpha\mu + \sigma^2)Y_t dt - 2\alpha Y_t^2 dt + 2\sigma Y_t \sqrt{Y_t} dW_t$$

Now, apply expectation on both sides and use Fubini's theorem to change the order of integration. Finally, use the fact that once again expectation of a stochastic integral is zero. Also, denote $y_t = E[r_t^2]$ and note that $y_0 = Y_0^2$ to obtain,

$$dy_t + 2\alpha y_t dt = (2\alpha\mu + \sigma^2)E[Y_t]dt.$$

Once again we multiply with the integrating factor which in this case is $e^{2\alpha t}$ to get:

$$d\left(e^{2\alpha t}y_t\right) = e^{2\alpha t}(2\alpha \overline{Y} + \sigma^2)\mathrm{E}[Y_t]dt$$

$$e^{2\alpha t}y_t - y_0 = \int_0^t e^{2\alpha s}(2\alpha \overline{Y} + \sigma^2)\mathrm{E}[Y_s]ds$$

All that remains is to calculate the integral in the right hand side:

$$\int_0^t e^{2\alpha s}(2\alpha \overline{Y} + \sigma^2)\mathrm{E}[Y_s]ds = \int_0^t e^{2\alpha s}(2\alpha \overline{Y} + \sigma^2)(\overline{Y} + (r_0 - \overline{Y})e^{-\alpha s})ds$$

$$= (2\alpha \overline{Y} + \sigma^2)\left(\frac{\overline{Y}}{2}(e^{2\alpha t} - 1) + (Y_0 - \overline{Y})(e^{\alpha t} - 1)\right)$$

Now going back we finally obtain the second moment as:

$$\mathrm{E}[Y_t^2] = Y_0^2 e^{-2\alpha t} + (2\alpha \overline{Y} + \sigma^2)\left(\frac{\overline{Y}}{2}(1 - e^{-2\alpha t}) + (Y_0 - \overline{Y})(e^{-\alpha t} - e^{-2\alpha t})\right) \quad (10.24)$$

Using the expression in (10.24) and the square of the mean in (10.23), we obtain the variance as follows:

$$\mathrm{Var}(Y_t) = \mathrm{E}[Y_t^2] - \left(\mathrm{E}[Y_t]\right)^2 = \frac{\sigma^2 \overline{Y}}{2\alpha} + \frac{\sigma^2}{\alpha}(Y_0 - \overline{Y})e^{-\alpha t}$$

$$+ \frac{4\alpha^2 Y_0 + \sigma^2(\overline{Y} - 2Y_0)}{2\alpha}e^{-2\alpha t} \quad (10.25)$$

One may find expressions for higher moments as well by just following the same procedures outlined earlier.

10.5 Problems

1. Describe any stochastic volatility model and explain its significance.
2. Suppose that the process X_t follows the stochastic volatility model:

$$\begin{cases} dS_t = rS_t dt + \sqrt{Y_t}S_t^\beta dW_t \\ dY_t = \alpha^2 Y_t dt + 2\alpha Y_t dZ_t \end{cases}$$

Estimate the expectation and variance using Monte Carlo simulations. Then calculate these quantities using formulas. Compare.
3. Suppose that a stock price is governed by the equation:

$$dS = aSdt + bSd\chi$$

Find the equation that governs the process S^α
4. Application of Ito's lemma
You have a stock price that is governed by the equation:

$$dS = aSdt + bSd\chi$$

Applying Ito's lemma on $Y = \ln S$, verify the equation
$$dY = (a - \frac{1}{2}b^2)dt + bd\chi$$

5. Prove that the second moment of a mean-reverting O-U process Y_t specified as in (10.15) is given by:

$$E[Y_t^2] = e^{-2\alpha t}\left[E[Y_0^2] + \frac{\beta^2}{2\alpha}\left(e^{2\alpha t} - 1\right) + 2mE[Y_0]\left(e^{\alpha t} - 1\right) + m^2\left(e^{\alpha t} - 1\right)^2\right].$$

(10.26)

Hint: The idea is to apply expectations in both sides of (10.19), substitute $E[Y_u]$ with the expression (10.17), and finally integrate with respect to u.

6. Prove that the integrated second moment of a mean-reverting O-U process Y_t specified as in (10.15) is given by:

$$\int_0^t E[Y_s^2]ds = t\left(m^2 + \frac{\beta^2}{2\alpha}\right) + \left(E[(Y_0 - m)^2] - \frac{\beta^2}{2\alpha}\right)\frac{1 - e^{-2\alpha t}}{2\alpha} + 2mE[Y_0 - m]\frac{1 - e^{-\alpha t}}{\alpha}.$$

(10.27)

Hint: The expected value in (10.26) is integrated to yield after another series of calculations the expression in (10.27).

11

Jump Diffusion Models

11.1 Introduction

Jump diffusion models are special cases of exponential Lévy models in which the frequency of jumps are finite. They are considered as prototypes for a large class of complex models such as the stochastic volatility plus jumps models (see [17]). They have been used extensively in finance to model option prices (see [127, 128, 151]). General Lévy processes will be studied in details in Chapter 12.

The jump diffusion models comprises two parts, namely, a jump part and a diffusion part. The diffusion term is determined by the driving Brownian motion and the jump term is determined by the Poisson process. The Poisson process causes price changes in the underlying asset and is determined by a distribution function. The jump part enables to model sudden and unexpected price jumps of the underlying asset. Examples of the jump diffusion model include the Merton model (see [151]), the Black–Scholes models with jumps (see [152]), the Kou double exponential jump diffusion model (see [128]), and several others. In this chapter we introduce these models by briefly discussing the Poisson process (jumps) and the compound Poisson process.

11.2 The Poisson Process (Jumps)

Definition 11.2.1 (Poisson Process). A stochastic process $N = \{N_t, t \geq 0\}$ with intensity parameter $\lambda > 0$ is a *Poisson process* if it fulfills the following conditions:

1) $N_0 = 0$.
2) The process has independent and stationary increments.
3) For $s < t$ the random variable $N_t - N_s$ has a Poisson distribution with parameter $\lambda(t - s)$ such that

Table 11.1 Moments of the poisson distribution with intensity λ.

	Poisson (λ)
Mean	λ
Variance	λ
Skewness	$\frac{1}{\sqrt{\lambda}}$
Kurtosis	$3 + \lambda^{-1}$

$$\mathbb{P}[N_t - N_s = n] = \frac{\lambda^n(t-s)^n}{n!} e^{-\lambda(t-s)}.$$

The Poisson process is the simplest of all the Lévy processes. It is based on the Poisson distribution, which depends on the parameter λ and has the following characteristic function:

$$\phi_{Poisson}(z; \lambda) = \exp(\lambda(\exp(iz) - 1)).$$

The domain of the Poisson distribution is the set of nonnegative integers $k = \{0, 1, 2, \ldots\}$, and the probability mass function at any given point k is given by:

$$f(k; \lambda) = \frac{\lambda^k e^{-\lambda}}{k!}.$$

Since the Poisson distribution is infinitely divisible, we can define a Poisson process $N = \{N_t, t \geq 0\}$ with intensity parameter λ as the process which starts at zero has independent and stationary increments and where the increments over a time interval of length $s > 0$ follow the Poisson distribution. The Poisson process is an increasing pure jump process with jump sizes equal to one. That is, the time between two consecutive jumps follows an exponential distribution with mean λ^{-1}. This is referred to as the $\Gamma(1, \lambda)$ law since the exponential distribution is a special case ($n = 1$) of the gamma distribution. The moments of the Poisson distribution are given in Table 11.1.

11.3 The Compound Poisson Process

Definition 11.3.1 (Compound Poisson Process). A *compound Poisson* process with intensity parameter λ and a jump size distribution L is a stochastic process $X = \{X_t, t \geq 0\}$ defined as:

$$X_t = \sum_{k=1}^{N_t} \chi_k \tag{11.1}$$

where $N_t, t \geq 0$ is a Poisson process with intensity parameter λ and $(\chi_k, k = 1, 2, \ldots)$ is an independent and identically distributed sequence.

The sample paths of X are piecewise constant and the value of the process at time t, X_t is the sum of N_t random numbers with law L. The jump times have the same law as those of the Poisson process N. The ordinary Poisson process corresponds to the case where $\chi_k = 1$ for $k = 1, 2, \ldots$. The characteristic function of X_t is given by

$$\mathbb{E}[\exp(izX_t)] = \exp\left(t\int_{-\infty}^{\infty}(\exp(izx) - 1)\nu(dx)\right) \quad \forall u \in \mathbb{R}, \quad (11.2)$$

where ν is called the *Lévy measure* of the process X. ν is a positive measure on $\mathbb{R}$ but not a probability measure since $\int \nu(dx) = \lambda \neq 1$. In the next section, we discuss the Black–Scholes models with jumps.

11.4 The Black–Scholes Models with Jumps

The Black–Scholes models with jumps arise from the fact that the Brownian motion is a continuous process, and so there are difficulties fitting the financial data presenting large fluctuations. The necessity of taking into account the large market movements and a great amount of information arriving suddenly (i.e. a jump) has led to the study of partial integro-differential equations PIDEs in which the integral term is modeling the jump.

In [5, 152], the following partial integro-differential equation (PIDE) in the variables t and S is obtained:

$$\frac{1}{2}\sigma^2 S^2 F_{SS} + (r - \lambda k)S F_S + F_t - rF + \lambda \epsilon\{F(SY, t) - F(S, t)\} = 0. \quad (11.3)$$

Here, r denotes the riskless rate, λ the jump intensity, and $p = \epsilon(P - 1)$, where ϵ is the expectation operator and the random variable $P - 1$ measures the percentage change in the stock price if the jump modeled by a Poisson process occurs. See [152] for the details.

The following PIDE is a generalization of (14.1) for d assets with prices $S_1, \ldots, S_d$:

$$\sum_{i=1}^{d}\frac{1}{2}\sigma_i^2 S_i^2\frac{\partial^2 F}{\partial S_i^2} + \sum_{i\neq j}\frac{1}{2}\rho_{ij}\sigma_i\sigma_j S_i S_j\frac{\partial^2 F}{\partial S_i \partial S_j} + \sum_{i=1}^{d}(r - \lambda k_i)S_i\frac{\partial F}{\partial S_i} + \frac{\partial F}{\partial t} - rF +$$

$$(11.4)$$

$$\lambda\int\left[F(S_1 Y_1, \ldots, S_d Y_d, t) - F(S_1, \ldots, S_d, t)\right]g(Y_1, \ldots, Y_d)dY_1 \ldots dY_d = 0$$

where

$$\rho_{ij}dt = \epsilon\{dz_i, dz_j\}$$

is the correlation coefficient.

In this section we model the *jump* given in (11.4) in a special way so that it admits an elegant solution when the domain is bounded (i.e. the stocks values cannot be arbitrarily large). We shall assume that the Brownian motions have no correlation, i.e., $\rho_{ij} = 0$, for all $i \neq j$, so the system (11.4) becomes weakly coupled. We also assume d assets, i.e. $S = (S_1, \ldots, S_d)$, and the *boundedness* of the stocks is given by $\sum_{i=1}^{d} \left(\ln \frac{S_i}{|E|} \right)^2 \leq R'^2$, for some constant R'. Let us define this region by $\mathbb{U}$. Define $\alpha_i = -\frac{1}{2}(\frac{r-\lambda p_i}{\sigma^2/2} - 1)$, for $i = 1, \ldots, d$ and $\omega = \sum_{i=1}^{d} \alpha_i - 1$. We consider the equation

$$\frac{\partial C}{\partial t} + \sum_{i=1}^{d} \frac{1}{2} \sigma_i^2 S_i^2 \frac{\partial^2 C}{\partial S_i^2} + \sum_{i=1}^{d} (r - \lambda p_i) S_i \frac{\partial C}{\partial S_i} - rC$$

$$+ \lambda |E|^\omega \int_{\mathbb{U}} G(S, P) C(P, t) \left(\prod_{i=1}^{d} P_i^{\alpha_i + 1} \right)^{-1} dP = 0, \qquad (11.5)$$

for some random variable $P = (P_1, \ldots, P_d) \in \mathbb{U}$, where λ is the jump intensity. We take $G(S, P) = g(\ln \frac{S_1}{P_1}, \ldots, \ln \frac{S_d}{P_d})$, where g is a probability density function of its variables, $p_i = \mathbb{E}(P_i - 1)$, where $\mathbb{E}$ is the expectation operator and the random variable $P_i - 1$ measures the percentage change in the stock price for S_i if a jump occurs. Further, we shall assume that the volatility is the same for all the assets, that is, $\sigma_i = \sigma, i = 1, \ldots, d$. So (11.5) becomes

$$\frac{\partial C}{\partial t} + \frac{1}{2} \sigma^2 \sum_{i=1}^{d} S_i^2 \frac{\partial^2 C}{\partial S_i^2} + \sum_{i=1}^{d} (r - \lambda p_i) S_i \frac{\partial C}{\partial S_i} - rC$$

$$+ \lambda |E|^\omega \int_{\mathbb{U}} G(S, P) C(P, t) \left(\prod_{i=1}^{d} P_i^{\alpha_i + 1} \right)^{-1} dP = 0. \qquad (11.6)$$

We set

$$S_i = |E| e^{x_i}, \quad P_i = |E| e^{y_i}, \quad t = T - \frac{\tau}{\sigma^2/2},$$

and

$$C(S_1, \ldots, S_d, t) = |E| \exp\left(\sum_{i=1}^{d} \alpha_i \right) u(x_1, \ldots, x_d, \tau).$$

Then after the change of variables, we get

$$-\frac{\partial u}{\partial \tau} + \gamma u + \Delta u + \lambda \int_{\Omega} g(x - Y) u(Y) dY = 0 \qquad (11.7)$$

where $\Omega := \mathbb{B}(R') = \{ x = (x_1, \ldots, x_d) \in \mathbb{R}^d | \sum_{i=1}^{d} x_i^2 \leq R'^2 \}$ and

$$\gamma = \left[\sum_{i=1}^{d} (\alpha_i^2 + (k_i - 1)\alpha_i) - k' \right], \quad k_i = \frac{r - \lambda p_i}{\sigma^2/2}, \quad k' = \frac{r}{\sigma^2/2}.$$

11.4 The Black–Scholes Models with Jumps

In order to present a solution to this PIDE, we choose a specific form of g, namely,

$$g(X) = \frac{1}{N_{R'}} \frac{J_\nu(c|X|)}{(c|X|)^\nu}, \tag{11.8}$$

where J_ν is the Bessel function for order ν, with $\nu = \frac{d-2}{2}$ and $N_{R'}$ is a normalizing constant such that $\int_{\mathbb{B}(R')} g(X) dX = 1$. To solve problem (11.7) with g given by (11.8), we need the following two theorems. Proof of the theorems are found in [185].

Theorem 11.4.1 Suppose $\mathbf{x} = (r, \eta)$ and $\mathbf{y} = (r', \xi)$ are in $\mathbb{R}^2$ where η and ξ are angular parts of $\mathbf{x}$ and $\mathbf{y}$, respectively. Then

$$\int_{S^1} J_0(c|\mathbf{x} - \mathbf{y}|) e^{ik\xi} d\xi = 2\pi J_k(cr) J_k(cr') e^{ik\eta}.$$

Theorem 11.4.2 Suppose $\mathbf{x} = (r, \eta)$ and $\mathbf{y} = (r', \xi)$ are in $\mathbb{R}^d$ where η and ξ are angular parts of $\mathbf{x}$ and $\mathbf{y}$, respectively, and $\nu = \frac{d-2}{2}$. Then

$$\int_{S^{d-1}} \frac{J_\nu(c|\mathbf{x} - \mathbf{y}|)}{(c|\mathbf{x} - \mathbf{y}|)^\nu} S_k^s(\cdot) d\xi = \frac{2^{3\nu+1}}{\pi^{\nu-1}} \Delta_n(\nu, cr) \Delta_d(\nu, cr') S_k^s(\eta),$$

where $\Delta_m(\nu, r) = \left(\frac{\pi}{2r}\right)^\nu J_{\nu+m}(r)$.

We consider here the case $d \geq 3$. The case $d = 2$ will be similar and simpler. The solution of problem (11.7) is presented in the theorem that follows. Before we present the results, we state the definition of spherical harmonics.

Definition 11.4.1 Spherical harmonics are a frequency-space basis for representing functions defined over the sphere.

Theorem 11.4.3 Let $\mathbf{H}_l$ denote the space of spherical harmonics of degree l on the d-sphere. If g is given by (11.8), then there exists a solution of (11.7) of the form

$$u(x, \tau) = \sum_{N=0}^{\infty} \sum_{l=1}^{h(N,p)} T_{Nl}(\tau) R_{Nl}(r) S_N^l(\eta), \tag{11.9}$$

where $x = (x_1, \ldots, x_d) = (r, \eta)$,

$$T'_{Nl}(\tau) = \Lambda T_{Nl}(\tau),$$

and

$$\gamma R_{Nl}(r) + r^{1-d} \frac{\partial}{\partial r}\left(r^{d-1} \frac{\partial R_{Nl}(r)}{\partial r}\right) - R_{Nl}(r) \frac{N(N+d-2)}{r^2} + \zeta \Delta_N(\nu, cr) I$$

$$= \Lambda R_{Nl}(r),$$

where Λ is a constant, $\zeta = \frac{2^{3\nu+1}\lambda}{\pi^{\nu-1}N_{R'}}$ and

$$I = \int_0^{R'} \Delta_N(\nu, cr')R_{Nl}(r')r'^{p+1}dr'.$$

Proof. With the substitution of (11.9) and using the fact that $Y = (Y_1, \ldots, Y_d) = (r', \xi)$ alongside applying Theorem 11.4.2, the integral term of (11.7) becomes

$$\lambda \int_{\mathbb{B}(R')} g(x-Y)u(Y)dY$$

$$= \frac{\lambda}{N_{R'}} \int_{\mathbb{B}(R')} \frac{J_\nu(c|x-Y|)}{(c|x-Y|)^\nu} u(Y)dY$$

$$= \frac{\lambda}{N_{R'}} \sum_{N=0}^{\infty} \sum_{l=1}^{h(N,p)} \int_0^{R'} r'^{p+1}dr' \int_{S^{d-1}} \frac{J_\nu(c|x-Y|)}{(c|x-Y|)^\nu} T_{Nl}(\tau)R_{Nl}(r')S_N^l(\xi)d\xi$$

$$= \frac{\lambda}{N_{R'}} \sum_{N=0}^{\infty} \sum_{l=1}^{h(N,p)} \int_0^{R'} r'^{p+1}dr' \frac{2^{3\nu+1}}{\pi^{\nu-1}} \Delta_N(\nu, cr)\Delta_N(\nu, cr')T_{Nl}(\tau)R_{Nl}(r')S_N^l(\eta)$$

$$= \sum_{N=0}^{\infty} \sum_{l=1}^{h(N,p)} \frac{2^{3\nu+1}\lambda}{\pi^{\nu-1}N_{R'}} \Delta_N(\nu, cr)T_{Nl}(\tau) \left(\int_0^{R'} \Delta_N(\nu, cr')R_{Nl}(r')r'^{p+1}dr' \right) S_N^l(\eta).$$

Therefore (11.7) becomes

$$-\sum_{N=0}^{\infty} \sum_{l=1}^{h(N,p)} T'_{Nl}(\tau)R_{Nl}(r)S_N^l(\eta) + \gamma \sum_{N=0}^{\infty} \sum_{l=1}^{h(N,p)} T_{Nl}(\tau)R_{Nl}(r)S_N^l(\eta)$$

$$+ \sum_{N=0}^{\infty} \sum_{l=1}^{h(N,p)} T_{Nl}(\tau)r^{1-d}\frac{\partial}{\partial r}\left(r^{d-1}\frac{\partial R_{Nl}(r)}{\partial r}\right)S_N^l(\eta)$$

$$- \sum_{N=0}^{\infty} \sum_{l=1}^{h(N,p)} T_{Nl}(\tau)R_{Nl}(r)\frac{N(N+d-2)}{r^2}S_N^l(\eta)$$

$$+ \sum_{N=0}^{\infty} \sum_{l=1}^{h(N,p)} \frac{2^{3\nu+1}\lambda}{\pi^{\nu-1}N_{R'}} \Delta_N(\nu, cr)T_{Nl}(\tau)$$

$$\left(\int_0^{R'} \Delta_N(\nu, cr')R_{Nl}(r')r'^{p+1}dr' \right) S_N^l(\eta) = 0.$$

Since $S_N^l(\eta)$ are linearly independent, comparing the coefficients we have the following equations for $N = 0, 1, \ldots$ and $l = 1, \ldots, h(N,p)$:

$$T'_{Nl}(\tau)R_{Nl}(r)$$

$$= \gamma T_{Nl}(\tau)R_{Nl}(r) + T_{Nl}(\tau)r^{1-d}\frac{\partial}{\partial r}\left(r^{d-1}\frac{\partial R_{Nl}(r)}{\partial r}\right) - T_{Nl}(\tau)R_{Nl}(r)\frac{N(N+d-2)}{r^2}$$

$$+ \frac{2^{3\nu+1}\lambda}{\pi^{\nu-1}N_{R'}} \Delta_N(\nu, cr)T_{Nl}(\tau) \left(\int_0^{R'} \Delta_N(\nu, cr')R_{Nl}(r')r'^{p+1}dr' \right).$$

Therefore we have the following equations
$$T'_{Nl}(\tau) = \Lambda T_{Nl}(\tau), \tag{11.10}$$
and
$$\gamma R_{Nl}(r) + r^{1-d}\frac{\partial}{\partial r}\left(r^{d-1}\frac{\partial R_{Nl}(r)}{\partial r}\right) - R_{Nl}(r)\frac{N(N+d-2)}{r^2} + \zeta \Delta_N(v,cr)I = \Lambda R_{Nl}(r), \tag{11.11}$$

where Λ is a constant, $\zeta = \frac{2^{3v+1}\lambda}{\pi^{v-1}N_{R'}}$ and

$$I = \int_0^{R'} \Delta_N(v,cr')R_{Nl}(r')r'^{p+1}dr'.$$

Thus we arrive at the desired results. Initial values for $T_{Nl}(\tau)$ and the boundary values of $R(r)$ are obtained from the initial-boundary conditions of the original given problem (11.7). □

The solution of (11.10) is given by $T_{Nl}(\tau) = T_{Nl}(0)e^{\Lambda \tau}$ and that of (11.11) can be obtained by standard techniques such as *homotopy perturbation method*. The homotopy perturbation method is a combination of homotopy in topology and classic perturbation techniques provides a convenient way to obtain analytic or approximate solutions for a wide variety of problems arising in different disciplines. Please refer to [51, 92] for a comprehensive understanding of the method.

Here we present an outline of the methodology. Observe that (11.11) can be rewritten as

$$\frac{\partial^2 R_{Nl}(r)}{\partial r^2} + \frac{d-1}{r}\frac{\partial R_{Nl}(r)}{\partial r} + \left(\gamma - \frac{N(N+d-2)}{r^2} - \Lambda\right)R_{Nl}(r)$$
$$+ \zeta \Delta_N(v,cr)\int_0^{R'} \Delta_N(v,cr')R_{Nl}(r')r'^{p+1}dr' = 0. \tag{11.12}$$

By homotopy perturbation technique, we construct a homotopy

$$H(v,p) = \frac{\partial^2 v(r)}{\partial r^2} - \frac{\partial^2 y_0(r)}{\partial r^2} + p\frac{\partial^2 y_0(r)}{\partial r^2} - p[\left(\frac{N(N+d-2)}{r^2} + \Lambda - \gamma\right)v(r)$$
$$- \frac{d-1}{r}\frac{\partial v(r)}{\partial r} - \zeta \Delta_N(v,cr)\int_0^{R'} \Delta_N(v,cr')v(r')r'^{p+1}dr'] = 0, \tag{11.13}$$

where $y_0(r)$ is the initial approximation. According to homotopy perturbation theory, we can first use the embedding parameter p as a *small parameter* and assume that the solution of (11.13) can be written as a power series in p. That is,

$$v(r) = v_0(r) + pv_1(r) + p^2 v_2(r) + \cdots . \tag{11.14}$$

Setting $p = 1$, we can get the solution for (11.12) as

$$R_{Nl}(r) = v_0(r) + v_1(r) + v_2(r) + \cdots . \tag{11.15}$$

Substituting (11.14) in (11.13) and equating coefficients, we obtain

$$p^0: \quad \frac{\partial^2 v_0(r)}{\partial r^2} - \frac{\partial^2 y_0(r)}{\partial r^2} = 0, \tag{11.16}$$

$$p^1: \quad \frac{\partial^2 v_1(r)}{\partial r^2} + \frac{\partial^2 y_0(r)}{\partial r^2} - \left[\left(\frac{N(N+d-2)}{r^2} + \Lambda - \gamma\right)v_0(r) - \frac{d-1}{r}\frac{\partial v_0(r)}{\partial r}\right.$$
$$\left. - \zeta \Delta_N(v, cr) \int_0^{R'} \Delta_N(v, cr') v_0(r') r'^{p+1} dr'\right] = 0, \tag{11.17}$$

$$p^k: \quad \frac{\partial^2 v_k(r)}{\partial r^2} - \left[\left(\frac{N(N+d-2)}{r^2} + \Lambda - \gamma\right)v_{k-1}(r) - \frac{d-1}{r}\frac{\partial v_{k-1}(r)}{\partial r}\right.$$
$$\left. -\zeta \Delta_N(v, cr) \int_0^{R'} \Delta_N(v, cr') v_{k-1}(r') r'^{p+1} dr'\right] = 0, \quad k \geq 2.$$

$$\tag{11.18}$$

Then starting with an initial approximation $y_0(r)$ and solving successively the aforementioned equations, we can find $v_k(r)$ for $k = 0, 1, 2, \ldots$. Therefore we can obtain the k-th approximation of the exact solution (11.15) as $R_{NI}^k(r) = v_0(r) + v_1(r) + \cdots + v_{k-1}(r)$.

In the section that follows, we describe methods to find solutions to the partial-integral differential systems. The typical differential equations obtained are of parabolic type. We will be discussing a more general model capable of producing realistic paths. The resulting option price may be found as the solution of a system of PIDEs.

11.5 Solutions to Partial-Integral Differential Systems

The problem of pricing derivatives in financial mathematics often leads to studying partial differential and/or integral equations. In recent years, the complexity of the equations studied has increased, due to the inclusion of stochastic volatility, stochastic interest rate, and jumps in the mathematical models governing the dynamics of the underlying asset prices. For instance, [68] considered a continuous time asset price model containing both stochastic volatility and discontinuous jumps. In this model, the volatility is driven by a second correlated Brownian motion, and the jump is modeled by a compound Poisson process.

Standard risk-neutral pricing principle is used to obtain a single second order PIDE for the prices of European options written on the asset. Motivated by this financial mathematics problem, a general integro-differential parabolic problem is posed and studied in the cited work [68]. The existence of solution is proved by employing a method of upper and lower solutions and a diagonal argument. Moreover, the proof can provide an approximation method for numerically finding the solution of the general type PIDE which was later implemented in [73].

A very significant result is found in Section 11.5.4 which is Theorem 11.5.1. The theorem provides conditions on the integral terms in the PIDE system which guarantee the existence of the solution to this system. The emphasis in this section is on the applied mathematical methods rather than the stochastic process due to the technical nature of this result.

We begin with an introduction and motivation for the regime-switching jump diffusion model.

11.5.1 Suitability of the Stochastic Model Postulated

From the beginning of the twentieth century starting with Louis Jean-Baptiste Alphonse Bachelier (1870–1946), researchers have been looking for mathematical models which are capable of capturing the main features of an observed price path. The most famous attempt is the Black–Scholes–Merton model [24, 150] which influenced so much of the literature on asset pricing. Of course, more sophisticated models are needed for high frequency data, and many attempts have been made in the last 20 years to capture the complexity exhibited by the evolution of asset prices. In recent years, considerable attention has been drawn to regime-switching models in financial mathematics aiming to include the influence of macroeconomic factors on the individual asset price behavior. See, for example, [84, 135, 136]. In this setting, asset prices are dictated by a number of stochastic differential equations coupled by a finite-state Markov chain, which represents various randomly changing economical factors. Mathematically, the regime-switching models generalize the traditional models in such a way that various coefficients in the models depend on the Markov chain. Consequently, a system (not a single one) of coupled PDEs (or PIDEs) is obtained for option prices. We present the one day evolution of all trades for a particular equity gathered from a single exchange. This image or sample path is generally representative for many traded assets in any markets during any given day. Looking at the image we recognize several characteristics which can be captured by using a regime-switching jump diffusion model. The price path seems to jump in several places during the day (either up or down), and in between these jumps it follows processes with different parameters. For example, the variability at the beginning of the day seems to be larger than the variability in the middle of the day.

In the next section of this chapter, we will describe the regime-switching jump diffusion model. In this model, the process jumps at random times by a random amount, and in between jumps, the process could follow diffusion with distinct coefficients. We believe such a model is appropriate for describing the observed features of the asset price during the day.

11.5.2 Regime-Switching Jump Diffusion Model

We assume that all the stochastic processes in this subsection are defined on some underlying complete probability space $(\mathcal{S}, \mathcal{F}, \mathcal{P})$. Let B_t be a one-dimensional standard Brownian motion. Let α_t be a continuous-time Markov chain with state space $\mathcal{M} := \{1, \ldots, m\}$. Let $Q = (q_{ij})_{m \times m}$ be the intensity matrix (or the generator) of α_t. In this context the generator q_{ij}, $i, j = 1, \ldots, m$ satisfies:

(I) $q_{ij} \geq 0$ if $i \neq j$;
(II) $q_{ii} = -\sum_{j \neq i} q_{ij}$ for each $i = 1, \ldots, m$.

We assume that the Brownian motion B_t and the Markov chain α_t are independent. Before we proceed, we state the definition of a Cox process.

Definition 11.5.1 A Cox process is a stochastic process which is a generalization of a Poisson process where the time-dependent intensity λt is itself a stochastic process.

Cox processes are very useful in financial mathematics because they produce a framework for modeling prices of financial instruments where credit risk is a significant factor. Please refer to [42] for a detailed study of Cox Processes.

Let N_t be a Cox process with regime-dependent intensity λ_{α_t}. Thus, when the current state is $\alpha_t = i$, the time until the next jump is given by an exponential random variable with mean $1/\lambda_i$. N_t models the number of the jumps in the asset price up to time t. Let the jump sizes be given by a sequence of iid random variables $Y_i, i = 1, 2, \ldots$, with probability density $g(y)$. Assuming that the jump sizes $Y_i, i = 1, 2, \ldots$, are independent of B_t and α_t, we can model the time evolution of the asset price S_t by using the regime-switching jump diffusion:

$$\frac{dS_t}{S_t} = \mu_{\alpha_t} dt + \sigma_{\alpha_t} dB_t + dJ_t, \quad t \geq 0, \tag{11.19}$$

where μ_{α_t} and σ_{α_t} are the appreciation rate and the volatility rate of the asset S_t, respectively, and the jump component J_t is given by:

$$J_t = \sum_{k=1}^{N_t} (Y_k - 1). \tag{11.20}$$

The $Y_i - 1$ values represent the percentage of the asset price by which the process jumps. Note that, in between switching times, the process follows a regular jump diffusion with constant coefficients. However, the coefficients are switching as governed by the corresponding state of the Markov chain. In the model setting (11.19) the volatility is modeled as a finite-state stochastic Markov chain σ_{α_t}. As further reference for the model usefulness, (11.19) may be considered as a discrete approximation of a continuous-time diffusion model for the stochastic volatility. An example is the Heston model. Please see [135] and references therein for more details.

Next we discuss the option pricing problem and the resulting system of PIDEs that we will study in Section 11.5.4.

11.5.3 The Option Pricing Problem

Given that the asset price process follows the hypothesized model (11.19), we look into the problem of derivative pricing written on the corresponding asset. To this end denote r_{α_t} as the risk-free interest rate corresponding to the state α_t of the Markov chain.

We consider a European type option written on the asset S_t with maturity $T < \infty$. Let $V_i(S, t)$ denote the option value functions at time to maturity t, when the asset price $S_t = S$ and the regime $\alpha_t = i$ (assuming that the regime α_t is observable). Under these assumptions the value functions $V_i(S, t), i = 1, \ldots, m$, satisfy the system of PIDEs:

$$\frac{1}{2}\sigma_i^2 S^2 \frac{\partial^2 V_i}{\partial S^2} + (r_i - \lambda_i \kappa) S \frac{\partial V_i}{\partial S} - r_i V_i - \frac{\partial V_i}{\partial t}$$
$$+ \lambda_i E[V_i(SY, t) - V_i(S, t)] + \sum_{j \neq i} q_{ij}[V_j - V_i] = 0, \quad (11.21)$$

where we use the notation $\kappa = E[Y - 1] = \int (y - 1)g(y)dy$. Recalling that $q_{ii} = -\sum_{j \neq i} q_{ij}$ and using the density $g(y)$, we can rewrite (11.21) as:

$$\frac{1}{2}\sigma_i^2 S^2 \frac{\partial^2 V_i}{\partial S^2} + (r_i - \lambda_i \kappa) S \frac{\partial V_i}{\partial S} - (r_i + \lambda_i - q_{ii}) V_i - \frac{\partial V_i}{\partial t}$$
$$= -\lambda_i \int V_i(Sy, t)g(y)dy - \sum_{j \neq i} q_{ij} V_j. \quad (11.22)$$

Standard risk-neutral pricing principle (i.e. used to compute the values of derivatives product) is used for the derivation of Eq. (11.21) from the dynamics (11.19); we refer for instance to [84, 118].

Such types of systems are complicated and hard to approach. In [68] the authors analyzed a single PIDE which appears when the process exhibit jumps and has stochastic volatility. The approach was further implemented and an algorithm to calculate the solution was provided in [73]. The current problem is more complex by involving a system of PIDEs. However, note that the system is coupled only through the final term in Eq. (11.22); the rest of the terms in each equation i are in the respective $V_i(\cdot,\cdot)$. This fact provides hope that an existence proof (and a potential solving algorithm) may be provided in the current situation as well.

William Feller (1906–1970) and his students developed the semigroup theory for Markov processes, and there is a well-known direct link through them with the resulting PDEs for option pricing (see e.g. [62] or [198] for more details). However, they worked with diffusion processes (and later jump diffusion processes) characterizing Markov processes and these models lead to simple PIDEs.

In the case presented here, while the regime switching is governed by a continuous-time Markov chain and while each process being switched is indeed a continuous-time Markov process (jump diffusion), the overall structure may not be described by a simple Markov process with a diffusion plus a density type infinitesimal generator. Instead, the resulting overall Markov process is complex and produces the type of coupled systems of PIDEs studied in this chapter. The analysis we present proves an existence of solution theorem for such systems. This system is very different from the work published by [168]. In Pitt's results, the analysis naturally followed a different technique.

11.5.4 The General PIDE System

In order to obtain a solution to the system (11.22), we formulate the problem using more general terms. This will provide a universal approach to the kind of PIDE systems arising when solving complex option pricing problems (see [72]).

We first recall that the Black–Scholes equation

$$\frac{\partial V}{\partial t} + \frac{1}{2}\sigma^2 S^2 \frac{\partial^2 V}{\partial S^2} + rS\frac{\partial V}{\partial S} - rV = 0$$

becomes a heat type equation after performing the classical (Euler type) change of variable: $S = Ee^x$ and $t = T - \frac{2\tau}{\sigma^2}$, where E, T, and σ are constants; see, for example, [211]. From now on, we assume that this classical change of variable for Black–Scholes type equations was performed.

To this end, let $\Omega \subset \mathbb{R}^d$ be an unbounded smooth domain, and we consider a collection of m functions $u_i(x,t)$, $i = 1, \ldots, m$, where $x = (x_1, x_2, \ldots, x_d)$ ($u_i : \mathbb{R}^d \times [0,T] \to \mathbb{R}$). Let the operator $\mathcal{L}_i$ be defined by:

$$\mathcal{L}_i u_i = \sum_{j=1}^{d}\sum_{k=1}^{d} a_{jk}^i(x,t)\frac{\partial u_i}{\partial x_j \partial x_k} + \sum_{j=1}^{d} b_j^i(x,t)\frac{\partial u_i}{\partial x_j} + c^i(x,t)u_i, \quad i = 1,\ldots,m,$$

(11.23)

where the coefficients a_{jk}^i, b_j^i, and c^i, $i \in \{1,\ldots,m\}$; $j,k \in \{1,\ldots,d\}$ belong to the Hölder space $C^{\delta,\delta/2}(\overline{\Omega} \times [0,T])$ and satisfy the following conditions:

11.5 Solutions to Partial-Integral Differential Systems

- There exist two constants Λ_1, Λ_2 with $0 < \Lambda_1 \leq \Lambda_2 < \infty$ such that

$$\Lambda_1 |v|^2 \leq \sum_{j=1}^{d} \sum_{k=1}^{d} a_{jk}^i(x,t) v_j v_k \leq \Lambda_2 |v|^2, \ \forall v = (v_1, \ldots, v_d)^T \in \mathbb{R}^d. \tag{11.24}$$

- There exists a constant $C > 0$ such that

$$|b_j^i(x,t)| \leq C. \tag{11.25}$$

- The functions

$$c^i(x,t) \leq 0. \tag{11.26}$$

This general formulation encompasses all models presented including the diffusion model of Black–Scholes and the jump diffusion of Merton. The conditions are needed to ensure the existence of solution for a system of the type (2). Generally, these conditions are satisfied by most option pricing equations arising in finance.

The generalized problem corresponding to the system of PIDEs in equation (2) on an unbounded smooth domain Ω is:

$$\mathcal{L}_i u_i - \frac{\partial u_i}{\partial t} = \mathcal{G}_i(t, u_i) - \sum_{j \neq i} q_{ij} u_j \quad \text{in } \Omega \times (0, T)$$

$$u_i(x, 0) = u_{i,0}(x) \quad \text{on } \Omega \times \{0\}$$

$$u_i(x, t) = h_i(x, t) \quad \text{on } \partial \Omega \times (0, T) \tag{11.27}$$

for $i = 1, \ldots, m$, where $\mathcal{G}_i$ are continuous integral operators. We assume that the boundary conditions $u_{i,0} \in C^{2+\delta}(\overline{\Omega})$ and $h_i \in C^{2+\delta, 1+\delta/2}(\overline{\Omega} \times [0, T])$ satisfy the compatibility condition

$$h_i(x, 0) = u_{i,0}(x), \quad \text{for any } x \in \partial \Omega, \ i = 1, \ldots, m. \tag{11.28}$$

The spaces $C^{2+\delta}(\overline{\Omega})$ and $C^{2+\delta, 1+\delta/2}(\overline{\Omega} \times [0, T])$ follow from the definitions in Section 4.3. We note that as applied to problem (2) the operators $\mathcal{L}_i$ and $\mathcal{G}_i$ differ in the parameter values only, not in functional form. However, the general problem formulation as described earlier contains the case when the option is written on a basket of assets (not only a single stock) which are all modeled by different jump-diffusion type processes, and they are all dependent on the same regime-switching Markov process α_t.

The goal is to establish the existence of a solution to the system (11.27) using the method of upper and lower solutions.

11.5.4.1 The Method of Upper and Lower Solutions

Definition 11.5.2 A collection of m smooth functions $u = \{u_i, 1 \leq i \leq m\}$ is called an upper (lower) solution of problem (11.27) if

$$\mathscr{L}_i u_i - \frac{\partial u_i}{\partial t} \leq (\geq) \mathscr{G}_i(t, u_i) - \sum_{j \neq i} q_{ij} u_j \quad \text{in } \Omega \times (0, T)$$

$$u_i(x, 0) \geq (\leq) u_{i,0}(x) \quad \text{on } \Omega \times \{0\}$$

$$u_i(x, t) \geq (\leq) h_i(x, t) \quad \text{on } \partial\Omega \times (0, T) \tag{11.29}$$

for $i = 1, \ldots, m$.

A very important result is stated in the following theorem.

Theorem 11.5.1 Let the operators $\mathscr{L}_i$ and $\mathscr{G}_i$, $1 \leq i \leq m$ be as defined earlier. Assume that either:

- for each $1 \leq i \leq m$, $\mathscr{G}_i$ is nonincreasing with respect to u_i or
- for each $1 \leq i \leq m$, there exists a continuous and increasing one-dimensional function f_i such that $\mathscr{G}_i(t, u_i) - f_i(u_i)$ is nonincreasing with respect to u_i.

Furthermore, assume there exist a lower solution $\alpha = \{\alpha_i, 1 \leq i \leq m\}$ and an upper solution $\beta = \{\beta_i, 1 \leq i \leq m\}$ of problem (1) satisfying $\alpha \leq \beta$ component-wise (i.e. $\alpha_i \leq \beta_i$, $1 \leq i \leq m$) in $\Omega \times (0, T)$. Then (1) admits a solution u such that $\alpha \leq u \leq \beta$ in $\Omega \times (0, T)$.

To this end, we first solve an analogous problem in a bounded domain and then extend the solution to the unbounded domain $\Omega \times (0, T)$. We note that we need this extension since general option problems are solved on $(S_1, \ldots, S_d, t) \in (0, \infty)^d \times [0, T]$. Please also note that the theory may be used just as well for perpetual options (when $T = \infty$).

Lemma 11.5.1 Let U be a smooth and bounded subset of Ω. Then, there exists a unique collection of functions $\varphi_U = \{\varphi_{U,i}, 1 \leq i \leq m\}$ with $\varphi_{U,i} \in C^{2+\delta, 1+\delta/2}(\overline{U} \times [0, T])$ such that

$$\mathscr{L}_i \varphi_{U,i} - \frac{\partial \varphi_{U,i}}{\partial t} = 0, \quad (x, t) \in U \times (0, T),$$

$$\varphi_{U,i}(x, 0) = u_{i,0}(x), \quad x \in U,$$

$$\varphi_{U,i}(x, t) = h_i(x, t), \quad (x, t) \in \partial U \times [0, T], \tag{11.30}$$

for $i = 1, \ldots, m$. Moreover, if α and β are, respectively, a lower and an upper solution of this reduced problem (11.30) with $\alpha \leq \beta$ in $U \times (0, T)$, then

$$\alpha(x, t) \leq \varphi_U(x, t) \leq \beta(x, t), \quad (x, t) \in \overline{U} \times [0, T]. \tag{11.31}$$

Proof. Note that the homogeneous system (11.30) is decoupled. Thus, solving the system means solving the individual PDEs. Applying Lemma 2.6.1 in [68] to each of the m component equations, we obtain the expected result. □

The next result is crucial and it is the lemma that makes the transition from simple PDEs to a complex system of PIDEs on a bounded domain.

Lemma 11.5.2 Let $U \in \mathbb{R}^d$ be a smooth and bounded domain. Let $0 < \tilde{T} < T$. Let φ_U be defined as in Lemma 11.5.1. Assume α and β are, respectively, a lower and an upper solution of the initial problem (11.27) on the bounded domain $\overline{U} \times [0, \tilde{T}]$ with $\alpha \leq \beta$. Then the problem

$$\mathscr{L}_i u_i - \frac{\partial u_i}{\partial t} = \mathscr{G}_i(t, u_i) - \sum_{j \neq i} q_{ij} u_j \quad \text{in } U \times (0, \tilde{T})$$

$$u_i(x, 0) = u_{i,0}(x) \quad \text{on } U \times \{0\}$$

$$u_i(x, t) = \varphi_{U,i}(x, t) \quad \text{on } \partial U \times (0, \tilde{T}) \tag{11.32}$$

for $i = 1, \ldots, m$ admits at least one solution u such that $\alpha(x, t) \leq u(x, t) \leq \beta(x, t)$ for $x \in U, 0 \leq t \leq \tilde{T}$.

Proof. Suppose first that for each $1 \leq i \leq m$, $\mathscr{G}_i$ is nonincreasing with respect to u_i. Let $V = U \times (0, \tilde{T})$. In this proof we use the following result provided by the existence and uniqueness of the solution for homogeneous PDEs (Lemma 11.5.1) and the extension to nonhomogeneous PDEs (this is a standard extension; see for example [134]):

Given a collection of m functions with $w_i \in W_p^{2,1}(V)$, the problem

$$\mathscr{L}_i v_i - \frac{\partial v_i}{\partial t} = \mathscr{G}_i(t, w_i) - \sum_{j \neq i} q_{ij} w_j \quad \text{in } U \times (0, \tilde{T})$$

$$v_i(x, 0) = u_{i,0}(x) \quad \text{on } U \times \{0\}$$

$$v_i(x, t) = \varphi_{U,i}(x, t) \quad \text{on } \partial U \times (0, \tilde{T}) \tag{11.33}$$

for $i = 1, \ldots, m$ has a unique solution $v = \{v_i, 1 \leq i \leq m\}$ with $v_i \in W_p^{2,1}(V)$.

The idea in the proof of the lemma is to construct a convergent sequence of functions and show that the limit is a solution to the general system (11.32). To this end we use an inductive construction starting with $u^0 = \alpha$ and constructing a sequence of solutions $\{u^n, n = 0, 1, 2, \ldots\}$ such that $u^{n+1} = \{u_i^{n+1}, 1 \leq i \leq m\}$ is the unique solution of the problem

$$\mathscr{L}_i u_i^{n+1} - \frac{\partial u_i^{n+1}}{\partial t} = \mathscr{G}_i(t, u_i^n) - \sum_{j \neq i} q_{ij} u_j^n \quad \text{in } U \times (0, \tilde{T})$$

$$u_i^{n+1}(x, 0) = u_{i,0}(x) \quad \text{on } U \times \{0\}$$

$$u_i^{n+1}(x, t) = \varphi_{U,i}(x, t) \quad \text{on } \partial U \times (0, \tilde{T}) \tag{11.34}$$

for $i = 1, \ldots, m$.

We claim that component-wise,

$$\alpha \leq u^n \leq u^{n+1} \leq \beta \quad \text{in } \overline{U} \times [0, \tilde{T}], \ \forall n \in \mathbb{N}. \tag{11.35}$$

Using the maximum principle (i.e. a property of solutions to certain partial differential equations, of the elliptic and parabolic types) we can show that $u^1 \geq \alpha$ (i.e. $u_i^1 \geq \alpha_i$, for all $1 \leq i \leq m$). If we assume this is not true, there would exist an index $1 \leq i_0 \leq m$ and a point $(x_0, t_0) \in \overline{U} \times [0, \tilde{T}]$ such that $u_{i_0}^1(x_0, t_0) < \alpha_{i_0}(x_0, t_0)$. Since $u_{i_0}^1|_{\partial \overline{U} \times [0,\tilde{T}]} \geq \alpha_{i_0}|_{\partial \overline{U} \times [0,\tilde{T}]}$, we deduce that $(x_0, t_0) \in U \times (0, \tilde{T})$ (interior of the domain), and furthermore we may assume that (x_0, t_0) is a maximum point of $\alpha_{i_0} - u_{i_0}^1$ since both functions are smooth. Since the point is a maximum, it follows that $\nabla(\alpha_{i_0} - u_{i_0}^1)(x_0, t_0) = 0$, $\Delta(\alpha_{i_0} - u_{i_0}^1)(x_0, t_0) < 0$, and $\frac{\partial(\alpha_{i_0} - u_{i_0}^1)}{\partial t}(x_0, t_0) = 0$. By the conditions imposed on its coefficients, we have that

$$\mathcal{L}_{i_0}(\alpha_{i_0} - u_{i_0}^1)(x_0, t_0) < 0. \tag{11.36}$$

since $\mathcal{L}_{i_0}$ is strictly elliptic. On the other hand, in view of the definition (11.29) for the lower solution α and the way u^1 is constructed in (11.34), we have

$$\mathcal{L}_{i_0} u_{i_0}^1(x_0, t_0) - \frac{\partial u_{i_0}^1}{\partial t}(x_0, t_0) = \mathcal{G}_{i_0}(t, \alpha_{i_0})(x_0, t_0) - \sum_{j \neq i_0} q_{i_0 j} \alpha_j(x_0, t_0)$$

$$\leq \mathcal{L}_{i_0} \alpha_{i_0}(x_0, t_0) - \frac{\partial \alpha_{i_0}}{\partial t}(x_0, t_0), \tag{11.37}$$

resulting in $\mathcal{L}_{i_0}(\alpha_{i_0} - u_{i_0}^1)(x_0, t_0) \geq 0$, a contradiction with (11.36). Therefore, we must have $u^1 \geq \alpha$.

Next, since for each $1 \leq i \leq m$, $\mathcal{G}_i$ is nonincreasing with respect to u_j, and $q_{ij} \geq 0$ whenever $i \neq j$, we have for each $1 \leq i \leq m$ that

$$\mathcal{L}_i u_i^1 - \frac{\partial u_i^1}{\partial t} = \mathcal{G}_i(t, \alpha_i) - \sum_{j \neq i} q_{ij} \alpha_j \geq \mathcal{G}_i(t, \beta_i) - \sum_{j \neq i} q_{ij} \beta_j \geq \mathcal{L}_i \beta_i - \frac{\partial \beta_i}{\partial t}. \tag{11.38}$$

Again, by the maximum principle we obtain that $u^1 \leq \beta$. If the inequality did not hold, there would exist an index $1 \leq i_0 \leq m$ and a point $(x_0, t_0) \in \overline{U} \times [0, \tilde{T}]$ such that $u_{i_0}^1(x_0, t_0) > \beta_{i_0}(x_0, t_0)$. Since $u_{i_0}^1|_{\partial \overline{U} \times [0,\tilde{T}]} \leq \beta_{i_0}|_{\partial \overline{U} \times [0,\tilde{T}]}$ by the maximum principle, we deduce that $(x_0, t_0) \in U \times (0, \tilde{T})$ and furthermore we may assume that (x_0, t_0) is a maximum point of $u_{i_0}^1 - \beta_{i_0}$. It follows that $\nabla(u_{i_0}^1 - \beta_{i_0})(x_0, t_0) = 0$, $\Delta(u_{i_0}^1 - \beta_{i_0})(x_0, t_0) < 0$, and $\frac{\partial(u_{i_0}^1 - \beta_{i_0})}{\partial t}(x_0, t_0) = 0$. Since $\mathcal{L}_{i_0}$ is strictly elliptic, we have

$$\mathcal{L}_{i_0}(u_{i_0}^1 - \beta_{i_0})(x_0, t_0) < 0. \tag{11.39}$$

On the other hand, (11.38) implies that at the maximum point (x_0, t_0), $\mathscr{L}_{i_0}(u_{i_0}^1 - \beta_{i_0})(x_0, t_0) \geq 0$, a contradiction with (11.39).

In the general induction step, given $\alpha \leq u^{n-1} \leq u^n \leq \beta$, we can use a similar argument to show that $\alpha \leq u^n \leq u^{n+1} \leq \beta$. First, we claim that $u^n \leq u^{n+1}$. If this is not true, there exists an index $1 \leq i_0 \leq m$ and a point $(x_0, t_0) \in U \times (0, \tilde{T})$ such that

$$\mathscr{L}_{i_0}(u_{i_0}^n - u_{i_0}^{n+1})(x_0, t_0) < 0. \tag{11.40}$$

On the other hand, from the way the sequence is defined in (11.34) and the fact that, $\mathscr{G}_i$ is nonincreasing with respect to u_i for each $1 \leq i \leq m$ and $q_{ij} \geq 0$ for $i \neq j$, we have

$$\mathscr{L}_i u_i^{n+1} - \frac{\partial u_i^{n+1}}{\partial t} = \mathscr{G}_i(t, u_i^n) - \sum_{j \neq i} q_{ij} u_j^n \leq \mathscr{G}_i(t, u_i^{n-1})$$

$$- \sum_{j \neq i} q_{ij} u_j^{n-1} = \mathscr{L}_i u_i^n - \frac{\partial u_i^n}{\partial t}. \tag{11.41}$$

It follows that at the maximum point (x_0, t_0), $\mathscr{L}_{i_0}(u_{i_0}^n - u_{i_0}^{n+1})(x_0, t_0) \geq 0$, a contradiction with (11.40). In a similar way, we can show that $u^{n+1} \leq \beta$.

We now define:

$$u(x, t) = \lim_{n \to \infty} u^n(x, t), \tag{11.42}$$

or componentwise,

$$u_i(x, t) = \lim_{n \to \infty} u_i^n(x, t), \quad \forall (x, t) \in \overline{U} \times [0, \tilde{T}], \ i = 1, \ldots m. \tag{11.43}$$

Since $u^n \leq \beta$ and $\beta \in L^p(V)$, by the Lebesgue's dominated convergence theorem, we obtain that $\{u_i^n\}_{n=1}^\infty$ is a convergent sequence, therefore a Cauchy sequence in the complete space $L^p(V)$ for each $i = 1, \ldots m$. Using the results in [[134], chapter 7], the $W_p^{2,1}$-norm of the difference $u_i^n - u_i^m$, i.e. $||u_i^n - u_i^m||_{W_p^{2,1}}$, can be controlled by its L^p-norm and the L^p-norm of its image under the operator $\mathscr{L}_i - \frac{\partial}{\partial t}$. Using these results, there exists a constant $C > 0$ such that

$$||u_i^n - u_i^m||_{W_p^{2,1}(V)}$$
$$= ||D^2(u_i^n - u_i^m)||_{L^p(V)} + ||(u_i^n - u_i^m)_t||_{L^p(V)}$$
$$\leq C \left(\mathscr{L}_i(u_i^n - u_i^m) - \frac{\partial(u_i^n - u_i^m)}{\partial t} \bigg|_{L^p(V)} + u_i^n - u_i^m \bigg|_{L^p(V)} \right). \tag{11.44}$$

By construction,

$$\mathscr{L}_i(u_i^n - u_i^m) - \frac{\partial(u_i^n - u_i^m)}{\partial t} = \mathscr{G}_i(\cdot, u_i^{n-1}) - \mathscr{G}_i(\cdot, u_i^{m-1}) - \sum_{j \neq i} q_{ij}(u_j^{n-1} - u_j^{m-1}). \tag{11.45}$$

Since $\mathscr{G}_i$ is a completely continuous operator (an operator is completely continuous if it maps every relatively weakly compact subset of one space into a relatively compact subset of another space), there is a constant $C_1 > 0$ such that

$$||\mathscr{G}_i(\cdot, u_i^{n-1}) - \mathscr{G}_i(\cdot, u_i^{m-1}) - \sum_{j \neq i} q_{ij}(u_j^{n-1} - u_j^{m-1})||_{L^p(V)}$$

$$\leq C_1 \sum_{j=1}^{m} ||u_j^{n-1} - u_j^{m-1}||_{L^p(V)}. \tag{11.46}$$

Combining (11.44)–(11.46), it follows that $\{u_i^n\}_{n=1}^{\infty}$ is a Cauchy sequence in $W_p^{2,1}(V)$ for each $i = 1, \ldots m$. Hence $u_i^n \to u_i$ in the $W_p^{2,1}$-norm, and thus $u = \{u_i, 1 \leq i \leq m\}$ is a strong solution of the problem (11.32).

Now suppose the condition on $\mathscr{G}_i(t, u_i)$ is that for each $1 \leq i \leq m$, there exists a continuous and increasing function f_i such that $\mathscr{G}_i(t, u_i) - f_i(u_i)$ is nonincreasing with respect to u_i. Starting with $\tilde{u}^0 = 0$, we define recursively a sequence $\{\tilde{u}^n, n = 0, 1, \ldots\}$ such that $\tilde{u}^{n+1} = \{\tilde{u}_i^{n+1} \in W_p^{2,1}(V), 1 \leq i \leq m\}$ is the unique solution of the problem

$$\mathscr{L}_i \tilde{u}_i^{n+1} - \frac{\partial \tilde{u}_i^{n+1}}{\partial t} - f_i(\tilde{u}_i^{n+1}) = \mathscr{G}_i(t, \tilde{u}_i^n) - f_i(\tilde{u}_i^n) - \sum_{j \neq i} q_{ij} \tilde{u}_j^n \quad \text{in } U \times (0, \tilde{T})$$

$$\tilde{u}_i^{n+1}(x, 0) = u_{i,0}(x) \quad \text{on } U \times \{0\}$$

$$\tilde{u}_i^{n+1}(x, t) = \varphi_{U,i}(x, t) \quad \text{on } \partial U \times (0, \tilde{T}) \tag{11.47}$$

for $i = 1, \ldots, m$. The same arguments as before may be repeated almost verbatim to show that

$$0 \leq \tilde{u}^n \leq \tilde{u}^{n+1} \leq \beta \quad \text{in } \overline{U} \times [0, \tilde{T}], \; \forall n \in \mathbb{N}. \tag{11.48}$$

This will imply that $\{\tilde{u}_i^n\}_{n=1}^{\infty}$ is a Cauchy sequence in $W_p^{2,1}(V)$ for each $i = 1, \ldots m$. If we denote with $\tilde{u}_i = \lim_{n \to \infty} \tilde{u}_i^n$, then $\tilde{u} = \{\tilde{u}_i, 1 \leq i \leq m\}$ is a strong solution of problem (11.32). Note that the function f is continuous and thus the solution of the modified problem (11.47) also solves the original system. □

Finally, all that remains is to extend the solution to the original unbounded domain.

Proof of Theorem 11.5.1. We first approximate the unbounded domain Ω by a nondecreasing sequence $(\Omega_N)_{N \in \mathbb{N}}$ of bounded smooth subdomains of Ω, which can be chosen in such a way that $\partial \Omega$ is also the union of the nondecreasing sequence $\partial \Omega_N \cap \partial \Omega$.

11.5 Solutions to Partial-Integral Differential Systems

In view of Lemma 11.5.2, we define $u^N = \{u_i^N, 1 \leq i \leq m\}$ as a solution of the problem

$$\mathscr{L}_i u_i - \frac{\partial u_i}{\partial t} = \mathscr{G}_i(t, u_i) - \sum_{j \neq i} q_{ij} u_j \quad \text{in } \Omega_N \times (0, T - \frac{1}{N})$$

$$u_i(x, 0) = u_{i,0}(x) \quad \text{on } \Omega_N \times \{0\}$$

$$u_i(x, t) = h_i(x, t) \quad \text{on } \partial\Omega_N \times (0, T - \frac{1}{N}) \tag{11.49}$$

for $i = 1, \ldots, m$, such that $0 = \alpha \leq u^N \leq \beta$ in $\Omega_N \times (0, T - \frac{1}{N})$. Define $V_N = \Omega_N \times (0, T - \frac{1}{N})$ and choose $p > d$. For $M > N$, we have:

$$\|D^2(u_i^M)\|_{L^p(V_N)} + \|(u_i^M)_t\|_{L^p(V_N)}$$

$$\leq C_1 \left(\|\mathscr{L}_i u_i^M - \frac{\partial u_i^M}{\partial t}\|_{L^p(V_N)} + \|u_i^M\|_{L^p(V_N)} \right)$$

$$\leq C_1 \left(\|\mathscr{G}_i(t, u_i^M) - \sum_{j \neq i} q_{ij} u_j^M\|_{L^p(V_N)} + \|\beta\|_{L^p(V_N)} \right) \leq C, \tag{11.50}$$

for some constant C depending only on N.

By the well-known Morrey imbedding $W_p^{2,1}(V_N) \hookrightarrow C(\overline{V}_N)$ (see e.g. [2]), there exists a subsequence that converges uniformly on $\overline{V}_N$.

Now, we apply the well-known Cantor diagonal argument (see [190]): For $N = 1$, we extract a subsequence of $u_i^M|_{\overline{\Omega}_1 \times [0, T-1]}$ (still denoted $\{u_i^M\}$) that converges uniformly to some function u_{i1} over $\overline{\Omega}_1 \times [0, T-1]$. Next, we extract a subsequence of $u_i^M|_{\overline{\Omega}_2 \times [0, T-\frac{1}{2}]}$ for $M \geq 2$ (still denoted $\{u_i^M\}$) that converges uniformly to some function u_{i2} over $\overline{\Omega}_2 \times [0, T-\frac{1}{2}]$ and so on. As the families $\{\Omega_N\}$ and $\{\partial\Omega_N \cap \partial\Omega\}$ are nondecreasing, it is clear that $u_{iN}(x, 0) = u_{iN}(x)$ for $x \in \Omega_N$ and that $u_{iN}(x, t) = h(x, t)$ for $x \in \partial\Omega \cap \partial\Omega_N$ and $t \in (0, T - \frac{1}{N})$. Moreover, as $u_{i(N+1)}$ is constructed as the limit of a subsequence of $u_i^M|_{\overline{\Omega}_{N+1} \times [0, T-\frac{1}{N+1}]}$, which converges uniformly to some function u_{iN} over $\overline{\Omega}_N \times [0, T - \frac{1}{N}]$, it follows that $u_{i(N+1)}|_{\overline{\Omega}_N \times [0, T-\frac{1}{N}]} = u_{iN}$ for every N.

Thus, the diagonal subsequence (still denoted $\{u_i^M\}$) converges uniformly over compact subsets of $\Omega \times (0, T)$ to the function u_i defined as $u_i = u_{iN}$ over $\overline{\Omega}_N \times [0, T - \frac{1}{N}]$. For $V = U \times (0, \tilde{T})$, $U \subset \Omega$, and $\tilde{T} < T$, taking $M, N \geq N_V$ for some N_V large enough, we have that

$$\|u_i^M - u_i^N\|_{W_p^{2,1}(V)}$$

$$= \|D^2(u_i^M - u_i^N)\|_{L^p(V)} + \|(u_i^M - u_i^N)_t\|_{L^p(V)}$$

$$\leq C \left(\|\mathscr{L}_i(u_i^M - u_i^N) - \frac{\partial(u_i^M - u_i^N)}{\partial t}\|_{L^p(V)} + \|u_i^M - u_i^N\|_{L^p(V)} \right). \tag{11.51}$$

By construction,

$$\mathcal{L}_i(u_i^M - u_i^N) - \frac{\partial(u_i^M - u_i^N)}{\partial t} = \mathcal{G}_i(\cdot, u_i^{M-1}) - \mathcal{G}_i(\cdot, u_i^{N-1}) - \sum_{j \neq i} q_{ij}(u_j^{M-1} - u_j^{N-1}). \tag{11.52}$$

As in the proof of Lemma 11.5.2, since $\mathcal{G}_i$ is continuous, $\alpha \leq u^N \leq \beta$, using the Lebesgue's dominated convergence theorem, it follows that $\{u_i^N\}$ is a Cauchy sequence in $W_p^{2,1}(V)$ for each $i = 1, \ldots m$. Hence $u_i^N \to u_i$ in the $W_p^{2,1}(V)$-norm, and then $u = \{u_i, 1 \leq i \leq m\}$ is a strong solution in V. It follows that u satisfies the equation on $\Omega \times (0, T)$. Furthermore, it is clear that $u_i(x, 0) = u_{i,0}(x)$. For $M > N$ we have that $u_i^M(x, t) = u_i^N(x, t) = h_i(x, t)$ for $x \in \partial\Omega_N \cap \partial\Omega$, $t \in (0, T - \frac{1}{N})$. It then follows that u satisfies the boundary conditions $u_i(x, t) = h_i(x, t)$, $1 \leq i \leq m$ on $\partial\Omega \times [0, T)$. This completes the proof. □

The existence proof of solution of the system of PIDEs was coupled in a very specific way. This coupling type arises in regime-switching models when the assets are all changing their stochastic dynamics according to the same continuous-time Markov chain α_t with intensity matrix $Q = (q_{ij})_{m \times m}$. The proof of Theorem 11.5.1 uses a construction that may be used in a numerical scheme implementing a PDE solver.

Theorem 11.5.1 is directly applicable to our motivating system (2), noticing that in this case $\mathcal{G}_i(t, u) = -\lambda_i \int u_i(Sy, t)g(y)dy$ is a nonincreasing continuous operator in u_i and that $\alpha = \{\alpha_i(S, t) = 0, \ 1 \leq i \leq m\}$ is a lower solution of the option problem since the boundary conditions $u_{i,0}$ and h_i are nonnegative functions (represent the monetary value of the option on the boundaries). The upper solution also exists in these cases but its specific form depends on the jump distribution $g(y)$ and needs to be derived in each case. Note that the construction in Theorem 11.5.1 does not use the upper solution at all but its existence guarantees the convergence of the final solution. For specific examples of upper solutions as depending on the distribution $g(y)$, we refer to [68, 73].

Remarks 11.5.1 The result of Theorem 11.5.1 is applicable whenever the jump diffusion process and the regime switching may be thought of as Markovian. In particular, a simple generalization is to make the distribution of jumps dependent on the state of the regime as in $g_{\alpha_t}(\cdot)$. This is directly solvable with the theory presented.

11.6 Problems

1. Prove that

$$\mathcal{L}_i u_i - \frac{\partial u_i}{\partial t} = \mathcal{G}_i(t, u_i) - \sum_{j \neq i} q_{ij} u_j \quad \text{in } \Omega \times (0, T)$$

$$u_i(x, 0) = u_{i,0}(x) \quad \text{on } \Omega \times \{0\}$$
$$u_i(x, t) = h_i(x, t) \quad \text{on } \partial\Omega \times (0, T)$$

for $i = 1, \ldots, m$, where $\mathcal{G}_i$, are continuous integral operators, is the generalized problem corresponding to the system of PIDEs

$$\frac{1}{2}\sigma_i^2 S^2 \frac{\partial^2 V_i}{\partial S^2} + (r_i - \lambda_i \kappa) S \frac{\partial V_i}{\partial S} - (r_i + \lambda_i - q_{ii}) V_i - \frac{\partial V_i}{\partial t}$$
$$= -\lambda_i \int V_i(Sy, t) g(y) dy - \sum_{j \neq i} q_{ij} V_j.$$

on an unbounded smooth domain Ω.

2. Suppose $V_i(S, t)$ denote the option value functions at time to maturity t, when the asset price $S_t = S$. For the following PIDE,

$$\frac{1}{2}\sigma_i^2 S^2 \frac{\partial^2 V_i}{\partial S^2} + (r_i - \lambda_i \kappa) S \frac{\partial V_i}{\partial S} - (r_i + \lambda_i - q_{ii}) V_i - \frac{\partial V_i}{\partial t}$$
$$= -\lambda_i \int V_i(Sy, t) g(y) dy - \sum_{j \neq i} q_{ij} V_j.$$

(a) Discuss how to solve the PIDE by using Monte Carlo method.
(b) Discuss how to write a numerical code to solve the PIDE by using the recursive method in the upper and lower solutions method.

3. Discuss the modeling of the evolution of a Put option near a crash using the jump diffusion model.
4. Discuss why in the regime-switching jump diffusion model the process jumps at random times by a random amount.
5. Explain the reason why in the construction of Theorem 11.5.1, the final solution converged even though the upper solution was not used.
6. Discuss why options written on a basket of stocks which follows a different jump diffusion are all dependent on the same regime-switching process α_t and also solve a system of PIDEs of the type analyzed in Theorem 11.5.1.

12

General Lévy Processes

12.1 Introduction and Definitions

In this chapter we give basic definitions concerning stochastic processes and Lévy processes and present the main properties and examples of the Lévy stochastic processes. We will conclude this chapter with an introduction to the range scale analysis and detrended fluctuation analysis (DFA). The range scale analysis and DFA are very important methodologies that can be used to analyze extreme events, like financial crashes and earthquakes.

12.2 Lévy Processes

Stochastic Lévy processes play a fundamental role in mathematical finance and in other fields such as physics (turbulence), engineering (telecommunications and dams), actuarial science (insurance risk), and several others. References on stochastic processes and Lévy processes can be found in [8, 21, 159, 177, 179, 180, 200].

We recall the definition of stochastic processes which was previously defined in Chapter 1 of this book.

Definition 12.2.1 (Stochastic Process). A *stochastic process* is a parametrized collection of random variables $\{X_t : t \geq 0\}_{t \in T}$ defined on a probability space $(\Omega, \mathcal{F}, P)$ and taking values in $\mathbb{R}^n$. The parameter space T is usually the interval $[0, 1)$, but it may also be an arbitrary interval $[a, b]$, the nonnegative integers and even subsets of $\mathbb{R}^n$ for $n \geq 1$. A stochastic process is usually written as $\{X_t\}$.

A stochastic process $\{Y_t\}$ is called a modification of a stochastic process $\{X_t\}$, if

$$\mathbb{P}[X_t = Y_t] = 1 \quad \text{for} \quad t \in [0, \infty). \tag{12.1}$$

Quantitative Finance, First Edition. Maria C. Mariani and Ionut Florescu.
© 2020 John Wiley & Sons, Inc. Published 2020 by John Wiley & Sons, Inc.

Two stochastic processes $\{X_t\}$ and $\{Y_t\}$ are identical in law, written as

$$\{X_t\} \stackrel{d}{=} \{Y_t\}, \tag{12.2}$$

if the systems of their finite-dimensional distributions are identical, where $\stackrel{d}{=}$ denotes the equality of the finite-dimensional distribution.

Stochastic processes model the time evolution of random phenomena; therefore the index t is usually taken for time. The most basic process modeled for continuous random motions is the Brownian motion, and the one for jumping random motions is the Poisson process.

Definition 12.2.2 (Stochastic Continuity). A stochastic process $\{X_t\}$ on $\mathbb{R}^n$ is *stochastically continuous* if for every $t \geq 0$ and $\epsilon > 0$,

$$\lim_{s \to t} \mathbb{P}[|X_s - X_t| > \epsilon] = 0. \tag{12.3}$$

Definition 12.2.3 (Characteristic Function). The *Characteristic Function* ϕ of a random variable X is the Fourier Stieltjes transform of the distribution function $F(x) = \mathbb{P}(X \leq x)$:

$$\phi_X(u) = \mathbb{E}[e^{iuX}] = \int_{-\infty}^{\infty} e^{iux} dF(x), \tag{12.4}$$

where i is the imaginary number.

One important property of the characteristic function is the fact that for any random variable X, it always exists, it is continuous, and it determines X unequivocally. If X and Y are independent random variables, then

$$\phi_{X+Y}(u) = \phi_X(u)\phi_Y(u). \tag{12.5}$$

We present some functions related to the characteristic function that will be used in this chapter.

- The cumulant function:

$$k(u) = \log \mathbb{E}[e^{-uX}] = \log \phi(iu).$$

- The cumulant characteristic function or characteristic exponent:

$$\psi(u) = \log \mathbb{E}[e^{iuX}] = \log \phi(u) \tag{12.6}$$

Definition 12.2.4 (Infinitely Divisible Distribution). Suppose $\phi(u)$ is the characteristic function of a random variable X. If for every positive integer n, $\phi(u)$ is also the nth power of a characteristic function, we say that the distribution is infinitely divisible.

Given the aforementioned definitions, we now provide a formal definition of the Lévy process.

Definition 12.2.5 (Lévy Process). A stochastic process $\{X_t : t \geq 0\}$ on $\mathbb{R}^n$ is a *Lévy process* if the following conditions are satisfied:

1) For any choice of $n \geq 1$ and $0 \leq t_0 < t_1 < \cdots < t_n$, the random variables $X_{t_0}, X_{t_1} - X_{t_0}, X_{t_2} - X_{t_1}, \ldots, X_{t_n} - X_{t_{n-1}}$ are independent. The process has independent increments.
2) $X_0 = 0$.
3) The distribution of $X_{s+t} - X_s$ does not depend on s. The process has stationary increments.
4) It is stochastically continuous.
5) There is $\Omega_0 \in \mathcal{F}$ (σ algebra) with $\mathbb{P}[\Omega_0] = 1$ such that for every $\omega \in \Omega_0$, $X_t(\omega)$ is right continuous on $t \geq 0$ and has left limits in $t > 0$.

Remark 12.2.1 A Lévy process defined on $\mathbb{R}^n$ is known as an *n*-dimensional Lévy process. The law at time *t* of a Lévy process is completely determined by the law of X_1. The only degree of freedom we have when specifying a Lévy process is to define its distribution at a single time.

The following theorem describes the one-to-one relationship between Lévy processes and infinitely divisible distributions.

Theorem 12.2.1 (Infinite Divisibility of Lévy Processes). Let $X = \{X_t, t \geq 0\}$ be a Lévy process. Then X has infinitely divisible distributions F for every t. Conversely if F is an infinitely divisible distribution, there exists a Lévy process X such that the distribution of X_1 is given by F.

Please refer to [177] for the proof of the theorem. We can write

$$\phi_{X_t}(z) = \mathbb{E}[e^{-izX_t}] = e^{t\psi(z)},$$

where $\psi(z) = \log(\phi(z))$ is the characteristic exponent as in Eq. (12.6). The characteristic exponent $\psi(z)$ of a Lévy process satisfies the following *Lévy–Khintchine formula* (see [21, 121]):

$$\psi(z) = -\frac{1}{2}\sigma^2 z^2 + i\gamma z + \int_{-\infty}^{\infty} \left(e^{izx} - 1 - izx\mathbb{1}_{\{|x|<1\}}\right) v(dx), \qquad (12.7)$$

where $\gamma \in \mathbb{R}, \sigma^2 \geq 0$, and v is a measure on $\mathbb{R} \setminus \{0\}$ with

$$\int_{-\infty}^{\infty} \inf\{1, x^2\} v(dx) = \int_{-\infty}^{\infty} (1 \wedge x^2) v(dx) < \infty. \qquad (12.8)$$

From Eq. (12.7), we observe that in general, a Lévy process consists of three independent parts, namely, linear deterministic part, Brownian part, and pure jump part. We say that the corresponding infinitely divisible distribution has a Lévy triplet $[\gamma, \sigma^2, v(dx)]$. The measure v is called the *Lévy measure* of X.

Definition 12.2.6 (Lévy measure). Let $X = \{X_t : t \geq 0\}$ be a Lévy process on $\mathbb{R}^n$. The measure v on $\mathbb{R}^n$ defined by

$$v(A) = \frac{1}{t}\mathbb{E}\left(\sum_{0<s\leq t} \mathbb{1}_{\{\Delta X_s \in A\}}\right), \quad A \in \mathcal{B}(\mathbb{R}) \tag{12.9}$$

is called a *Lévy measure*, where $v(A)$ dictates how jumps occur. Jumps of sizes in the set A occur according to a Poisson process with parameter $v(A) = \int_A v(dx)$. In other words, $v(A)$ is the expected number of jumps per unit time, whose size belongs to A.

A Lévy measure has no mass at the origin, but singularities may occur near the origin (small jumps). Special attention has to be considered on small jumps since the sum of all jumps smaller than some $\epsilon > 0$ may not converge. For instance, consider the example where the Lévy measure $v(dx) = \frac{dx}{x^2}$. As we move closer to the origin, there is an increasingly large number of small jumps and $\int_{-1}^{1} |x|v(dx) = +\infty$. But for the integral

$$\int_{-1}^{1} x^2 v(dx),$$

substituting $v(dx) = \frac{dx}{x^2}$ leads to the integral

$$\int_{-1}^{1} x^2 \frac{dx}{x^2} = \int_{-1}^{1} dx = 2$$

which is finite. However, as we move away from the origin, $v([-1,1]^c)$ is finite and we do not experience any difficulties with the integral in Eq. (12.8) being finite. Brownian motion has continuous sample paths with no jumps and as such $\Delta X_t = 0$.

A Poisson process with rate parameter λ and jump size equal to 1 is a pure jump process with $\Delta X_t = 1$ and Lévy measure,

$$v(A) = \begin{cases} \lambda & \text{if } \{1\} \in A \\ 0 & \text{if } \{1\} \notin A \end{cases}$$

If the Lévy measure is of the form $v(dx) = u(x)dx$, then $u(x)$ is known as the Lévy density. The Lévy density has similar properties to a probability density. However, it does not need to be integrable and must have zero mass at the origin.

Consider the class of models with risk-neutral dynamics of the underlying asset given by $S_t = \exp(rt + X_t)$, where X_t is a Lévy process and X_0 is set to zero. The characteristic function of X_t has the following Lévy–Khintchine representation

$$\mathbb{E}[e^{izX_t}] = \exp t\phi(z),$$

where

$$\phi(z) = -\frac{\sigma^2 z^2}{2} + i\gamma z + \int_{-\infty}^{\infty} (e^{izx} - 1 - izx 1_{|x|\leq 1}) v(dx),$$

$\sigma \geq 0, \gamma$ are real constants, and v is a positive random measure on $\mathbb{R} - \{0\}$ verifying

$$\int_{-1}^{+1} x^2 v(dx) < \infty \quad \text{and} \quad \int_{|x|>1} v(dx) < \infty.$$

12.3 Examples of Lévy Processes

In this section we present examples of some popular Lévy processes. We will begin with the so-called subordinators. A subordinator is a one-dimensional Lévy process that is nondecreasing almost surely (a.s). They will be discussed in details in Section 12.4. Next, we will present some examples of Lévy processes defined on the real line. For each Lévy process defined, we will analyze their density function, characteristic function, and other important properties. We will compute their respective moments, variance, skewness, and kurtosis if possible. In Chapter 11 of this book, we discussed the Poisson and compound Poisson processess which are examples of Lévy processes. For more examples of Lévy processes, see [8, 179, 180].

12.3.1 The Gamma Process

Definition 12.3.1 (Gamma Process). A stochastic process $X = \{X_t, t \geq 0\}$ with parameters a and b is a *Gamma process* if it fulfills the following conditions:

1) $X_0 = 0$.
2) The process has independent and stationary increments.
3) For $s < t$ the random variable $X_t - X_s$ has a $\Gamma(a(t-s), b)$ distribution.

A random variable X has a gamma distribution $\Gamma(a, b)$ with rate and shape parameters, $a > 0$ and $b > 0$, respectively, if its density function is given by:

$$f_X(x; a, b) = \frac{b^a}{\Gamma(a)} x^{a-1} e^{-bx}, \quad \forall x > 0. \tag{12.10}$$

The moments of the $\Gamma(a, b)$ distribution are presented in Table 12.1.

The Gamma process is a nondecreasing Lévy process and its characteristic function is given by:

$$\phi(z; a, b) = (1 - \frac{iz}{b})^{-a}.$$

Table 12.1 Moments of the $\Gamma(a,b)$ distribution.

	$\Gamma(a,b)$
Mean	$\frac{a}{b}$
Variance	$\frac{a}{b^2}$
Skewness	$2a^{-\frac{1}{2}}$
Kurtosis	$3(1 + 2a^{-1})$

12.3.2 Inverse Gaussian Process

The inverse Gaussian (IG) process $\{Y(t); t \geq 0\}$ is defined as the stochastic process satisfying the following properties:

- $Y(t)$ has independent increments, i.e. $Y(t_2) - Y(t_1)$ and $Y(s_2) - Y(s_1)$ are independent for all $t_2 > t_1 \geq s_2 > s_1$
- $Y(t) - Y(s)$ follows an IG distribution $\mathscr{IG}(\Lambda(t) - \Lambda(s), \eta[\Lambda(t) - \Lambda(s)]^2)$, for all $t > s \geq 0$

where $\Lambda(t)$ is a monotone increasing function and $\mathscr{IG}(a,b), a,b > 0$ denotes the IG distribution with probability density function,

$$f_{\mathscr{IG}}(y; a, b) = \sqrt{\frac{b}{2\pi y^3}} \exp\left[-\frac{b(y-a)^2}{2a^2 y}\right], y > 0.$$

Define $T^{(a,b)}$ to be the first time a standard Brownian motion with drift $b > 0$ reaches a positive level $a > 0$. Then the time follows the IG law and has a characteristic function:

$$\phi(z; a, b) = \exp(-a(\sqrt{-2iz + b^2} - b)).$$

The IG distribution is infinitely divisible and we can simply redefine the IG process $IG(a,b)$ to be a stochastic process X with parameters a, b as the process which starts at zero and has independent and stationary increments such that

$$\mathbb{E}[\exp(izX_t)] = \phi(z; at, b)$$
$$= \exp(-at(\sqrt{-2iz + b^2} - b)). \quad (12.11)$$

12.3.3 Exponential Lévy Models

Suppose $(\Omega, F, F_t, \mathbb{P})$ is a filtered probability space and let $(S_t)_{t \in [0,T]}$ be the price of a financial asset modeled as a stochastic process on that space. Here F_t is taken to be the price history up to t. As mentioned in [41] under the hypothesis of absence of arbitrage, there exists a measure $\mathbb{Q}$ equivalent to $\mathbb{P}$ under which the discounted prices of all traded financial assets are ℚ-martingales. In particular the discounted underlying $(e^{-rt}S_t)_{t\in[0,T]}$ is a martingale under ℚ.

In exponential Lévy models the dynamics of S_t under ℚ is represented as the exponential of a Lévy process $S_t = S_0 e^{rt+X_t}$, where X_t is a Lévy process (under ℚ) with characteristic triplet (σ, γ, ν), and the interest rate is given by r. As observed in [41], the absence of arbitrage then imposes that $\hat{S}_t = S_t e^{-rt} = \exp X_t$ is a martingale, which is equivalent to the following conditions on the triplet (σ, γ, ν):

$$\int_{|y|>1} \nu(dy)e^y < \infty, \quad \gamma = \gamma(\sigma,\nu) = -\frac{\sigma^2}{2} - \int (e^y - 1 - y\mathbf{1}_{|y|\le 1})\nu(dy). \tag{12.12}$$

Assuming this the infinitesimal generator L of (14.3) becomes

$$Lu(x) = \frac{\sigma^2}{2}\left[\frac{\partial^2 u}{\partial x^2} - \frac{\partial u}{\partial x}\right](x) + \int_{-\infty}^{\infty} \nu(dy)\left[u(x+y) - u(x) - (e^y - 1)\frac{\partial u}{\partial x}(x)\right]. \tag{12.13}$$

The Poisson process and compound Poisson process defined in Sections 11.2 and 11.3, respectively, in Chapter 11 are all examples of Lévy processes.

12.4 Subordination of Lévy Processes

Subordination is a transformation of a stochastic process to a new stochastic process through random time change by increasing a Lévy process (subordinator) independent of the original process. The new process is a subordinate to the original one. The idea of subordination was introduced by Bochner in 1949; see [25]. In probability theory, a subordinator is a very important concept related to stochastic processes. A subordinator is itself a stochastic process of the evolution of time within another stochastic process that is the subordinated stochastic process. A formal definition of a subordinator is as follows:

Definition 12.4.1 (Subordinator). A *subordinator* is a one-dimensional Lévy process that is nondecreasing almost surely (a.s). As mentioned earlier, a subordinator can be thought of as a random model of time evolution. This is due to the fact that if $T = (T(t), t \ge 0)$ is a subordinator, we have $T(t) \ge 0$ a.s for each $t > 0$ and $T(t_1) \le T(t_2)$ a.s whenever $t_1 \le t_2$. Thus a subordinator will determine the random number of time steps that occur within the subordinated process for a given unit of chronological time.

We proceed by stating the following theorem.

Theorem 12.4.1 If T is a subordinator, then its Lévy symbol takes the form

$$\eta(z) = ibz + \int_0^\infty (e^{izy} - 1)\lambda(dy), \tag{12.14}$$

where $b \geq 0$ and the Lévy measure λ satisfies the additional requirements

$$\lambda(-\infty, 0) = 0 \quad \text{and} \quad \int_0^\infty (y \wedge 1)\lambda(dy) < \infty.$$

Conversely, any mapping from $\mathbb{R}^d \to \mathbb{C}$ of the form (12.14) is the Lévy symbol of a subordinator.

The proof of Theorem 12.4.1 can be found in [21]. The pair (b, λ) is the characteristic of the subordinator T.

Some classical examples of subordinators are the Poisson process, compound Poisson process (if and only if the $\chi_k, k = 1, 2, \ldots$ in Eq. (11.1) are all $\mathbb{R}^+$- valued), and Gamma process. For more examples of subordinators, please refer to the book by [8].

12.5 Rescaled Range Analysis (Hurst Analysis) and Detrended Fluctuation Analysis (DFA)

The DFA and rescaled range analysis (Hurst Analysis) are important techniques that allow detecting the presence of long-range correlations in nonstationary temporal series. We begin our discussion with the rescaled range analysis.

12.5.1 Rescaled Range Analysis (Hurst Analysis)

Rescaled range analysis is a statistical technique designed to assess the nature and magnitude of variability in data over time. It describes how the apparent variability of a series changes with the length of the time-period being considered. Rescaled range analysis has been used to detect and evaluate the amount of persistence, randomness, or mean reversion in financial markets time series data.

The author in [103] initially developed the rescaled range analysis (R/S analysis). He observed many natural phenomena that followed a biased random walk, i.e. every phenomenon showed a pattern. He measured the trend using an exponent now called the Hurst exponent. The authors [141, 142] later introduced a generalized form of the Brownian motion model, the fractional Brownian motion to model the Hurst effect. Closely associated with rescaled range analysis is the Hurst exponent also known as the "index of dependence" or the "index of long-range dependence." The Hurst exponent ranges between 0 and

1 and measures three types of trends in a time series: persistence, randomness, or mean reversion.

The numerical procedure to estimate the Hurst exponent by using the R/S analysis is presented next (for more details please see [187] and references therein).

1) Let N be the length of time series $(y_1, y_2, y_3 \cdots, y_N)$. The logarithmic ratio of the time series is obtained. The length of the new time series $M(t)$ will be $N - 1$.

$$M(t) = \log\left(\frac{y_{t+1}}{y_t}\right)$$

for $t = 1, 2, \cdots N - 1$.

2) The time series is then divided into m subseries of length n. n represents the number of elements in the series and m represents the number of subseries. Thus $m \cdot n = N - 1$. Each subseries can be labeled as Q_a where $a = 1, 2, \cdots, m$ and each element in Q_a can be labeled as $L_{k,a}$ for $k = 1, 2, \cdots, n$.

3) For each Q_a, the average value is calculated:

$$Z_a = \frac{1}{n}\sum_{k=1}^{n} L_{k,a}$$

4) The cumulative deviation in each Q_a is calculated:

$$C_{k,a} = \sum_{j=1}^{k}(L_{j,a} - Z_a)$$

$k = 1, 2, \cdots n$.

5) Thus the range of each sub-series Q_a is given as:

$$R(Q_a) = \max(C_{k,a}) - \min((C_{k,a}))$$

6) The standard deviation of each subseries Q_a is calculated as follows:

$$S(Q_a) = \sqrt{\left(\frac{1}{n}\sum_{j=1}^{n}(L_{j,a} - Z_a)^2\right)}$$

7) Each subseries is normalized by dividing the range, $R(Q_a)$, by the standard deviation, $S(Q_a)$. The average value of R/S for subseries of length n is obtained by:

$$(R/S)_n = \frac{1}{m}\sum_{a=1}^{m}\frac{R(Q_a)}{S(Q_a)}$$

8) Steps 2 through 7 are repeated for all possible values of n, thus obtaining the corresponding R/S values for each n. The relationship between length of the subseries n and the rescaled range R/S is:

$$(R/S) = (cn)^H$$

where R/S is the rescaled range, n is the length of the subseries of the time series, and H is the Hurst exponent. Taking logarithms yields,

$$\log(R/S) = H \log c + H \log n$$

9) An ordinary least squares regression is performed using $\log(R/S)$ as a dependent variable and $\log(n)$ as an independent variable. The slope of the equation is the estimate of the Hurst exponent H.

If the Hurst exponent H for the investigated time series is 0.5, then it implies that the time series follows a random walk which is a process with independent increments. For data series with long memory effects, H would lie between 0.5 and 1. It suggests all the elements of the observation are dependent. This means that what happens now would have an impact on the future. Time series that exhibit this property are called persistent time series and this character enables prediction of any time series as it shows a trend. Lastly, if H lies between 0 and 0.5, it implies that the time series possess antipersistent behavior (negative autocorrelation), i.e. it is likely to reverse trend over the time frame considered.

12.5.2 Detrended Fluctuation Analysis

The DFA method is an important technique in revealing long-range correlations in nonstationary time series. This method was developed by [165, 166] and has been successfully applied to the study of cloud breaking [111], DNA [31, 164, 165], cardiac dynamics [163, 166], climatic studies [125, 126], solid-state physics [116, 201], and economic time series [12, 37, 137]. The advantages of DFA over conventional methods are that it permits the detection of intrinsic self-similarity embedded in a seemingly nonstationary time series (i.e. time series whose means, variances, and covariances change over time) and also avoids the detection of apparent self-similarity, which may be an artifact of extrinsic trends. In general, self-similarity occurs when the shape of an object is similar or approximately similar to the part of itself. That is, each portion of the self-similar object can be considered as a reduced scale of the whole. More details of self-similarity will be presented in Chapter 13 subSection 13.2.3.

Next we present the DFA methodology as follows.

To begin, the absolute value of $M(t)$, i.e. the logarithmic returns of the indices calculated in the R/S analysis, is integrated:

$$y(t) = \sum_{i=1}^{t} |M(i)|$$

Then the integrated time series of length N is divided into m boxes of equal length n with no intersection between them. As the data is divided into equal-length intervals, there may be some leftover at the end. In order to take account of these leftover values, the same procedure is repeated but beginning from the end, obtaining $2N/n$ boxes. Then, a least squares line is fitted to each box, representing the trend in each box, thus obtaining $y_n(t)$. Finally the root mean square fluctuation is calculated by using the formula:

$$F(n) = \sqrt{\frac{1}{2N} \sum_{t=1}^{2N} (y(t) - y_n(t))^2}$$

This computation is repeated over all box sizes to characterize a relationship between the box size n and $F(n)$. A linear relationship between the $F(n)$ and n (i.e. box size) in a log–log plot reveals that the fluctuations can be characterized by a scaling exponent α, the slope of the line relating $\log F(n)$ to $\log n$.

For data series with no correlations or short-range correlation, α is expected to be 0.5. For data series with long-range power law correlations, α would lie between 0.5 and 1 and for power law anticorrelations, α would lie between 0 and 0.5. This method was used to measure correlations in financial series of high frequencies and in the daily evolution of some of the most relevant indices.

12.5.3 Stationarity and Unit Root Test

In order to study the fractional behavior of a times series using the R/S or the DFA analysis, it is important to investigate whether the underlying time series is stationary or not. In this subsection, we discuss a unit root methodology that is often used to test long memory behavior namely, Augmented Dickey Fuller Test (ADF) (see [217]).

The ADF is an extension of the basic autoregressive unit root test to accommodate general autoregressivemoving-average (ARMA(p, q)) models with unknown orders. The test is very powerful and can handle more complex models. The ADF test tests the null hypothesis that a time series y_t is a unit root against the alternative and that it is stationary, assuming that the dynamics in the data have an ARMA structure. The ADF test is based on estimating the test regression:

$$y_t = \beta D_t + \phi y_{t-1} + \sum_{j=1}^{p} \psi_j \Delta y_{t-j} + \epsilon_t \tag{12.15}$$

where ϵ_t are i.i.d. with mean 0 and constant variance σ^2 and D_t is a vector of deterministic term. The p lagged difference terms, Δy_{t-j}, are used to approximate the ARMA structure of the errors, and the value of p is set so that the error ϵ_t is serially uncorrelated. An alternative formulation of (12.15) is

$$\Delta y_t = \beta D_t + \pi \phi y_{t-1} + \sum_{j=1}^{p} \psi_j \Delta y_{t-j} + \epsilon_t \tag{12.16}$$

where $\pi = \phi - 1$. In practice, the test regression (12.16) is often used since the ADF t-statistic is the usual t-statistic reported for testing the significance of the coefficient y_{t-1}. An important practical issue for the implementation of the ADF test is the specification of the lag length p. If p is too small then the remaining serial correlation in the errors will bias the test. If p is too large then the power of the test will suffer. For the ADF t-statistic test, small p-values suggest the data is stationary. Usually, small p-values are less than 5 % level of significance.

12.6 Problems

1. Present examples of stochastic processes that are not Lévy processes.
2. Present examples of stochastic processes that are Lévy processes.
3. Present examples of Lévy processes that are not subordinators.
4. Present examples of Lévy processes that are subordinators.
5. Prove directly that, for each $t \geq 0$,

$$\mathbb{E}(e^{-zT(t)}) = \int_0^\infty e^{-zs} f_{T(t)}(s) ds = e^{-tz^{1/2}},$$

where $(T(t), t \geq 0)$ is the Lévy subordinator.
6. Present examples of exponential Lévy models that have infinite number of jumps in every compact interval.
7. Discuss and compare the detrended fluctuation analysis and rescaled range analysis.
8. What is a unit root test and what are its consequences?
9. If we have unit root, how can we transform the data, so that we can use the autoregressive-moving-average (ARMA) methodology?
10. Download monthly, weekly, and daily equity data for the past 10 years. Please use any equity of your choice. For each time frequency calculate a sequence of continuously compounded returns. Apply R/S range analysis and obtain an estimate for the H parameter. What do you observe about the estimate when the sampling frequency increases? Repeat the analysis for the DFA estimator.

13

Generalized Lévy Processes, Long Range Correlations, and Memory Effects

13.1 Introduction

In this chapter we introduce the stable distributions and generalized Lévy processes. In order to introduce generalized Lévy models, we first present some background in stable distributions. In probability theory, a distribution is said to be stable if a linear combination of two independent random variables with this distribution has the same distribution, up to location and scale parameters. The family of stable distributions is also referred to as the Lévy alpha-stable distribution named after Paul Lévy. In this chapter, we will present some forms of stable distributions which includes the Lévy flight models.

The Black–Scholes is not appropriate for the study of high frequency data or for the study of financial indices or asset prices when a Market Crash takes place. For these financial data, other models are more appropriate, such as the Lévy-like stochastic processes. The generalized Lévy models like the Range Scale Analysis, Detrended Fluctuation Analysis (DFA) (see Chapter 12 of this book) and Diffusion Entropy Analysis (DEA) (see [178]) are very convenient methodologies for the analysis of extreme events, like financial crashes and earthquakes. The DEA method determines the correct scaling exponent in a time series even when the statistical properties, as well as the dynamic properties, are irregular.

We will conclude this chapter by presenting some applications to financial high frequency data and analyze the presence of long range correlations and long memory effects.

13.1.1 Stable Distributions

Consider the sum of n independent identically distributed (i.i.d.) random variables x_i:

$$S_n = x_1 + x_2 + x_3 + \cdots + x_n = x(n\Delta t).$$

Quantitative Finance, First Edition. Maria C. Mariani and Ionut Florescu.
© 2020 John Wiley & Sons, Inc. Published 2020 by John Wiley & Sons, Inc.

Observe that S_n can be regarded as the sum of n random variables or as the position of a single walker at time $t = n\Delta t$ where n is the number of steps performed and Δt the time required to perform one step.

As the variables are independent, the sum can be obtained as the convolution, namely,

$$P[x(2\Delta t)] = P(x_1) \otimes P(x_2)$$

or more generally,

$$P[x(n\Delta t)] = P(x_1) \otimes P(x_2) \cdots \otimes P(x_n).$$

We will say that the distribution is stable if the functional form of $P[x(n\Delta t)]$ is the same as the functional form of $P[x(\Delta t)]$. Specifically, given a random variable X, if we denote with $Law(X)$ its probability density function (p.d.f.) (for example, for a Gaussian random variable, we write $Law(X) = N(\mu, \sigma^2)$), then we will say that the random variable X is stable or that it has a stable distribution if for any $n \geq 2$ there exists a positive number C_n and a number D_n so that

$$Law(X_1 + X_2 + \cdots + X_n) = Law(C_n X + D_n)$$

where $X_1, X_2, \cdots X_n$ are independent random copies of X; this means that $Law(X_i) = Law(X)$ for $i = 1, 2, \cdots n$. If $D_n = 0$, X is said to be a strictly stable variable.

The authors in [176] showed that $C_n = n^{\frac{1}{\alpha}}$ for some parameter α, $0 < \alpha \leq 2$. For example, X is a Lorentzian random variable with distribution

$$P(x) = \frac{\gamma}{\pi} \frac{1}{\gamma^2 + x^2}$$

and its characteristic function (that is, its Fourier transform) is given by

$$\varphi(q) = \int_{I\!R} \exp(iqx) f(x) dx = E(\exp(iqx))$$

where $f(x)$ is the p.d.f. associated to the distribution $P(x)$. In this case we obtain the characteristic function to be:

$$\varphi(q) = \exp(-\gamma |q|)$$

Suppose X_1 and X_2 are two i.i.d. Lorentzian random variables, we have that

$$P[X(2\Delta t)] = P(X_1) \otimes P(X_2) = \frac{2\gamma}{\pi} \frac{1}{4\gamma^2 + x^2}.$$

Also, as the Fourier transform of a convolution is the product of the Fourier transforms, we obtain:

$$\varphi_2(q) = \exp(-2\gamma |q|) = (\varphi(q))^2$$

and so in general we conclude that
$$\varphi_n(q) = (\varphi(q))^n.$$
As a second example, if X is a Gaussian random variable,
$$P(X) = \frac{1}{\sqrt{2\pi}\sigma} \exp\left(\frac{-x^2}{2\sigma^2}\right)$$
its characteristic function is
$$\varphi(q) = \exp\left(-\frac{\sigma^2}{2}|q|^2\right) = \exp\left(-\gamma|q|^2\right)$$
where $\gamma = \frac{\sigma^2}{2}$, and again, we have that
$$\varphi_2(q) = (\varphi(q))^2.$$
By performing the inverse Fourier transform we obtain that
$$P_2[X(2\Delta t)] = \frac{1}{\sqrt{8\pi\gamma}} \exp\left(\frac{-x^2}{8\gamma}\right) = \frac{1}{\sqrt{2\pi}\left(\sqrt{2}\sigma\right)} \exp\left(\frac{-x^2}{2\left(\sqrt{2}\sigma\right)^2}\right)$$
That is, the variance is now $\sigma_2^2 = 2\sigma^2$. Therefore if two stable stochastic processes exist, in this case the Lorentzian and Gaussian, then their Fourier transform has the form:
$$\varphi(q) = \exp(-\gamma|q|^\alpha)$$
where $\alpha = 1$ for the Lorentzian and $\alpha = 2$ for the Gaussian.

Characteristic functions
$$\varphi(q) = \exp(-\gamma|q|^\alpha)$$
with $1 \leq \alpha \leq 2$ are stable see [112].

In the next section, we will see the form of all stable distributions also called Lévy distributions.

13.2 The Lévy Flight Models

13.2.1 Background

The authors in [121, 161] solved the problem of determining the functional form that all the stable distributions must follow. They found that the most general representation is through the characteristic functions $\varphi(q)$, which are defined by the following equations:
$$\ln(\varphi(q)) = i\mu q - \gamma|q|^\alpha \left[1 - i\beta\frac{q}{|q|}\tan\left(\frac{\pi\alpha}{2}\right)\right] \text{ if } \alpha \neq 1,$$

and

$$\ln(\varphi(q)) = \mu q - \gamma |q| \left[1 + i\beta \frac{q}{|q|} \frac{2}{\pi} \log(q)\right] \text{ if } \alpha = 1,$$

where $0 < \alpha \leq 2$ (which is the same parameter mentioned before) is called the stability exponent or the characteristic parameter, γ is a positive scale factor, μ is a real number called the location parameter, and β is an asymmetry parameter ranging from -1 to 1, which is called the skewness parameter.

The analytical form of the Lévy stable distribution is known only for a few values of α and β:

1) $\alpha = \frac{1}{2}$, $\beta = 1$ (Lévy–Smirnov). This distribution is also called the one-sided stable distribution on (μ, ∞) with density

$$\left(\frac{\gamma}{2\pi}\right)^{\frac{1}{2}} \frac{1}{(x-\mu)^{\frac{3}{2}}} \exp\left(-\frac{\gamma}{2(x-\mu)}\right).$$

2) $\alpha = 1$, $\beta = 0$ (Lorentzian). This distribution is also called the Cauchy stable distribution with density

$$\frac{\gamma}{\pi} \frac{1}{\gamma^2 + (x-\mu)^2}.$$

3) $\alpha = 2$, $\beta = 0$ (Gaussian). This distribution is also called the normal distribution with density

$$\frac{1}{\sqrt{2\pi\gamma}} \exp\left(\frac{-(x-\mu)^2}{2\gamma^2}\right).$$

From the previous discussion it is clear that μ will be (in same cases) the mean and that γ^2 will coincide in same cases with the variance σ^2.

We consider the symmetric distribution ($\beta = 0$) with a zero mean ($\mu = 0$). In this case the characteristic function takes the form:

$$\varphi(q) = \exp(-\gamma |q|^\alpha)$$

As the characteristic function of a distribution is its Fourier transform, the stable distribution of index α and scale factor γ is

$$P_L(x) = \frac{1}{\pi} \int_0^\infty \exp(-\gamma |q|^\alpha) \cos(qx) dq.$$

The asymptotic behavior of the distribution for big values of the absolute value of x is given by:

$$P_L(x) \simeq \frac{\gamma \Gamma(1+\alpha) \sin \frac{\pi \alpha}{2}}{\pi |x|^{1+\alpha}} \simeq |x|^{-(1+\alpha)}$$

and the value at zero $P_L(x = 0)$ by:

$$P_L(x = 0) = \frac{\Gamma(1/\alpha)}{\pi \alpha \gamma^{1/\alpha}}.$$

The fact that the asymptotic behavior for big values of x is a power law has as a consequence that $E[|x|^n]$ diverges for $n \geq \alpha$ when $\alpha < 2$. In particular, all the stable Lévy processes with $\alpha < 2$ have infinite variance. In order to avoid the problems arising in the infinite second moment, *Mantegna and Stanley* considered a stochastic process with finite variance that follows scale relations called truncated Lévy flight (TLF) [143].

The *TLF* distribution is defined by

$$T(x) = cP(x)\chi_{(-l,l)}(x)$$

where $P(x)$ is a symmetric Lévy distribution.

The *TLF* distribution is not stable, but it has finite variance; thus independent variables from this distribution satisfy a regular Central Limit Theorem. However, depending on the size of the parameter l (the cutoff length), the convergence may be very slow [143]. If the parameter l is small (so that the convergence is fast), the cut that it presents in its tails is very abrupt. In order to have continuous tails, *Koponen* [124] considered a *TLF* in which the cut function is a decreasing exponential characterized by a parameter l.

The characteristic function of this distribution is defined as:

$$\varphi(q) = \exp\left\{c_0 - c_1 \frac{(q^2 + 1/l^2)^{\frac{\alpha}{2}}}{\cos(\pi\alpha/2)} \cos(\alpha \arctan(l|q|))\right\}$$

where the scale factor c_1 is given as

$$c_1 = \frac{2\pi \cos(\pi\alpha/2)}{\alpha\Gamma(\alpha)\sin(\pi\alpha)} At$$

and

$$c_0 = \frac{l^{-\alpha}}{\cos(\pi\alpha/2)} c_1$$

$$= \frac{2\pi}{\alpha\Gamma(\alpha)\sin(\pi\alpha)} A l^{-\alpha} t.$$

The variance can be calculated from the characteristic function as follows:

$$\sigma^2(t) = \frac{\partial^2 \varphi(q)}{\partial q^2}\bigg|_{q=0}$$

$$= t\frac{2A\pi(1-\alpha)}{\Gamma(\alpha)\sin(\pi\alpha)} l^{2-\alpha}$$

If we discretize in time with steps Δt, we obtain that $T = N\Delta t$. Thus we deduce that at the end of each interval, we must calculate the sum of N stochastic variables that are independent and identically distributed. Therefore, the new characteristic function will be:

$$\varphi(q, N) = \varphi(q)^N = \exp\left\{c_0 N - c_1 \frac{N(q^2 + 1/l^2)^{\alpha/2}}{\cos(\pi\alpha/2)} \cos(\alpha \arctan(l|q|))\right\}.$$

13 Generalized Lévy Processes, Long Range Correlations, and Memory Effects

For small values of N the probability will be very similar to the stable Lévy distribution:

$$P_L(x=0) = \frac{\Gamma(1/\alpha)}{\pi\alpha(\gamma N)^{1/\alpha}}.$$

In this case, we observe that the parameter γ has been changed by $N\gamma$. In fact this model can be improved by standardizing as follows:

$$\sigma^2 = -\left.\frac{\partial^2\varphi(q)}{\partial q^2}\right|_{q=0}$$

we have that

$$-\left.\frac{\partial^2\varphi(q/\sigma)}{\partial q^2}\right|_{q=0} = -\frac{1}{\sigma^2}\left.\frac{\partial^2\varphi(q)}{\partial q^2}\right|_{q=0} = 1.$$

Thus our standardized model is:

$$\ln\varphi_S(q) = \ln\varphi\left(\frac{q}{\sigma}\right)$$

$$= c_0 - c_1\frac{((q/\sigma)^2 + 1/l^2)^{\alpha/2}}{\cos(\pi\alpha/2)}\cos\left(\alpha\arctan\left(l\frac{|q|}{\sigma}\right)\right)$$

Substituting the constants c_0 and c_1 into the above equation and simpifying, we obtain the standardized model:

$$\ln\varphi_S(q) = \frac{2\pi A l^{-\alpha}t}{\alpha\Gamma(\alpha)\sin(\pi\alpha)}\left[1 - \left(\left(\frac{ql}{\sigma}\right)^2 + 1\right)^{\alpha/2}\cos\left(\alpha\arctan\left(\frac{ql}{\sigma}\right)\right)\right].$$

(13.1)

Before we proceed, we state a well-known theorem applicable in our case.

Theorem 13.2.1 Let $\hat{p}(k)$ be the characteristic function (Fourier transform) of $p(x)$. Let $\log\hat{p}(k) = f_a(k) + f_s(k)$, where $f_a(k)$ is analytic and $f_s(k) \sim -a|k|^\alpha$ as $k \to 0$ with $a \in \mathbb{R}$ and finite. Then

$$p(x) \sim \frac{A}{|x|^{1+\alpha}}, \quad \text{as } x \to \infty,$$

where

$$A = \frac{a\sin(\alpha\pi/2)\Gamma(\alpha)\alpha}{\pi}.$$

$\square$

Please see [218] and references therein for the proof of the theorem.

Going back to the TLF distribution, for small q, our standardized model (13.1) is approximated as

$$\log\varphi_S(q) \sim \frac{2\pi A l^{-\alpha}t}{\alpha\Gamma(\alpha)\sin(\pi\alpha)}\left[1 - \left(1 + \frac{\alpha}{2}\left(\frac{ql}{\sigma}\right)^2\right)\cos\left(\alpha\arctan\left(\frac{ql}{\sigma}\right)\right)\right].$$

Therefore using the aforementioned theorem,

$$p(x) \sim \frac{1}{|x|^3} \tag{13.2}$$

We can justify (13.2) from the viewpoint of the cutoff function used in [124] given by:

$$f(x) = \begin{cases} A_- e^{-|x|/l}|x|^{-1-\alpha} & x < 0 \\ A_+ e^{-|x|/l}|x|^{-1-\alpha} & x \geq 0. \end{cases}$$

If we consider the following cases of $p(x)$,

$$p(x) = \begin{cases} 0 & x > l \\ cL(x) & -l \leq x \leq l \\ 0 & x < -l. \end{cases}$$

where $L(x)$ is the Lévy distribution. In this instance, the decay is faster than the *heavy tail* decay due to the presence of the exponential function in $f(x)$. However, the main advantage of using this form of density $p(x)$ is that we can keep the *heavy tail* of the Lévy for arbitrarily large cutoff values (by adjusting the parameter l) and then have a sharp decay (faster than *heavy tail*). In other words, we are keeping the features of the heavy tail of the Lévy distribution as well as the finite variance feature given by exponential decay and truncation.

13.2.2 Kurtosis

Kurtosis is a measure of the "tailedness" of the probability distribution of a real-valued random variable. It is a descriptor of the shape of a probability distribution. In general, the nth moment of a distribution, if it is finite, can be computed from the characteristic function $\varphi(q)$ in a similar way we would compute the nth moment from the moment generating function. The nth moment computed from the characteristic function φ will be:

$$\mu_n = E[X^n] = \frac{1}{n} \frac{P^n \varphi}{Pq^n}\bigg|_{q=0}.$$

For all parameters of α except $\alpha = 2$ of the Lévy stable distribution, we have infinite 4th moment. For a truncated Lévy distribution, we can compute the 4th moment of (13.1). A lengthy but straightforward computation gives us

$$\mu_4 = -\frac{\alpha c_0 l^4 (\alpha - 1)(3\alpha c_0 - 5\alpha - 3\alpha^2 c_0 + \alpha^2 + 6)}{\sigma^4}$$

Since (13.1) is normalized, the kurtosis will be [33]:

$$Kurt[X] = \gamma_2 = \mu_4 - 3$$
$$= -\frac{\alpha c_0 l^4 (\alpha - 1)(3\alpha c_0 - 5\alpha - 3\alpha^2 c_0 + \alpha^2 + 6)}{\sigma^4} - 3 \tag{13.3}$$

Kurtosis is useful as a measure of the degree to which there is "excess" probability density in comparison to a normal distribution. Specifically, a leptokurtic ($Kurt[X] > 0$) distribution is often referred to as a "fat tailed" distribution since it has more area in tails than the normal which decays very fast. In finance this is important since ALL return distributions are leptokurtic. For financial data a truncated Levy distribution may provide a better fit.

In the previous chapter, we mentioned that a random variable Y is infinitely divisible if it can be written as the sum of n independent copies of some random variable X. This type of decomposition of a distribution is used in probability and statistics to find families of probability distributions that might be natural choices for certain models.

To see whether TLF is infinitely divisible, we introduce the following theorem.

Theorem 13.2.2 Let X_i be independent and identically distributed random variables with characteristic function φ_X. Then the random variable Y given by

$$Y = \sum_{i=1}^{n} X_i$$

has characteristic function φ_Y given by

$$\varphi_Y = \prod_{i=1}^{n} \varphi_X = (\varphi_X)^n.$$

Next we define a characteristic function that suggests that the TLF is infinitely divisible.

Given that

$$\log \varphi_{S,N} = \frac{2\pi A l^{-\alpha} t}{N \alpha \Gamma(\alpha) \sin(\pi \alpha)} \left[1 - \left(\left(\frac{ql}{\sigma} \right)^2 + 1 \right)^{\alpha/2} \cos\left(\alpha \arctan\left(\frac{ql}{\sigma} \right) \right) \right],$$

if we consider the sum of random variables given by $\varphi_{S,N}$, then by using the aforementioned theorem, we obtain:

$$\varphi_S = (\varphi_{S,N})^N$$

This means that the TLF is infinitely divisible.

Remark 13.2.1 It is very remarkable the fact that the stable Lévy processes (also called Lévy flight) have independent increments but are designated as long memory processes. Indeed, this is the case for these processes due to the fact that the increments are heavy tailed. Heavy-tailed distributions are probability distributions whose tails are not exponentially bounded, that is, they have heavier tails than the exponential distribution.

13.2.3 Self-Similarity

Definition 13.2.1 A stochastic process is said to be *self-similar* if there exists a constant $H > 0$ such that for any scaling factor $a > 0$, the process $\{X_{at}\}_{t \geq 0}$ and $\{a^H X_t\}_{t \geq 0}$ have the same law in the sense of finite dimensional distributions. The constant H is called the self-similarity exponent or the Hurst exponent of the process X.

Self-similar processes occur in several areas of applied mathematics. They are used to describe fractals and chaos theory, long memory processes, and spectral analysis. Hurst exponent estimation has been applied in areas ranging from biophysics to computer networking. Estimation of the Hurst exponent was originally developed in hydrology. However, the modern techniques for estimating the Hurst exponent come from fractal mathematics.

As observed in [40], from the definition we obtain that for any c, t positive, X_{ct} and $c^H X_t$ have the same distribution. Choosing $c = 1/t$ yields that $X_t = t^H X_1$ in distribution, for any positive t. Therefore,

$$F_t(x) = P(X_t \leq x) = P(t^h X_1 \leq x) = F_1\left(\frac{x}{t^H}\right).$$

If F_t has a density ρ_t, that is, if ρ_t denotes the density or distribution function of X_t, then we obtain by differentiating the previous formula and Definition 13.2.1 implies that

$$\rho_t(x) = \frac{1}{t^H} \rho_1\left(\frac{x}{t^H}\right). \tag{13.4}$$

If $\Phi_t(q)$ is the characteristic function of X_t and furthermore $\Phi_t(q)$ is an even function, then

$$\rho_t(x) = \frac{1}{\pi} \int_0^\infty \Phi_t(q) \cos(qx) dq, \tag{13.5}$$

and

$$\rho_1\left(\frac{x}{t^H}\right) = \frac{1}{\pi} \int_0^\infty \Phi_1(q) \cos\left(q \frac{x}{t^H}\right) dq. \tag{13.6}$$

Thus in that case (13.4) gives

$$\int_0^\infty \Phi_t(q) \cos(qx) dq = \frac{1}{t^H} \int_0^\infty \Phi_1(q) \cos\left(q \frac{x}{t^H}\right) dq. \tag{13.7}$$

The Lévy process (where the increments are independent and follow the Lévy distribution) is self-similar. For this process $\Phi_t(q) = \exp(-t|q|^\alpha)$, and we obtain from (13.7) that

$$\int_0^\infty \exp(-t|q|^\alpha) \cos(qx) dq = \frac{1}{t^H} \int_0^\infty \exp(-|q|^\alpha) \cos\left(\frac{qx}{t^H}\right) dq$$

by using the change of variable: $q = t^H y$; the second integral becomes:

$$\int_0^\infty \exp(-t^{\alpha H}|q|^\alpha)\cos(qx)dq$$

so the equality holds only if

$$\alpha = \frac{1}{H}.$$

This result is well known and has been used widely in literature.

13.2.4 The H - α Relationship for the Truncated Lévy Flight

We now consider the standardized truncated Lévy model. This model is not self-similar due to truncation and standardization. However, we would like to find out how close this model is from a self-similar model. We will do this by looking at the relationship between the α characterizing the *TLF* and the resulting H parameter characterizing the self-similar property.

As observed in previous subsection for standardized *TLF* model,

$$\Phi_t(q) = \varphi_S(q) = \varphi(\frac{q}{\sigma}) = \exp\left[G(\alpha)t\left[1 - \left(\left(\frac{q}{\sigma}\right)^2 + 1\right)^{\frac{\alpha}{2}}\cos\left(\alpha\arctan\left(\frac{ql}{\sigma}\right)\right)\right]\right], \tag{13.8}$$

where

$$G(\alpha) = \frac{2\pi A l^{-\alpha}}{\alpha\Gamma(\alpha)\sin(\pi\alpha)}. \tag{13.9}$$

Therefore

$$p_t(x) = \frac{1}{\pi}\int_0^\infty \exp\left[G(\alpha)t\left[1 - \left(\left(\frac{q}{\sigma}\right)^2 + 1\right)^{\frac{\alpha}{2}}\cos\left(\alpha\arctan\left(\frac{ql}{\sigma}\right)\right)\right]\right]\cos(qx)dq. \tag{13.10}$$

Observe that when $\frac{q}{\sigma}$ is large enough (in other words q is large enough), $\cos\left(\alpha\arctan\left(\frac{ql}{\sigma}\right)\right) \approx \cos\left(\alpha\frac{\pi}{2}\right)$ because $\arctan(\infty) = \frac{\pi}{2}$ and $\left(\left(\frac{q}{\sigma}\right)^2 + 1\right)^{\frac{\alpha}{2}} \approx \left(\frac{q}{\sigma}\right)^\alpha$. Choose B large enough so that when $q > \frac{B}{t^{1/\alpha}}$,

$$1 - \left(\left(\frac{q}{\sigma}\right)^2 + 1\right)^{\frac{\alpha}{2}}\cos\left(\alpha\arctan\left(\frac{ql}{\sigma}\right)\right) \approx -\left(\frac{q}{\sigma}\right)^\alpha\cos\left(\alpha\frac{\pi}{2}\right).$$

Then

$$p_t(x) = \frac{1}{\pi}\int_0^{\frac{B}{t^{1/\alpha}}} \exp\left[G(\alpha)t\left[1 - \left(\left(\frac{q}{\sigma}\right)^2 + 1\right)^{\frac{\alpha}{2}}\cos\left(\alpha\arctan\left(\frac{ql}{\sigma}\right)\right)\right]\right]\cos(qx)dq$$

$$+ \frac{1}{\pi}\int_{\frac{B}{t^{1/\alpha}}}^\infty \exp\left[G(\alpha)t\left[1 - \left(\left(\frac{q}{\sigma}\right)^2 + 1\right)^{\frac{\alpha}{2}}\cos\left(\alpha\arctan\left(\frac{ql}{\sigma}\right)\right)\right]\right]\cos(qx)dq$$

$$\approx I_0^{\frac{B}{t^{1/\alpha}}}(t,x) + \frac{1}{\pi}\int_{\frac{B}{t^{1/\alpha}}}^\infty \exp\left[-G(\alpha)t\cos\left(\alpha\frac{\pi}{2}\right)\left(\frac{q}{\sigma}\right)^\alpha\right]\cos(qx)dq$$

$$= I_0^{\frac{B}{t^{1/\alpha}}}(t,x) + \frac{1}{t^{1/\alpha}}\frac{1}{\pi}\int_B^\infty \exp\left[-G(\alpha)\cos\left(\alpha\frac{\pi}{2}\right)\left(\frac{u}{\sigma}\right)^\alpha\right]\cos\left(u\frac{x}{t^{1/\alpha}}\right)du, \tag{13.11}$$

where

$$I_0^{\frac{B}{t^{1/\alpha}}}(t,x) = \frac{1}{\pi}\int_0^{\frac{B}{t^{1/\alpha}}} \exp\left[G(\alpha)t\left[1-\left(\left(\frac{q}{\sigma}\right)^2+1\right)^{\frac{\alpha}{2}}\cos\left(\alpha\arctan\left(\frac{ql}{\sigma}\right)\right)\right]\right]\cos(qx)dq,$$
(13.12)

and the change of variable $q = \frac{u}{t^{\frac{1}{\alpha}}}$ is done in the second term to obtain the last equality.

So we obtain the formula:

$$\rho_t(x) = I_0^{\frac{B}{t^{1/\alpha}}}(t,x) + \frac{1}{t^{1/\alpha}}\frac{1}{\pi}\int_B^\infty \exp\left[-G(\alpha)\cos\left(\alpha\frac{\pi}{2}\right)\left(\frac{u}{\sigma}\right)^\alpha\right]\cos\left(u\frac{x}{t^{1/\alpha}}\right)du$$

Taking $t = 1$ in the last formula, we obtain

$$\rho_1(x) \approx I_0^B(1,x) + \frac{1}{\pi}\int_B^\infty \exp\left[-G(\alpha)\cos\left(\alpha\frac{\pi}{2}\right)\left(\frac{u}{\sigma}\right)^\alpha\right]\cos(ux)du.$$

Thus, in the last formula, evaluating in $\frac{x}{t^{\frac{1}{\alpha}}}$ and multiplying by $t^{\frac{1}{\alpha}}$ both sides, we have:

$$\frac{1}{t^{1/\alpha}}\rho_1\left(\frac{x}{t^{1/\alpha}}\right) \approx \frac{1}{t^{1/\alpha}}I_0^B\left(1,\frac{x}{t^{1/\alpha}}\right)$$
$$+ \frac{1}{t^{1/\alpha}}\frac{1}{\pi}\int_B^\infty \exp\left[-G(\alpha)\cos\left(\alpha\frac{\pi}{2}\right)\left(\frac{u}{\sigma}\right)^\alpha\right]\cos\left(u\frac{x}{t^{1/\alpha}}\right)du.$$

Therefore

$$\rho_t(x) - \frac{1}{t^{1/\alpha}}\rho_1\left(\frac{x}{t^{1/\alpha}}\right) \approx I_0^{\frac{B}{t^{1/\alpha}}}(t,x) - \frac{1}{t^{1/\alpha}}I_0^B\left(1,\frac{x}{t^{1/\alpha}}\right). \quad (13.13)$$

So when the quantity $\left|I_0^{\frac{B}{t^{1/\alpha}}}(t,x) - \frac{1}{t^{1/\alpha}}I_0^B(1,\frac{x}{t^{1/\alpha}})\right|$ becomes small, the standardized truncated Lévy model tends to show the self-similar structure with corresponding Hurst exponent H approximately equal to $\frac{1}{\alpha}$. The choice of B in (13.13) depends on the numerical computation.

Remark 13.2.2 For the case of the present analysis, empirical evidence presented in [15] shows that when $T = 1$, $H \approx \frac{1}{\alpha}$. In general because of the truncation and standardization of the Lévy model, we should not expect $H = \frac{1}{\alpha}$.

13.3 Sum of Lévy Stochastic Variables with Different Parameters

In this section we will be discussing the sum of Lévy stochastic variables with different parameters. We will prove and present for similar distributions which will be used later in this chapter for the analysis of time series data. We begin with a discussion of the sum of exponential random variables with different parameters.

13.3.1 Sum of Exponential Random Variables with Different Parameters

Suppose X and Y follow exponential distributions with parameters β_1, β_2. Let X has p.d.f. $f_X(x) = \frac{e^{-x/\beta_1}}{\beta_1}$ and Y has p.d.f. $f_Y(y) = \frac{e^{-y/\beta_2}}{\beta_2}$, where X and Y independent. Applying the change of variables $u = x + y$ and $v = x$, we obtain the Jacobian of this transformation that is 1. Thus the joint pdf in new variables is given by

$$f(u,v) = \frac{1}{\beta_1 \beta_2} \exp\left(-v\left(\frac{1}{\beta_1} - \frac{1}{\beta_2}\right)\right) \exp\left(-\frac{u}{\beta_2}\right).$$

Therefore when $\beta_1 \neq \beta_2$,

$$f_U(u) = \int_0^u f(u,v) dv = \frac{1}{\beta_1 - \beta_2}\left[e^{-u/\beta_1} - e^{-u/\beta_2}\right].$$

When $\beta_1 = \beta_2$,

$$f_U(u) = \frac{u e^{-u/\beta_1}}{\beta_1^2} = \Gamma(2, \beta_1).$$

So $u \sim \Gamma(2, \beta_1)$. Similarly with random variables X, Y, and Z with p.d.f. $f_X(x) = \frac{e^{-x/\beta_1}}{\beta_1}, f_Y(y) = \frac{e^{-y/\beta_2}}{\beta_2}$, and $f_Z(z) \frac{e^{-y/\beta_3}}{\beta_3}$, then using a change of variable

$$u = x + y + z, \quad v = x + y, \quad w = x,$$

we arrive to

$$f(u,v) = \frac{e^{-u/\beta_3}}{\beta_3(\beta_1 - \beta_2)}\left[\exp(-v(1/\beta_1 - 1/\beta_3)) - \exp(-v(1/\beta_2 - 1/\beta_3))\right],$$

so that

$$\int_0^u f(u,v) dv = \frac{\beta_1}{\beta_1 - \beta_2}\left[\frac{e^{-u/\beta_1} - e^{-u/\beta_3}}{\beta_1 - \beta_3}\right] - \frac{\beta_2}{\beta_1 - \beta_2}\left[\frac{e^{-u/\beta_2} - e^{-u/\beta_3}}{\beta_2 - \beta_3}\right] = f_U(u),$$

and if $\beta_1 = \beta_2 = \beta_3, f(u,v) = \frac{1}{\beta^3} e^{-u/\beta} v$, then

$$f_U(u) = \int_0^u f(u,v) dv = \frac{1}{\beta^3} e^{-u/\beta} \frac{u^2}{2} = \Gamma(3, \beta).$$

Generally, if we have n random variables X_i with p.d.f. $f_{X_i}(x)$, then the p.d.f. of $X = \sum_{i=1}^n X_i$ will have the p.d.f. given by a convolution of p.d.f. [33].

$$f_X(x) = (f_{X_1} * (f_{X_2} * (f_{X_3} * \ldots)))[x]. \tag{13.14}$$

In the case of n independent variables following p.d.f. $f(x) = \frac{e^{-x/\beta}}{\beta}$ with different coefficient β, it is more convenient to work with the *characteristic function* which is just a Fourier transform of a p.d.f. The convolution theorem states that

13.3 Sum of Lévy Stochastic Variables with Different Parameters

Fourier transform of convolution of two functions is the product of their Fourier transform.

$$\mathcal{F}\{f * g\} = \mathcal{F}\{f\} \cdot \mathcal{F}\{g\}.$$

Thus if we use this theorem on (13.14) we get,

$$\mathcal{F}\{f_X\} = \mathcal{F}\{(f_{X_1} * (f_{X_2} * (f_{X_3} * \ldots)))\} = \prod_{i=1}^{n} \mathcal{F}\{f_{X_i}\}$$

but $\mathcal{F}\{f_{X_i}\}$ is just characteristic function of X_i denoted by φ_{X_i}. We showed that

$$\varphi_X = \prod_{i=1}^{n} \varphi_{X_i}. \tag{13.15}$$

In other words, characteristic function of a sum of independent random variables is the product of their characteristic functions.

Suppose we consider the general version for n random variables. We will assume the following parametrization, using scale parameter β, of the exponential random variable p.d.f.

$$f(x) = \frac{1}{\beta} e^{-x/\beta}, x \geq 0. \tag{13.16}$$

The characteristic function for this random variable will be [33]

$$\varphi_X(t) = (1 - it\beta)^{-1}.$$

We also know, using (13.15), that sum of independent identically distributed exponential random variables will have characteristic function

$$\varphi_{\sum X_k}(t) = \prod_{k=1}^{n} \varphi_{X_k}(t) = (1 - it\beta)^{-n} \tag{13.17}$$

and this defines the gamma distribution.

Similarly, the sum of independent exponential random variables $X_1, \ldots, X_n$ with scale parameters $\beta_1, \ldots, \beta_n$ will have the characteristic function

$$\varphi_{\sum X_k}(t) = \prod_{k=1}^{n} \varphi_{X_k}(t) = \prod_{k=1}^{n} (1 - it\beta_i)^{-1} \tag{13.18}$$

Definition 13.3.1 (Hypoexponential Distribution). [[196], section 7.6.3.] We say that the random variable X is hypoexponentially distributed if its characteristic function is of form (13.18). Sometimes it is also called a generalized Erlang distribution.

We will show that the sum of exponential random variables with different parameters β_k (13.18), where β_k differs by a small variation from some $\overline{\beta}$, can be reasonably approximated by the gamma distribution (13.17). We will assume the following scale parameters:

$$\beta_k = \overline{\beta} + \Delta\beta_k, k \in \{1, \ldots, n\}$$

$$\overline{\beta} = \frac{1}{n}\sum_{j=1}^{n}\beta_j$$

$$0 = \sum_{k=0}^{n}\Delta\beta_k \tag{13.19}$$

where $\Delta\beta_k$ is a small perturbation from the mean value $\overline{\beta}$. If we substitute β_k into (13.16), we get

$$\varphi_{X_k} = (1 - it\overline{\beta} - it\Delta\beta_k)^{-1}. \tag{13.20}$$

This function is holomorphic in $\Delta\beta_k$ for small values of $\Delta\beta_k$; thus it has power series expansion in $\Delta\beta_k$ at zero, i.e.

$$\varphi_{X_k} = (1 - it\overline{\beta})^{-1} + it(1 - it\overline{\beta})^{-2}\Delta\beta_k + \mathcal{O}(\Delta\beta_k^2). \tag{13.21}$$

Using this result and substituting (13.21) into (13.18), we get

$$\varphi_{\sum X_k}(t) = \left((1 - it\overline{\beta})^{-1} + it(1 - it\overline{\beta})^{-2}\Delta\beta_1 + \mathcal{O}(\Delta\beta_1^2)\right) \cdot$$
$$\left((1 - it\overline{\beta})^{-1} + it(1 - it\overline{\beta})^{-2}\Delta\beta_2 + \mathcal{O}(\Delta\beta_2^2)\right) \cdot$$
$$\left((1 - it\overline{\beta})^{-1} + it(1 - it\overline{\beta})^{-2}\Delta\beta_3 + \mathcal{O}(\Delta\beta_3^2)\right) \cdot$$
$$\left((1 - it\overline{\beta})^{-1} + it(1 - it\overline{\beta})^{-2}\Delta\beta_4 + \mathcal{O}(\Delta\beta_4^2)\right) \cdot$$
$$\vdots$$
$$\left((1 - it\overline{\beta})^{-1} + it(1 - it\overline{\beta})^{-2}\Delta\beta_n + \mathcal{O}(\Delta\beta_n^2)\right).$$

Multiplying the previous expression gives us

$$\varphi_{\sum X_k}(t) = (1 - it\overline{\beta})^{-n} + it(1 - it\overline{\beta})^{-(n+1)}\sum_{k=1}^{n}\Delta\beta_k + \mathcal{O}(\Delta\beta_k^2) \tag{13.22}$$

and from $\sum_{k=0}^{n}\Delta\beta_k = 0$ from (13.19), we finally get

$$\varphi_{\sum X_k}(t) = (1 - it\overline{\beta})^{-n} + \mathcal{O}(\Delta\beta_k^2). \tag{13.23}$$

Therefore, the characteristic function differs from that of the gamma distribution (13.17) by a term depending on a square of perturbation $\Delta\beta_k$. This justifies our fit using the gamma distribution with a scale parameter equal to the mean value $\overline{\beta}$. Next we discuss the sum of Lévy random variables with different parameters.

13.3.2 Sum of Lévy Random Variables with Different Parameters

We present a result for the Lévy distribution that is analogous to the one found in the previous sections for convolutions of exponential distributions. Let us assume random variables $X_1, \ldots, X_n$ with characteristic function (13.1) and coefficients $\beta = 0, \mu = 0, \gamma_k = \gamma$. The characteristic function for X_k will be

$$\varphi_{X_k}(q) = e^{-|\gamma q|^{\alpha_k}}. \tag{13.24}$$

We will use a transformation of parameters α_k:

$$\alpha_k = \overline{\alpha} + \Delta\alpha_k, k \in \{1, \ldots, n\}$$

$$\overline{\alpha} = \frac{1}{n}\sum_{j=1}^{n} \alpha_j$$

$$0 = \sum_{j=0}^{n} \Delta\alpha_k \tag{13.25}$$

where $\Delta\alpha_k$ is a small perturbation from the mean value $\overline{\alpha}$. If we substitute α_k into (13.24), we get

$$\varphi_{X_k} = e^{-|\gamma q|^{\overline{\alpha}+\Delta\alpha_k}}. \tag{13.26}$$

The function (13.26) is holomorphic (it is also entire) in $\Delta\alpha_j$ since it is differentiable in a neighborhood of every point in its domain and has a power series expansion in terms of $\Delta\alpha_j$ at zero, i.e.

$$\varphi_{X_k} = e^{-|\gamma q|^{\overline{\alpha}}} + e^{-|\gamma q|^{\overline{\alpha}}}(-|\gamma q|^{\overline{\alpha}})\ln(|\gamma q|) \cdot \Delta\alpha_k + \mathcal{O}(\Delta\alpha_k^2) \tag{13.27}$$

Again, the sum of random variables X_k will have the characteristic function given by a product of characteristic functions:

$$\varphi_{\sum X_k} = \prod_{k=1}^{n} \varphi_{X_k} = \prod_{k=1}^{n} e^{-|\gamma q|^{\overline{\alpha}+\Delta\alpha_k}}$$

Substituting (13.27) into the equation above, we obtain:

$$\varphi_{\sum X_k} = \left(e^{-|\gamma q|^{\overline{\alpha}}} + e^{-|\gamma q|^{\overline{\alpha}}}(-|\gamma q|^{\overline{\alpha}})\ln(|\gamma q|) \cdot \Delta\alpha_1 + \mathcal{O}(\Delta\alpha_1^2)\right) \cdot$$
$$\left(e^{-|\gamma q|^{\overline{\alpha}}} + e^{-|\gamma q|^{\overline{\alpha}}}(-|\gamma q|^{\overline{\alpha}})\ln(|\gamma q|) \cdot \Delta\alpha_2 + \mathcal{O}(\Delta\alpha_2^2)\right) \cdot$$
$$\left(e^{-|\gamma q|^{\overline{\alpha}}} + e^{-|\gamma q|^{\overline{\alpha}}}(-|\gamma q|^{\overline{\alpha}})\ln(|\gamma q|) \cdot \Delta\alpha_3 + \mathcal{O}(\Delta\alpha_3^2)\right) \cdot$$
$$\left(e^{-|\gamma q|^{\overline{\alpha}}} + e^{-|\gamma q|^{\overline{\alpha}}}(-|\gamma q|^{\overline{\alpha}})\ln(|\gamma q|) \cdot \Delta\alpha_4 + \mathcal{O}(\Delta\alpha_4^2)\right) \cdot$$
$$\vdots$$
$$\left(e^{-|\gamma q|^{\overline{\alpha}}} + e^{-|\gamma q|^{\overline{\alpha}}}(-|\gamma q|^{\overline{\alpha}})\ln(|\gamma q|) \cdot \Delta\alpha_n + \mathcal{O}(\Delta\alpha_n^2)\right).$$

Therefore multiplying these expressions and simplifying the terms, we get:

$$\varphi_{\sum X_k} = e^{-n|\gamma q|^{\bar{\alpha}}} + e^{-n|\gamma q|^{\bar{\alpha}}}(-|\gamma q|^{\bar{\alpha}})\ln(|\gamma q|) \cdot \sum_{k=1}^{n} \Delta \alpha_k + \mathcal{O}(\Delta \alpha_k^2) \quad (13.28)$$

and using that $\sum_{j=0}^{n} \Delta \alpha_k = 0$ from (13.25), we finally get

$$\varphi_{\sum X_k} = e^{-n|\gamma q|^{\bar{\alpha}}} + \mathcal{O}(\Delta \alpha_k^2) \quad (13.29)$$

which has similar properties or same structure as the characteristic function for the Lévy stable distribution with scale factor $N\gamma$.

13.4 Examples and Applications

In the this section, we present examples and applications of applying the Lévy model, DFA, and Rescaled Range Analysis discussed in Chapter 12 of this book to financial time series.

We will begin the examples by first considering the standardized truncated Lévy models.

13.4.1 Truncated Lévy Models Applied to Financial Indices

Example 13.4.1 In this example, we studied the behavior of stock prices comprising the Dow Jones Industrial Average (DJIA) index, along with the index itself. Specifically, we have studied the shares' values for City group, JPMorgan, and Microsoft Corporation and the values of the DJIA index. The values are from 2 January 1985 to 1 May 2010. The analyzed stochastic variable is the return r_t, defined as the difference of the logarithm of two consecutive stock (or index) prices.

In this example, we plot on the same grid the cumulative distribution of the observed returns for different time lags T in order to visualize how good the fitting is. The time lag $T = 1$ means the returns are calculated by using two consecutive observations; for a general T, the returns are calculated by using: $r_t = \log(Xt/Xt - T)$ with $X_t = I_t$, where I_t denotes the stock (or index) price at time t and T is the difference (in labor days) between two values of the stock or index.

The results for the estimated Lévy flight parameter are presented in Figures 13.1–13.4.

Example 13.4.2 On Sunday, 16 March 2008 Bear Stearns signed a merger agreement with JP Morgan Chase essentially selling the company for $2 a share (price revised on March 24 to $10/share). The same stock was traded at $172 in January 2007 and $93 a share in February 2007.

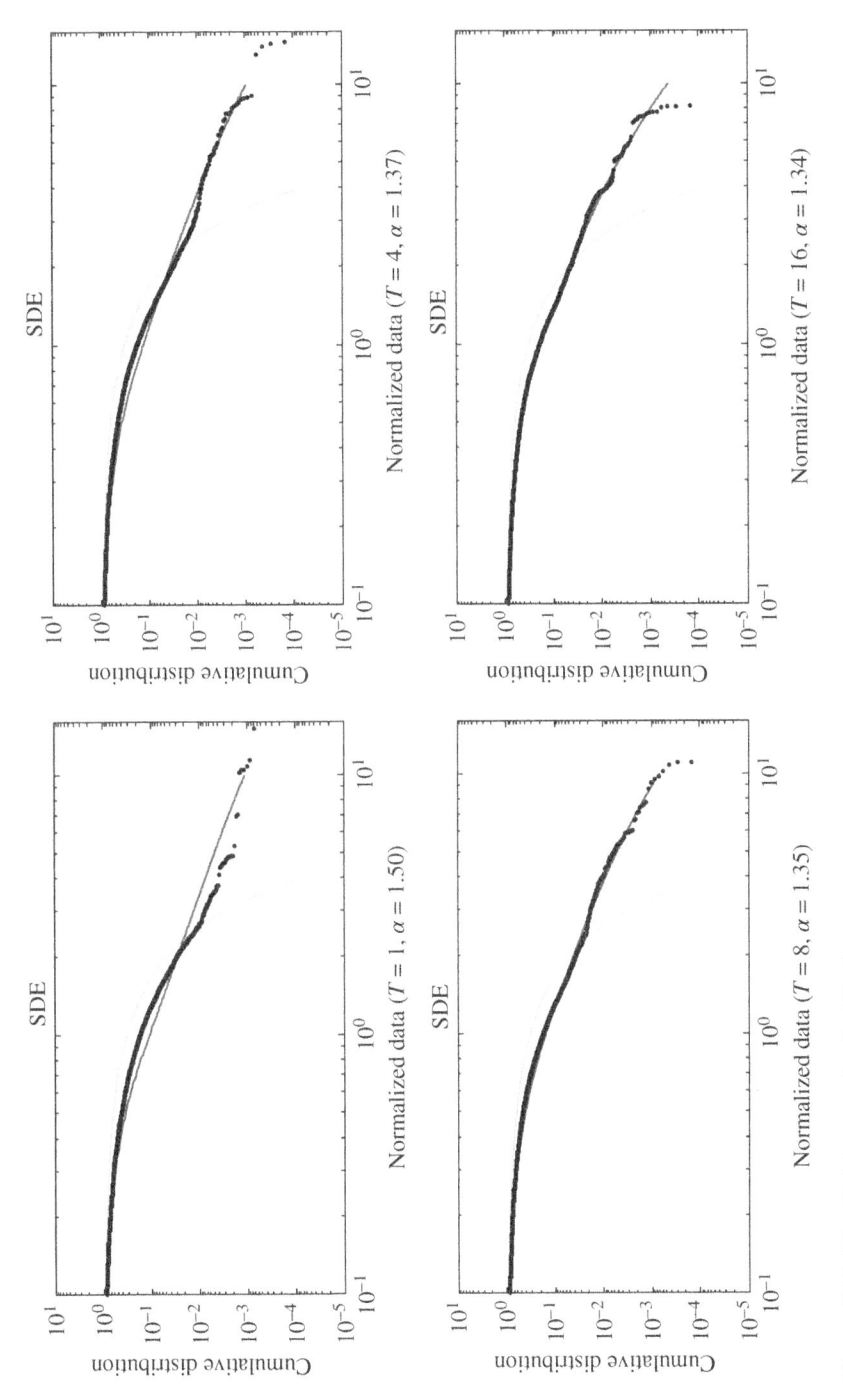

Figure 13.1 Lévy flight parameter for City Group.

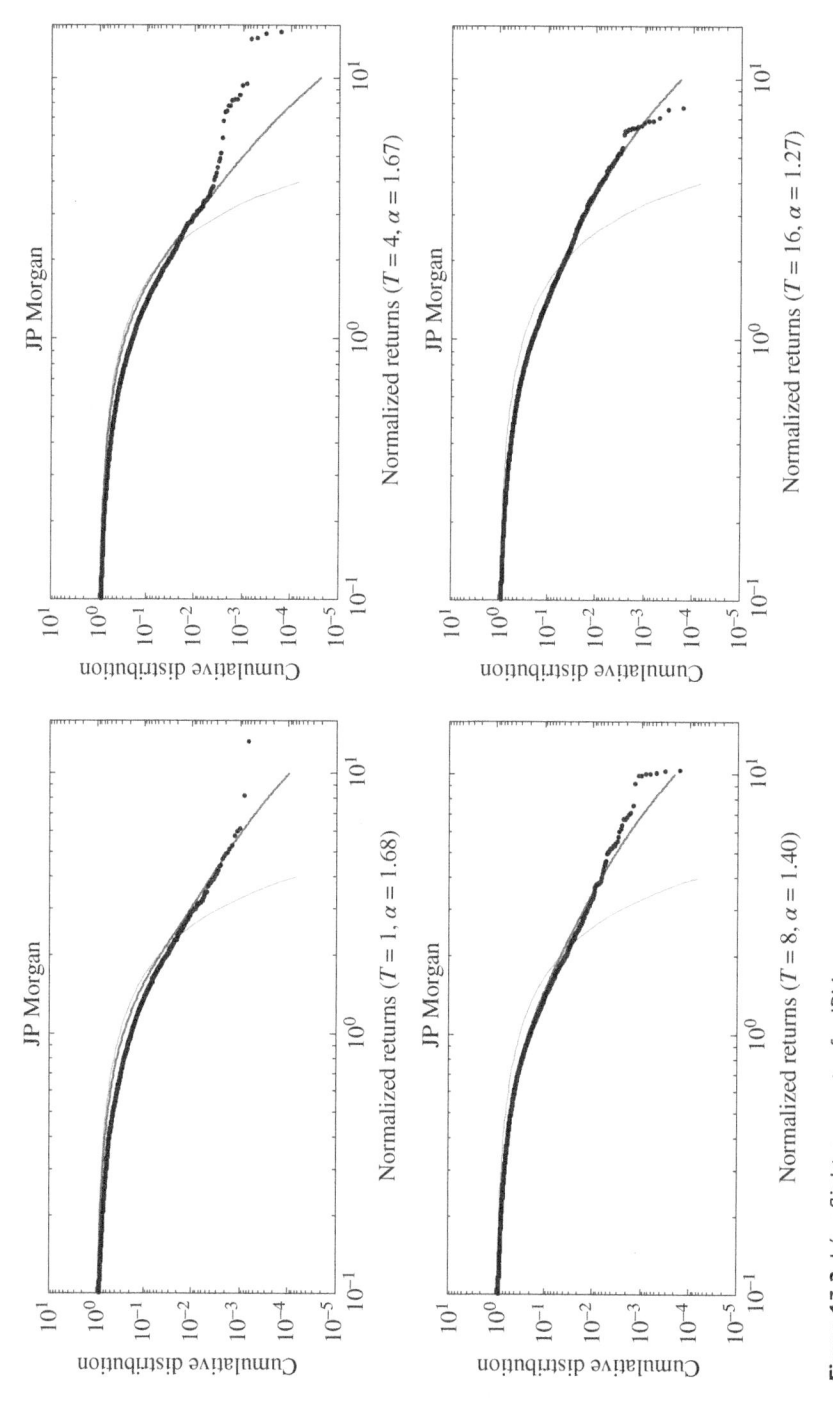

Figure 13.2 Lévy flight parameter for JPMorgan.

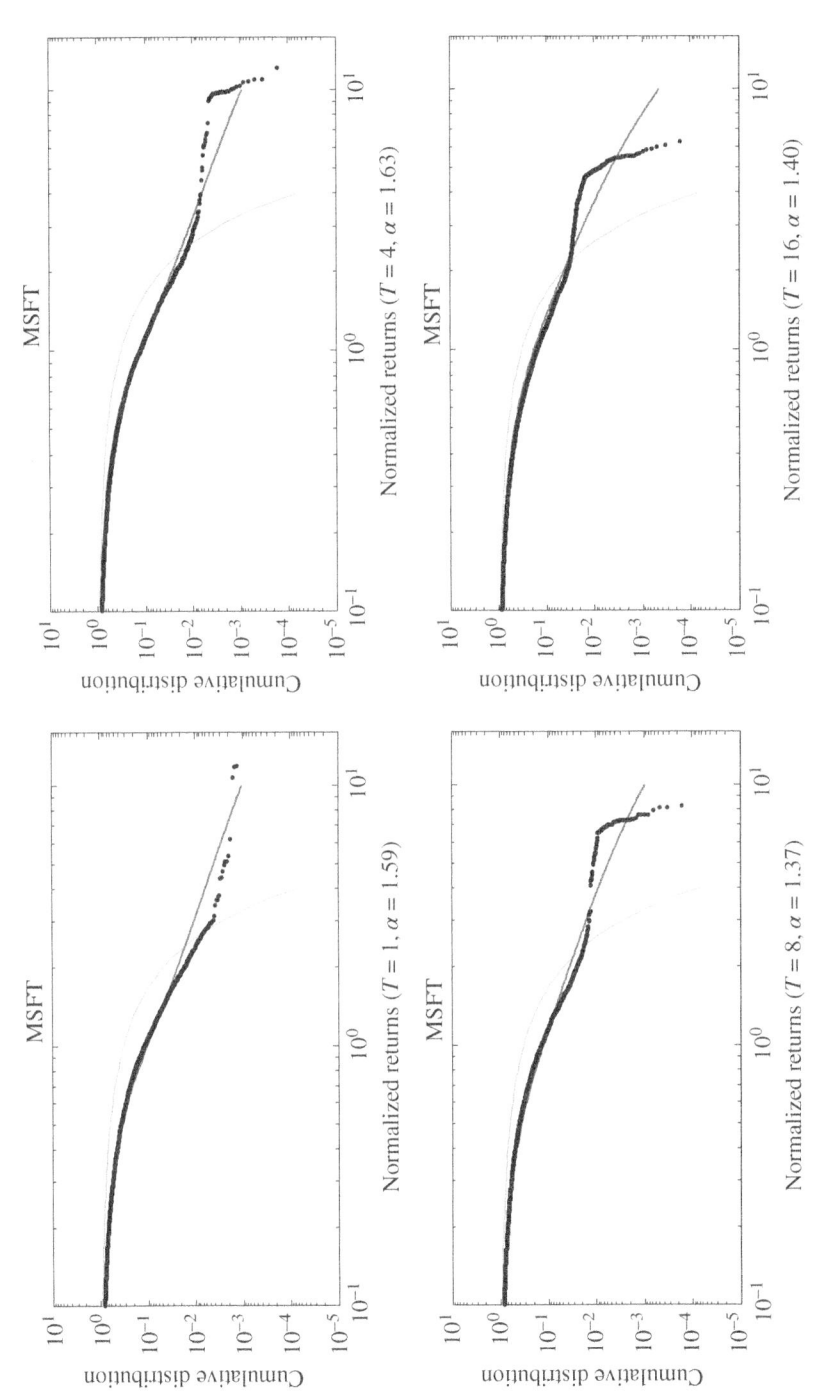

Figure 13.3 Lévy flight parameter for Microsoft Corporation.

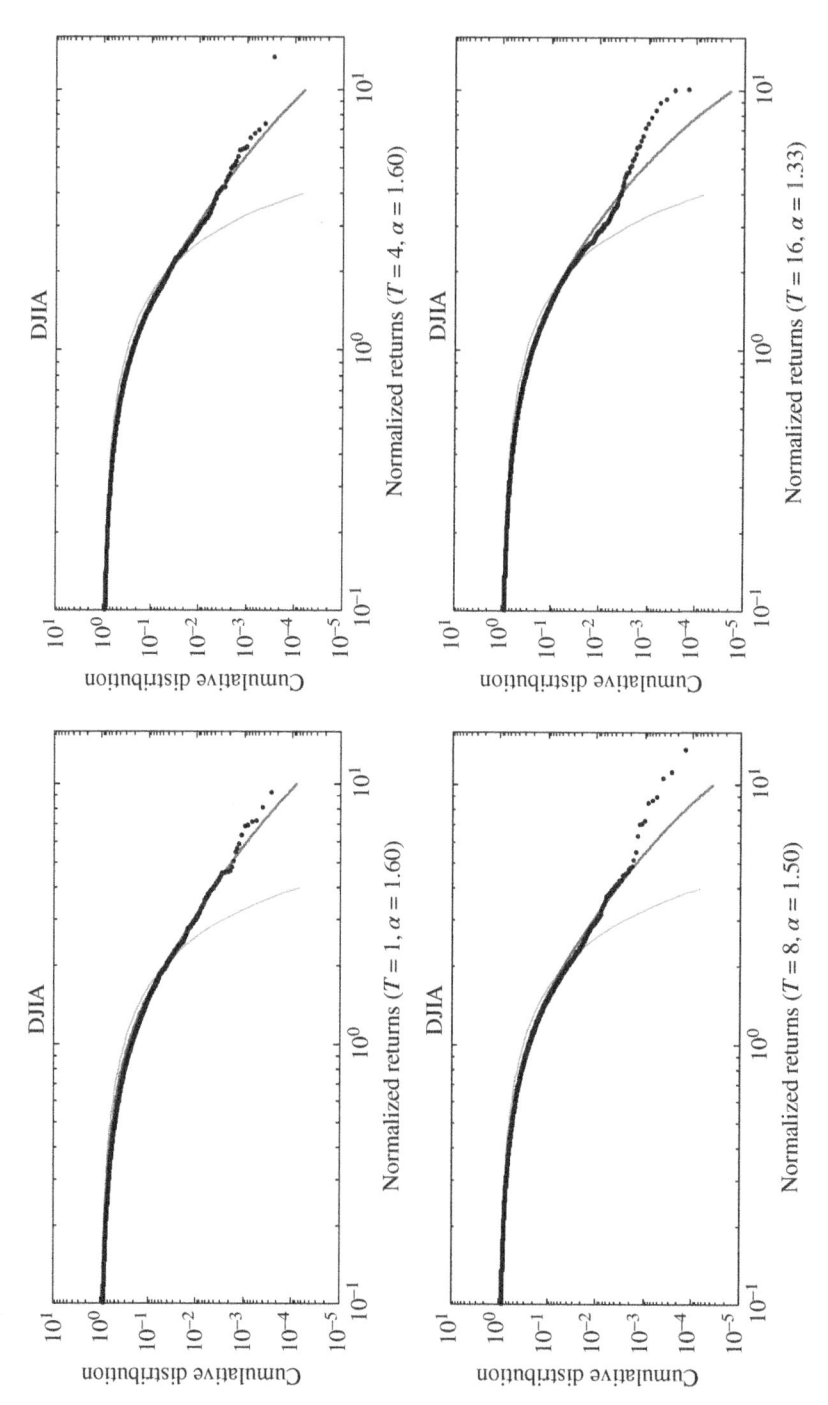

Figure 13.4 Lévy flight parameter for the DJIA index.

We analyzed data comprising the five trading days between 10 and 14 March 2008 predating the merging announcement over the weekend as well as the two trading days 17 and 18 March following the event. These data is a high frequency(minute) data. The companies studied are as follows: International Business Machines (IBM), Google, Walmart, and Walt Disney Company.

The estimated Lévy flight parameter for the four analyzed companies are given in Figures 13.5–13.8.

In the two examples, we obtained that the evolution of data of the financial index along with the specific companies and the high frequency data can be described by the model. From the plots shown, we can see that all the values obtained for the exponent α are lower than 2. We also observe that the Lévy exponents are more closer to 1 for the high frequency data. Based on the results obtained, we conclude that the (TLF) is suitable for modeling financial time series.

13.4.2 Detrended Fluctuation Analysis (DFA) and Rescaled Range Analysis Applied to Financial Indices

The premise we make in this analysis is a simple one. In the normal market conditions, all the participating agents have diverse views, and accordingly the price process should not exhibit large long-term memory effects. Of course even in normal market conditions when working with high frequency data, the price process is far from the lognormal specification [14]. On the other hand, when a crisis situation is anticipated, all the agents start to behave in a similar way, and accordingly the resulting price process starts to exhibit large memory effects.

We estimate the Hurst parameter (H) as well as the Detrended Fluctuation parameter (α) and we compare with 0.5. The further these parameters are from 0.5, the stronger the evidence of a crash event waiting to happen. The reason why we estimate both parameters is that the α parameter works better with nonstationary data than H. On the other hand when working with stationary data H is much more relevant [15].

Example 13.4.3 The data studied comprises of the five trading days 10–14 March 2008 predating the merging announcement over the weekend as well as the two trading days 17 and 18 March following the event. The data used consists of the closing value within each minute of the trading day (8.5 hours). We analyze four very large companies, namely, IBM, Google (GOOG), Microsoft (MSFT) and Walmart (WMT).

Table 13.1 gives the values of the Hurst parameter H using an R/S analysis, α of the Lévy process, and DFA. Looking at these numbers we recognize a much closer behavior with what we expected to see. The estimated values for the slopes suggests that the results obtained using DFA and R/S methods agree.

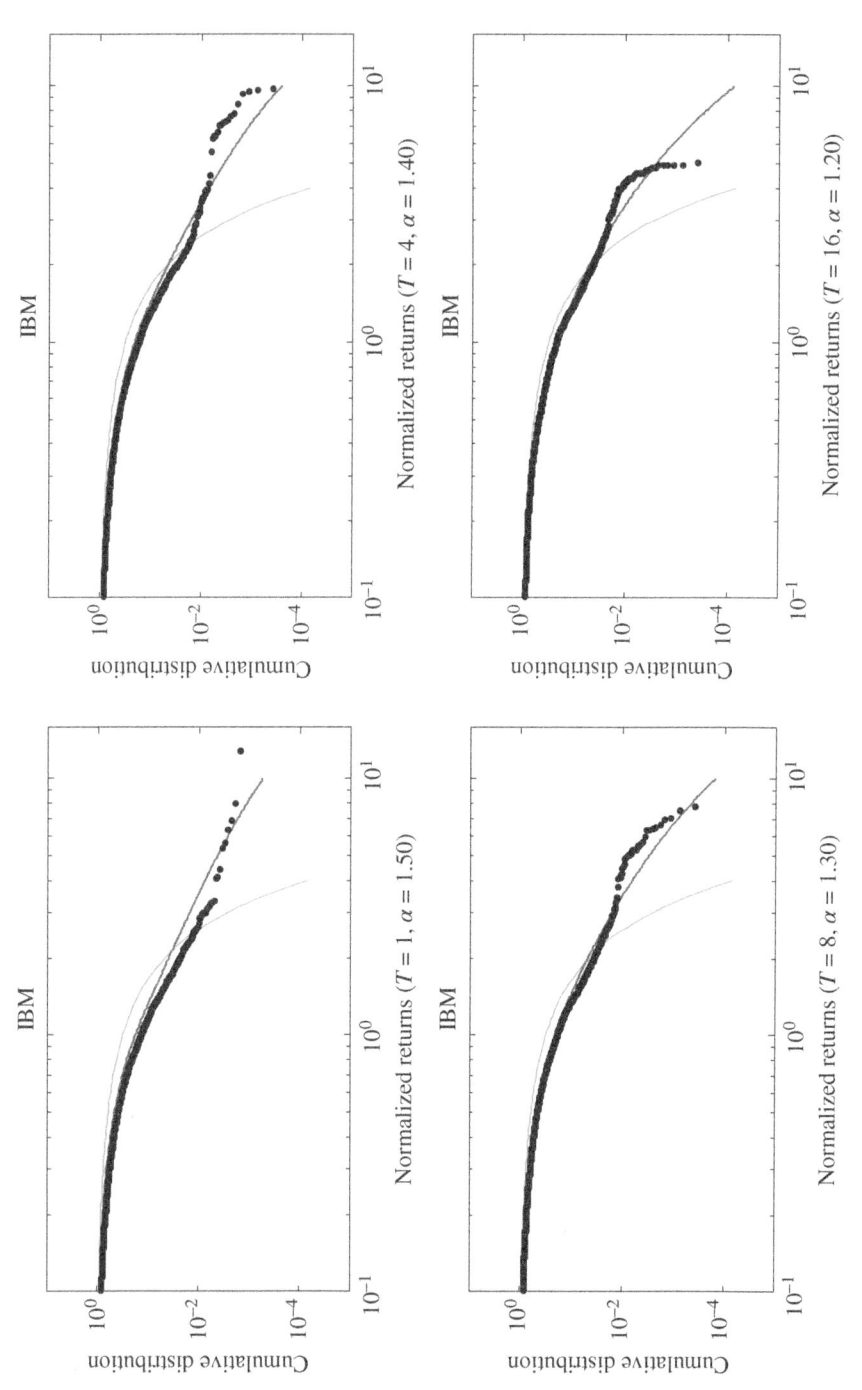

Figure 13.5 Lévy flight parameter for IBM, high frequency (tick) data.

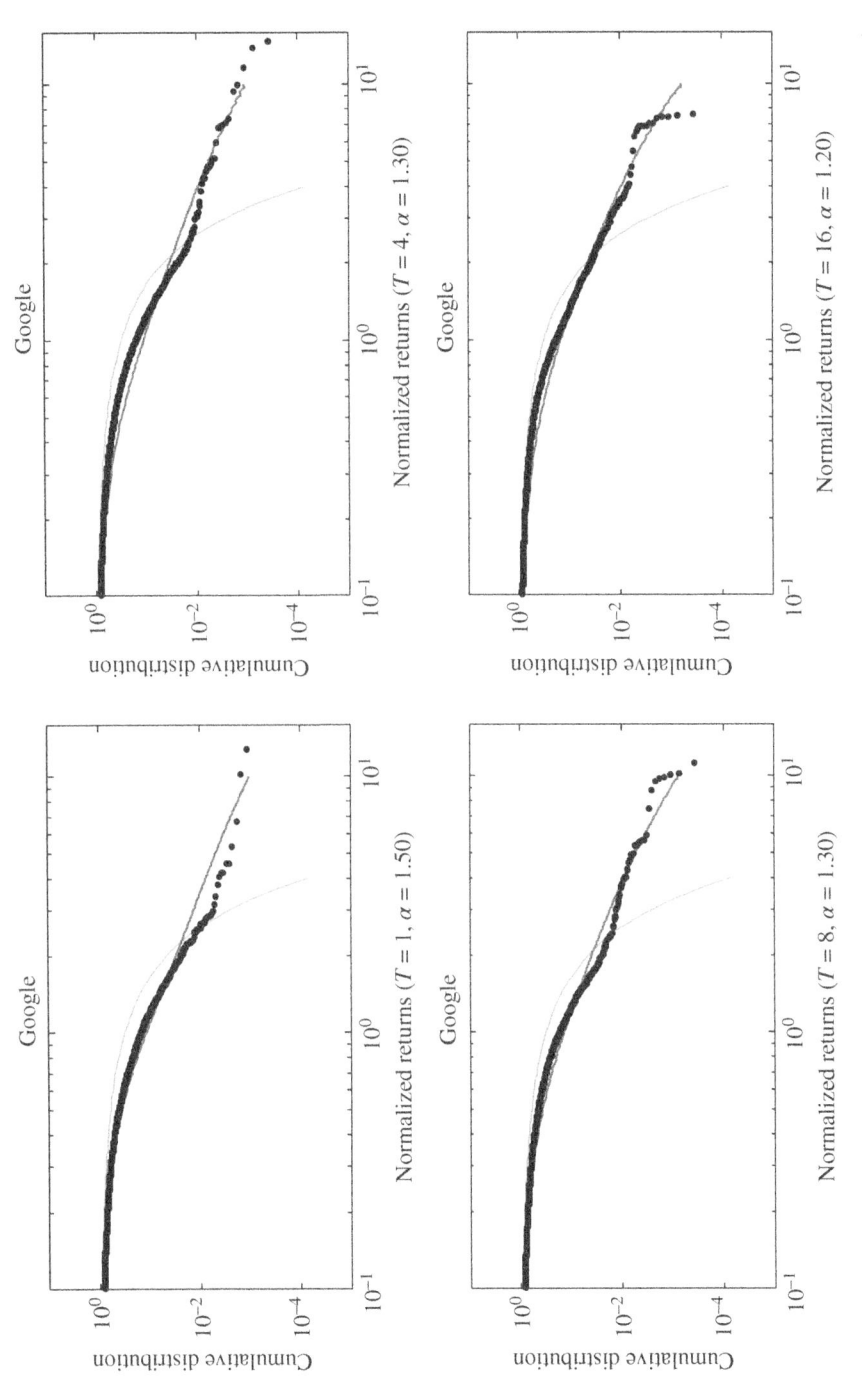

Figure 13.6 Lévy flight parameter for Google, high frequency (tick) data.

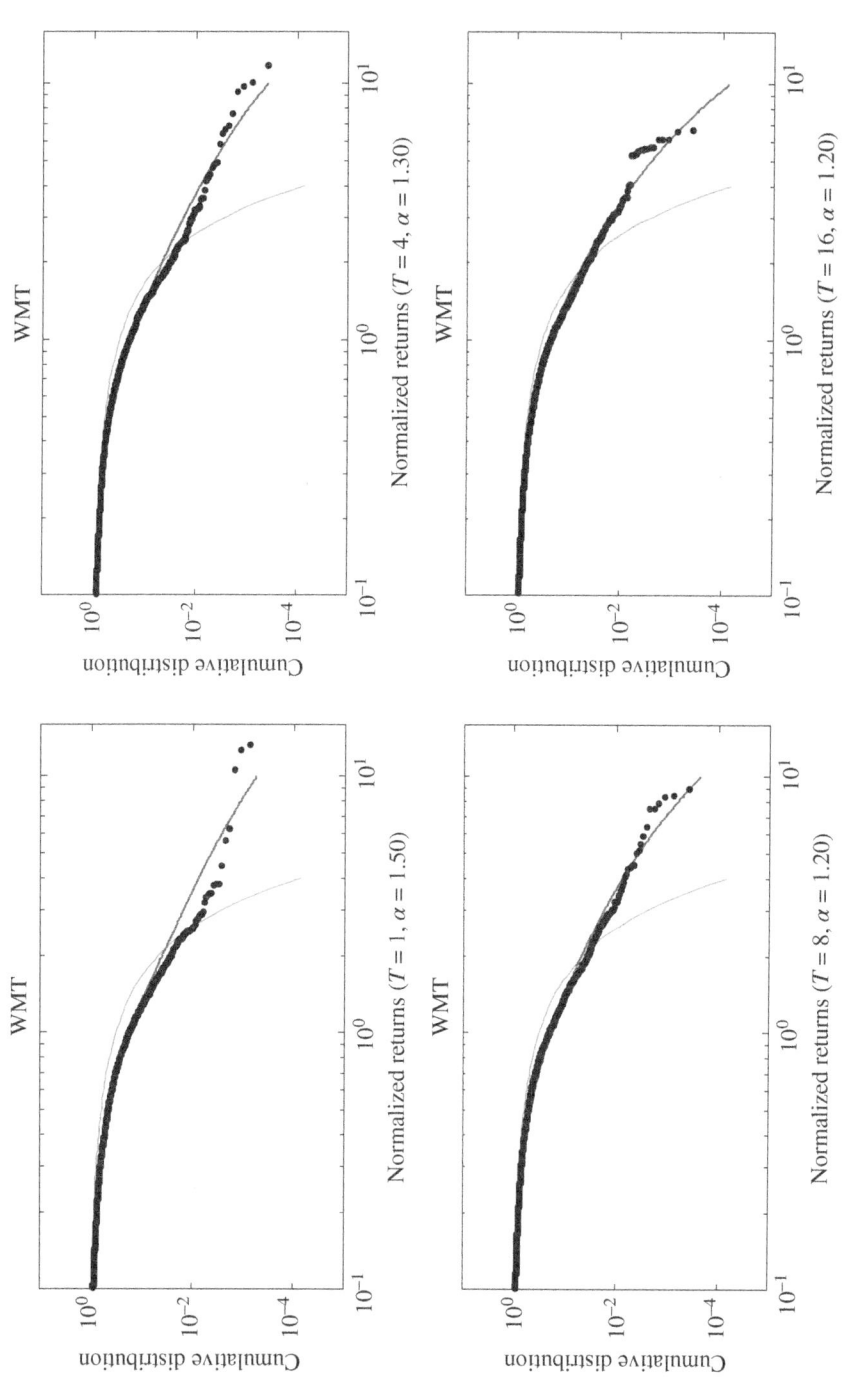

Figure 13.7 Lévy flight parameter for Walmart, high frequency (tick) data.

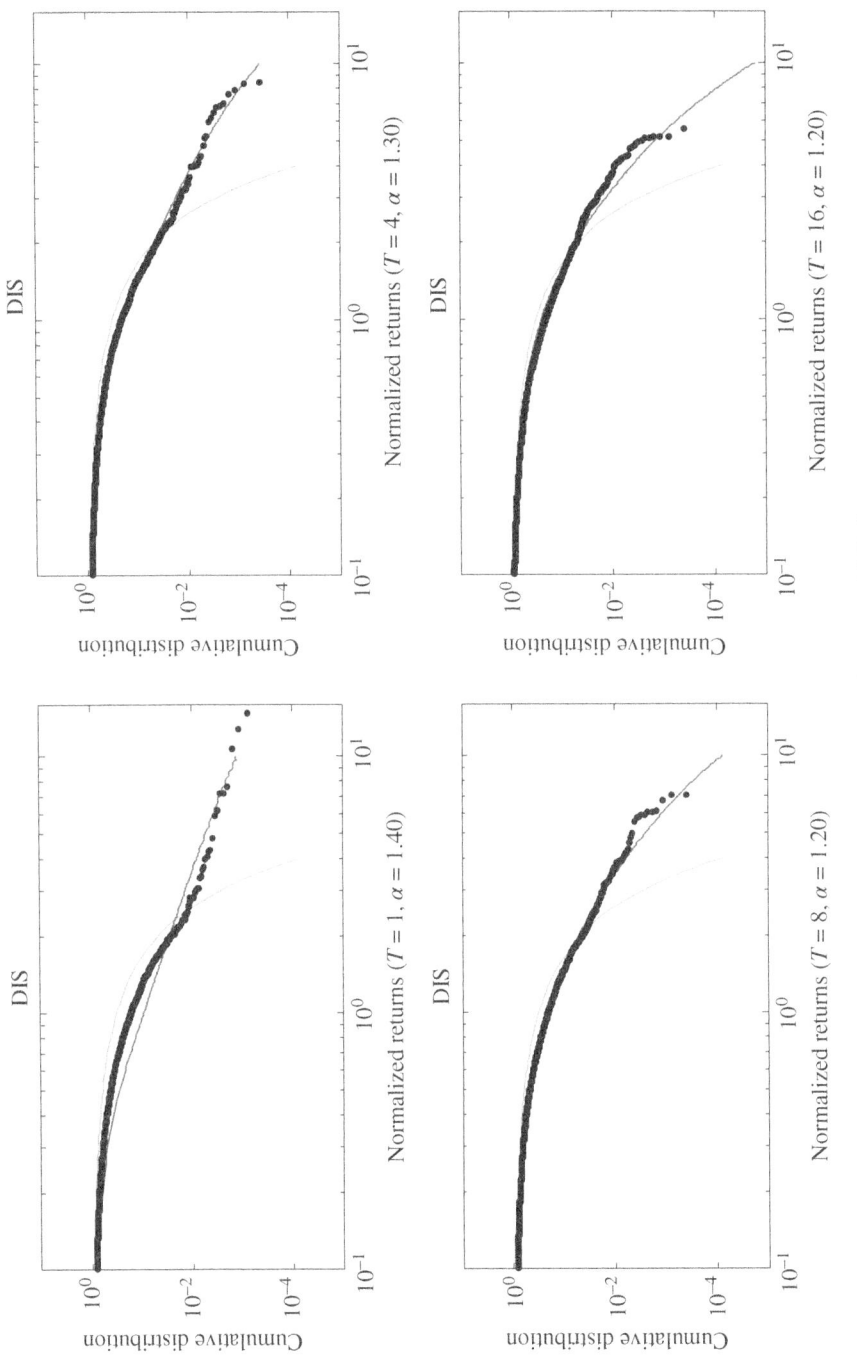

Figure 13.8 Lévy flight parameter for the Walt Disney Company, high frequency (tick) data.

Table 13.1 Lévy parameter for the high frequency data corresponding to the crash week.

Companies	DFA	Hurst	Levy coefficient (α)			
			T=1	T=2	T=3	T=4
IBM	0.71	0.62	1.50	1.40	1.30	1.20
IBM 3/10	0.57	0.62	1.70	1.80	2.00	1.70
IBM 3/11	0.83	0.67	1.40	1.70	1.80	1.99
IBM 3/12	0.60	0.59	1.70	1.85	1.80	1.70
IBM 3/13	0.57	0.59	1.75	1.70	1.60	1.70
IBM 3/14	0.47	0.62	1.40	1.30	1.30	1.60
IBM 3/17	0.60	0.67	1.70	1.70	1.70	2.00
IBM 3/18	0.70	0.56	1.30	1.30	1.20	1.20
Google	0.64	0.62	1.50	1.30	1.30	1.20
MSFT	0.75	0.66	1.50	1.40	1.30	1.20
WMT	0.66	0.62	1.50	1.30	1.20	1.20
WMT 3/10	0.73	0.68	1.65	1.75	1.60	1.70
WMT 3/11	0.63	0.60	1.65	1.90	1.90	1.99
WMT 3/12	0.56	0.60	1.90	1.90	1.90	1.99
WMT 3/13	0.64	0.60	1.60	1.50	1.70	1.75
WMT 3/14	0.51	0.62	1.80	1.50	1.50	1.60
WMT 3/17	0.49	0.64	1.99	1.99	1.99	1.99
WMT 3/18	0.63	0.59	1.75	1.85	1.70	1.99

Thus the data analyzed exhibited long memory effects. For the R/S analysis, we observe that the Hurst parameter estimates are all close to 0.6. This is expected since during a normal day when analyzing data sampled with high frequency, the prices exhibit long memory effects.

Figures 13.9 and 13.10 show the plots of the DFA and Hurst analysis applied to IBM and Google, respectively.

Next, we investigate the TLF distributions and self-similarity. For the case of the present analysis, if we consider, for example, the companies Google and Microsoft (MSFT), we see that at $T = 1$ and the Hurst parameter $H \approx \frac{1}{\alpha}$, where α is the Lévy coefficient. The closeness of the values of H with the values of $\frac{1}{\alpha}$ suggests a very good self-similar structure for the present analysis.

13.5 Problems

1. Present example of stable distributions and of nonstable distributions.
2. Explain the differences between the Lévy processes, the truncated Lévy processes, and the standardized Lévy processes.

DFA analysis: IBM

[Plot: log(Fn) vs log(n), ($\alpha = 0.715$)]

Hurst analysis: IBM

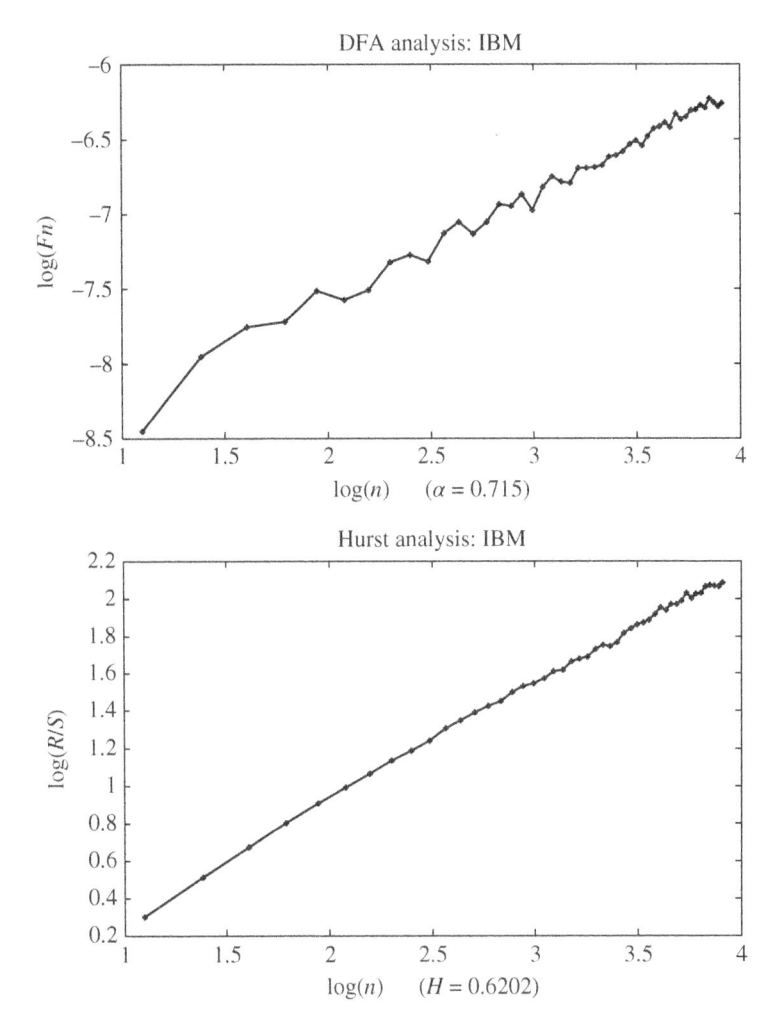

[Plot: log(R/S) vs log(n), ($H = 0.6202$)]

Figure 13.9 DFA and Hurst methods applied to the data series of IBM, high frequency data corresponding to the crash week.

3. The standardized model (13.1) can be approximated as

$$\log \varphi_S(q) \sim \frac{2\pi A l^{-\alpha} t}{\alpha \Gamma(\alpha) \sin(\pi\alpha)} \left[1 - \left(1 + \frac{\alpha}{2} \left(\frac{ql}{\sigma} \right)^2 \right) \cos\left(\alpha \arctan\left(\frac{ql}{\sigma} \right) \right) \right]$$

for small q. Therefore using Theorem 13.2.1, show that $p(x) \sim \frac{1}{|x|^3}$.

4. Explain the differences between Range Scale Analysis and Detrended Fluctuation Analysis.
5. Develop a Matlab module for simulating the Lévy–Smirnov and the Lorentzian distribution.

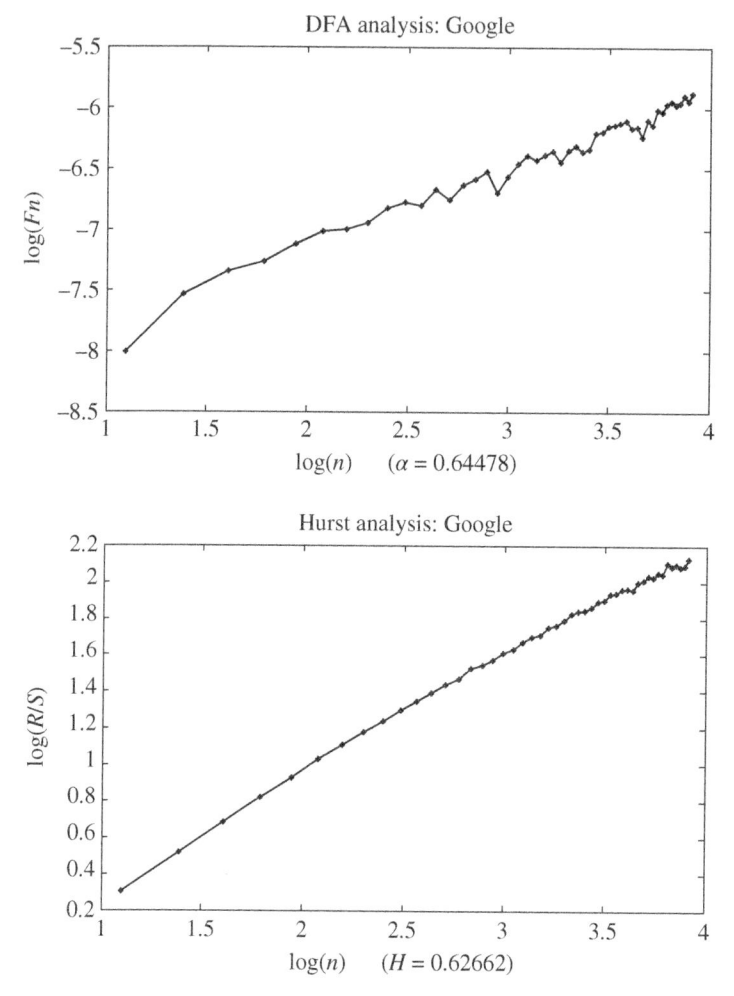

Figure 13.10 DFA and Hurst methods applied to the data series of Google, high frequency data corresponding to the crash week.

6. Present examples of *self-similar* stochastic processes and of a stochastic processes that are not *self-similar*.
7. Explain why a random variable X is hypoexponentially distributed if its characteristic function is of form (13.18).

14

Approximating General Derivative Prices

Gradient Method

14.1 Introduction

In this chapter, we will study integro-differential parabolic problems arising in financial mathematics. The solution of these problems models processes with jumps and stochastic volatility. Under suitable conditions, we will prove the existence of solutions in a general domain using the method of upper and lower solutions and a diagonal argument. This type of methodology can be extended to other cases which can be applied to similar problems. We begin this section with some background.

In recent years there has been an increasing interest in solving PDE problems arising in financial mathematics and in particular on option pricing. The standard approach to this problem leads to the study of equations of parabolic type.

Usually the Black–Scholes model (see e.g. [24, 57, 101, 104, 152]) is used for pricing derivatives, by means of a backward parabolic differential equation. In this model, an important quantity is the volatility which is a measure of the fluctuation (risk) in the asset prices and corresponds to the diffusion coefficient in the Black–Scholes equation.

In the standard Black–Scholes model, a basic assumption is that the volatility is constant. However, several models proposed in recent years, such as the model found in [95], have allowed the volatility to be nonconstant or a stochastic variable. In this model, the underlying security S follows, as in the Black–Scholes model, a stochastic process

$$dS_t = \mu S_t dt + \sigma_t S_t dZ_t,$$

where Z is a standard Brownian motion. Unlike the classical model, the variance $v(t) = \sigma_t^2$ also follows a stochastic process given by

$$dv_t = \kappa(\theta - v(t))dt + \gamma \sqrt{v_t} dW_t,$$

Quantitative Finance, First Edition. Maria C. Mariani and Ionut Florescu.
© 2020 John Wiley & Sons, Inc. Published 2020 by John Wiley & Sons, Inc.

where W is another standard Brownian motion. The correlation coefficient between W and Z is denoted by ρ:

$$\text{Cov}(dZ_t, dW_t) = \rho\, dt.$$

This leads to a generalized Black–Scholes equation:

$$\frac{1}{2}vS^2\frac{\partial^2 U}{\partial S^2} + \rho\gamma vS\frac{\partial^2 U}{\partial v \partial S} + \frac{1}{2}v\gamma^2\frac{\partial^2 U}{\partial v^2} + rS\frac{\partial U}{\partial S}$$
$$+ [\kappa(\theta - v) - \lambda v]\frac{\partial U}{\partial v} - rU + \frac{\partial U}{\partial t} = 0.$$

Similar models have been considered in [4].

More general models with stochastic volatility have been proposed, where the following problem is derived using the Feynman-Kac lemma [159]:

$$u_t = \frac{1}{2}\text{Tr}\left(M(x,\tau)D^2 u\right) + q(x,\tau)\cdot Du,$$
$$u(x,0) = u_0(x),$$

where M is some diffusion matrix and u_0 is the payoff function. The Feynman–Kac formula was derived in Chapter 6 of this textbook.

In Chapter 11, we showed that the Black–Scholes model with jumps arise from the fact that the driving Brownian motion is a continuous process, and so there are difficulties fitting the financial data presenting large fluctuations. The necessity of taking into account the large market movements and a great amount of information arriving suddenly (i.e. a jump) has led to the study of partial integro-differential equations (PIDE) in which the integral term is modeling the jump.

In [5, 152], the following PIDE in the variables t and S is obtained:

$$\frac{1}{2}\sigma^2 S^2 F_{SS} + (r - \lambda k)S F_S + F_t - rF + \lambda\epsilon\{F(SY, t) - F(S, t)\} = 0. \quad (14.1)$$

Here, r denotes the riskless rate, λ the jump intensity, and $k = \epsilon(Y - 1)$, where ϵ is the expectation operator and the random variable $Y - 1$ measures the percentage change in the stock price if the jump, modeled by a Poisson process, occurs. See [152] for the details.

The following PIDE is a generalization of (14.1) for N assets with prices $S_1, \ldots, S_N$:

$$\sum_{i=1}^{N} \frac{1}{2}\sigma_i^2 S_i^2 \frac{\partial^2 F}{\partial S_i^2} + \sum_{i \neq j} \frac{1}{2}\rho_{ij}\sigma_i\sigma_j S_i S_j \frac{\partial^2 F}{\partial S_i \partial S_j} + \sum_{i=1}^{N}(r - \lambda k_i)S_i \frac{\partial F}{\partial S_i} + \frac{\partial F}{\partial t} - rF$$
$$+ \lambda \int \left[F(S_1 Y_1, \ldots, S_d Y_d, t) - F(S_1, \ldots, S_d, t)\right] g(Y_1, \ldots, Y_d) dY_1 \ldots dY_d = 0$$

with the correlation coefficients

$$\rho_{ij} dt = \epsilon\{dz_i, dz_j\}.$$

14.1 Introduction

The case in which F is increasing and all jumps are negative corresponds to the evolution of a Call option near a crash. Please see [5, 152] for more details.

As described in [144], when modeling high frequency data in applications, a Lévy-like stochastic process appears to be the best fit. When using these models, option prices are found by solving the resulting PIDE. For example, integro-differential equations appear in exponential Lévy models, where the market price of an asset is represented as the exponential of a Lévy stochastic process. We will outline such a model, but first, we will discuss the Black–Scholes model with stochastic volatility.

When the volatility is stochastic, we may consider the following process:

$$dS = S\sigma dZ + S\mu dt$$
$$d\sigma = \beta\sigma dW + \alpha\sigma dt,$$

where Z and W are two standard Brownian motions with correlation coefficient ρ. If $F(S, \sigma, t)$ is the price of an option depending on the price of the asset S, then by Ito's lemma (see [104]), we have

$$dF(S, \sigma, t) = F_S dS + F_\sigma d\sigma + \mathscr{L} F dt,$$

where $\mathscr{L}$ is given by

$$\mathscr{L} = \frac{\partial}{\partial t} + \frac{1}{2}\sigma^2 S^2 \frac{\partial^2}{\partial S^2} + \frac{1}{2}\beta^2\sigma^2 \frac{\partial^2}{\partial \sigma^2} + \rho\sigma^2 S\beta \frac{\partial^2}{\partial S \partial \sigma}.$$

Under an appropriate choice of the portfolio, the stochastic term of the equation vanishes.

A generalized tree process has been developed in [71] that approximates any stochastic volatility model. Unlike the nonrandom volatility case, the tree construction is stochastic every time, since that is the only way we are able to deal with the huge complexity involved.

If in this model we add a jump component modeled by a compound Poisson process to the process S and we follow Merton [152], we obtain the following PIDE:

$$\frac{\partial F}{\partial t} + \frac{1}{2}\sigma^2 S^2 \frac{\partial^2 F}{\partial S^2} + \frac{1}{2}\sigma^2\beta^2 \frac{\partial^2 F}{\partial \sigma^2} + \rho\sigma^2 \beta S \frac{\partial^2 F}{\partial S \partial \sigma} + (r - \lambda k)S\frac{\partial F}{\partial S} - \frac{1}{2}\rho\sigma^2\beta\frac{\partial F}{\partial \sigma}$$
$$+ \lambda \int_\mathbb{R} [F(SY, \sigma, t) - F(S, \sigma, t)] g(Y) \, dY - rF = 0 \quad (14.2)$$

Once again, r is the riskless rate, λ is the jump intensity, and $k = \epsilon(Y - 1)$. Here, ϵ is the expectation operator, and the random variable $Y - 1$ measures the percentage change in the stock price if the jump, modeled by a Poisson process, occurs. See section 9.2 of [152] for more details. The authors in [68] generalized Eq. (14.2) and proved the existence and uniqueness of a classical solution to a more general problem in a parabolic domain $Q_T = \Omega \times (0, T)$, where Ω is an open, unbounded subset of $\mathbb{R}^d$, with a smooth boundary $\partial\Omega$.

In this chapter we consider a problem similar to (14.2) with constant volatility. We will begin by describing the statement of the problem. Next, we will discuss a more general integro-differential equation and give a proof of existence of a *strong solution*. The results will be generalized from a bounded domain to an unbounded domain. We begin the next section with the statement of the problem.

14.2 Statement of the Problem

A Lévy process is a (strong) Markov process where the associated semigroup is a convolution semigroup and its infinitesimal generator is an integro-differential operator given by

$$Lu(x) = \lim_{t \to 0} \frac{\mathbb{E}[u(x + X_t)] - u(x)}{t}$$

$$= \frac{\sigma^2}{2} \frac{\partial^2 u}{\partial x^2} + \gamma \frac{\partial u}{\partial x} + \int v(dy) \left[u(x + y) - u(x) - y \mathbf{1}_{|y| \le 1} \frac{\partial u}{\partial x}(x) \right], \quad (14.3)$$

which is well defined for $u \in C^2(\mathbb{R})$ with compact support because its support is a compact set. We now present the problem considered in this section.

Consider the following integro-differential model for a European option, where the market price of an asset is represented as the exponential of a Lévy stochastic process [41]:

$$\frac{\partial C}{\partial t}(S, t) + rS \frac{\partial C}{\partial S}(S, t) + \frac{\sigma^2 S^2}{2} \frac{\partial^2 C}{\partial S^2}(S, t) - rC(S, t)$$

$$+ \int v(dy) \left[C(Se^y, t) - C(S, t) - S(e^y - 1) \frac{\partial C}{\partial S}(S, t) \right] = 0, \quad (14.4)$$

with the final payoff

$$C(S, T) = \mathbb{E}[\max(S_T - K), 0], \quad (14.5)$$

where $K > 0$ is the strike price. If we introduce the change of variables

$$\tau = T - t,$$

$$x = \ln\left(\frac{S}{K}\right) + r\tau,$$

$$u(x, \tau) = \frac{e^{r\tau}}{K} C\left(Ke^{x-r\tau}, T - \tau\right),$$

then (14.4) becomes

$$\frac{\partial u}{\partial \tau}(x, \tau) = \frac{\sigma^2}{2} \left[\frac{\partial^2 u}{\partial x^2}(x, \tau) - \frac{\partial u}{\partial x}(x, \tau) \right] + \mathcal{F}(u, u_x) \quad (14.6)$$

with the initial condition

$$u(x, 0) = u_0(x) \quad \text{for all} \quad x \in \mathbb{R}. \quad (14.7)$$

The term

$$F(u, u_x) = \int \left[u(x+y, \tau) - u(x, \tau) - (e^y - 1)\frac{\partial u}{\partial x}(x, \tau) \right] v(dy) \qquad (14.8)$$

is an integro-differential operator modeling the jump.

The above equation is similar to the results obtained by [41]. We are interested in generalizing this equation and presenting a solution technique. In Section 14.4 we will discuss the result for the existence problem of the one stated in [41] in a generalized setting and with some different boundedness condition. This gives an approach to solve the problem in [41] with growth condition that is different from [16] where they considered nonlocal operators as given in (14.3).

Now we show how to obtain Eqs. (14.6)–(14.8). First, notice that to convert back to the original variables, we use the change of variables:

$$t = T - \tau,$$

$$S = Ke^{x-r\tau},$$

$$C(S, t) = C\left(Ke^{x-r\tau}, T - \tau\right) = Ke^{-r\tau} u(x, \tau).$$

Next, we will compute each partial derivative in (14.4). We do this by using the chain and product rules repeatedly and the expression $Ke^{-r\tau} u(x, \tau)$ for $C(S, t)$:

$$\frac{\partial C}{\partial t} = \frac{\partial C}{\partial \tau}\frac{\partial \tau}{\partial t} = -\frac{\partial C}{\partial \tau} = rKe^{-r\tau} u(x, \tau) - rKe^{-r\tau}\frac{\partial u}{\partial x} - Ke^{-r\tau}\frac{\partial u}{\partial \tau},$$

$$\frac{\partial C}{\partial S} = \frac{\partial C}{\partial x}\frac{\partial x}{\partial S} = \frac{1}{S}\frac{\partial C}{\partial x} = \frac{1}{S}Ke^{-r\tau}\frac{\partial u}{\partial x},$$

$$\frac{\partial^2 C}{\partial S^2} = \frac{\partial}{\partial S}\left(\frac{1}{S}Ke^{-r\tau}\frac{\partial u}{\partial x}\right) = -\frac{1}{S^2}Ke^{-r\tau}\frac{\partial u}{\partial x} + \frac{1}{S}Ke^{-r\tau}\frac{\partial^2 u}{\partial x^2}\frac{\partial x}{\partial S},$$

$$= -\frac{1}{S^2}Ke^{-r\tau}\frac{\partial u}{\partial x} + \frac{1}{S^2}Ke^{-r\tau}\frac{\partial^2 u}{\partial x^2}.$$

Furthermore, notice that the first term in the integral operator of (14.4) can be expressed as

$$C(Se^y, t) = C\left(Ke^{x+y-r\tau}, T - \tau\right) = Ke^{-r\tau} u(x+y, \tau).$$

If we substitute everything into (14.4) and divide through by $Ke^{-r\tau}$, we obtain Eqs. (14.6) and (14.8). It is clear that $S > 0$ implies $x \in \mathbb{R}$ and $t = T$ implies $\tau = 0$. Using this two facts, Eq. (14.5) becomes

$$C(S, T) = Ku(x, 0) = \tilde{u}_0(x).$$

This justifies the initial condition in (14.7), where $u_0(x) = \tilde{u}_0(x)/K$.

For the classical Black–Scholes model and for any other Black–Scholes models, such as models which take into account jumps, it follows that $C(S, T) \sim 0$ when $S \sim 0$ and $C(S, T) \sim S$ when S is very large. This observation will justify the boundary conditions we will be using later on in this chapter. That is the boundary condition for the Black–Scholes model with jumps should be the same boundary condition used for the classical Black–Scholes model whenever the spatial domain for S is bounded and sufficiently large. In the next section, we present a more general integro-differential parabolic problem.

14.3 A General Parabolic Integro-Differential Problem

The discussion of statement of problem in Section 14.2 motivates us to consider more general integro-differential parabolic problems. First, we will consider the following initial-boundary value problem in the bounded parabolic domain $Q_T = \Omega \times (0, T)$, $T > 0$:

$$
\begin{aligned}
u_t - Lu &= \mathcal{F}(x, t, u, \nabla u) &&\text{in } Q_T, \\
u(x, 0) &= u_0(x) &&\text{on } \Omega, \\
u(x, t) &= g(x, t) &&\text{on } \partial\Omega \times (0, T).
\end{aligned}
\tag{14.9}
$$

Then, we will extend our results to the corresponding initial-value problem in the unbounded domain $\mathbb{R}_T^{d+1} = \mathbb{R}^d \times (0, T)$:

$$
\begin{aligned}
u_t - Lu &= \mathcal{F}(x, t, u, \nabla u) &&\text{in } \mathbb{R}_T^{d+1}, \\
u(x, 0) &= u_0(x) &&\text{on } \mathbb{R}^d.
\end{aligned}
\tag{14.10}
$$

Here, $L = L(x, t)$ is a second order elliptic operator in divergence form, namely,

$$
L(x, t) := \sum_{i,j=1}^{d} \frac{\partial}{\partial x_j}\left(a^{ij}(x, t)\frac{\partial}{\partial x_i}\right) - \sum_{i=1}^{d} b^i(x, t)\frac{\partial}{\partial x_i} - c(x, t).
$$

The integro-differential operator is defined by

$$
\mathcal{F}(x, t, u, \nabla u) = \int_\Omega f(x, t, y, u(x, t), \nabla u(x, t))\, dy.
\tag{14.11}
$$

This integro-differential operator will be a continuous integral operator as the ones defined in (14.1)–(14.4) and (14.8) modeling the jump. The case in which f is decreasing with respect to u and all jumps are positive corresponds to the evolution of a call option near a crash. Throughout this section, we impose the following assumptions:

A(1) The coefficients $a^{ij}(x, t)$, $b^i(x, t)$, and $c(x, t)$ belong to the Hölder space $C^{\delta, \delta/2}(\overline{Q_T})$.

A(2) The operator $\frac{\partial}{\partial t} - L$ is (uniformly) parabolic. That is, there exists a constant $\theta > 0$ such that

$$
\theta|v|^2 \leq \sum_{i,j=1}^{d} a^{ij}(x, t) v_i v_j
$$

for all $(x, t) \in Q_T, v \in \mathbb{R}^d$.

A(3) There exists a constant $B \in \mathbb{R}$, such that for all $v \in H_0^1(\Omega)$, $Bv_{x_i}v \leq b^i(x,t)v_{x_i}v$, $\forall (x,t) \in Q_T$. For all $(x,t) \in Q_T$, $c(x,t) \geq 0$.

A(4) $u_0(x)$ and $g(x,t)$ belong to the Hölder spaces $C^{2+\delta}(\mathbb{R}^d)$ and $C^{2+\delta, 1+\delta/2}(\overline{Q_T})$, respectively.

A(5) The two consistency conditions

$$g(x,0) = u_0(x) \text{ and}$$
$$g_t(x,0) - L(x,0)u_0(x) = 0$$

are satisfied for all $x \in \partial\Omega$.

A(6) $f(x,t,y,z,p)$ is nonnegative and belongs to $C^1(Q_T \times \Omega \times \mathbb{R}^{d+1})$.

A(7) For some $C_0 > 0$, f satisfies the estimate

$$|f(x,t,y,z,p)| \leq C_0(1 + |z| + |p|),$$

for all $(x,t,y,z,p) \in \overline{Q_T} \times \overline{\Omega} \times \mathbb{R}^{d+1}$, where C_0 is independent of parameters of f.

A(8) For $u \in H_{loc}^2(\mathbb{R})$, $\mathcal{F}(x,t,u,\nabla u) \in L^2(0,T; L_{loc}^2(\mathbb{R}))$.
If $w_k \to w$ in $L^2(0,T; H_0^1(\Omega))$, then $\mathcal{F}(x,t,w_k, \nabla w_k) \to \mathcal{F}(x,t,w, \nabla w)$ in $L^2(0,T; L^2(\Omega))$.

In the next subsection, we state the Schaefer's fixed point theorem that will be used to show the existence of the solution for our present problem. The theorem is an example of the mathematical principle saying "a priori estimates implies existence."

14.3.1 Schaefer's Fixed Point Theorem

Let X be a real Banach space.

Definition 14.3.1 The space $C([0,T]; X)$ consists of all continuous functions $u : [0,T] \to X$ with

$$\|u\|_{C([0,T];X)} := \max_{0 \leq t \leq T} \|u(t)\|_X < \infty.$$

$C([0,T]; X)$ is a Banach space when endowed with the norm $\|u\|_{C([0,T];X)}$.

Definition 14.3.2 A mapping $A : X \to X$ is said to be compact if and only if for each bounded sequence $\{u_k\}_{k=1}^\infty$, the sequence $\{A[u_k]\}_{k=1}^\infty$ is precompact, that is, there exists a subsequence $\{u_{k_j}\}_{j=1}^\infty$ such that $\{A[u_{k_j}]\}_{j=1}^\infty$ converges in X.

Theorem 14.3.1 (Schaefer's fixed point theorem). Suppose $A : X \to X$ is a continuous and compact mapping. Assume further that the set

$$\{u \in X : u = \lambda A[u], \text{ for some } 0 \leq \lambda \leq 1\}$$

is bounded. Then A has a fixed point.

Please refer to [191] for details of the theorem.

We shall prove the existence of a weak solution to (14.13) using Schaefer's fixed point theorem. First we prove the result for bounded regions. Afterwards, we will use a standard argument to show that our solution can be extended to give us a solution to the initial-value problem in $\mathbb{R}_T^{d+1}$.

14.4 Solutions in Bounded Domains

In this subsection, $Q_T = \Omega \times (0, T)$ will always denote a bounded parabolic domain, where $\Omega \subset \mathbb{R}^d$ is open and has smooth boundary $\partial \Omega$. We will find a weak solution for (14.13). The following lemma follows immediately from theorem 10.4.1 in [129].

Lemma 14.4.1 There exists a unique solution $\varphi \in C^{2+\delta, 1+\delta/2}\left(\overline{Q}_T\right)$ to the problem

$$\begin{aligned} u_t - Lu &= 0 && \text{in } Q_T, \\ u(x, 0) &= u_0(x) && \text{on } \Omega, \\ u(x, t) &= g(x, t) && \text{on } \partial \Omega \times (0, T). \end{aligned} \qquad (14.12)$$

As we have already mentioned in previous subsection, Lemma 14.4.1 can be thought of as generalization of the classical Black–Scholes model where the stock price S satisfies

$$\varepsilon < S < S_{\max}$$

i.e. the spatial domain for S is bounded and sufficiently large. In practice, one should not assume that S is bounded away from 0.

The problem in the next theorem can be regarded as a generalization of (14.6) and (14.7), where the stock price S is bounded above and bounded below away from 0 as in Lemma 14.4.1. We take the same boundary condition as in Lemma 14.4.1 because of our earlier comment regarding the behavior of the option value when S is really small or really large for any Black–Scholes model.

Theorem 14.4.1 Let φ be defined as in Lemma 14.4.1. Then if **A(1)–A(8)** are satisfied, there exists a weak solution $u \in L^2(0, T; H_0^1(\Omega)) \cap C([0, T]; L^2(\Omega))$ to the problem

$$\begin{aligned} u_t - Lu &= \mathcal{F}(x, t, u, \nabla u) && \text{in } Q_T, \\ u(x, 0) &= u_0(x) && \text{on } \Omega, \\ u(x, t) &= \varphi(x, t) = g(x, t) && \text{on } \partial \Omega \times (0, T). \end{aligned} \qquad (14.13)$$

14.4 Solutions in Bounded Domains

Proof. First, we introduce a change of variables to transform our problem into one with a zero boundary condition. If we let

$$v(x,t) = u(x,t) - \varphi(x,t),$$
$$v_0(x) = u_0(x) - \varphi(x,0) = 0,$$

then v will satisfy the initial-boundary value problem

$$\begin{aligned} v_t - Lv &= F_g(x,t,v+\varphi,\nabla(v+\varphi)) &&\text{in } Q_T, \\ v(x,0) &= 0 &&\text{on } \Omega, \\ v(x,t) &= 0 &&\text{on } \partial\Omega \times (0,T). \end{aligned} \quad (14.14)$$

If problem (14.14) has a strong solution, then (14.13) will have a strong solution since $u = v + \varphi$. We use an iteration procedure to construct the solution to (14.14). Consider the problem

$$\begin{aligned} \beta_t - L\beta &= F_g(x,t,\alpha+\varphi,\nabla(\alpha+\varphi)) &&\text{in } Q_T, \\ \beta(x,0) &= 0 &&\text{on } \Omega, \\ \beta(x,t) &= 0 &&\text{on } \partial\Omega \times (0,T), \end{aligned} \quad (14.15)$$

where $\alpha \in C^{2+\delta,1+\delta/2}(\overline{Q}_{T,U})$ is arbitrary. It is clear that $F_g(x,t,\alpha+\varphi,\nabla(\alpha+\varphi)) \in C^{\delta,\delta/2}(\overline{Q}_T)$. By Theorem 10.4.1 in [129], there exists a unique solution $\beta \in C^{2+\delta,1+\delta/2}(\overline{Q}_T)$ to problem (14.15). Moreover the solution can be extended analytically in $C^{2+\delta,1+\delta/2}(\overline{2Q}_T)$. Using this result, we can now define $v^n \in C^{2+\delta,1+\delta/2}(\overline{Q}_T)$, $n \geq 1$ to be the unique solution to the linearized problem

$$\begin{aligned} \partial_t v^n - L v^n &= F_g(x,t,v^{n-1}+\varphi,\nabla(v^{n-1}+\varphi)) &&\text{in } Q_T, \\ v^n(x,0) &= 0 &&\text{on } \Omega, \\ v^n(x,t) &= 0 &&\text{on } \partial\Omega \times (0,T), \end{aligned} \quad (14.16)$$

where $v^0 = v_0(x) = 0 \in C^{2+\delta,1+\delta/2}(\overline{Q}_{T,U})$. To prove the existence of a solution to problem (14.14), we will show that this sequence converges. From [130], there exists a Green's function [193], i.e. $G(x,y,t,\tau)$ for problem (14.16). Green's function is an integral kernel that is used to solve difficult differential equations with initial or boundary value conditions. For $n \geq 1$, the solution v^n can be written as

$$\begin{aligned} v^n(x,t) &= \int_0^t \int_\Omega G(x,y,t,\tau) \, F_g(y,\tau,v^{n-1}+\varphi,\nabla(v^{n-1}+\varphi)) \, dy \, d\tau \\ &\quad + \int_\Omega G(x,y,t,0) v_0(y) \, dy \\ &= \int_0^t \int_\Omega G(x,y,t,\tau) \, F_g(y,\tau,v^{n-1}+\varphi,\nabla(v^{n-1}+\varphi)) \, dy \, d\tau, \end{aligned}$$

because $v_0(y) = 0$. Here,

$$F_g(y, \tau, v^{n-1} + \varphi, \nabla(v^{n-1} + \varphi))$$
$$= \int_\Omega \left[f\left(y, \tau, z, g\left(v^{n-1}(x) + \varphi(x), v^{n-1}(x+z) + \varphi(x+z)\right), \nabla(v^{n-1} + \varphi)(y, \tau)\right) \right] dz.$$

For convenience, we will write

$$\mathcal{F}^{n-1}(y, \tau) = F_g(y, \tau, v^{n-1} + \varphi, \nabla(v^{n-1} + \varphi))$$
$$= \int_\Omega f\left(y, \tau, z, g\left(v^{n-1}(x) + \varphi(x), v^{n-1}(x+z) + \varphi(x+z)\right), \nabla(v^{n-1} + \varphi)(y, \tau)\right) dz.$$

Now we take the first and second derivatives of $v^n(x, t)$ with respect to x:

$$v^n_{x_i}(x, t) = \int_0^t \int_\Omega G_{x_i}(x, y, t, \tau) \, \mathcal{F}^{n-1}(y, \tau) \, dy \, d\tau,$$

$$v^n_{x_i x_j}(x, t) = \int_0^t \int_\Omega G_{x_i x_j}(x, y, t, \tau) \, \mathcal{F}^{n-1}(y, \tau) \, dy \, d\tau.$$

From Chapter IV.16 in [130], we have the estimates

$$|G(x, y, t, \tau)| \leq c_1 (t - \tau)^{-\frac{d}{2}} \exp\left(-C_2 \frac{|x-y|^2}{t - \tau}\right), \tag{14.17}$$

$$|G_{x_i}(x, y, t, \tau)| \leq c_1 (t - \tau)^{-\frac{d+1}{2}} \exp\left(-C_2 \frac{|x-y|^2}{t - \tau}\right), \tag{14.18}$$

$$|G_{x_i x_j}(x, y, t, \tau)| \leq c_1 (t - \tau)^{-\frac{d+2}{2}} \exp\left(-C_2 \frac{|x-y|^2}{t - \tau}\right), \tag{14.19}$$

where $t > \tau$ and the constants c_1 and C_2 are independent of all parameters of G. If we combine everything together, we get

$$||v^n(\cdot, t)||_{W^2_\infty(\Omega)} = ||v^n(\cdot, t)||_{L^\infty(\Omega)} + \sum_{i=1}^d ||v^n_{x_i}(\cdot, t)||_{L^\infty(\Omega)} + \sum_{i,j=1}^d ||v^n_{x_i x_j}(\cdot, t)||_{L^\infty(\Omega)}$$

$$\leq \int_0^t \int_\Omega ||G(\cdot, y, t, \tau)||_{L^\infty(\Omega)} |\mathcal{F}^{n-1}(y, \tau)| dy d\tau$$

$$+ \sum_{i=1}^d \int_0^t \int_\Omega ||G_{x_i}(\cdot, y, t, \tau)||_{L^\infty(\Omega)} |\mathcal{F}^{n-1}(y, \tau)| dy d\tau$$

$$+ \sum_{i,j=1}^d \int_0^t \left|\left| \int_\Omega G_{x_i x_j}(\cdot, y, t, \tau) \mathcal{F}^{n-1}(y, \tau) dy \right|\right|_{L^\infty(\Omega)} d\tau.$$

The aim is to show that $||v^n(\cdot, t)||_{W^2_\infty(\Omega)}$ is uniformly bounded on the interval $[0, T]$ so that we can use the Arzelà–Ascoli theorem and a weak compactness argument (Theorem 3 of Appendix D in [63]). From **A(6)**, we have $|\mathcal{F}^{n-1}| = \mathcal{F}^{n-1}$. We obtain the following estimates by using **A(7)** and **A(8)**:

14.4 Solutions in Bounded Domains

$F^{n-1}(y, \tau)$

$$\leq \int_\Omega \left| f\left(y, \tau, z, g\left(v^{n-1}(x) + \varphi(x), v^{n-1}(x+z) + \varphi(x+z)\right), \nabla(v^{n-1} + \varphi)(y, \tau)\right) \right| dz$$

$$\leq \int_\Omega C_0 \left(1 + \left|g\left(v^{n-1}(x) + \varphi(x), v^{n-1}(x+z) + \varphi(x+z)\right)\right|\right.$$
$$\left. + |\nabla v^{n-1}(y, \tau)| + |\nabla \varphi(y, \tau)| \right) dz$$

$$\leq \int_\Omega C_0 \left(2||v^{n-1}(\cdot, \tau)||_{L^\infty(\Omega)} + \sum_{i=1}^d ||v^{n-1}_{y_i}(\cdot, \tau)||_{L^\infty(\Omega)} \right) dz$$

$$+ C_0 \left(1 + 2 \sup_{Q_T} |\varphi(y, \tau)| + \sum_{i=1}^d \sup_{Q_T} |\varphi_{y_i}(y, \tau)| \right) dz$$

$$\leq C_3 ||v^{n-1}(\cdot, \tau)||_{W^2_\infty(\Omega)} + C_T,$$

where C_3 is a constant independent of T, whereas C_T is a constant which depends on T. By a direct calculation, we can easily see that (with $|x - y|^2 = (x_1 - y_1)^2 + \cdots + (x_d - y_d)^2$),

$$\int_\Omega (t - \tau)^{-\frac{d}{2}} \exp\left(-C_2 \frac{|x - y|^2}{t - \tau}\right) dy$$
$$\leq \int_{\mathbb{R}^d} (t - \tau)^{-\frac{d}{2}} \exp\left(-C_2 \frac{|x - y|^2}{t - \tau}\right) dy$$
$$= C_2^{-\frac{d}{2}} \int_{\mathbb{R}^d} e^{-\sigma^2} d\sigma$$
$$= \left(\frac{\pi}{C_2}\right)^{\frac{d}{2}}.$$

We can see this by computing the integral in one dimension:

$$\int_{-\infty}^\infty (t - \tau)^{-\frac{1}{2}} \exp\left(-C_2 \frac{(x_1 - y_1)^2}{t - \tau}\right) dy_1$$
$$= \int_{-\infty}^\infty (t - \tau)^{-\frac{1}{2}} \sqrt{\frac{t - \tau}{C_2}} e^{-\omega_1^2} d\omega_1$$
$$= C_2^{-\frac{1}{2}} \int_{-\infty}^\infty e^{-\omega_1^2} d\omega_1$$
$$= \left(\frac{\pi}{C_2}\right)^{\frac{1}{2}},$$

where we use

$$\omega_1 = \sqrt{\frac{C_2}{t - \tau}} (x_1 - y_1).$$

The integral in $\mathbb{R}^d$ is a product of these one-dimensional integrals. This gives us the desired result. The Green's function estimate

$$\left\| \int_\Omega G_{x_i x_j}(\cdot, y, t, \tau) \, dy \right\|_{L^\infty(\Omega)} \leq C_4 (t - \tau)^{-\beta}, \tag{14.20}$$

where C_4 is a constant independent of T, $0 < \beta < 1$ and $t > \tau$ can be found in [213].

We also need to estimate the integral

$$\int_0^t \left\| \int_\Omega G_{x_i x_j}(x, y, t, \tau) F^{n-1}(y, \tau) \, dy \right\|_{L^\infty(\Omega)}$$

$$= \int_0^t \left\| \int_\Omega G_{x_i x_j}(x, y, t, \tau) \left(F^{n-1}(y, \tau) - F^{n-1}(x, \tau) + F^{n-1}(x, \tau) \right) dy \right\|_{L^\infty(\Omega)}$$

$$\leq \int_0^t \left\| \int_\Omega G_{x_i x_j}(x, y, t, \tau) \left(F^{n-1}(y, \tau) - F^{n-1}(x, \tau) \right) dy \right\|_{L^\infty(\Omega)}$$

$$+ \int_0^t \left\| \int_\Omega G_{x_i x_j}(x, y, t, \tau') \left(F^{n-1}(x, \tau) \right) dy \right\|_{L^\infty(\Omega)}. \tag{14.21}$$

To estimate the second term in the right-hand side of (14.21), we use (14.20). Since F^{n-1} is in a Hölder space, for the term

$$\left(F^{n-1}(y, \tau) - F^{n-1}(x, \tau) \right),$$

we have the inequality

$$\left| F^{n-1}(y, \tau) - F^{n-1}(x, \tau) \right| \leq C_5 \left(\sqrt{(y-x)^2} \right)^\delta, \delta \in (0, 1),$$

where C_5 is only a function of T and is independent of the spatial variable. Using the aforementioned result and the fact that F^{n-1} is in the Hölder space for t, the time variable, we can find some constant C_6 which is independent of both space and time variable so that

$$\left| F^{n-1}(y, \tau) - F^{n-1}(x, \tau) \right| \leq C_6 \left(\sqrt{(y-x)^2} \right)^\delta, \delta \in (0, 1). \tag{14.22}$$

The first term in (14.21) is bounded by using the inequality (14.22) to obtain

$$\left| \int_\Omega G_{x_i x_j}(x, y, t, \tau) \left(F^{n-1}(y, \tau) - F^{n-1}(x, \tau) \right) dy \right|$$

$$\leq \tilde{C}(t - \tau)^{-1+\delta/2},$$

where $\tilde{C}$ is a constant independent of T. Combining the two previous results together, putting them under the same exponent $\gamma = \min(\beta, 1 - \frac{\delta}{2})$, and using all of our previous estimates, we obtain

14.4 Solutions in Bounded Domains

$$||v^n(\cdot,t)||_{W^2_\infty(\Omega)} = ||v^n(\cdot,t)||_{L^\infty(\Omega)} + \sum_{i=1}^{d}||v^n_{x_i}(\cdot,t)||_{L^\infty(\Omega)} + \sum_{i,j=1}^{d}||v^n_{x_ix_j}(\cdot,t)||_{L^\infty(\Omega)}$$

$$\leq \int_0^t \left(A + B(t-\tau)^{-\frac{1}{2}} + D(t-\tau)^{-\gamma}\right)\left(C_3||v^{n-1}(\cdot,\tau)||_{W^2_\infty(\Omega)} + C_T\right) d\tau$$

$$= C_T\left(At + 2Bt^{1/2} + D\frac{t^{1-\gamma}}{1-\gamma}\right)$$

$$+ C_3 \int_0^t \left(A + B(t-\tau)^{-\frac{1}{2}} + D(t-\tau)^{-\gamma}\right)||v^{n-1}(\cdot,\tau)||_{W^2_\infty(\Omega)} d\tau$$

$$\leq C(T,\gamma) + C\int_0^t \left(A + B(t-\tau)^{-\frac{1}{2}} + D(t-\tau)^{-\gamma}\right)||v^{n-1}(\cdot,\tau)||_{W^2_\infty(\Omega)} d\tau$$

where the constants $A, B, D,$ and C are independent of T. The constant $C(T,\gamma)$ depends only on T and γ. Therefore we have

$$||v^n(\cdot,t)||_{W^2_\infty(\Omega)} \leq C(T,\gamma)$$

$$+ C\int_0^t \left(A + B(t-\tau)^{-\frac{1}{2}} + D(t-\tau)^{-\gamma}\right)||v^{n-1}(\cdot,\tau)||_{W^2_\infty(\Omega)} d\tau. \tag{14.23}$$

Observe that there exists an upper bound (ϵ) of the integral

$$\int_0^\tau \left(A + B(\tau-\tau')^{-\frac{1}{2}} + D(\tau-\tau')^{-\gamma}\right) d\tau',$$

for $\tau \in [0, T_1]$, with $T_1 \leq T$, so that $|\epsilon C| < 1$. This is possible as C depends on T. We choose this T_1 to be the initial time for the next time step. We will follow exactly the same computation as follows to obtain a solution in the interval and we move on to the next interval (until we reach T). When we solve the problem in the interval $[0, T_1]$ by the method described in the following, we will find a solution given by v. $v(T_1)$ will denote the initial value of the same problem in the next interval. If $v(T_1) \neq 0$, in order to get (14.14), we need to use $v - v(T_1)$ as the new variable. This will lead to a constant term in the right-hand side of the first equation in (14.14). But that will not change any other subsequent derivations.

Thus by dividing the interval $[0, T]$ properly, we can obtain the required solution. Next, we present the proof for obtaining a solution in the interval $\tau \in [0, T_1]$.

We observe from (14.23) that

$$||v^1(\cdot,t)||_{W^2_\infty(\Omega)} \leq C(T,\gamma),$$

and so we can deduce that

$$||v^2(\cdot,t)||_{W^2_\infty(\Omega)} \leq C(T,\gamma) + C\int_0^t \left(A + B(t-\tau)^{-\frac{1}{2}} + D(t-\tau)^{-\gamma}\right)||v^1(\cdot,t)||_{W^2_\infty(\Omega)} d\tau$$

$$\leq C(T,\gamma) + C(T,\gamma)C\epsilon.$$

Again,

$$\|v^3(\cdot,t)\|_{W^2_\infty(\Omega)} \leq C(T,\gamma) + C\int_0^t \left(A + B(t-\tau)^{-\frac{1}{2}} + D(t-\tau)^{-\gamma}\right)\|v^2(\cdot,\tau)\|_{W^2_\infty(\Omega)}\, d\tau$$
$$\leq C(T,\gamma) + C(C(T,\gamma) + C(T,\gamma)C\epsilon)\epsilon$$
$$= C(T,\gamma) + C(T,\gamma)C\epsilon + C(T,\gamma)C^2\epsilon^2.$$

By induction we obtain,

$$\|v^n(\cdot,t)\|_{W^2_\infty(\Omega)} \leq C(T,\gamma)\left(1 + C\epsilon + \cdots + C^{n-1}\epsilon^{n-1}\right).$$

Since $|\epsilon C| < 1$, we obtain $\|v^n(\cdot,t)\|_{W^2_\infty(\Omega)} \leq \frac{C(T,\gamma)}{1-\epsilon C}$, where $n = 0, 1, 2, \ldots$. Consequently $\|v^n(\cdot,t)\|_{W^2_\infty(\Omega)}$ is uniformly bounded on the closed interval $[0, T]$. Using this result along with (14.14), we can easily show that $\|v^n_t(\cdot,t)\|_{L^\infty(\Omega)}$ is also uniformly bounded on $[0, T]$.

Since $\|v^n(\cdot,t)\|_{W^2_\infty(\Omega)}$ and $\|v^n_t(\cdot,t)\|_{L^\infty(\Omega)}$ are continuous functions of t on the closed interval $[0, T]$, it follows that $|v^n|$, $|v^n_{x_i}|$, $|v^n_{x_i x_j}|$ and $|v^n_t|$ are uniformly bounded on $\overline{Q}_T$. Thus $v^n(\cdot, t)$ is equicontinuous in $C(\overline{Q}_T)$. By the Arzelà–Ascoli theorem, there exists a subsequence $\{v^{n_k}\}_{k=0}^\infty$ such that as $k \to \infty$,

$$v^{n_k} \to v \in C(\overline{Q}_T) \quad \text{and}$$
$$v^{n_k}_{x_i} \to v_{x_i} \in C(\overline{Q}_T),$$

where the convergence is uniform. Furthermore, by Theorem 3 in Appendix D of [63],

$$v^{n_k}_{x_i x_j} \to v_{x_i x_j} \in L^\infty(\overline{Q}_T) \quad \text{and}$$
$$v^{n_k}_t \to v_t \in L^\infty(\overline{Q}_T),$$

as $k \to \infty$. Here, the convergence is in the weak sense. Therefore, v^{n_k} converges uniformly on the compact set $\overline{Q}_T$ to a function $v \in C^{1+1,0+1}(\overline{Q}_T)$. By **A(7)** and **A(8)**, we have

$$\left|f\left(y,\tau,z,g\left(v^{n_k-1}(x) + \varphi(x), v^{n_k-1}(x+z) + \varphi(x+z)\right), \nabla(v^{n_k-1} + \varphi)(y,\tau)\right)\right|$$
$$\leq C_0\left(1 + 2\sup_{Q_T}|v^{n_k-1}(y,\tau)| + \sum_{i=1}^d \sup_{Q_T}|v^{n_k-1}_{y_i}(y,\tau)|\right.$$
$$\left. + 2\sup_{Q_T}|\varphi(y,\tau)| + \sum_{i=1}^d \sup_{Q_T}|\varphi_{y_i}(y,\tau)|\right) \leq C',$$

where C' is independent of n_k. Hence, by the dominated convergence theorem and the continuity of f,

$$\lim_{k\to\infty} F^{n_k-1}(y,\tau)$$
$$= \int_\Omega \lim_{k\to\infty} f\left(y,\tau,z,g\left(v^{n_k-1}(x)+\varphi(x), v^{n_k-1}(x+z)+\varphi(x+z),\right)\right.$$
$$\left.\nabla(v^{n_k-1}+\varphi)(y,\tau)\right) dz$$
$$= \int_\Omega f(y,\tau,z,g(v(x)+\varphi(x), v(x+z)+\varphi(x+z)), \nabla(v+\varphi)(y,\tau)) dz.$$

Therefore, by passing to the limit $k \to \infty$ in (14.16), we see that v is a classical solution to the initial-boundary value problem (14.14). Consequently, u is a strong solution to (14.13). $\square$

Now, we present that we can extend this solution in order to obtain a classical solution on the unbounded domain $\mathbb{R}_T^{d+1} = \mathbb{R}^d \times (0,T)$.

Theorem 14.4.2 There exists a classical solution $u \in C^{2,1}(\mathbb{R}_T^{d+1})$ to the problem

$$\begin{aligned} u_t - Lu &= F_g(x,t,u,\nabla u) &&\text{in } \mathbb{R}_T^{d+1} \\ u(x,0) &= u_0(x) &&\text{on } \mathbb{R}^d \end{aligned} \qquad (14.24)$$

such that the solution $u(x,t) \to v(x,t)$ as $|x| \to \infty$.

Proof. We approximate the domain $\mathbb{R}^d$ by a nondecreasing sequence $\{\Omega_N\}_{N=1}^\infty$ of bounded smooth subdomains of Ω. For simplicity, we will let $\Omega_N = B(0,N)$ be the open ball in $\mathbb{R}^d$ centered at the origin with radius N. Also, we let $V_N = \Omega_N \times (0,T)$.

Using the previous theorem, we let $u_M \in C^{2,1}(\overline{V}_M)$ be a solution to the problem

$$\begin{aligned} u_t - Lu &= F_g(x,t,u,\nabla u) &&\text{in } V_M \\ u(x,0) &= u_0(x) &&\text{on } \Omega_M \\ u(x,t) &= v(x,t) &&\text{on } \partial\Omega_M \times (0,T). \end{aligned} \qquad (14.25)$$

Since $M \geq 1$ is arbitrary, we can use a standard diagonal argument (for details see [145]) to extract a subsequence that converges to a solution u to the problem on the whole unbounded space $\mathbb{R}_T^{d+1}$. Clearly, $u(x,0) = u_0(x)$ and $u(x,t) \to v(x,t)$ as $|x| \to \infty$. $\square$

For the rest of this subsection we will prove the aforementioned theorem. First, we introduce a change of variables to transform our problem into one with a zero boundary condition. If we let

$$v(x,t) = u(x,t) - \varphi(x,t) \text{ and}$$
$$v_0(x) = u_0(x) - \varphi(x,0) = 0,$$

then v will satisfy the initial-boundary value problem
$$\begin{aligned} v_t - Lv &= F(x, t, v + \varphi, \nabla(v + \varphi)) & \text{in } & Q_T, \\ v(x, 0) &= 0 & \text{on } & \Omega, \\ v(x, t) &= 0 & \text{on } & \partial\Omega \times (0, T). \end{aligned} \qquad (14.26)$$

Definition 14.4.1 v is said to be a weak solution of (14.26) if
$$v \in L^2(0, T; H_0^1(\Omega)), \quad \frac{\partial v}{\partial t} \in L^2(0, T; H^{-1}(\Omega))$$
and
$$\int_\Omega \left(\frac{\partial v}{\partial t} g + \sum_{i,j=1}^d a^{ij}(x, t) v_{x_i} g_{x_j} + \sum_{i=1}^d b^i(x, t) g_{x_i} \phi + c(x, t) v g \right) dx$$
$$= \int_\Omega F(x, t, v + \varphi, \nabla(v + \varphi)) g \, dx, \qquad (14.27)$$
for all $g \in H_0^1(\Omega)$.

Lemma 14.4.2 (See [63] Theorem 3 in Section 5.9.2). If $v \in L^2(0, T; H_0^1(\Omega))$ and $\frac{\partial v}{\partial t} \in L^2(0, T; H^{-1}(\Omega))$, then:

i) $v \in C([0, T]; L^2(\Omega))$
ii) The mapping $t \to \|v(t)\|_{L^2(\Omega)}^2$ is absolutely continuous with
$$\frac{d}{dt} \|v(t)\|_{L^2(\Omega)}^2 = 2 \int_\Omega \frac{\partial v}{\partial t} v \, dt, \qquad (14.28)$$
for a.e $0 \le t \le T$.

Before we proceed, we state the important theorem that will be used in the next lemma.

Theorem 14.4.3 (Gauss–Green theorem). Assume that U is a bounded, open subset of $\mathbb{R}^n$ and ∂U is C^1. Let $v = (v^1, \ldots, v^n)$ denote the unit outward normal vector:

1) Let $u : U \to \mathbb{R}$ such that u is C^1. Then,
$$\int_U u_{x_i} \, dx = \int_{\partial U} u v^i \, dS \text{ for } i = 1, \ldots, n.$$

2) For a vector field $\mathbf{u} \in C^1(U; \mathbb{R}^n)$ we have,
$$\int_U \operatorname{div} \mathbf{u} \, dx = \int_{\partial U} \mathbf{u} \cdot v \, dS.$$

Please see [65] and references therein for the proof.

Lemma 14.4.3 If v is a weak solution of (14.26), then there exits a positive constant C independent of v such that

$$||v(t)||_{L^2(0,T;H^1_0(\Omega))} \leq C. \tag{14.29}$$

Proof. Choose $v(t) \in H^1_0(\Omega)$ as the test function in (14.27). Then we obtain

$$\int_\Omega \left(\frac{\partial v}{\partial t} v + \sum_{i,j=1}^d a^{ij}(x,t) v_{x_i} v_{x_j} + \sum_{i=1}^d b^i(x,t) v_{x_i} v + c(x,t) v^2 \right) dx$$

$$= \int_\Omega \mathcal{F}(x,t,v+\varphi, \nabla(v+\varphi)) v dx. \tag{14.30}$$

Observe that by **A(7)** we have

$$|\mathcal{F}(x,t,u,\nabla u)| \leq \int_\Omega |f(x,t,y,u,\nabla u)| dy \leq C'(1+|u|+|\nabla u|), \tag{14.31}$$

where the constant C' depends only on the region Ω. Also observe from **A(2)** that

$$\theta |\nabla v|^2 \leq \sum_{i,j=1}^d a^{ij} v_{x_i} v_{x_j}. \tag{14.32}$$

We obtain from **A(3)**

$$\int_\Omega \sum_{i=1}^d b^i(x,t) v_{x_i} v dx \geq B \sum_{i=1}^d \int_\Omega v_{x_i} v dx$$

$$= \frac{B}{2} \sum_{i=1}^d \int_\Omega \frac{\partial}{\partial x_i}(v^2) dx$$

$$= 0,$$

where in the last step we used the boundary condition of v and the Gauss–Green theorem. Therefore

$$\int_\Omega \sum_{i=1}^d b^i(x,t) v_{x_i} v dx \geq 0. \tag{14.33}$$

Finally using **A(3)** we obtain

$$\int_\Omega c(x,t) v^2 dx \geq 0. \tag{14.34}$$

We use Lemma 14.4.2 and (14.31)–(14.34) in (14.30) to get

$$\frac{1}{2} \frac{d}{dt} ||v(t)||^2_{L^2(\Omega)} + \theta \int_\Omega |\nabla v|^2 dx \leq C' \int_\Omega (1+|v|+|\varphi|+|\nabla v|+|\nabla \varphi|)|v| dx.$$

$$\tag{14.35}$$

By using the Cauchy inequality with $\epsilon > 0$, we get

$$\frac{1}{2}\frac{d}{dt}||v(t)||^2_{L^2(\Omega)} + \theta \int_\Omega |\nabla v|^2 dx$$
$$\leq C' \left[\left(\epsilon \int_\Omega dx + \frac{1}{4\epsilon}\int_\Omega |v|^2 dx \right) + \int_\Omega |v|^2 dx + \left(\epsilon \int_\Omega |\nabla v|^2 dx + \frac{1}{4\epsilon}\int_\Omega |v|^2 dx\right) \right.$$
$$\left. + \left(\epsilon \int_\Omega |\varphi|^2 dx + \frac{1}{4\epsilon}\int_\Omega |v|^2 dx\right) + \left(\epsilon \int_\Omega |\nabla \varphi|^2 dx + \frac{1}{4\epsilon}\int_\Omega |v|^2 dx\right) \right].$$

Choosing $\epsilon \ll 1$ we obtain

$$\frac{1}{2}\frac{d}{dt}||v(t)||^2_{L^2(\Omega)} + C_1 ||v||^2_{H^1_0(\Omega)} \leq C_2 ||v||^2_{L^2(\Omega)} + \tilde{C}(t),$$

for almost everywhere $0 \leq t \leq T$, C_1 and C_2 are some positive constants and

$$\tilde{C}(t) = C'\epsilon \int_\Omega (1 + |\varphi(x,t)|^2 + |\nabla \varphi(x,t)|^2) dx.$$

Since $\varphi \in C^{2+\delta, 1+\delta/2}\left(\overline{Q_T}\right)$, therefore $\tilde{C}(t) \leq C_3$, for some constant C_3 where $0 \leq t \leq T$. Thus

$$\frac{1}{2}\frac{d}{dt}||v(t)||^2_{L^2(\Omega)} + C_1 ||v||^2_{H^1_0(\Omega)} \leq C_2 ||v||^2_{L^2(\Omega)} + C_3, \quad \text{for a.e. } 0 \leq t \leq T.$$
(14.36)

We write $\eta(t) := ||v(t)||^2_{L^2(\Omega)}$, then (14.36) gives

$$\eta'(t) \leq 2C_2 \eta(t) + 2C_3, \quad \text{for a.e. } 0 \leq t \leq T.$$

The differential form of Gronwall inequality [98] thus gives

$$\eta(t) \leq e^{2C_2 t}(\eta(0) + 2C_3 t), \quad \text{for a.e. } 0 \leq t \leq T.$$

But $\eta(0) = ||v(0)||^2_{L^2(\Omega)} = 0$. Thus we have

$$||v(t)||^2_{L^2(\Omega)} \leq 2e^{2C_2 t} C_3 t, \quad \text{for a.e. } 0 \leq t \leq T.$$

Thus

$$\max_{0 \leq t \leq T} ||v(t)||^2_{L^2(\Omega)} \leq C_4, \quad (14.37)$$

for some constant C_4 independent of v. We consider now (14.36) and integrate from 0 to T (with $||v(0)||^2_{L^2(\Omega)} = 0$) to get

$$C_1 \int_0^T ||v||^2_{H^1_0(\Omega)} dt \leq C_2 \int_0^T ||v||^2_{L^2(\Omega)} dt + C_3 T.$$

Use inequality (14.37) to obtain

$$C_1 ||v||_{L^2(0,T;H^1_0(\Omega))} \leq C_2 C_4 T + C_3 T.$$

Define a constant $C > 0$ such that $CC_1 = C_2C_4T + C_3T$. Thus we have

$$||v(t)||_{L^2(0,T;H_0^1(\Omega))} \leq C.$$

Hence the proof. □

Before we prove the existence theorem, we state the following lemma from the linear theory of parabolic partial differential equations. The lemma follows directly from [63], Theorem 2, page 354.

Lemma 14.4.4 (Energy estimates). Consider the problem

$$\begin{cases} \frac{\partial v}{\partial t} - L(v) = f(x,t) & \text{in } \Omega \times (0,T) \\ v(x,0) = 0 & \text{on } \Omega \times \{0\} \\ v(x,t) = 0 & \text{on } \partial\Omega \times [0,T] \end{cases} \quad (14.38)$$

with $f \in L^2(0,T;L^2(\Omega))$. Then there exists a unique $u \in L^2(0,T;H_0^1(B_R)) \cap C([0,T];L^2(B_R))$ solution of (14.38) that satisfies

$$\max_{0 \leq t \leq T} ||u(t)||_{L^2(\Omega)} + ||u||_{L^2(0,T;H_0^1(\Omega))} + \left\|\frac{\partial u}{\partial t}\right\|_{L^2(0,T;H^{-1}(\Omega))} \leq C'||f||_{L^2(0,T;L^2(\Omega))},$$

(14.39)

where C' is a positive constant depending only on Ω and T and the coefficients of the operator L.

Lemma 14.4.5 follows directly from [63], Theorem 5, page 360.

Lemma 14.4.5 (Improved regularity). Consider the problem

$$\begin{cases} \frac{\partial v}{\partial t} - L(v) = f(x,t) & \text{in } \Omega \times (0,T) \\ v(x,0) = 0 & \text{on } \Omega \times \{0\} \\ v(x,t) = 0 & \text{on } \partial\Omega \times [0,T] \end{cases}$$

with $f \in L^2(0,T;L^2(\Omega))$. Then there exists a unique weak solution of the problem $u \in L^2(0,T;H_0^1(B_R)) \cap C([0,T];L^2(B_R))$, with $\frac{\partial u}{\partial t} \in L^2(0,T;H^{-1}(\Omega))$. Moreover

$$u \in L^2(0,T;H^2(\Omega)) \cap L^\infty(0,T;H_0^1(\Omega)), \quad \frac{\partial u}{\partial t} \in L^2(0,T;L^2(\Omega)).$$

We also have the estimate

$$\underset{0 \leq t \leq T}{\text{ess sup}} ||u(t)||_{H_0^1(\Omega)} + ||u||_{L^2(0,T;H^2(\Omega))} + \left\|\frac{\partial u}{\partial t}\right\|_{L^2(0,T;L^2(\Omega))} \leq C''||f||_{L^2(0,T;L^2(\Omega))},$$

(14.40)

where C'' is a positive constant depending only on Ω and T and the coefficients of the operator L.

Now we proceed to prove Theorem 14.4.1. We will actually prove the existence of a weak solution for the equivalent problem of (14.13) given by (14.26).

Proof. (of Theorem 14.4.1)
Given $w \in L^2(0, T; H_0^1(\Omega))$, set $f_w(x, t) := \mathcal{F}(x, t, w + \varphi, \nabla(w + \varphi))$. By **A(8)**, $f_w \in L^2(0, T; L^2(\Omega))$. By Lemma (14.4.4) there exists a unique $v \in L^2(0, T; H_0^1(\Omega)) \cap C([0, T]; L^2(\Omega))$ which is the solution of

$$\begin{cases} \frac{\partial v}{\partial t} - Lv = f_w(x, t) & \text{in} \quad \Omega \times (0, T) \\ v(x, 0) = 0 & \text{on} \quad \Omega \times \{0\} \\ v(x, t) = 0 & \text{on} \quad \partial\Omega \times [0, T] \end{cases} \quad (14.41)$$

Define the mapping

$$A : L^2(0, T; H_0^1(\Omega)) \to L^2(0, T; H_0^1(\Omega))$$
$$w \mapsto A(w) = v$$

where v which is the solution of (14.41) is derived from w. We will show that the mapping A is continuous and compact.

We begin by showing that A is continuous.
Continuity: Let $\{w_k\}_k \subset L^2(0, T; H_0^1(\Omega))$ be a sequence such that

$$w_k \to w \text{ in } L^2(0, T; H_0^1(\Omega)). \quad (14.42)$$

By the *improved regularity* (14.40), there exists a constant C'', independent of $\{w_k\}_k$ such that

$$\sup_k \|v_k\|_{L^2(0,T;H^2(\Omega))} \leq C'' \|f_{w_k}\|_{L^2(0,T;L^2(\Omega))}, \quad \text{for } v_k = A[w_k], k = 1, 2, \ldots. \quad (14.43)$$

But $f_{w_k}(x, t) := \mathcal{F}(x, t, w_k + \varphi, \nabla(w_k + \varphi))$, and thus by **A(8)** as $w_k \to w$ in $L^2(0, T; H_0^1(\Omega))$, we must have $f_{w_k}(x, t) \to f_w(x, t)$ in $L^2(0, T; L^2(\Omega))$ and thus $\|f_{w_k}(x, t)\|_{L^2(0,T;L^2(\Omega))} \to \|f_w(x, t)\|_{L^2(0,T;L^2(\Omega))}$.
Therefore the sequence $\{\|f_{w_k}\|_{L^2(0,T;L^2(\Omega))}\}_k$ is bounded and

$$\sup_k \|f_{w_k}\|_{L^2(0,T;L^2(\Omega))} \leq C''', \quad (14.44)$$

for some constant C'''. Thus by (14.43) and (14.44) the sequence $\{v_k\}_k$ is bounded uniformly in $L^2(0, T; H^2(\Omega))$. Similarly it can be proved that $\{\frac{\partial v_k}{\partial t}\}_k$ is uniformly bounded in $L^2(0, T; H^{-1}(\Omega))$. Thus by Rellich's theorem (see [74]) there exists a subsequence $\{v_{k_j}\}_j \in L^2(0, T; H_0^1(\Omega))$ and a function $v \in L^2(0, T; H_0^1(\Omega))$ with

$$v_{k_j} \to v \text{ in } L^2(0, T; H_0^1(\Omega)), \text{ as } j \to \infty. \quad (14.45)$$

Therefore

$$\int_\Omega \left(\frac{\partial v_{k_j}}{\partial t}\phi - Lv_{k_j}\right) dx = \int_\Omega f_{w_{k_j}}(x,t)\phi \, dx$$

for each $\phi \in H_0^1(\Omega)$. Then using (14.42) and (14.45), we see that

$$\int_\Omega \left(\frac{\partial v}{\partial t}\phi - L\phi\right) dx = \int_\Omega f_w(x,t)\phi \, dx.$$

Thus $v = A[w]$. Therefore

$$A[w_k] \to A[w] \text{ in } L^2(0,T;H_0^1(\Omega)).$$

Compactness: For compactness, we need to show that every compact subset of A is closed and bounded. The compactness result follows from similar arguments.

Finally, to apply Schaefer's fixed point theorem with $X = L^2(0,T;H_0^1(\Omega))$, we need to show that the set $\{w \in L^2(0,T;H_0^1(\Omega)) : w = \lambda A[w] \text{ for some } 0 \leq \lambda \leq 1\}$ is bounded. This follows directly from Lemma 14.4.3 with $\lambda = 1$. This proves the theorem. $\square$

14.5 Construction of the Solution in the Whole Domain

Next we construct a solution of problem (14.26) in $\mathbb{R}_T^{d+1} = \mathbb{R}^d \times [0,T]$. The construction is presented in the following theorem.

Theorem 14.5.1 *If assumptions* **A(1)–A(8)** *are satisfied, then there exists a weak solution for*

$$\begin{aligned} u_t - Lu &= \mathcal{F}(x,t,u,\nabla u) &&\text{in } \mathbb{R}_T^{d+1}, \\ u(x,0) &= u_0(x) &&\text{on } \mathbb{R}^d. \end{aligned} \tag{14.46}$$

Proof. Let us consider $\Omega = \mathbb{B}_N$, where $\mathbb{B}_N = \{x \in \mathbb{R}^d : |x| < N\}$ is a ball in $\mathbb{R}^d$ center at zero and radius N. We approximate $\mathbb{R}^d$ as $\mathbb{R}^d = \cup_{N \in \mathbb{N}} \mathbb{B}_N = \lim_{N \to \infty} \mathbb{B}_N$. For $N \in \mathbb{N}$, by the previous subsection, there exists $w_N \in L^2(0,T;H_0^1(\mathbb{B}_N)) \cap C([0,T];L^2(\mathbb{B}_N))$ with $\frac{\partial w_N}{\partial t} \in L^2(0,T;H^{-1}(\mathbb{B}_N))$, weak solution of

$$\begin{aligned} (w_N)_t - Lw_N &= \mathcal{F}(x,t,w_N+\varphi,\nabla(w_N+\varphi)) &&\text{in } \mathbb{B}_N \times [0,T], \\ w_N(x,0) &= 0 &&\text{on } \mathbb{B}_N, \\ w_N(x,t) &= 0 &&\text{on } \partial \mathbb{B}_N \times (0,T). \end{aligned} \tag{14.47}$$

For any given $\rho > 0$, the following sequences are bounded uniformly in $N > 2\rho$ because it is possible to find one constant that bounds all sequences, i.e. 2ρ:

$$\{w_N\}_N \text{ in } L^2(0, T; H_0^1(\mathbb{B}_\rho))$$

$$\left\{\frac{\partial w_N}{\partial t}\right\}_N \text{ in } L^2(0, T; H^{-1}(\mathbb{B}_\rho))$$

Since these spaces are compactly embedded in $L^2(\mathbb{B}_\rho \times (0, T))$, then the sequence $\{w_N\}_N$ is relatively compact in $L^2(\mathbb{B}_\rho \times (0, T))$. Suppose we take $\rho \in \mathbb{N}$. Then using the compactness just described, we can construct a sequence consisting of diagonal elements of the converging subsequences with each $\rho \in \mathbb{N}$. Denote this sequence also by $\{w_N\}_N$. Then there exists $w \in L^2(0, T; L^2_{loc}(\mathbb{R}^d)) \cap L^2(0, T; H^1_{loc}(\mathbb{R}^d))$ with $w(x, 0) = 0$ so that

$$w_N \to w \text{ a.e and in } L^2(0, T; L^2_{loc}(\mathbb{R}^d))$$

and also

$$w_N \to w \text{ weakly in } L^2(0, T; H^1_{loc}(\mathbb{R}^d)).$$

By **A(8)**, passing to the limit in (14.47) yields that w is a weak solution of the problem (14.26) in $\mathbb{R}^d \times [0, T]$. That is, there exists a weak solution for Eq. (14.46). Hence the proof. □

14.6 Problems

1. State and prove Schaefer's fixed point theorem.
2. Consider a stochastic process such that the underlying security S follows the model:

$$dS_t = \mu S_t dt + \sigma_t S_t dZ_t,$$

where Z is a standard Brownian motion. Suppose the variance $v(t) = \sigma_t^2$ also follows a stochastic process given by

$$dv_t = \kappa(\theta - v(t))dt + \gamma\sqrt{v_t}dW_t,$$

where W is a standard Brownian motion. If the correlation coefficient between W and Z is denoted by ρ,

$$\text{Cov}\left(dZ_t, dW_t\right) = \rho\, dt,$$

prove that the generalized Black–Scholes equation is

$$\frac{1}{2}vS^2\frac{\partial^2 U}{\partial S^2} + \rho\gamma vS\frac{\partial^2 U}{\partial v \partial S} + \frac{1}{2}v\gamma^2\frac{\partial^2 U}{\partial v^2} + rS\frac{\partial U}{\partial S}$$
$$+ [\kappa(\theta - v) - \lambda v]\frac{\partial U}{\partial v} - rU + \frac{\partial U}{\partial t} = 0.$$

3. Derive the Black–Scholes equation if the volatility is time dependent.

4. For a model with several assets, show how to obtain the Black–Scholes PIDE for n assets.
5. Explain why the boundary condition for the Black–Scholes model with jumps should be the same boundary condition used for the classical Black–Scholes model whenever the spatial domain for S is bounded and sufficiently large.
6. Explain how to find a solution of (14.1) subject to the boundary conditions
$$F(0,t) = 0, \quad F(S,0) = \max(0, S - E)$$
and outline a numerical scheme.
7. Explain why there exists a weak solution for the problem:
$$u_t - Lu = F(x, t, u, \nabla u) \quad \text{in} \quad \mathbb{R}_T^{d+1},$$
$$u(x, 0) = u_0(x) \quad \text{on} \quad \mathbb{R}^d,$$
in $\mathbb{R}_T^{d+1} = \mathbb{R}^d \times [0, T]$.

15

Solutions to Complex Models Arising in the Pricing of Financial Options

15.1 Introduction

In this chapter we will first analyze a market model where the assets are driven by stochastic volatility models, and trading the assets involves paying proportional transaction costs. The stochastic volatility model is an enhancement of the Black–Scholes model for the pricing of financial options. In the Black–Scholes model, the volatility of the underlying security is assumed to be constant. However in stochastic volatility models, the price of the underlying security is a random variable. Allowing the price to vary helps improve the accuracy of calculations and forecasts.

In this chapter, we will show how the price of an option written on this type of equity may be obtained as a solution to a partial differential equation (PDE). We will obtain the option pricing PDE for the scenario where the volatility is a traded asset. In this case all option prices may be found as solutions to the resulting nonlinear PDE. Furthermore, hidden within this scenario is the case where the option depends on two separate assets, and the assets are correlated in the same form as S and σ. The treatment of the option in this case is entirely equivalent with the case discussed in this chapter.

15.2 Option Pricing with Transaction Costs and Stochastic Volatility

This is a problem with a long history in mathematical finance. In a complete frictionless (i.e. without transaction costs) financial market, Black and Scholes (see [24]) provide a hedging strategy for any European type contingent claim, i.e. a payout that is dependent on the realization of some uncertain future event. One needs to trade continuously to rebalance the hedging portfolio, and therefore, such an operation tends to be infinitely expensive in a market with transaction costs (the cost of participating in a market). For example, [192] shows

Quantitative Finance, First Edition. Maria C. Mariani and Ionut Florescu.
© 2020 John Wiley & Sons, Inc. Published 2020 by John Wiley & Sons, Inc.

that the best hedging strategy in this case is to simply buy the asset and hold it for the duration of the Call or Put option. This is the reason why the requirement of replicating the value of the option continuously and exactly has to be relaxed.

As described by [55], the approaches taken were local in time and global in time. The former approach (which is also the one taken in this chapter) pioneered by [132] (continued e.g. [27, 97]), considers risk and return over a short interval of time. The later approach pioneered by [96] (continued, for example, in [50]) adopts "optimal strategies," in which risk and return are considered over the lifetime of the option.

In the seminal work Leland [132] introduced the idea of using expected transaction costs for a small interval. He assumed that the portfolio is rebalanced at deterministic discrete times, δt units apart, and that the transaction costs are proportional to the value of the underlying. Specifically, the cost incurred is $\kappa|v|S$, where v is the number of shares of the underlying asset bought ($v > 0$) or sold ($v < 0$) at price S and κ is a proportionality constant characteristic to the individual investor. Leland proposed a hedging strategy based on replicating an option with an adjusted volatility, i.e.

$$\hat{\sigma} = \sigma\left(1 + \sqrt{\frac{2}{\pi}\frac{\kappa}{\sigma\sqrt{\delta t}}}\right)^{1/2}.$$

Leland claimed in his paper that the hedging error approaches zero using this strategy when the length of revision intervals goes to zero, a claim later disproved by many, first being Kabanov and Safarian [114].

Notwithstanding the claim of the hedging error approaching zero for this modified strategy, the idea of using (conditional) expectations when calculating transaction costs proved valuable. This idea was continued by [27] in discrete time and by [97] (and further [55, 206]) in continuous time. Later, this influential line of work derives a nonlinear PDE whose solution provides the option value. For the reader's convenience we replicate the derivation of the PDE using modern notation in Section 15.3. This section is illustrating the idea of modeling transaction costs using conditional expectations in a simple model. In particular, we will extend the transaction costs model when the asset price is approximated using stochastic volatility models.

15.3 Option Price Valuation in the Geometric Brownian Motion Case with Transaction Costs

Suppose ˙ is the value of the hedging portfolio and $C(S, t)$ is the value of the option. The asset follows a geometric Brownian motion. Using a discrete time approximation [97], assume the underlying asset follows the process:

$$\delta S = \mu S \delta t + \sigma S \Phi \sqrt{\delta t}, \qquad (15.1)$$

where Φ is a standard normal random variable, μ is a measure of the average rate of growth of the asset price also known as the drift, and the volatility σ is a measure of the fluctuation (risk) in the asset prices and corresponds to the diffusion coefficient. The quantities involving δ denote the increment of processes over the time step δt. If the portfolio is given by $\Pi = C - \Delta S$, then the change in portfolio value is given by

$$\delta \Pi = \sigma S \left(\frac{\partial C}{\partial S} - \Delta \right) \Phi \sqrt{\delta t} + \left(\frac{1}{2} \sigma^2 S^2 \frac{\partial^2 C}{\partial S^2} + \mu S \frac{\partial C}{\partial S} + \frac{\partial C}{\partial t} - \mu \Delta S \right) \delta t - \kappa S |\nu|$$

To derive the portfolio change in the previous equation, one may assume that the quantity of shares Δ is kept constant. This however should not be the case. In fact in the derivation earlier Δ is stochastic. To obtain the expression we use the fact that the constructed portfolio Π is self-financing, i.e. at all time steps when the portfolio is rebalanced, no extra funds are added to the portfolio or consumed from the portfolio value. Portfolio rebalancing means buying and selling investments in order to restore a portfolio to its original asset allocation model. The basic derivation when the asset follows a geometric Brownian motion and when the pricing of vanilla option is desired may be found in [188]. In Section 15.5.2 we provide a generalization of this result under any dynamics for the stock price. In fact, the same rule applies for a portfolio constructed using any number of assets S_i.

The dynamic discussed earlier leads to the delta hedging strategy. Specifically, let the quantity of asset be held short at time t, $\Delta = \frac{\partial C}{\partial S}(S, t)$. The time step is assumed to be small; thus the number of assets traded after a time δt is

$$\nu = \frac{\partial C}{\partial S}(S + \delta S, t + \delta t) - \frac{\partial C}{\partial S}(S, t)$$

$$= \delta S \frac{\partial^2 C}{\partial S^2} + \delta t \frac{\partial^2 C}{\partial t \partial S} + \dots$$

Since $\delta S = \sigma S \Phi \sqrt{\delta t} + \mathcal{O}(\delta t)$, keeping only the leading term yields

$$\nu \simeq \frac{\partial^2 C}{\partial S^2} \sigma S \Phi \sqrt{\delta t}.$$

Thus, the expected transaction cost over a time step is

$$E[\kappa S |\nu|] = \sqrt{\frac{2}{\pi}} \kappa \sigma S^2 \left| \frac{\partial^2 C}{\partial S^2} \right| \sqrt{\delta t},$$

where $\sqrt{2/\pi}$ is the expected value of $|\Phi|$. Therefore, the expected change in the value of the portfolio is

$$E(\delta \Pi) = \left(\frac{\partial C}{\partial t} + \frac{1}{2} \sigma^2 S^2 \frac{\partial^2 C}{\partial S^2} - \kappa \sigma S^2 \sqrt{\frac{2}{\pi \delta t}} \left| \frac{\partial^2 C}{\partial S^2} \right| \right) \delta t.$$

In [211] the authors used the standard no arbitrage arguments to deduce that the portfolio will earn the risk-free interest rate r,

$$E(\delta \Pi) = r\left(C - S\frac{\partial C}{\partial S}\right)\delta t.$$

They derived the PDE for option pricing with transaction costs as:

$$\frac{\partial C}{\partial t} + \frac{1}{2}\sigma^2 S^2 \frac{\partial^2 C}{\partial^2 S} + rS\frac{\partial C}{\partial S} - rC - \kappa\sigma S^2 \sqrt{\frac{2}{\pi \delta t}}\left|\frac{\partial^2 C}{\partial S^2}\right| = 0, \qquad (15.2)$$

on the domain $(S, T) \in (0, \infty) \times (0, T)$ with terminal condition

$$C(S, T) = \max(S - E, 0), \quad S \in (0, \infty) \qquad (15.3)$$

for European Call options with strike price E and a suitable terminal condition for European Puts.

The portfolio is considered to be revised every δt where δt is a noninfinitesimal fixed time step. This approach is now classified into the so-called local in time hedging strategy. Equation (15.2) is claimed as one of the first nonlinear PDEs in finance [97]. It is also one of the most studied in finance; we refer to [105] for analytical solution and numerical implementation and to [206] for asymptotic analysis for this model and two other models in the presence of transaction costs. Theorem 1 in [105] proves that under the condition $2\kappa\sqrt{\frac{2}{\sigma^2 \pi \delta t}} < 1$, Eq. (15.2) has a solution for any option V with payoff $V(S, T) \approx \alpha S$ when $S \to \infty$. In general, all the option types used in practice have this kind of payoff.

The asset model used is presented in Section 15.4. When working with stochastic volatility models, the market is incomplete and contingent claims do not have unique prices. This is because every contingent claim can be replicated by means of a trading strategy. The classical approach is to "complete the market" by fixing a related tradeable asset and deriving the option price in terms of this asset as well as the underlying equity.

15.4 Stochastic Volatility Model with Transaction Costs

A basic assumption in modeling the equity using a geometric Brownian motion as described earlier is that the volatility is constant. Much of the literature today shows this is an unrealistic assumption. Any model where the volatility is random is called a stochastic volatility model. A possible alternative approach to stochastic volatility models is to use jump diffusion processes or more general Lévy processes. We do not consider jumps in this section as they will lead to nonlinear PDEs with an integral term, which are very hard to work with.

In this section we consider the stochastic volatility model

$$dS_t = \mu(S_t)dt + \sigma_t S_t dX_1(t), \tag{15.4}$$

$$d\sigma_t = \alpha(\sigma_t)dt + \beta\sigma_t dX_2(t). \tag{15.5}$$

where the two Brownian motions $X_1(t)$ and $X_2(t)$ are correlated with correlation coefficient ρ:

$$E(dX_1(t)dX_2(t)) = \rho\, dt \tag{15.6}$$

The stochastic volatility model considered is a modified Hull–White process [102, 209], which contains general drift terms in S and σ. These general drift terms do not influence the PDE derivation. We note that the process mentioned earlier may also be viewed as a generalization of the SABR process [86] which is the stochastic volatility model mostly used in the financial industry to capture the volatility smile in derivatives markets.

The market is arbitrage free and incomplete when using stochastic volatility models. The fundamental theorem of asset pricing [51, 88] guarantees no arbitrage if an equivalent martingale measure exists and completeness of the market if the equivalent martingale measure is unique.

In the case of stochastic volatility models (with the exception of the trivial case when the Brownian motions are perfectly correlated $\rho = \pm 1$), there exists an infinite number of equivalent martingale measures (see [71]), and therefore the market is arbitrage free but not complete, i.e. the traded asset price does not uniquely determine the derivative prices.

In [71], the authors fixed the price of a particular derivative as a given asset and expressed all the other derivative prices in terms of the price of the given derivative. For the present analysis an alternative is discussed in Section 15.5. This is the case when the volatility is a traded asset, e.g. for S&P500 and its associated volatility index (VIX). In this case we present a market completion solution, i.e. we will form a portfolio using this asset as well as the underlying and we will derive a PDE which may explicitly give the price of options in the option chain. However, the PDE obtained is nonlinear with a very different nonlinear structure from the classical market completion approach.

15.5 The PDE Derivation When the Volatility is a Traded Asset

The results in this section are applicable when there exists a proxy for the stochastic volatility which is actively traded. Proxies such as the intraday high–low range or the realized volatility are important objects for modeling financial asset prices and volatility. Good proxies increase forecast accuracy and improve parameter estimation for discrete time volatility models. An example of such a case in today's financial derivative market is the Standard and Poor 500 equity index (in fact the exchange traded fund that replicates it: either SPX or SPY) and the VIX. The VIX is a traded asset, supposed to represent the implied volatility of an option with strike price exactly at the money (i.e. strike price equal with the spot value of SPX) and with maturity exactly one month from the current date. The VIX is calculated using an interpolating formula from the (out of money, i.e. Call option with a strike price that is higher than the market price of the underlying asset or a Put option with a strike price that is lower than the market price of the underlying asset) options available and traded on the market. In our setting we view the VIX as a traded asset, a proxy for the value of the stochastic volatility process in the model we propose here. Using the traded volatility index as a proxy provides a further advantage. The problem of parameter estimation in the stochastic volatility specification (15.5) is much simpler since the volatility process becomes observable. The volatility distribution may be further estimated using a filtering methodology as described for an example in [71].

In the future, it is possible that more volatility indices will be traded on the market, and we denote in what follows S as the spot equity price and with σ the matching spot volatility. It is important that this σ be traded (sold and bought). In the present subsection we are considering the volatility index σ as a perfect proxy for the stochastic volatility. In depth analysis about the suitability of this assumption is beyond the scope of the book.

We consider a portfolio Π that contains one option, with value $V(S, \sigma, t)$ and quantities Δ and Δ_1 of S and σ, respectively. That is,

$$\Pi = V - \Delta S - \Delta_1 \sigma. \tag{15.7}$$

We apply Itô's formula to get the dynamics of V. A derivation to find the portfolio dynamics is presented in Section 15.5.2. Applying this derivation we obtain the change in value of the portfolio Π as

$$d\Pi = \left(\frac{\partial V}{\partial t} + \frac{1}{2} \sigma^2 S^2 \frac{\partial^2 V}{\partial S^2} + \frac{1}{2} \beta^2 \sigma^2 \frac{\partial^2 V}{\partial \sigma^2} + \rho \sigma^2 \beta S \frac{\partial^2 V}{\partial S \partial \sigma} \right) dt$$
$$+ \left(\frac{\partial V}{\partial S} - \Delta \right) dS + \left(\frac{\partial V}{\partial \sigma} - \Delta_1 \right) d\sigma - \kappa S |v| - \kappa_1 \sigma |v_1|,$$

where $\kappa S |v|$ and $\kappa_1 \sigma |v_1|$ represent the transaction cost associated with trading v of the main asset S and v_1 of the volatility index σ during the time step δt. It is important to note (see Section 15.5.2) that this equation is an approximation to the exact dynamics of the portfolio. Nevertheless, even though Δ and Δ_1 are treated as constants, this derivation is correct and based on the self-financing property of the portfolio.

15.5 The PDE Derivation When the Volatility is a Traded Asset

The costs for trading S and σ are different and proportional with quantity transacted. We use k, k_1 to denote cost and v and v_1 to denote quantity transacted, respectively, for S and σ. We choose Δ and Δ_1 which are the quantities of stock volatility to be owned every time portfolio rebalancing is performed as the solutions of

$$\left(\frac{\partial V}{\partial S} - \Delta\right) = 0,$$

and

$$\left(\frac{\partial V}{\partial \sigma} - \Delta_1\right) = 0, \text{ respectively.}$$

This choice once again eliminates the drift terms and the portfolio dynamics become

$$d\Pi = \left(\frac{\partial V}{\partial t} + \frac{1}{2}\sigma^2 S^2 \frac{\partial^2 V}{\partial S^2} + \frac{1}{2}\beta^2 \sigma^2 \frac{\partial^2 V}{\partial \sigma^2} + \rho\sigma^2 \beta S \frac{\partial^2 V}{\partial S \partial \sigma}\right) dt - \kappa S|v| - \kappa_1 \sigma |v_1|.$$

(15.8)

15.5.1 The Nonlinear PDE

In Section 15.2 we pointed out that transaction cost is the cost incurred in making an economic exchange of some sort, or in other words the cost of participating in a market. In this subsection, we will investigate the costs associated with trading assets present in the market, perform a detailed analysis of the cost associated with trading S, and find an approximate value for the quantities traded, i.e. v and v_1.

If the number of assets held short at time t is

$$\Delta_t = \frac{\partial V}{\partial S}(S, \sigma, t),$$

(15.9)

after a time step δt and rehedging, the number of assets we hold short is

$$\Delta_{t+\delta t} = \frac{\partial V}{\partial S}(S + \delta S, \sigma + \delta \sigma, t + \delta t).$$

Since the time step δt is assumed small, the changes in asset and the volatility are also small, and applying the Taylor's formula to expand $\Delta_{t+\delta t}$ yields

$$\Delta_{t+\delta t} \simeq \frac{\partial V}{\partial S}(S, \sigma, t) + \delta t \frac{\partial^2 V}{\partial t \partial S}(S, \sigma, t) + \delta S \frac{\partial^2 V}{\partial S^2}(S, \sigma, t) + \delta \sigma \frac{\partial^2 V}{\partial \sigma \partial S}(S, \sigma, t) + \ldots$$

Since $\delta S = \sigma S \delta X_1 + \mathcal{O}(\delta t)$ and $\delta \sigma = \beta \sigma \delta X_2 + \mathcal{O}(\delta t)$,

$$\Delta_{t+\delta t} \simeq \frac{\partial V}{\partial S} + \sigma S \delta X_1 \frac{\partial^2 V}{\partial S^2} + \beta \sigma \delta X_2 \frac{\partial^2 V}{\partial \sigma \partial S}.$$

(15.10)

Subtracting (15.9) from (15.10), we find the number of assets traded during a time step:

$$v = \sigma S \delta X_1 \frac{\partial^2 V}{\partial S^2} + \beta \sigma \delta X_2 \frac{\partial^2 V}{\partial \sigma \partial S}.$$

(15.11)

Note that v is a random variable. We base our estimation of quantity traded on the expectation of this variable, and we use it to calculate the expected transaction cost. Since X_1 and X_2 are correlated Brownian motions, we consider Z_1 and Z_2 two independent normal variables with mean 0 and variance 1, and thus we may write the distribution of X_1, X_2 as

$$\delta X_1 = Z_1 \sqrt{\delta t}$$
$$\delta X_2 = \rho Z_1 \sqrt{\delta t} + \sqrt{1-\rho^2} Z_2 \sqrt{\delta t}.$$

Substituting these expressions in v and denoting

$$\alpha_1 = \sigma S \sqrt{\delta t} \frac{\partial^2 V}{\partial S^2} + \beta \sigma \rho \sqrt{\delta t} \frac{\partial^2 V}{\partial \sigma \partial S}$$
$$\beta_1 = \beta \sigma \sqrt{1-\rho^2} \sqrt{\delta t} \frac{\partial^2 V}{\partial \sigma \partial S}, \tag{15.12}$$

we write the change in the number of shares over a time step δt as

$$v = \alpha_1 Z_1 + \beta_1 Z_2.$$

We calculate the expected value of the transaction costs associated with trading the asset S:

$$E[\kappa S|v| \mid S] = \sqrt{\frac{2}{\pi}} \kappa S \sqrt{\alpha_1^2 + \beta_1^2}.$$

Analyzing the transaction costs associated with trading the volatility index σ proceeds in an entirely similar way and produces a similar formula to (15.11):

$$v_1 = \sigma S \delta X_1 \frac{\partial^2 V}{\partial S \partial \sigma} + \beta \sigma \delta X_2 \frac{\partial^2 V}{\partial \sigma^2}. \tag{15.13}$$

Therefore (15.8) leads to the nonlinear PDE

$$\frac{\partial V}{\partial t} + \frac{1}{2}\sigma^2 S^2 \frac{\partial^2 V}{\partial S^2} + \frac{1}{2}\beta^2 \sigma^2 \frac{\partial^2 V}{\partial \sigma^2} + \rho \sigma^2 \beta S \frac{\partial^2 V}{\partial S \partial \sigma} + rS \frac{\partial V}{\partial S} + r\sigma \frac{\partial V}{\partial \sigma} - rV$$
$$- \kappa S \sqrt{\frac{2}{\pi \delta t}} \sqrt{\sigma^2 S^2 \left(\frac{\partial^2 V}{\partial S^2}\right)^2 + 2\rho \beta \sigma^2 S \frac{\partial^2 V}{\partial S^2} \frac{\partial^2 V}{\partial S \partial \sigma} + \beta^2 \sigma^2 \left(\frac{\partial^2 V}{\partial S \partial \sigma}\right)^2}$$
$$- \kappa_1 \sigma \sqrt{\frac{2}{\pi \delta t}} \sqrt{\sigma^2 S^2 \left(\frac{\partial^2 V}{\partial S \partial \sigma}\right)^2 + 2\rho \beta \sigma^2 S \frac{\partial^2 V}{\partial S \partial \sigma} \frac{\partial^2 V}{\partial \sigma^2} + \beta^2 \sigma^2 \left(\frac{\partial^2 V}{\partial \sigma^2}\right)^2} = 0. \tag{15.14}$$

The two final radical terms in the resulting PDE mentioned earlier are coming from transaction costs. As noted in the first section, in this equation δt is a noninfinitesimal fixed time step not to be taken very close to zero. It is the time period for rebalancing and again if it is too small, this term will explode the solution of the equation.

In general for PDE models in finance, this is a terminal value problem. The specific boundary condition depends on the particular type of option priced but in all cases is expressed at $t = T$ the maturity of the option. For example, for a European Call the condition is $V(S, \sigma, T) = \max(S - K, 0)$ for all σ, where K is the particular option's strike. The general treatment presented in the next subsection is applicable to any option with boundary condition at maturity T, i.e. a function of S_T and T only. In fact the theorems stated apply to options whose payoff value is a function of σ_T as well. This is very valuable for certain types of nonvanilla options such as variance swaps.

The next subsection is devoted to the study of this type of nonlinear equations. In particular, we transform the final value boundary problem (FVBP) to an initial value boundary problem (IVBP) by changing the time variable from t to $\tau = T - t$. Note that this change will only modify the time derivative $\partial V/\partial t$ which becomes negative in the IVBP.

15.5.2 Derivation of the Option Value PDEs in Arbitrage Free and Complete Markets

In this subsection we present the correct derivation of portfolio dynamics used when deriving the PDE (15.8), i.e.

$$d\Pi = \left(\frac{\partial V}{\partial t} + \frac{1}{2}\sigma^2 S^2 \frac{\partial^2 V}{\partial S^2} + \frac{1}{2}\beta^2 \sigma^2 \frac{\partial^2 V}{\partial \sigma^2} + \rho \sigma^2 \beta S \frac{\partial^2 V}{\partial S \partial \sigma} \right) dt - \kappa S |v| - \kappa_1 \sigma |v_1|$$

and the nonlinear PDE (15.14), i.e.

$$\frac{\partial V}{\partial t} + \frac{1}{2}\sigma^2 S^2 \frac{\partial^2 V}{\partial S^2} + \frac{1}{2}\beta^2 \sigma^2 \frac{\partial^2 V}{\partial \sigma^2} + \rho \sigma^2 \beta S \frac{\partial^2 V}{\partial S \partial \sigma} + rS \frac{\partial V}{\partial S} + r\sigma \frac{\partial V}{\partial \sigma} - rV$$

$$- \kappa S \sqrt{\frac{2}{\pi \delta t}} \sqrt{\sigma^2 S^2 \left(\frac{\partial^2 V}{\partial S^2}\right)^2 + 2\rho \beta \sigma^2 S \frac{\partial^2 V}{\partial S^2} \frac{\partial^2 V}{\partial S \partial \sigma} + \beta^2 \sigma^2 \left(\frac{\partial^2 V}{\partial S \partial \sigma}\right)^2}$$

$$- \kappa_1 \sigma \sqrt{\frac{2}{\pi \delta t}} \sqrt{\sigma^2 S^2 \left(\frac{\partial^2 V}{\partial S \partial \sigma}\right)^2 + 2\rho \beta \sigma^2 S \frac{\partial^2 V}{\partial S \partial \sigma} \frac{\partial^2 V}{\partial \sigma^2} + \beta^2 \sigma^2 \left(\frac{\partial^2 V}{\partial \sigma^2}\right)^2} = 0.$$

Suppose that we want to price a claim V which at time t is dependent on S, σ, and t. We note that the same approach works if the contingent claim V is contingent on any set of n traded assets $S_1(t), \ldots, S_n(t)$, but for clarity we use the specific case presented in this chapter.

The market contains two traded assets $S(t)$ and $\sigma(t)$, which have some specific dynamics irrelevant to this derivation as well as a risk-free account that earns the risk-free interest rate r. This risk-free account is available from the moment $t = 0$ when the portfolio is constructed. Specifically, one share of this money market account solves the ODE

$$dM(t) = rM(t)dt.$$

Integrating both sides of the aforementioned ODE and solving for $M(t)$, we obtain the solution:

$$M(t) = Ae^{rt},$$

where A is a constant.

Suppose we form a portfolio (any portfolio) containing shares in these assets and in the money account (i.e. an interest-bearing account that typically pays a higher interest rate than a savings account and which provides the account holder with limited check-writing ability). Divide the interval $[0, t]$ into intervals with endpoints $0 = t_0 < t_1 < \cdots < t_N = T$, and for simplicity assume that the times are equally spaced at intervals δt wide. Suppose that at one of these times t_k our portfolio has value:

$$X(k) = \Delta(k)S(k) + \Delta_1(k)\sigma(k) + \Gamma(k)M(k),$$

where $\Delta(k)$ and $\Delta_1(k)$ are the number of shares of respective assets while $\Gamma(k)$ is the number of shares of the riskless asset we own.

Suppose that at the next time t_{k+1} we need to rebalance this portfolio to contain exactly some other weights $\Delta(k+1)$ and $\Delta_1(k+1)$. To this purpose, we need to trade the assets, and thus we pay transaction costs depending on the differences of the type $\Delta(k+1) - \Delta(k)$ as well as on the price of the specific asset traded.

Here we make the assumption that the portfolio is self-financing. This means that any extra or missing monetary value resulting from rebalancing the portfolio will be put or borrowed from the money account. Mathematically, we need to have the following two quantities equal:

$$X_{k+1} = \Delta(k)S(k+1) + \Delta_1(k)\sigma(k+1) + \Gamma(k)M(k+1)$$
$$X(k+1) = \Delta(k+1)S(k+1) + \Delta_1(k+1)\sigma(k+1)$$
$$+ \Gamma(k+1)M(k+1) - v(k+1).$$

In this expression all transaction costs incurred at time t_{k+1} are lumped into the term $v(k+1)$. Setting the two quantities equal and rearranging the terms gives

$$(\Delta(k+1) - \Delta(k))S(k+1) + (\Delta_1(k+1) - \Delta_1(k))\sigma(k+1)$$
$$+ (\Gamma(k+1) - \Gamma(k))M(k+1) - v(k+1) = 0.$$

In this equation we add and subtract $S(k)(\Delta(k+1) - \Delta(k))$ and $\sigma(k)(\Delta_1(k+1) - \Delta_1(k))$, which gives the following self-financing condition:

$$S(k)(\Delta(k+1) - \Delta(k)) + (\Delta(k+1) - \Delta(k))(S(k+1) - S(k))$$
$$+ \sigma(k)(\Delta_1(k+1) - \Delta_1(k)) + (\Delta_1(k+1) - \Delta_1(k))(\sigma(k+1) - \sigma(k))$$
$$+ (\Gamma(k+1) - \Gamma(k))M(k) + (\Gamma(k+1) - \Gamma(k))(M(k+1) - M(k)) - v(k+1) = 0$$
(15.15)

15.5 The PDE Derivation When the Volatility is a Traded Asset

The next step requires some explanation. We plan to sum these expressions over k and to take the mesh of partition $\max_k |t_{k+1} - t_k|$ to converge to zero. However, we need to deal with the transaction costs term. If the rebalancing length of the interval goes to 0, then the transaction costs become infinite. This is why it is important to realize that the actual rebalancing needs to be done at fixed points in time length with stepsize δt. Because of this, the limit is an approximation of the PDE dynamic.

If we take the limit while at the same time bounding the transaction costs, we obtain the stochastic integrals for all the expressions in (15.15). Expressing the integrals in differential form for the sake of compactness, we obtain the continuous time self-financing condition

$$S(t)d\Delta(t) + d\langle\Delta, S\rangle_t + \sigma(t)d\Delta_1(t) + d\langle\Delta_1, \sigma\rangle_t$$
$$+ M(t)d\Gamma(t) + d\langle\Gamma, M\rangle_t - \nu(\delta t) \approx 0 \tag{15.16}$$

For no transaction costs (last term zero), this condition is exact for any portfolio which is self-financing and for any stochastic dynamics of the weights and assets. In the presence of transaction costs, we need to bound the total transaction costs over the interval δt when the subinterval length approaches 0. For this reason Eq. (15.16) is only approximately satisfied when dealing with transaction costs. The notation $\nu(\delta t)$ was used to bound the transaction costs over the interval δt. The expected value of this term will be calculated.

The condition (15.16) is valid for any self-financing portfolios. Next we form a specific portfolio, one that will replicate the payoff of the contingent claim V at time T. Such a portfolio involves stochastic weights and has the form

$$\Pi(t) = V(t) - \Delta(t)S(t) - \Delta_1(t)\sigma(t),$$

since we replicate using only the underlying assets S and σ. The weights are stochastic, and they are suitably chosen to replicate the contingent claim V. The dynamics of the portfolio $\Pi(t)$ may be derived by applying the Itô's lemma to obtain:

$$d\Pi(t) = dV(t) - \Delta(t)dS(t) - S(t)d\Delta(t) - d\langle\Delta, S\rangle_t$$
$$- \Delta_1(t)d\sigma(t) - \sigma(t)d\Delta_1(t) - d\langle\Delta_1, \sigma\rangle_t \tag{15.17}$$

Now, the idea is that the latter two terms will disappear when we use the self-financing condition. More specifically, since the suitable choices of Δ terms allow us to replicate the option value $V(t)$, the amount in the money account at time t is exactly $\Pi(t)$. Therefore, the number of shares held in the money account at any time is $\Gamma(t) = \frac{\Pi(t)}{M(t)}$. The last two terms in (15.17) are substituted using the self-financing condition (15.16). In so doing, we need to calculate the terms $M(t)d\Gamma(t) + d\langle\Gamma, M\rangle_t$ for this specific $\Gamma(t) = \frac{\Pi(t)}{M(t)}$. Here it pays to know that $M(t)$ as well as $1/M(t) = e^{-rt}$ are deterministic, and therefore the terms $d\langle\Pi/M, M\rangle_t$ and $d\langle\Pi, 1/M\rangle_t$ vanish in the resulting expression. Furthermore,

$$M(t)d\Gamma(t) = M(t)d\frac{\Pi(t)}{M(t)} = d\Pi(t) + \Pi(t)(-r)dt + M(t)d\left\langle\Pi, \frac{1}{M}\right\rangle_t,$$

and as mentioned the last term is zero. After we perform the calculations and some simple algebras, we end up with the expression

$$dV(t) - \sum_{i=1}^{n} \Delta_i(t)dS_i(t) - r\Pi(t)dt - v(\delta t) = 0. \tag{15.18}$$

The next step is to use suitable replicating weights to make the stochastic integrals disappear by equating all the terms multiplying dt which gives the PDEs used in Section 15.5.

The derivation presented in this subsection is clearly valid if we use discrete time. However, the self-financing condition needs to hold at all times when we rebalance the portfolio. In this subsection we used the continuous time condition (15.16) instead of the discrete one (15.15) simply due to the convenience of working with Itô's lemma and thus vanishing quadratic variations in (15.17), instead of higher dt terms in the Taylor expansion. However, if we go the long route and replace (15.16) with its discrete counterpart from the proof of Itô's lemma, the same terms as in the continuous version will also disappear in the discrete time expression. But, since we do not rebalance inside intervals of width δt, there are no further transaction costs while taking subpartitions of these original intervals. Therefore when the mesh of the subpartitions is converging to zero, we will obtain the final Eq. (15.18).

15.6 Problems

1. Assume that Π is the value of the hedging portfolio and $C(S, t)$ is the value of the option and assuming that the underlying asset follows the process:

   $$\delta S = \mu S \delta t + \sigma S \Phi \sqrt{\delta t}, \tag{15.19}$$

   where Φ is a standard normal random variable, μ is the drift term, and σ is the diffusion coefficient. If the portfolio is given by $\Pi = C - \Delta S$, then show that the change in portfolio value is given by:

   $$\delta\Pi = \sigma S\left(\frac{\partial C}{\partial S} - \Delta\right)\Phi\sqrt{\delta t} + \left(\frac{1}{2}\sigma^2 S^2 \frac{\partial^2 C}{\partial S^2} + \mu S\frac{\partial C}{\partial S} + \frac{\partial C}{\partial t} - \mu\Delta S\right)\delta t - \kappa S|\nu|.$$

   (Hint: Assume that the quantity of shares Δ is kept constant)
2. Discuss why a market is arbitrage free and incomplete when using stochastic volatility models.
3. Using the same approach as in Section 15.2,
   (a) Obtain the PDE corresponding to a model that includes only transaction costs.

(b) The PDE corresponding to a model that includes stochastic volatility and transaction cost.

4. Using the same approach as in Section 11.5, obtain the system of equations corresponding to a model where we have n option value functions with no jumps.

5. Explain with an example why when working with stochastic volatility models, the market is incomplete and contingent claims do not have unique prices.

6. Given a portfolio that replicates the payoff of the contingent claim V at time T, i.e.

$$\Pi(t) = V(t) - \Delta(t)S(t) - \Delta_1(t)\sigma(t),$$

applying the Itô's lemma, prove that the dynamics of the portfolio $\Pi(t)$ is:

$$d\Pi(t) = dV(t) - \Delta(t)dS(t) - S(t)d\Delta(t) - d\langle \Delta, S \rangle_t$$
$$- \Delta_1(t)d\sigma(t) - \sigma(t)d\Delta_1(t) - d\langle \Delta_1, \sigma \rangle_t$$

7. In Section 15.5.2, it was mentioned that mathematically, we need to have the following two quantities equal:

$$X_{k+1} = \Delta(k)S(k+1) + \Delta_1(k)\sigma(k+1) + \Gamma(k)M(k+1)$$
$$X(k+1) = \Delta(k+1)S(k+1) + \Delta_1(k+1)\sigma(k+1) + \Gamma(k+1)M(k+1)$$
$$- v(k+1).$$

Explain the reasoning behind this idea.

16

Factor and Copulas Models

16.1 Introduction

In this chapter we will discuss factor and copulas models. Factor models are very useful for analyzing high-dimensional response data with dependence coming from unobservable variables or factors. There are several types of factor models, but they all are constructed using factor analysis techniques and can be divided into three basic categories: statistical, macroeconomic, and fundamental. In the first part of this chapter, we will discuss all three categories. The second half of this chapter will focus on copulas models. A copula is a cumulative distribution function connecting multivariate marginal distributions in a specified form. Copulas are popular in high-dimensional statistical applications since they allow one to easily model and estimate the distribution of random vectors by estimating marginals and copula separately. An advantage of the copula approach is the fact that it helps achieve the goal of data reduction by locating the common factor variables affecting the selected data series. The copulas can better assist in identifying the dependence relationships and model the chosen financial data.

16.2 Factor Models

A factor model relates the return on an asset (be it a stock, bond, or mutual fund) to the values of a limited number of factors, with the relationship described by a linear equation. In a general form, such a model can be written as:

$$\begin{aligned} R_{it} &= \alpha_i + \beta_{1i} f_{1t} + \ldots + \beta_{ki} f_{kt} + \epsilon_{it} \\ &= \alpha_i + \beta_i^T f_t + \epsilon_{it} \end{aligned} \quad (16.1)$$

Quantitative Finance, First Edition. Maria C. Mariani and Ionut Florescu.
© 2020 John Wiley & Sons, Inc. Published 2020 by John Wiley & Sons, Inc.

where,

$$f_t = \begin{pmatrix} f_{1t} \\ \vdots \\ f_{kt} \end{pmatrix} \text{ and } \beta_i = \begin{pmatrix} \beta_{1i} \\ \vdots \\ \beta_{ki} \end{pmatrix}$$

The variables $R_{it}, f_{1t}, \ldots, f_{kt}$ and ϵ_{it} are generally not known in advance. The values of such stochastic variables are uncertain. Thus we do not know what the return on the asset R_{it} will be, since we do not know the values that the factors $f_{1t}, \ldots, f_{kt}$ will take on nor do we know the amount of the asset's return that will come from other sources ϵ_{it}. On the other hand, we do know (or at least assume that we know) the sensitivities of the return on the asset to each of the factors $\beta_{1i}, \ldots, \beta_{ki}$ – these are deterministic (not subject to uncertainty).

We will assume that f_t are stationary with the following properties:

$$E(f_t) = \mu_f$$
$$Cov(f_t) = \Omega_f$$
$$Cov(f_{jt}, \epsilon_{it}) = 0, \text{ for all } j, i$$
$$Cov(\epsilon_{it}, \epsilon_{js}) = \sigma_i^2, \text{ if } i = j, \text{ else } 0.$$

These properties make the linear factor model powerful in the sense that it rules out many possible combinations of outcomes. But greater power comes at a cost. The more restrictive a model, (it has to satisfy the aforementioned properties), the greater the chance that it may be inconsistent with reality.

Before we proceed, we define some notations that will be used throughout this section.

$$R_t = \begin{pmatrix} R_{1t} \\ \vdots \\ R_{Nt} \end{pmatrix}, \quad R_i = \begin{pmatrix} R_{i1} \\ \vdots \\ R_{iT} \end{pmatrix} \text{ and } R_t = \begin{pmatrix} R_{11} & \cdots & \cdots & R_{N1} \\ \vdots & \cdots & \cdots & \vdots \\ R_{1T} & \cdots & \cdots & R_{NT} \end{pmatrix} \text{ all data.}$$

16.2.1 Cross-Sectional Regression

Linear regression is one of the most basic tools in statistics and is widely used throughout finance and economics. Linear regressions success is owed to two important features, namely, the availability of simple closed-form estimators and the ease and directness of interpretation.

Linear regression expresses a dependent variable as a linear function of independent variables and an error term.

$$R_t = \beta_1 f_{1,t} + \beta_2 f_{2,t} + \ldots + \beta_k f_{k,t} + \epsilon_t \tag{16.2}$$

where R_t is known as the regressand or dependent variable. The k variables $f_{1,t}, \ldots, f_{k,t}$ are known as the regressors or independent variables. The constants $\beta_1, \ldots, \beta_k$ are the regression coefficients and ϵ_t is known as the error term.

Linear regression allows coefficients to be interpreted. Specifically, the effect of a change in one variable can be examined without changing the others. Regression analysis helps determine if models containing all of the information relevant for determining R_t is of significant interest or not. This feature provides the mechanism to interpret the coefficient on an independent variable as the unique effect of that regressor, a feature that makes linear regression very attractive.

The concepts of linear regression will be explored in the context of a cross-sectional regression of returns on a set of factors thought to capture systematic risk. Cross-sectional regressions in financial econometrics date back at least to the Capital Asset Pricing Model (CAPM) (see [147, 186]) a model formulated as a regression of individual assets excess returns of the market. The CAPM also describes the relationship between systematic risk and expected return for assets, particularly stocks. It is used throughout finance for the pricing of risky securities, generating expected returns for assets given the risk of those assets, and calculating costs of capital.

A cross-sectional regression is defined as a type of regression in which the explained and explanatory variables are associated with one period or point in time. The cross-sectional regression analysis is different from the time-series regression or longitudinal regression in which the variables are considered to be associated with a sequence of points in time. Let

$$R_t = \alpha + Bf_t + \epsilon_t \tag{16.3}$$

where the dependent variable f_t is defined as

$$f_t = \begin{pmatrix} f_{1t} \\ \vdots \\ \vdots \\ f_{kt} \end{pmatrix},$$

the constant B is also defined as

$$B = \begin{pmatrix} \beta_1^T \\ \vdots \\ \beta_N^T \end{pmatrix} = \begin{pmatrix} \beta_{11} & \cdots & \cdots & \beta_{1k} \\ \vdots & \cdots & \cdots & \vdots \\ \beta_{N1} & \cdots & \cdots & \beta_{Nk} \end{pmatrix}$$

and finally,

$$\mathrm{Var}(f_t) = D = \begin{pmatrix} \sigma_1^2 & 0 & \cdots & 0 \\ 0 & \sigma_2^2 & \cdots & 0 \\ \vdots & \vdots & \ddots & \vdots \\ 0 & 0 & \cdots & \sigma_N^2 \end{pmatrix}$$

A cross-sectional regression would have as each data point an observation on a particular individual's money holdings, income, and perhaps other variables at a single point in time, and different data points would reflect different individuals at the same point in time.

16.2.2 Expected Return

Expected returns are profits or losses that investors expect to earn based on anticipated rates of return. Often, the realized returns are different than the expected returns due to the volatility of the markets. Given

$$E[R_{it}] = \alpha_i + \beta_i^T E[f_t] \tag{16.4}$$

where $\beta_i^T E[f_t]$ is the explained return due to systematic factors, Equation (16.4) can be rewritten as:

$$\alpha_i = E[R_{it}] - \beta_i^T E[f_t]$$

which is the unexpected return specific to a given asset.

The covariance structure for $R_t = \alpha + Bf_t + \epsilon_t$ which implies that $\mathrm{Var}(R_t) = B\Omega_f B^T + D$. Thus individually,

$$\mathrm{Var}(R_{it}) = \beta_i^T \Omega_f \beta_i + \sigma_i^2$$
$$\mathrm{Cov}(R_{it}, R_{jt}) = \beta_i^T \Omega_f \beta_j$$

We recall that the expected return is the profit that an investor anticipates on investment. For example, it may be the expected return on a bond if the bond pays out the maximum return at maturity. In fact, the expected return is a possible return on investment, and the outcomes are continuous, i.e. between 0 and infinity.

We consider a portfolio analysis as follows. Let $w = (w_1, \ldots, w_N)$ be a vector of weights such that $\sum w_i = 1$. If R_t is a vector or simple returns, then

$$R_t = \alpha + Bf_t + \epsilon_t$$
$$R_{P,t} = w^T R_t = w^T \alpha + w^T Bf_t + w^T \epsilon_t$$
$$\mathrm{Var}(R_{P,t}) = w^T B\Omega_f B^T w + w^T Dw$$

In portfolio analysis, the returns of a portfolio are the sum of each potential return multiplied by the probability of occurrence or weight. With this in mind, one may look at the expected return to be the weighted average of potential returns of the portfolio, weighted by the potential returns of each asset class included in the portfolio.

We remark that for active portfolios, the weights are rebalanced at fixed intervals. However for static portfolios, the weights are fixed over time.

In the next subsection we study three different types of factor models, namely, macroeconomic factor model, fundamental factor model, and statistical factor model.

16.2.3 Macroeconomic Factor Models

Macroeconomic factor models are the simplest and most intuitive type of factor models. They use observable economic time series as measures of the pervasive factors in security returns. Some macroeconomic variables commonly used as factors are inflation, the percentage change in industrial production, and the excess return to long-term government bonds. The random return of each security is assumed to respond linearly to the macroeconomic shocks. The economic problems that we are interested in are as follows:

1) Choice of factors
2) Estimate β_i and σ_i^2 using regression
3) Estimate factor covariance Ω_f from history of factors

Example 16.2.1 (CAPM single factor model). As mentioned earlier in this section, CAPM describes the relationship between systematic risk and expected return for assets, particularly stocks. The CAPM single factor model can be defined as:

$$f_t = R_{Mt} - R_{Ft} \tag{16.5}$$

where R_{Mt} is the return on a market portfolio at time t and R_{Ft} is the risk-free rate at time t.

Example 16.2.2 (Multifactor models). Multifactor models are used to construct portfolios with certain characteristics, such as risk, or to track indexes. When constructing a multifactor model, it is difficult to decide how many and which factors to include. The authors [35] in their study extended the single factor model to multiple factors used for equity valuation. In [138] Lo extended the single factor model to hedge funds.

In all these instances, the factors are observable so they are very easy to estimate. A drawback to macroeconomic factor models is that they require identification and measurement of all the pervasive shocks affecting security returns.

16.2.4 Fundamental Factor Models

Fundamental factor models are the type of models that do not require time series regression. They rely on the empirical finding that the company attributes such as firm size, dividend yield, and book to market ratio.

For fundamental factor models, the following facts arise:

- Observe asset characteristics (for example, market capitalization, debt value, growth, etc.).
- Factor betas are constructed using the asset characteristics (i.e. B is known).
- f_t are calculated from R_t given B.

There are two approaches to address the concerns raised, namely, Bar Rosenberg (BARRA) approach and Eugene Fama and Kenneth French (FAMA-FRENCH) approach.

The BARRA approach is a multifactor model that is used to measure the overall risk associated with a security relative to the market. The procedure for the BARRA approach are as follows:

1) Treat observable asset-specific attributes as factor betas which are assumed to be time invariant.
2) Factor realizations f_t are unobservable but are estimated. Thus the problem is to estimate them using T cross-sectional regression.

In the FAMA-FRENCH approach for all assets, a portfolio sorts them by some characteristic (high and low return) and then forms a portfolio by using the long top 1/5 performers and short bottom 1/5 performers and finally creates two factors, namely, high minus low (HML) and small minus big (SMB). These two factors are created in addition to the market factor: $R_{Mt} - R_{Ft}$. Thus we have the FAMA and FRENCH three-factor model.

The FAMA and FRENCH three-factor model is an asset pricing model that expands on the CAPM by adding size and value factors to the market risk factor in CAPM. This model takes into account the fact that the value and small-cap stocks outperform markets on a regular basis. By including these two additional factors, the model adjusts for the out-performance tendency, which is a better tool for evaluating manager performance.

16.2.5 Statistical Factor Models

Statistical factor models use various maximum likelihood and principal component-based factor analysis and cross-sectional/time series samples of security returns to identify the pervasive factors in returns. Given

$$R_t = \mu + Bf_t + \epsilon_t$$

with the properties that

$$\text{Cov}(f_t, \epsilon_t) = 0, \text{ for all } t$$
$$\text{Var}(f_t) = I_k(\text{uncorrelated})$$
$$\text{Var}(\epsilon_t) = D(\text{diagonal})$$
$$\text{Cov}(R_t) = BB^T + D$$

then

$$\text{Var}(R_{it}) = \sum_{j=1}^{k} \beta_{ij}^2 + \sigma_i^2$$

where $\sum_{j=1}^{k} \beta_{ij}^2$ is the variance due to common factors and σ_i^2 is the variance due to specific return.

Macroeconomic and statistical factor models both estimate a firm's factor beta by time series regression. Given the nature of security returns data, this limitation is substantial. The time series regression requires a long and stable history of returns for a security to estimate the factor betas accurately (please see [39] for more details).

We remark that factors and loadings are not unique. In order to demonstrate this fact, we take an orthogonal matrix H such that $H^T = H^{-1}$. Then we can write:

$$R_t = \mu + BHH^T f_t + \epsilon_t \tag{16.6}$$

Equation (16.6) can be rewritten as:

$$R_t = \mu + B^* f_t^* + \epsilon_t$$

where $B^* = BH$ and $f_t^* = H^T f_t$. So we totally have different factors and loadings. Therefore interpreting the factors may not be apparent until we find the proper notation. We define the following steps:

1) Estimate the factors using Principal Components Analysis (PCA).
2) Construct the factor realizations f_t.
3) Rotate coordinates to enhance interpretation.

Recent literature provides testing procedures for the number of factors required. VARIMAX is a rotation technique where each factor has a small number of large loadings and a large number of small loadings.

In the next section we discuss other type of factor models such as the copula models.

16.3 Copula Models

Let X and Y be continuous random variables with distribution functions $F(x) = P(X \leq x)$ and $G(y) = P(Yy)$ and joint distribution function

$H(x,y) = P(X \leq x, Y \leq y)$. For every (x,y) in $[\infty, \infty]^2$ consider the point in $\mathbb{I}^3$ ($\mathbb{I} = [0,1]$) with coordinates $(F(x), G(y), H(x,y))$. This mapping from $\mathbb{I}^2$ to $\mathbb{I}$ is a copula. Copulas are also known as dependence functions or uniform representations.

Definition 16.3.1 A two-dimensional copula is a function $C : \mathbb{I}^2 \to \mathbb{I}$ such that:

(C1) $C(0,x) = C(x,0) = 0$ and $C(1,x) = C(x,1) = x$ for all $x \in \mathbb{I}$
(C2) C is 2-increasing: for $a, b, c, d \in \mathbb{I}$ with $a \leq b$ and $c \leq d$,

$$V_C([a,b] \times [c,d]) = C(b,d) - C(a,d) - C(b,c) + C(a,c) \geq 0.$$

The function V_C in item (C2) of Definition 16.3.1 is called the C-volume of the rectangle $[a,b] \times [c,d]$. Equivalently, a copula is the restriction to the unit square $\mathbb{I}^2$ of a bivariate distribution function whose margins are uniform on $\mathbb{I}$. We remark that a copula C induces a probability measure on $\mathbb{I}^2$ via $V_C([0,u] \times [0,v]) = C(u,v)$. Please see [156] for more details.

We recall that one may obtain any marginal distribution from knowing the joint distribution $F(x_1, \ldots, x_k)$ or $f(x_1, \ldots, x_k)$. So given historical observations of a vector $(x_1, \ldots, x_k)$, one could construct a k-dimensional histogram. A problem that normally arises is the fact that the required data should be large before one can obtain a precise histogram. For example if we have 100 assets to construct a simple 5 level histogram for each asset, we will need a total of 500 bins. Assume at least 5 observations for each bin that is 2500 observations or 10 years of daily data and this is very expensive. On the other hand marginals are easy to estimate.

Next we state Sklar's theorem which enables one to use the marginals to construct the joint distribution. Sklar's theorem states that any multivariate joint distribution can be written in terms of univariate marginal distribution functions and a copula which describes the dependence structure between the variables. Below is a formal presentation of Sklar's theorem.

Theorem 16.3.1 (Sklar's theorem). Given $X_1, \ldots, X_k$ with cumulative distribution function $F(x_1, \ldots, x_k)$ and marginals $F_1, \ldots, F_k$, there exists a function $C : [0,1]^k \to [0,1]$ such that for all $(x_1, \ldots, x_k)$ in $\mathbb{R}^k$

$$F(x_1, \ldots, x_k) = C(F_1(x_1), \ldots, F_k(x_k))$$

where C is a copula function.

The theorem suggests that such a function exists and it is unique. It does not actually say what is the function. Clearly, C must have the following properties:

1) $C(x_1, \ldots, x_k)$ is increasing in each component.
2) $C(1, \ldots, \mu_i, 1, \ldots, 1) = \mu_i$ for all i.

An example of the copula function is the Normal copula, i.e.

$$C(x_1, \ldots, x_k) = \Phi_k(\Phi^{-1}(x_1), \ldots, \Phi^{-1}(x_k))$$

where Φ^{-1} is the inverse Normal (0,1) CDF and Φ_k is the CDF of a k-dimensional multihandle Normal with mean 0 and covariance matrix R.

In the next subsection, we briefly discuss two families of copulas, namely, Normal copulas and Archimedean copulas.

16.3.1 Families of Copulas

If one has a collection of copulas, then using Sklars theorem, we can construct bivariate distributions with arbitrary margins. Thus, for the purposes of statistical modeling, it is desirable to have a collection of copulas at ones disposal. Many examples of copulas can be found in literature; most are members of families with one or more real parameters (members of such families are often denoted $C_\theta, C_{\alpha,\beta}, C_{\alpha,\beta,\gamma}$, etc.). We now present a very brief overview of some parametric families of copulas.

16.3.1.1 Gaussian Copulas

Let $N_\rho(x, y)$ denote the standard bivariate normal joint distribution function with correlation coefficient ρ. Then C_ρ, the copula corresponding to N_ρ, is given by $C_\rho(u, v) = N_\rho(\Phi^{-1}(u), \Phi^{-1}(v))$, where Φ denotes the standard normal distribution function. Since there is no closed form expression for Φ^{-1}, there is no closed form expression for N_ρ. However, N_ρ can be evaluated approximately in order to construct bivariate distribution functions with the same dependence structure as the standard bivariate normal distribution function but with non-normal marginals. Please refer to [155] and references therein for more details.

16.3.1.2 Archimedean Copulas

Let ϕ be a continuous strictly decreasing function from $\mathbb{I}$ to $[0, \infty]$ such that $\phi(1) = 0$, and let $\phi^{[-1]}$ denote the "pseudo-inverse" of ϕ: $\phi^{[-1]}(t) = \phi^{-1}(t)$ for $t \in [0, \phi(0)]$ and $\phi^{[-1]}(t) = 0$ for $t \geq \phi(0)$. Then $C(u, v) = \phi^{[-1]}(\phi(u) + \phi(v))$ satisfies condition (C1) for copulas. If, in addition, ϕ is convex, then it can be shown [154] that C also satisfies the two-increasing condition (C2) and is thus a copula. Such copulas are called Archimedean. When $\phi(0) = \infty$, we say that C is strict, and when $\phi(0) < \infty$, we say that C is nonstrict. When C is strict, $C(u, v) > 0$ for all (u, v) in $(0, 1]^2$.

Archimedean copulas are widely used in finance and insurance due to their simple form and nice properties. Procedures exist for choosing a particular member of a given family of Archimedean copulas to fit a data set ([79, 204]). However, there does not seem to be a natural statistical property for random variables with an associative copula.

16.4 Problems

1. Discuss the implications of assuming that asset returns are determined according to a three-factor model in which each factor is the rate of return on a given portfolio. What considerations should determine how the three given factors are defined?
2. Suppose the total volatility of returns on a stock is 25%. A linear model with two risk factors indicates that the stock has betas of 0.8 and 1.2 on the two risk factors. The factors have volatility 15 and 20%, respectively, with correlation of 0.5. How much of the stocks volatility can be attributed to the risk factors and how large is the stocks specific risk?
3. Suppose a portfolio is invested in only three assets with weights $0.25, 0.75$, and 0.50. Each asset has a factor model representation with the same two risk factors as in problem 2, and the betas with respect to the two factors are: $(0.2; 1.2)$ for asset 1, $(0.9; 0.2)$ for asset 2, and $(1.3; 0.7)$ for asset 3. What is the volatility due to the risk factors (i.e. the systematic risk) for this portfolio?
4. For the three types of factor models, discuss and compare their explanatory power.
5. In portfolio analysis, explain the reason why in active portfolios, the weights are rebalanced at fixed intervals.
6. Show that the function $\Pi(x, y) = xy$ is a copula. Hint: Show that it satisfies conditions ($C1$) and ($C2$).

Part IV

Fixed Income Securities and Derivatives

17

Models for the Bond Market

17.1 Introduction and Notations

This chapter serves as an introduction to bond market models. The main instrument studied is a bond price which is entirely different from stock price. A bond is an investment product that is issued by corporate and governmental entities to raise capital to finance and expand their projects and operations. The bond market is where debt securities are issued and traded. The differences in the bond and stock market lie in the manner in which the different products are sold and the risk involved in dealing with both markets. For instance, the stock market has stock exchanges where stocks are bought and sold. However, the bond market does not have a central trading place for bonds; rather bonds are sold mainly over the counter (OTC). The other difference between the stock and bond market is the risk involved in investing in both. Investing in bond market is usually less risky than investing in a stock market because the bond market is not as volatile as the stock market is. Throughout this chapter the notional principal value is taken as 1 (unit). The notional principal amount is the predetermined dollar amounts on which the exchanged interest payments are based. The notional principal never changes hands in the transaction, which is why it is considered notional, or theoretical.

17.2 Notations

We begin this section with some definitions and notations that would be used throughout this chapter.

Short Rate Let $r(t)$ be the (annualized) instantaneous return rate for \$1 invested today. In other words it is the rate at which an entity (bank primarily) can borrow money for an infinitesimally short period of time from the current time t. In practice this is really hard to estimate. Typically, its realization is directly replaced with some Treasury bill (T-Bill) (i.e. a short-term debt

Quantitative Finance, First Edition. Maria C. Mariani and Ionut Florescu.
© 2020 John Wiley & Sons, Inc. Published 2020 by John Wiley & Sons, Inc.

obligation) rate for the option pricing or the current (daily) London Interbank Offered Rate (LIBOR) rate. LIBOR is a benchmark rate that some of the world's leading banks charge each other for short-term loans.

Zero Coupon Bond Price $P(t, s)$ (with $t < s$) is the current price at time t of an instrument that pays 1 at maturity s (and nothing during its lifetime – no coupon payments), in other words, a security with a single cash flow equal to face value at maturity.

Spot Rate Associated with a Zero Coupon Bond $R(t, s)$ is the associated rate that the zero coupon bond yields throughout its lifetime:

$$P(t, s) = e^{-(s-t)R(t,s)} \quad \text{or} \quad R(t, s) = -\frac{1}{s-t} \ln P(s, t). \tag{17.1}$$

It is also called the yield rate for short.

Forward Contract on the Zero Coupon Bond A forward contract is an agreement to buy an asset at a future settlement date at a forward price specified today. The forward contract on the zero coupon bond $P(t, T, s)$ is the current price of an instrument that at time $T < s$ delivers a bond that pays 1 at maturity s. Because of no arbitrage we must have:

$$P(t, T, s) = \frac{P(t, s)}{P(t, T)}, \tag{17.2}$$

where $P(t, s) = e^{-(s-t)R(t,s)}$ and $P(t, T) = e^{-(T-t)R(t,T)}$ are regular zero coupon bond prices.

Forward Rate $f(t, T, s)$, is the predetermined delivery price for an underlying asset as decided by the buyer and the seller of the forward contract. Mathematically,

$$P(t, T, s) = e^{-(s-T)f(t,T,s)}. \tag{17.3}$$

But this has no meaning since it is some sort of average yield rate as viewed from today over the interval $[T, s]$. It is much more convenient to work with a similar notion with $r(t)$ but into the future.

Instantaneous Forward Rate $f(t, \tau)$ is the yield rate of a fake forward contract (i.e. forward contract is a customized contract between two parties to buy or sell an asset at a specified price on a future date) that starts at τ and matures very quickly afterwards at say $\tau + \Delta \tau$. Given this notion we can unify the price of the zero coupon and the forward price of a zero coupon in the following way:

$$P(t, s) = e^{-\int_t^s f(t,\tau)d\tau}$$

$$P(t, T, s) = e^{-\int_T^s f(t,\tau)d\tau}.$$

Since everything is calculated at time $t = 0$, we note that it is enough to model $f(\tau) = f(0, \tau)$. Furthermore, since in fact the rate $r(t)$ depends on the current time t, it is not an issue if it is recalibrated (and thus different) at every time t. This $f(t) = f(0, t)$ *is the main object of study in short rate models.* Also note that $f(t, t) = r(t)$.

Options on Bonds These are simply the equivalents of the stock options. We shall use:

$c(t, T, s)$ - the price of a European Call option with exercise date T on a pure discount bond[1] with maturity s and strike K.
$p(t, T, s)$ - the price of a European Put option with exercise date T on a pure discount bond with maturity s and strike K.

In the next section we will show how every instrument of interest will be expressed as a function of $P(t, s)$ and the vanilla options (i.e. a contract which gives the holder of that option the right but not the obligation to buy or sell this contract at a given price within a set time frame) described earlier.

17.3 Caps and Swaps

In this section we do not assume any model instead we show how to express price of complicated instruments in terms of the basic instruments described in the previous section. Thus if one can calculate the price of those basic instruments, one is finished with the calculation.

1) *Coupon bond option.* This is the price of a European type option with maturity T which if exercised at the maturity gives the right to buy or sell at the prespecified strike price K a bond that pays coupons (cash amounts) c_i at times s_i.
2) *Swaption.* This is a European option on a swap. A swap is a derivative contract through which two parties exchange financial instruments. In this chapter the swap considered is only an interest rate swap fixed versus floating of the reverse. Accordingly there are two types of options: you have the option to swap fixed payments with floating payments or the other way around. Both can be expressed in terms of basic instruments.

Swap Rate Calculation To calculate the swaption you will need to know how to calculate the fixed preagreed upon rate of the underlying swap. This calculation is detailed in Hull's book (Chapter 6). We are given the term structure of interest rates: $R(t, \tau)$ for all τ (the spot rates).

---
[1] This is another name for the zero coupon bond.

17 Models for the Bond Market

The swap rate R_{swap} is calculated so that at the beginning of the swap, the present value (PV) of the floating payments is equal to the PV of the fixed payments. In other words, nobody pays anyone anything.

Denote $f(t, t_{k-1}, t_k)$ the forward rate for the interval $[t_{k-1}, t_k]$ (the average of the instantaneous forward rate over that interval or $f(t, t_{k-1}, t_k) = \frac{1}{t_k - t_{k-1}} \int_{t_{k-1}}^{t_k} f(t, \tau) d\tau$). We recall from Eq. (17.3) that for any times $T < s$, $P(t, T, s) = e^{-(s-T)f(t,T,s)}$. Now the question that arises is how do we calculate these values in terms of the given values $R(t, \tau)$? From Eq. (17.2) we know the relationship between zero coupon bonds and forward contract on a zero coupon, that is, $P(t, t_{k-1}, t_k) = \frac{P(t, t_k)}{P(t, t_{k-1})}$, and also from Eq. (17.1) any zero coupon price is $P(t, s) = e^{-(s-t)R(t,s)}$. Therefore substituting the aforementioned into the forward rate, we obtain the following:

$$f(t, t_{k-1}, t_k) = \frac{(t_k - t)R(t, t_k) - (t_{k-1} - t)R(t, t_{k-1})}{t_k - t_{k-1}}. \tag{17.4}$$

Thus, we can calculate this rate for any times t_k and t_{k-1}.

Back to the calculation of the R_{swap}. This rate is calculated so that the PV of the fixed payments is equal to the present value of the floating payments. We take as notional principal 1. Suppose that there are n payments at $t_1, \ldots, t_n$ and at t_n there is also the principal exchange back. The current time is $t = t_0$. We can then calculate immediately the present value of the fixed and floating payments:

$$\text{PV of floating} = \sum_{k=1}^{n} f(t, t_{k-1}, t_k) P(t, t_k) + P(t, t_n) \tag{17.5}$$

$$\text{PV of fixed} = \sum_{k=1}^{n} (t_k - t_{k-1}) R_{swap} P(t, t_k) + P(t, t_n). \tag{17.6}$$

where PV is the present value and

$$R_{swap} = \frac{\sum_{k=1}^{n} f(t, t_{k-1}, t_k) P(t, t_k)}{\sum_{k=1}^{n} (t_k - t_{k-1}) P(t, t_k)}$$

The present value is the current worth of a future sum of money or stream of cash flows given a specified rate of return. Note that $f(t, t, t_1) = R(t, t_1)$ according to the calculation earlier and that the last term is just the principal swap. The rest of the terms are the actual payments (from a principal of 1) discounted back by multiplying them with the value of a zero coupon bond – also called the discount bond for this reason.

The swap rate R_{swap} is calculated as follows. First, we assume that the payments are equidistant in time (which is the usual case) or $t_k - t_{k-1} = \Delta t$. If we now substitute in Eq. (17.4), we obtain:

$$f(t, t_{k-1}, t_k) = \frac{k \Delta t R(t, t_k) - (k-1) \Delta t R(t, t_{k-1})}{\Delta t} = kR(t, t_k) - (k-1)R(t, t_{k-1}).$$

And if we put this back into the formula (17.5), we get:

$$\text{PV of floating} = \sum_{k=1}^{n}(kR(t,t_k) - (k-1)R(t,t_{k-1}))P(t,t_k) + P(t,t_n).$$

Intuitively assuming we have $1 today, we can generate all these payments into the future, so by no arbitrage the present value of the aforementioned should be 1. If this is true then the swap rate formula simplifies to:

$$R_{swap} = \frac{1 - P(t,t_n)}{\sum_{k=1}^{n}(t_k - t_{k-1})P(t,t_k)}$$

3) *Caps, floors, and collars.* An Interest rate cap is an agreement between two counterparties that limit the buyer's interest rate exposure to a maximum rate. The cap is actually the purchase of a call option on an interest rate. An interest rate floor is an agreement between two counterparties that limits the buyer's interest rate exposure to a minimum rate. Thus the floor is actually the purchase of a Put option on an interest rate. The combination of setting a cap rate and floor rate is called a collar.

The cap places a limit on the borrowers floating rate for some period of time. A floor places a lower limit on the interest rate charged and finally, a collar places both a upper and lower limit on the floating rate charged.

At any time the cap is in place, one does not need to use it if – in fact – the floating rate is lower than the agreed upon cap level. Therefore, this cap for the period considered can be thought of as a sum of options (caplets) each with maturity exactly the dates when the payments are supposed to happen. The mathematical deduction shows that in fact a cap is a sum of Put options and a floor is a sum of Call options.

One has to pay close attention when applying these formulas. In particular, we have to be cautious with the numbers and remember to price all the caplets that make a cap and sum them at the end.

17.4 Valuation of Basic Instruments: Zero Coupon and Vanilla Options on Zero Coupon

In the previous section we explained why we only have to give models capable of pricing the basic instruments presented in Section 17.2. Now we present two of such models in the following subsections.

17.4.1 Black Model

In the Black model one chooses the underlying modeled as a geometric Brownian motion (BM) so that the derivative of interest can be calculated using the

Black–Scholes formula. This is the most popular model in practice since it is the simplest but has some limitations. Before proceeding, we present concepts useful to the Black model.

1) *Options on zero coupon bond.* These depend directly on the forward bond price $P(t, T, s)$. So, for these options it is assumed that the forward bond price follows a geometric BM, i.e.:

$$dP(t, T, s) = \mu P(t, T, s)dt + \sigma_F P(t, T, s)dW_t.$$

We recall that this is simply the Black–Scholes formula. $P(t, T)$ is directly observable from the yield curve. A yield curve is a line that plots the interest rates, at a set point in time, of bonds having equal credit quality but differing maturity dates.

2) *Cap, floor, and collar.* Only cap pricing (in fact caplet pricing) is presented but the formulas for floorlet and floor are similar. To price a caplet the Black model assumes that $f(t, t_k, t_{k+1})$ is lognormal where this quantity represents the forward rate over the interval of interest $[t_k, t_{k+1}]$. The caplet formula is:

$$\text{caplet}(t, t_k, t_{k+1}) = P(t, t_{k+1}) \left[f(t, t_k, t_{k+1}) N(d_1) - R_{\text{cap}} N(d_2) \right] \Delta \tau L,$$

where R_{cap} is the agreed upon cap level and $\Delta \tau = t_{k+1} - t_k$.

3) *Swaption.* A swaption is an option on an interest rate swap. For the swaption, an instrument is assumed to follow a geometric BM, namely, the forward rate for the swap or R_{fswap}. The formula for swaption is:

$$\text{swaption}(t) = \Delta \tau \sum_{i=1}^{n} P(t, T_i) \left[R_{\text{fswap}} N(d_1) - KN(d_2) \right],$$

where $d_1 = \dfrac{\ln\left(\frac{F}{K}\right) + \left(\frac{\sigma^2}{2}\right)}{\sigma \sqrt{T}}$ and $d_2 = d_1 - \sigma \sqrt{T}$, $F-$ forward rate, $K-$ cap rate, and $\sigma-$ annualized volatility.

17.4.2 Short Rate Models

Short rate models are models in which the stochastic state variable is taken to be the instantaneous forward rate $f(t, \tau)$. Since usually this is done at the current time $t = 0$, one usually needs to model only $f(0, \tau)$ denoted with $r(t)$. The short rate $r(t)$ represents the interest rate of a loan emitted at t with immediate expiration $t + \Delta t$ as perceived now at time 0. If we have a model for $r(t)$, then the zero coupon bond price can be calculated as:

$$P(t, s) = \mathbf{E}_t \left[e^{\int_t^s r(u) du} \right] = \mathbf{E} \left[e^{\int_t^s r(u) du} \Big| \mathcal{F}_t \right]$$

and the price of an option that pays $C(T)$ at time T is:

$$C(t) = \mathbf{E}_t \left[e^{\int_t^s r(u) du} C(T) \right] = \mathbf{E} \left[e^{\int_t^s r(u) du} C(T) \Big| \mathcal{F}_t \right].$$

17.4 Valuation of Basic Instruments: Zero Coupon and Vanilla Options on Zero Coupon

Please refer to [133] for explicit formulas for these values. We shall focus on a tractable Gaussian model, namely, Vasicek model, Cox–Ingersoll–Ross (CIR) and the two factor models.

17.4.2.1 Vasicek Model (1977)
This was the first short rate model introduced by Oldrich Vasicek in 1977. The specification for the short rate is:

$$dr(t) = \alpha(\bar{r} - r(t))dt + \sigma dW_t$$

where $\bar{r}$ is the long-term rate and α, σ are parameters. The technical name of this model is mean-reverting Ørnstein–Uhlenbeck model.

With the Vasicek model, we can find explicit formulas for the zero coupon bond price and vanilla options. However the generated interest rates can become negative. Another secondary issue is the fact that it ignores the relationship between the volatility of the rate and the value of the rates. In fact empirical evidence suggests that volatility is high when interest rates are high.

17.4.2.2 Cox–Ingersoll–Ross (CIR) Model (1985)
The CIR model fixes the drawbacks in the previous model. Specifically, the rate is now modeled as:

$$dr(t) = \alpha(\bar{r} - r(t))dt + \sigma\sqrt{r(t)}dW_t$$

The introduction of the square root term in the stochastic integral fixes both problems with the Vasicek model since the generated interest rates can never be negative.

With the CIR, we can find explicit formulas for the zero coupon bond price and vanilla options. One major drawback is that the produced term structure is not flexible enough to fit the observed term structure (yield curve). Furthermore, under this model all the bond prices are perfectly correlated (not empirically true). This drawback is also true for the Vasicek model.

17.4.2.3 Two Factor Models
Fong and Vasicek (1992) and Longstaff and Schwartz (1992) tried to fix the drawback in the CIR model (statistically more complex models). In this model the instantaneous rate is represented as the sum of

a) the current rate $r_0(t)$
b) two stochastic state variables $r_1(t)$ and $r_2(t)$

A natural interpretation of these variables is that $r_1(t)$ controls the levels of the rates, while $r_2(t)$ controls the steepness of the forward curve.

This new model is capable to fitting the yield curve at one fixed moment in time. However, they still cannot fit two different yield curves (say today and tomorrow) without drastically changing the calibration parameters. Furthermore, for these models there is no known formula for the vanilla options (only for the zero coupon bond price).

17.5 Term Structure Consistent Models

We are faced typically with two important issues when modeling $r(t)$.

1) Finding the values of the parameters in the model. For example, for Vasiček model what exactly are the numbers $\alpha, \bar{r}$, and σ present in the model?
2) Once we know the numbers what are the prices of derivatives which depend on $r(t)$. This includes bond prices, as well as vanilla option prices on bonds.

We can see that the first problem is more important because we can't solve part 2 without first finding the answer for part 1, but as it turns out finding parameter values goes hand in hand with efficiently finding zero coupon bond pricing formulas. Next we justify the use of risk-neutral probability measure.

Given the probability space $(\Omega, \mathcal{F}, \mathbf{P})$, naturally, we model $r(t)$ under the assumption that everything is calculated with respect to this probability space, and the corresponding probability measure $\mathbf{P}$ is called the objective probability measure. Specifically, we assume the model for r under this measure is:

$$dr(t) = \mu(t, r(t))dt + \sigma(t, r(t))d\overline{W}(t) \tag{17.7}$$

where $\overline{W}$ is a BM under the probability measure $\mathbf{P}$. Furthermore, we assume the existence of a bank account earning interest at the rate $r(t)$ specifically:

$$dB(t) = r(t)B(t)dt.$$

Also assume that zero coupon bond prices are traded for all maturities T. In fact the bond prices are not uniquely determined by the $\mathbf{P}$- dynamics of $r(t)$?

The theorem that follows has many names in finance. It is called the fundamental theorem of finance, the fundamental theorem of arbitrage-free pricing, the Harrison–Pliska theorem, the Delbaen–Schachermayer theorem, and the Cameron-Martin theorem.

Theorem 17.5.1 (Metatheorem [88].). Let M = the number of underlying traded assets in the model, excluding the risk-free rate and let R = the number of random sources in the model. Then we have:

The model is arbitrage free if and only if $M \leq R$.
The model is complete if and only if $M \geq R$.
The model is complete and arbitrage free if and only if $M = R$.

17.5 Term Structure Consistent Models

A complete model is a model under which every contingent claim (think options) is replicable (think its value is attainable). As a consequence any contingent claim in a complete market has a unique price. For example, in the Black–Scholes model for an asset S_t, we had one traded asset (i.e. the equity) and one source of randomness (i.e. BM); therefore that model is arbitrage free and complete.

Now stepping back to our interest rate models, we recognize that in this case $M = 0$ and $R \geq 1$. In any model presented there exists at least one BM so $R \geq 1$. We also have $M = 0$ because the only external asset is the risk-free asset and we cannot form portfolios with it (there is no asset on the market whose price is $r(t)$).

Now we model $r(t)$ not under the probability measure **P** (the objective measure) but instead under **Q** (the equivalent martingale measure).

Assume that $r(t)$ has the dynamics given by (17.7) under **P** and we will shorten the notation to $dr(t) = \mu dt + \sigma d\overline{W}(t)$ while keeping in mind that both μ and σ are not constants but functions of t and $r(t)$.

Suppose the price of a zero coupon bond $P(t, T)$ is a smooth function of t, T and $r(t)$, specifically a twice derivable function in the first two variables and continuous in T:

$$P(t, T) = F(t, r(t), T)$$

Clearly, $F(T, r, T) = 1$ for any value of the short rate r.

Because we will use Itô's lemma in t, we will use the notation $F^T(t, r) = F(t, r, T)$.

Applying Itô's lemma to find the dynamics of F^T, we obtain:

$$\begin{aligned} dF^T(t) &= \frac{\partial F^T}{\partial t} dt + \frac{\partial F^T}{\partial r} dr + \frac{1}{2} \frac{\partial^2 F^T}{\partial r^2} (dr)^2 \\ &= F_t^T dt + F_r^T (\mu dt + \sigma d\overline{W}(t)) + \frac{1}{2} F_{rr}^T \sigma^2 dt \\ &= \left(F_t^T + \mu F_r^T + \frac{1}{2} F_{rr}^T \sigma^2 \right) dt + \sigma F_r^T d\overline{W}(t), \end{aligned}$$

where F_r^T is a notation for the first partial derivative of the function F^T with respect to r and so on the other derivatives. We now rewrite this dynamics as:

$$dF^T = F^T \alpha_T dt + F^T \sigma_T d\overline{W}(t), \tag{17.8}$$

$$\text{where} \begin{cases} \alpha_T = \frac{F_t^T + \mu F_r^T + \frac{1}{2} F_{rr}^T \sigma^2}{F^T} \\ \sigma_T = \frac{\sigma F_r^T}{F^T} \end{cases}$$

Now suppose we have two different bonds, one with maturity at S and the other with maturity at T. We form a portfolio with these two bonds by buying u_S units of the one with expiry at S and u_T units of the other one. Then the value of the portfolio is

$$V = u_T F^T + u_S F^S.$$

Since we do not modify the portfolio, we have the change in value depending only on the change in value of the two bonds, that is,

$$\frac{dV}{V} = u_T \frac{dF^T}{F^T} + u_S \frac{dF^S}{F^S}.$$

Now substituting the two terms on the right using Eq. (17.8) and solving for dV, we obtain:

$$dV = V(u_T \alpha_T + u_S \alpha_S)dt + V(u_T \sigma_T + u_S \sigma_S)d\overline{W}$$

In order to make this portfolio nonrandom, we take the weights in such a way that the coefficient of $\overline{W}$ disappears or the weights solve the system:

$$\begin{cases} u_T \sigma_T + u_S \sigma_S = 0 \\ u_T + u_S = 1 \end{cases} \qquad (17.9)$$

We selected the second condition for convenience so that we know how to divide the sum of money we have. We can easily solve the system (17.9) to obtain:

$$\begin{cases} u_T = -\dfrac{\sigma_S}{\sigma_T - \sigma_S} \\ u_S = \dfrac{\sigma_T}{\sigma_T - \sigma_S} \end{cases}$$

Substituting these particular weights into the dynamics of V we obtain the dynamics of this particular portfolio as:

$$dV = V\left(\frac{\alpha_S \sigma_T - \alpha_T \sigma_S}{\sigma_T - \sigma_S}\right) dt,$$

which is a nonrandom portfolio. The model is arbitrage free so any nonrandom portfolio earns the same as the bank account. Thus it should increase at the rate $r(t)$ and its dynamics should be:

$$dV = r(t)V dt.$$

Equating the two rates we obtain:

$$\frac{\alpha_S \sigma_T - \alpha_T \sigma_S}{\sigma_T - \sigma_S} = r(t), \quad \forall \ t \ (a.s.),$$

and rearranging this expression we get:

$$\frac{\alpha_S - r(t)}{\sigma_S} = \frac{\alpha_T - r(t)}{\sigma_T}, \quad \forall \ t$$

But our maturity times S and T were arbitrary so the aforementioned must happen for all the maturities. The only way this is true is if the expression $\frac{\alpha_T - r(t)}{\sigma_T}$ does not in fact depend on T. So, we must have:

17.5 Term Structure Consistent Models

$$\frac{\alpha_T - r(t)}{\sigma_T} = \lambda(t), \quad \forall \ t, \forall \ T. \tag{17.10}$$

$\lambda(t)$ is the risk market price which does not depend on T, $\alpha_T - r(t)$ is called the risk premium or *the excess rate of return*, and σ_T is called *the local volatility*.

Finally, if we substitute (17.10) into (17.8) and using the equations for α_T and σ_T after a little rearranging of terms, we obtain the following PDF for F^T:

$$\begin{cases} F_t^T + (\mu - \lambda\sigma)F_r^T + \frac{1}{2}\sigma^2 F_{rr}^T - rF^T = 0 \\ F^T(T, r) = 1 \end{cases} \tag{17.11}$$

The last is a terminal condition determined from $F(T, r, T) = F^T(T, r) = 1$ for any value of the short rate r.

Observe that the function $\lambda = \lambda(t, r)$ which is the external value determined by the market is not specified by the model. Therefore to solve the equation above, we need to specify the values for μ, σ, and λ. The first two are given by the model specification under the probability measure **P**. The probability density function (PDF) is the same as the value of an option in the lognormal case when the underlying dynamics of the return of the asset is:

$$dr(t) = (\mu - \lambda\sigma)dt + \sigma dW_t$$

We have only proven this when μ, σ, and λ are constants. The probability measure **P** - dynamics given in (17.7) is a little different than the expression discussed earlier, i.e.:

$$dr(t) = \mu dt + \sigma d\overline{W}_t.$$

Using Girsanov's theorem [159], we specifically rewrite the dynamics of $r(t)$ under probability measure **P** as follows:

$$dr(t) = (\mu - \lambda\sigma)dt + \lambda\sigma dt + \sigma d\overline{W}_t,$$

which can be rewritten as:

$$dr(t) = (\mu - \lambda\sigma)dt + \sigma\left(d\overline{W}_t + \lambda dt\right).$$

We recall that Girsanov theorem describes how the dynamics of stochastic processes change when the original measure is changed to an equivalent probability measure.

Next we construct an equivalent measure **Q** under which the process $W_t = \overline{W}_t + \int_0^t \lambda(s)ds$ is a BM. Note that for this process we have $dW_t = d\overline{W}_t + \lambda(t)dt$, exactly the process we need. We perform a change of measure where the measure **Q** is defined as:

$$\left.\frac{d\mathbf{Q}}{d\mathbf{P}}\right|_{F_t} = e^{X_t - \frac{1}{2}[X]_t}$$

and $X_t = \int_0^t \lambda(s)d\overline{W}_s$, so the quadratic variation of X_t is: $[X]_t = \int_0^t \lambda(s)ds$. More precisely, the measure $\mathbf{Q}$ is defined on the sigma algebra $\mathcal{F}_t$ and the measure of any set $B \in \mathcal{F}_t$ is:

$$Q(B) = \int_B e^{\int_0^t \lambda(s)d\overline{W}_s - \frac{1}{2}\int_0^t \lambda(s)ds} dP$$

As a consequence of the Girsanov theorem, the process $W_t = \overline{W}_t + \int_0^t \lambda(s)ds$ is a BM under $\mathbf{Q}$. So the dynamic of $r(t)$ under $\mathbf{Q}$ is exactly what we need, that is,

$$dr(t) = (\mu - \lambda\sigma)dt + \sigma dW_t.$$

We note that if we specify the drift of $r(t)$ under $\mathbf{Q}$ directly, we implicitly include in that specification the λ process.

If we specify directly the dynamic of $r(t)$ under the equivalent measure $\mathbf{Q}$ as

$$dr(t) = \mu dt + \sigma dW_t$$

then the price of a zero coupon bond with maturity T: $P(t, T) = F^T(t, r)$ is the solution to the following PDF:

$$\begin{cases} F_t^T + \mu F_r^T + \frac{1}{2}\sigma^2 F_{rr}^T - rF^T = 0 \\ F^T(T, r) = 1 \end{cases} \tag{17.12}$$

We note that all the models discussed, i.e Vasiček, CIR, and two factor models, specify the $r(t)$ dynamics under the probability measure $\mathbf{Q}$.

17.6 Inverting the Yield Curve

In this section, we find the parameters arising in the models discussed in the previous sections.

To begin, we assume that the model for $r(t)$ contains a vector of parameters generically denoted θ, i.e.

$$dr(t) = \mu(t, r(t), \theta)dt + \sigma(t, r(t), \theta)dW_t$$

Then the steps to find the solution of the PDF (17.12) for any given parameter vector θ is as follows:

- Denote this solution $P(t, T, \theta)$.
- Calculate the empirical term structure from available data, i.e. $P^*(0, T)$.
- Choose θ so that the empirical term structure fits best the theoretical curve:

$$P(0, T, \theta) = P^*(0, T)$$

This produces the optimal parameter vector θ^*. Once we determine this parameter, we are done. We can find the derivative prices and zero coupon bond prices because we determined the parameter values. The step that may bring some difficulties is to find a quick way to find the solution of the PDF (17.12). This is addressed in the next subsection.

17.6.1 Affine Term Structure

It turns out that the complications underlined earlier are easier to fix for some specific models. These models are said to posses an affine structure. More specifically, if for the zero coupon bond price $P(t, T)$, the solution of Eq. (17.12) looks like this:

$$P(t, T) = F^T(t, r(t)) = e^{A(t,T)+B(t,T)r(t)},$$

then the corresponding model for $r(t)$ is said to possess an affine term structure.

A question that arises is which of the models for $r(t)$ possesses an affine term structure? This question is not answered in general. However, if the functions μ and σ in the dynamic of $r(t)$ are of this form:

$$\begin{cases} \mu(t, r) = \alpha(t)r + \beta(t) \\ \sigma(t, r) = \sqrt{\gamma(t)r + \delta(t)}, \end{cases} \qquad (17.13)$$

then the corresponding $r(t)$ model has an affine term structure.

We note that all the models presented herein have an affine term structure.[2]

Finally, if the model possesses an affine term structure of the form (17.13), then the price of the zero coupon bond $P(t, T) = e^{A(t,T)+B(t,T)r(t)}$ is determined by finding A and B as solutions to the following system of ordinary differential equations:

$$\begin{cases} B_t + \alpha(t)B - \frac{1}{2}\gamma(t)B^2 = -1 \\ B(T, T) = 0 \\ A_t = \beta(t)B - \frac{1}{2}\delta(t)B^2 \\ A(T, T) = 0 \end{cases} \qquad (17.14)$$

The first equation is a Riccati equation with terminal condition. Once solved the solution B is plugged into the equation in A_t and integrated to find A.

Remember that when we match the theoretical and empirical term structure, we do so at the present time $t = 0$. But the present time changes all the time. This means that the optimal parameter vector θ^* may change its value dramatically from day to day.

---

[2] These two models do not need an affine term structure since they are lognormal processes, and the solution of (17.12) is found easily by applying the Black–Scholes formula.

This fact creates a demand for models that can be made to fit the observed term structure. This is what created the Hull–White class models in which the parameters change in time $\theta = \theta(t)$.

However, even these models are not generic and the estimates (the $\theta(t)$ is supposed to be smooth function) created are not stable. As a consequence if the observed curve for the term structure $P^*(0, T)$ changes dramatically (as it happens quite often in practice), these models will not be able to cope with the changes.

So, Heat–Jarrow–Morton [94] came up with the most general model in which the instantaneous forward rate actually depends on T as well (as it should and as we mentioned from the beginning). Specifically, under $\mathbf{P}$ the model for the instantaneous forward rate is now:

$$dr(t, T) = \mu(t, T)dt + \sigma(t, T)d\overline{W}(t)$$

where μ and σ are adapted and $\overline{W}(t)$ is a BM under the risk-neutral measure.

The approach for this framework is different than the one we have seen thus far. However, some of the more complex models presented earlier can be solved using this framework.

17.7 Problems

1. Consider two identical call options, with strikes K_1 and K_2; show that:
   (a) If $K_2 > K_1$ then $C(K_2) > C(K_1)$
   (b) If $K_2 > K_1$ then $K_2 - K_1 > C(K_2) - C(K_1)$
2. Consider a European Call option with strike K and time to expiration T. Denote the price of the call for $C(S, T)$ and let $B(T)$ the price of one unit of a zero coupon bond maturing at time T.

$$C(S, T) \geq \max\{0, S - KB(T)\}$$

3. A perpetual option is an option that never expires.
   (a) Describe such an option. Is it European or American?
   (b) Show that the value of a perpetual call on a non-dividend-paying stock is $C = S$.
   (c) Find the value of a perpetual Put option. In particular find the time-independent Black–Scholes equation.
4. A happy Call option is a Call option with payoff $\max(aS, S - K)$. So we always get something with a happy Call! If C_1 and C_2 are the prices of two call options with strikes nK and mK, show that the price of a happy Call is given by the formula

$$C_H = a_1 C_1 + a_2 C_2 + a_3 S$$

and find the constants $a_1, a_2, 2_3$.

5. If we have a model for the short rate $r(t)$, show that
   (a) the zero coupon bond price can be calculated as:
   $$P(t,s) = \mathrm{E}_t\left[e^{\int_t^s r(u)du}\right] = \mathrm{E}\left[e^{\int_t^s r(u)du}\Big| \mathcal{F}_t\right]$$
   (b) the price of an option that pays $C(T)$ at time T is:
   $$C(t) = \mathrm{E}_t\left[e^{\int_t^s r(u)du} C(T)\right] = \mathrm{E}\left[e^{\int_t^s r(u)du} C(T)\Big| \mathcal{F}_t\right].$$
   where $r(t)$ is the interest rate of a loan emitted at t with immediate expiration $t + \Delta t$ as perceived now at time zero.

18

Exchange Traded Funds (ETFs), Credit Default Swap (CDS), and Securitization

18.1 Introduction

In the previous chapter, we discussed models for pricing bonds. We recall that the bond market is where debt securities are issued and traded. Although the bond market appears complex, it is driven by the same risk and return tradeoffs as the stock market. In this chapter, we will begin by discussing Exchange Traded Funds (ETFs) which are marketable securities. Marketable securities are defined as any unrestricted financial instrument that can be bought or sold on a public stock exchange or a public bond exchange. Next we will describe the Credit Default Swap (CDS) which is a financial swap agreement where the seller of the CDS will compensate the buyer in the event of a loan default (by the debtor) or other credit event. Finally, we will end this chapter by discussing securitization, namely, Mortgage Backed Securities (MBS) and Collateralized Debt Obligation (CDO). Securitization is the procedure whereby an issuer designs a financial instrument by pooling various financial assets together and then markets tiers of the repackaged instruments to investors.

18.2 Exchange Traded Funds (ETFs)

In this section, we introduce and describe ETFs and present some examples.

An exchange traded fund (ETF) is an investment fund traded on stock exchanges [108, 183]. An ETF holds assets such as stocks, commodities, or bonds and generally operates with an arbitrage mechanism designed to keep it trading close to its net asset value, although deviations can occasionally occur. ETFs may be attractive as investments because of their low costs, tax efficiency, and stock-like features. Most ETFs track an index, such as a stock index or bond index. In recent time, ETFs has become a very popular form of exchange-traded product. Unlike mutual funds, an ETF trades like a common stock on a stock exchange. ETFs experience price changes throughout the

Quantitative Finance, First Edition. Maria C. Mariani and Ionut Florescu.
© 2020 John Wiley & Sons, Inc. Published 2020 by John Wiley & Sons, Inc.

day as they are bought and sold. ETFs are usually preferred by individual investors since they have lower fees and higher daily liquidity than other exchange-traded product such as the mutual fund shares.

ETF distributors only buy or sell ETFs directly from or to authorized participants, which are large broker-dealers with whom they have entered into agreements and then, only in creation units, which are large blocks of tens of thousands of ETF shares, usually exchanged in kind with baskets of the underlying securities. Payment in kind is the use of goods or services as payment instead of cash. Authorized participants may wish to invest in the ETF shares for the long term, but they usually act as market makers on the open market, using their ability to exchange creation units with their underlying securities to provide liquidity of the ETF shares and help ensure that their intraday market price approximates the net asset value of the underlying assets. Other investors, such as individuals using a retail broker, trade ETF shares on this secondary market.

In practice, an ETF combines the valuation feature of a mutual fund or unit investment trust, which can be bought or sold at the end of each trading day for its net asset value, with the tradability feature of a closed end fund, which trades throughout the trading day at prices that may be more or less than its net asset value. The net asset value represents the net value of an entity, which is usually calculated as the total value of the entitys assets minus the total value of its liabilities.

In the next subsection, we briefly explain the various types of ETFs.

18.2.1 Index ETFs

Most ETFs are index funds that attempt to replicate or reproduce the performance of a specific index. Indexes may be based on stocks, bonds, commodities, or currencies. An index fund seeks to track the performance of an index by holding in its portfolio either the contents of the index or a representative sample of the securities in the index (see [182] for more details). As of June 2012, in the United States, about 1200 index ETFs exist, with about 50 actively managed ETFs. Index ETF assets are about $1.2 trillion, compared with about $7 billion for actively managed ETFs [171]. Some index ETFs, known as leveraged ETFs or inverse ETFs, use investments in derivatives to seek a return that corresponds to a multiple of, or the inverse (opposite) of, the daily performance of the index. Leveraged ETFs will be studied later in this chapter.

Some index ETFs invest 100% of their assets proportionately in the securities underlying an index, a manner of investing called replication. Other index ETFs use representative sampling, investing 80–95% of their assets in the securities of an underlying index and investing the remaining 5–20% of their assets in other holdings, such as futures, option and swap contracts, and securities not in the underlying index. There are various ways the ETF can be weighted, such as equal weighting or revenue weighting. For index ETFs that invest in indices with thousands of underlying securities, some employs "aggressive sampling" and invest in only a tiny percentage of the underlying securities (see[182]).

18.2.2 Stock ETFs

A stock ETF is an asset that tracks a particular set of equities, similar to an index ETF. It trades just as a normal stock would on an exchange, but unlike a mutual fund, prices adjust throughout the day rather than at market close. These ETFs can track stocks in a single industry, such as energy, or an entire index of equities like the S&P 500. Stock ETFs can have different styles, such as large market capitalization (large-cap), small market capitalization (small-cap), growth, value, and several others. For example, the S&P 500 index is large and midmarket capitalization, so the SPDR S&P 500 ETF will not contain small-cap stocks. On the other hand, others such as iShares Russell 2000 are mainly for small-cap stocks. There are many style ETFs such as iShares Russell 1000 Growth and iShares Russell 1000 Value. ETFs focusing on dividends have been popular in the first few years of the 2010s decade, such as iShares Select Dividend.

18.2.3 Bond ETFs

Bond ETFs are exchange-traded funds that invest in bonds. They thrive during economic recessions because investors pull their money out of the stock market and into bonds (for example, government treasury bonds or those issued by companies regarded as financially stable). Because of this cause and effect relationship, the performance of bond ETFs may be indicative of broader economic conditions. There are several advantages of bond ETFs such as the reasonable trading commissions, but this benefit can be negatively offset by fees if bought and sold through a third party.

Bond ETFs have defined risk and reward characteristics. One should determine the risk and reward profile for the ETF portion of their portfolio and then choose bond ETFs that correspond to that profile. For example, if you want to minimize risk, one should choose a short-duration government bond fund. If on the other hand, one is willing to take on maximum risk (within the bond ETF market) in exchange for a higher return, consider choosing a high-yield corporate bond ETF. Please refer to [106] for more details.

18.2.4 Commodity ETFs

Commodities are known as the raw materials for production processes. Commodity ETFs (ETCs) invest in commodities, such as precious metals, agricultural products, or hydrocarbons. Commodity ETFs trade just like shares, are simple and efficient, and provide exposure to an ever-increasing range of commodities and commodity indices. However, it is important for an investor to realize that there are often other factors that affect the price of a commodity ETF that might not be immediately apparent. Gold is the first commodity ETFs which have been offered in a number of countries. The idea of a gold ETF was first officially conceptualised by Benchmark Asset Management Company Private Ltd in India when they filed a proposal with the SEBI in May 2002. The first gold exchange-traded fund was Gold Bullion Securities launched on the ASX in 2003, and the first silver exchange-traded fund was iShares Silver Trust launched on the NYSE in 2006. As of November 2010 a commodity ETF, namely, SPDR Gold Shares, was the second-largest ETF by market capitalization. Next we present some examples of commodity ETFs.

Precious metals like gold and silver are popular ETFs because the underlying commodity can't go bad or spoil. The SPDR Gold Shares and iShares Silver Trust are two of the largest gold and silver ETFs. The SPDR Gold Shares ETF has an expense ratio of 0.4%, and the iShares Silver Trust has an expense ratio of 0.5% [107].

Other popular type of commodity ETFs are oil and natural gas. However, since oil and gas can't be stockpiled like precious metals, these ETFs invest in futures contracts instead of the commodity itself. Generally commodity ETFs are index funds tracking nonsecurity indices. This is due to the fact that they do not invest in securities. Most ETCs implement a future trading strategy, which may produce quite different results from owning the commodity.

Commodity ETFs can also be referred to as exchange-traded notes (ETNs), which are not exchange-traded funds. An exchange-traded note (ETN) is a senior, unsecured, unsubordinated debt security issued by an underwriting bank. Similar to exchange-traded funds, ETNs are traded on major exchanges.

18.2.5 Currency ETFs

Currency ETFs are invested in a single currency or basket of currencies. They aim to replicate movements in currency in the foreign exchange market by holding currencies either directly or through currency-denominated short-term debt instruments. They offer investors exposure to a single currency or a basket of currencies. The funds are comprised of currency futures contracts. Some of these ETFs implement popular currency strategies, such as focusing on currencies of commodity producing economies.

In 2005, Rydex Investments launched the first currency ETF called the Euro Currency Trust in New York. Since then Rydex has launched a series of funds tracking all major currencies under their brand CurrencyShares. In 2007 Deutsche Bank's db X-trackers launched Euro OverNight Index Average (EONIA) Total Return Index ETF in Frankfurt tracking the euro and later in 2008 the Sterling Money Market ETF (LSE: XGBP) and US Dollar Money Market ETF (LSE: XUSD) in London. The funds are total return products where the investor gets access to the FX spot change, local institutional interest rates, and collateral yield.

18.2.6 Inverse ETFs

An inverse exchange-traded fund is an ETF, traded on a public stock market, which is designed to perform as the inverse of whatever index or benchmark it is designed to track. They are constructed by using various derivatives for the purpose of profiting from a decline in the value of the underlying benchmark. It is similar to holding several short positions or using a combination of advanced investment strategies to profit from falling prices. Many inverse ETFs use daily futures as their underlying benchmark.

An inverse S&P 500 ETF, for example, seeks a daily percentage movement opposite that of the S&P. If the S&P 500 rises by 5%, the inverse ETF is designed to fall by 5%, and if the S&P falls by 5%, the inverse ETF should rise by 5%.

18.2.7 Leverage ETFs

Leverage is the ratio of a company's debt to the value of its common stock. Leveraged exchange-traded funds (LETFs) (or leveraged ETFs) are a type of ETF that attempt to achieve returns that are more sensitive to market movements than nonleveraged ETFs. Leveraged index ETFs are often marketed as bull or bear funds. A bull or bear fund is a mutual fund that looks to provide higher returns amid market downturns. A leveraged bull ETF fund might, for example, attempt to achieve daily returns that are two times or three times more pronounced than the Dow Jones Industrial Average or the S&P 500. A leveraged inverse (bear) ETF fund on the other hand may attempt to achieve returns that are two times or three times the daily index return, meaning that it will gain double or triple the loss of the market. Leveraged ETFs require the use of financial engineering techniques, including the use of equity swaps, derivatives and rebalancing, and reindexing to achieve the desired return (see [175]). The most common way to construct leveraged ETFs is by trading future contracts.

The rebalancing and reindexing of leveraged ETFs may have considerable costs when markets are volatile ([149]). The rebalancing problem is that the fund manager incurs trading losses because he needs to buy when the index goes up and sell when the index goes down in order to maintain a fixed leverage ratio. A 2.5% daily change in the index will, for example, reduce value of a double bear fund by about 0.18% per day, which means that about a third of the fund may be wasted in trading losses within a year. Investors may however overcome this problem by buying or writing futures directly and accepting a varying leverage ratio. A more reasonable estimate of daily market changes is 0.5%, which leads to a 2.6% yearly loss of principal in a triple leveraged fund [149]. A triple-leveraged fund will return three times the daily performance of that index. The effect of leverage is also reflected in the pricing of options written on leveraged ETFs. In particular, the terminal payoff of a leveraged ETF European/American Put or Call depends on the realized variance (hence the path) of the underlying index. The impact of leverage ratio can also be observed from the implied volatility surfaces of leveraged ETF options. For instance, the implied volatility curves of inverse leveraged ETFs are commonly observed to be increasing in strike, which is characteristically different from the implied volatility smiles or skews seen for index options or nonleveraged ETF options.

Leveraged ETFs are available for most indexes, such as the Nasdaq 100 and the Dow Jones Industrial Average. These funds aim to keep a constant amount of leverage during the investment time frame [109].

In the section that follows, we will be focusing on credit derivatives. A credit derivative consists of privately held negotiable bilateral contracts that allow users to manage their exposure to credit risk. Credit derivatives are financial assets such as forward contracts, swaps, and options for which the price is driven by the credit risk of economic agents, such as private investors or governments. The financial assets mentioned were discussed in Chapter 17 of this book. The discussion begins with a financial instrument for swapping the risk of debt default that is CDS.

18.3 Credit Default Swap (CDS)

A CDS is a financial swap agreement where the seller of the CDS will compensate the creditor of the reference loan in the event of a loan default (by the debtor) or other credit event. In other words, CDS contract provides insurance against the risk of a default by a particular company. The company is known as the reference entity and a default by the company is known as a credit event. The buyer of the insurance obtains the right to sell a particular bond issued by the company for its par value when a credit event occurs. The bond is known as the reference obligation and the total par value of the bond that can be sold is known as the swap's notional principal. The buyer of the CDS makes periodic payments to the seller until the end of the life of the CDS or until a credit event occurs. A credit event usually requires a final accrual payment by the buyer. The swap is then settled by either physical delivery or in cash. If the terms of the swap require physical delivery, the swap buyer delivers the bonds to the seller in exchange for their par value (i.e. the face value of a bond). When there is cash settlement, the calculation agent polls dealers to determine the midmarket price, Z, of the reference obligation some specified number of days after the credit event. The cash settlement is then $(100 - Z)\%$ of the notional principal.

CDS are the most widely used type of credit derivative and a powerful force in the world markets. Credit derivatives are financial assets such as forward contracts, swaps, and options for which the price is driven by the credit risk of economic agents, such as governments or private investors.

Next we present two examples of CDS.

18.3.1 Example of Credit Default Swap

Example 18.3.1 Suppose that a bank has lent money to a firm in the form of a $1000 bond. The bank may then purchase a CDS from another company, e.g. a hedge fund. If the firm defaults on the loan, then the hedge fund will pay the bank the value of the loan. Thus the bank has insurance against loan default. The hedge fund has the opportunity to make profit, so long as the firm does not default on the loan. The more risky the loan, the higher will be the premium required on buying a CDS.

Example 18.3.2 Suppose John holds a 10-year bond issued by Modern Digital Electronics Inc. with a par value of $4000 and a coupon interest amount of $400 each year. Fearful that Modern digital electronics Inc will default on its bond obligations, Chris enters into a CDS with Thomas and agrees to pay him income payments of $80 (similar to an insurance premium) each year commensurate with the annual interest payments on the bond. In return, Thomas agrees to pay John the $4000 par value of the bond in addition to any remaining interest on the bond ($400 multiplied by the number of years remaining). If Modern Digital Electronics Inc. fulfills its obligation on the bond through maturity after 10 years, Thomas will make a profit on the annual $80 payments.

18.3.2 Valuation

We now move on to consider the valuation of a CDS. For convenience we assume that the notional principal is $1. We assume that default events, interest rates, and recovery rates are mutually independent. We also assume that the claim in the event of default is the face value (the nominal value of a security stated by the issuer) plus accrued interest (i.e. interest that has been earned but not collected). Suppose that default can only occur at times $t_1, t_2, \ldots, t_n$, we define the following:

T : Life of CDS in years
p_i : Risk-neutral probability of default at time t

$\hat{R}$: Expected recovery rate on the reference obligation in a risk-neutral world (this is assumed to be independent of the time of the default)

$u(t)$: Present value (PV) of payments at the rate of $1 per year on payment dates between time zero and time t

$e(t)$: PV of a payment at time t equal to $t - t^*$ dollars, where t^* is the payment date immediately preceding time t (both t and t^* are measured in years)

$v(t)$: PV of $ 1 received at time t

w : Payments per year made by CDS buyer per dollar

s : Value of w that causes the CDS to have a value of zero

π : The risk-neutral probability of no credit event during the life of the swap

$A(t)$: Accrued interest on the reference obligation at time t as a percent face value

The value of π is one minus the probability that a credit event will occur. It can be calculated from the p_i as follows:

$$\pi = 1 - \sum_{i=1}^{n} p_i$$

The payments last until a credit event or until time T, whichever is sooner. The PV of the payments is therefore

$$PV = w \sum_{i=1}^{n} \left[u(t_i) + e(t_i) \right] p_i + w\pi u(T).$$

If a credit event occurs at time t_i, the risk-neutral expected value of the reference obligation, as a percent of its face value, is $[1 + A(t_i)]\hat{R}$. Thus the risk-neutral expected payoff from a CDS is:

$$1 - [1 + A(t_i)]\hat{R} = 1 - \hat{R} - A(t_i)\hat{R}$$

The present value of the expected payoff from the CDS is

$$\text{PV of the expected payoff} = \sum_{i=1}^{n} [1 - \hat{R} - A(t_i)\hat{R}] p_i v(t_i)$$

and the value of the CDS to the buyer is the present value of the expected payoff minus the present value of the payments made by the buyer, i.e.

$$\text{PV of the expected payoff} - w \sum_{i=1}^{n} \left[u(t_i) + e(t_i) \right] p_i + w\pi u(T)$$

The CDS spread, s, is the value of w that makes this expression zero:

$$s = \frac{\text{PV of the expected payoff}}{\sum_{i=1}^{n} \left[u(t_i) + e(t_i) \right] p_i + \pi u(T)} \tag{18.1}$$

The variable s is referred to as the CDS spread. It is the payment per year, as a percent of the notional principal, for a newly issued CDS. We can extend

18.3 Credit Default Swap (CDS)

the analysis to allow defaults at any time. For instance suppose $g(t)$ is the risk-neutral default density at time t. Then the CDS spread s is defined as

$$s = \frac{\int_0^T [1 - \hat{R} - A(t)\hat{R}] g(t) v(t)}{\int_0^T [u(t) + e(t)] g(t) + \pi u(T)} \qquad (18.2)$$

which follows from (18.1).

18.3.3 Recovery Rate Estimates

The only variable necessary for valuing a CDS that cannot be observed directly in the market is the expected recovery rate. Fortunately the pricing of a plain vanilla CDS depends on the recovery rate to only a small extent. This is because the expected recovery rate affects CDS prices in two ways. It affects the estimates of risk-neutral default probabilities and the estimates of the payoff that will be made in the event of a default. These two effects largely offset each other.

18.3.4 Binary Credit Default Swaps

Nonstandard CDS can be quite sensitive to the expected recovery rate estimate. Consider a binary CDS. This is structured similarly to a regular CDS except that the payoff is a fixed dollar amount. In this case the expected recovery rate affects the probability of default but not the payoff. As a result the credit default spread is quite sensitive to the recovery rate. For example, a binary CDS spread when the expected recovery rate is 50% is typically about 80% higher than when the expected recovery rate is 10% [101].

18.3.5 Basket Credit Default Swaps

A basket default swap is similar to a single entity default swap except that the underlying is a basket of entities rather than one single entity. An add-up basket CDS provides a payoff when any of the reference entities default. It is equivalent to a portfolio of credit default swaps, one on each reference entity. A first-to-default basket credit default swap provides a payoff only when the first reference entity defaults. After that, there are no further payments on the swap and it ceases to exist. First-to-default swaps can be valued using Monte Carlo simulation. On each trial each reference entity is simulated to determine when it defaults. We calculate as follows:

a) the present value of the payoff (if any)
b) the present value of payments until the time of the first default or the end of the contract (whichever is earlier) at the rate of $1 per year

The swap spread is the average value of the calculations in (a) divided by the average value of the calculations in (b). First-to-default swaps are sensitive to the default correlation between reference entities. The higher the correlation, the lower the value. A conservative assumption for the seller of the swap is that all correlations are zero [101].

In the next section, we discuss a type of asset-backed security that is secured by a mortgage, i.e. a financial security collateralized by a pool of assets where the underlying securities are mortgage-based.

18.4   Mortgage Backed Securities (MBS)

A mortgage-backed security is created when a financial institution decides to sell part of its residential mortgage portfolio to investors. The mortgages sold are put into a pool and investors acquire a stake in the pool by buying units. The units are known as mortgage-backed securities. A secondary market is usually created for the units so that investors can sell them to other investors as desired. An investor who owns units representing $y\%$ of a certain pool is entitled to $y\%$ of the principal and interest cash flows received from the mortgages in the pool.

The mortgages in a pool are generally guaranteed by a government-related agency, such as the Government National Mortgage Association (GNMA) or the Federal National Mortgage Association (FNMA) so that investors are protected against defaults. This makes an MBS sound like a regular fixed-income security issued by the government. A fixed-income security is an investment that provides a return in the form of fixed periodic payments and the eventual return of principal at maturity. There is a critical difference between an MBS and a regular fixed-income security. The difference is the fact that the mortgages in an MBS pool have prepayment privileges. These prepayment privileges can be quite valuable to the householder. Most often, prepayments on mortgages occur for a variety of reasons. Sometimes interest rates have fallen and the owner of the house decides to refinance at a lower rate of interest. On other occasions, a mortgage is prepaid simply because the house is being sold. A critical element in valuing an MBS is the determination of what is known as the prepayment function. This is a function describing expected prepayments on the underlying pool of mortgages at a time t in terms of the yield curve at time t and other relevant variables.

For a comprehensive study of the valuation of MBS, please refer to [100] for more details.

A prepayment function is very unreliable as a predictor of actual prepayment experience for an individual mortgage. When many similar mortgage loans are combined in the same pool, there is a "law of large numbers" effect at work and prepayments can be predicted more accurately from an analysis of historical data. As mentioned, prepayments are not always motivated by pure interest rate considerations. Nevertheless, there is a tendency for prepayments to be more likely when interest rates are low than when they are high. This means that investors require a higher rate of interest on an MBS than on other fixed-income securities to compensate for the prepayment options they have written.

In the next section, we describe other type of bonds which are backed by mortgage, loan, and bonds.

18.5 Collateralized Debt Obligation (CDO)

A CDO is a way of packaging credit risk in much the same way as a collateralized mortgage obligation (see Section 18.5.1) is a way of packaging prepayment risk. The idea is that CDO pools together cash flow-generating assets and repackages this asset pool into discrete tranches (pieces, portions, or slices of debt or structured financing) that can be sold to investors. A cash flow CDO is one for which the collateral portfolio is not subjected to active trading by the CDO manager, which means that the uncertainty regarding interest and principal payments to the CDO tranches is induced mainly by the number and timing of defaults of the collateral securities. CDOs are backed by portfolios of assets that may include a combination of bonds, loans, securitized receivables, asset-backed securities, tranches of other collateralized debt obligations, or credit derivatives referencing any of the former. Please refer to [101, 139] for more details on CDOs.

In the subsections that follow, we describe the three types of CDOs.

18.5.1 Collateralized Mortgage Obligations (CMO)

Collateralized mortgage obligation (CMO) refers to a type of mortgage-backed security that contains a pool of mortgages bundled together and sold as an investment. In a CMO the investors are divided into a number of classes and rules are developed for determining how principal repayments are channeled to different classes.

As an example of a CMO, consider an MBS where investors are divided into three classes: class A, class B, and class C. All the principal repayments (both those that are scheduled and those that are prepayments) are channeled to class A investors until investors in this class have been completely paid off. Principal repayments are then channeled to class B investors until these investors have been completely paid off. Finally, principal repayments are channeled to class C investors. In this situation, class A investors bear the most prepayment risk. The class A securities can be expected to last for a shorter time than the class B

securities, and these, in turn, can be expected to last less long than the class C securities. The objective of this type of structure is to create classes of securities that are more attractive to institutional investors than those created by the simpler pass-through MBS. The prepayment risks assumed by the different classes depend on the par value in each class. For example, class C bears very little prepayment risk if the par values in classes A, B, and C are 400, 300, and 100, respectively. Class C bears rather more prepayment risk in the situation where the par values in the classes are 100, 200, and 500.

18.5.2 Collateralized Loan Obligations (CLO)

A collateralized loan obligation (CLO) is a security backed by a pool of debt, often low-rated corporate loans. CLO is a type of CDO which is similar to collateralized mortgage obligations, except for the different type of underlying loan. With a CLO, the investor receives scheduled debt payments from the underlying loans, assuming most of the risk in the event borrowers default, but is offered greater diversity and the potential for higher-than-average returns.

CLOs are complicated structures that combine multiple elements with the goal of generating an above-average return via income and capital appreciation. They consist of tranches that hold the underlying loans, which typically account for about 90% of total assets. The tranches are ranked from highest to lowest in order of credit quality, asset size, and income stream and thus from lowest to highest in order of riskiness.

18.5.3 Collateralized Bond Obligations (CBO)

A collateralized bond obligation (CBO) is another type of CDO. CBO is similar in structure to a CMO but different in that CBOs represent different levels of credit risk, not different maturities. In 1988, the first rated collateralized bond obligation backed by high yield bonds was brought to market. CBOs are backed by a portfolio of secured or unsecured senior or junior bonds issued by a variety of corporate or sovereign obligors. Often a collateralized bond obligation will include loans in the portfolio, but the majority of the collateral usually consists of bonds. Collateralized bond obligation transactions realize the positive spread between a portfolio of higher-return, higher-risk assets, often rated below BBB and lower-cost, highly rated collateralized bond obligation securities issued to purchase the portfolio of bonds backing the collateralized bond obligation.

For comprehensive studies on the risk analysis and valuation of CDOs, please refer to [58] and references therein for more details. In the valuation model proposed by [58], the authors did not deal directly with the effects of market imperfections. However, it takes as given the default risk of the underlying loans and assumes that investors are symmetrically informed.

18.6 Problems

1. Briefly explain the main difference between exchange-traded funds (ETFs) and mutual funds?
2. Explain the difference between Inverse ETFs and Leverage ETFs.
3. Briefly describe what a credit default swap is and how it is generally used.
4. A credit default swap requires a premium of 70 basis points per year paid semiannually. The principal is $350 million and the credit default swap is settled in cash. A default occurs after five years and two months, and the calculation agent estimates that the price of the reference bond is 50% of its face value shortly after the default. Please list the cash flows and their timing for the seller of the credit default swap.
5. Describe the risks of investing in MBS.
6. Explain some factors that affect the price of mortgage-backed securities.
7. Suppose you sell a CDO that insures the buyer against the event that Toyota Motor Corporation declares bankruptcy within a given calendar year. The likelihood of a declared bankruptcy depends on the state of the auto industry as a whole. It has been forecasted that the state of the industry next year will be "strong" with probability 0.70 and "weak" with probability 0.30. The probability of bankruptcy given that the auto industry is strong is 0.20. The probability of bankruptcy, given that the auto industry is weak, increases dramatically to 0.40.
   (a) What is the probability that the company will declare bankruptcy next year?
   (b) What is the probability that the auto industry is strong next year, given that the company covered by the CDO does not declare bankruptcy?
8. A typical debtor i has liablities of $100. His bank estimates that he will pay back the full face value with probability 60%. With probability 30%, he only pays back $50 and with probability 10% he defaults completely. The distribution of repayments is

$$R(i) = \begin{cases} 0, & \text{with probability } 0.05 \\ 50, & \text{with probability } 0.45 \\ 100, & \text{with probability } 0.50 \end{cases}$$

   (a) Consider a second debtor j with identical default risk and identical face value of debt. Suppose that the default risks of the two debtors are independent. Consider a portfolio which consists of the two credits. It has a face value of $200. Write down the probability distribution of the portfolios payoffs and calculate the expected repayment.

(b) The bank wants to build a collateralized debt obligation (CDO) out of this portfolio. Therefore, slice the portfolio into three tranches: an equity tranche with face value $50, a mezzanine tranche with face value $80, and a senior tranche with face value $70. Write down the probability distribution of repayments for each tranche and calculate the expected repayment for each tranche.

Bibliography

1 J. Abate and W. Whitt. The fourier-series method for inverting transforms of probability distributions. *Queueing Systems*, 10(1):5–87, 1992.
2 R.A. Adams. *Sobolev Spaces*. Academic Press, 1975.
3 Y. Aït-Sahalia. Maximum likelihood estimation of discretely sampled diffusions: A closed-form approximation approach. *Econometrica*, 70(1):223–262, 2002.
4 P. Amster, C. Averbuj, P. De Napoli, and M. C. Mariani. A parabolic problem arising on financial mathematics. *Nonlinear Analysis Series B: Real World Applications*, 11:759–763, 2010.
5 L. Andersen and J. Andreasen. Jump-Diffusion process: Volatility smile fitting and numerical methods for option pricing. *Review of Derivatives Research*, 4:231–262, 2000.
6 L.B.G. Andersen and V. V. Piterbarg. Moment explosions in stochastic volatility models. *Finance and Stochastics*, 11:29–50, 2007.
7 B. Anderson, J. Jackson, and M. Sitharam. Descartes' rule of signs revisited. *American Mathematical Monthly*, 105:447–451, 1998.
8 D. Applebaum. *Lévy Processes and Stochastic Calculus*. Cambridge University Press, 2004.
9 S. Asmussen and P.W. Glynn. *Stochastic Simulation: Algorithms and Analysis. Stochastic Modelling and Applied Probability*. Springer, 2007.
10 K. Atkinson. *An Introduction to Numerical Analysis*. John Wiley & Sons, Inc., 2nd edition, 1989.
11 K. Atkinson and W. Han. *Elementary Numerical Analysis*. John Wiley & Sons, Inc., 3rd edition, 2004.
12 M. Ausloos and K. Ivanova. Introducing false eur and false eur exchange rates. *Physica A: Statistical Mechanics and Its Applications*, 286:353–366, 2000.
13 M. Avellaneda, C. Friedman, R. Holmes, and D. Samperi. Calibrating volatility surfaces via relative-entropy minimization. *Applied Mathematical Finance*, 4(1):34–64, 1997.

Quantitative Finance, First Edition. Maria C. Mariani and Ionut Florescu.
© 2020 John Wiley & Sons, Inc. Published 2020 by John Wiley & Sons, Inc.

14 E. Barany, M.P. Beccar Varela, and I. Florescu. Long correlations applied to the study of memory effects in high frequency (tick) data, the down jones index and international indices. *Handbook of Modeling High Frequency Data*, Wiley, 1:327–346, 2011.

15 E. Barany, M.P. Beccar Varela, I. Florescu, and I. SenGupta. Detecting market crashes by analyzing long memory effects using high frequency data. *Quantitative Finance*, 12(4):623–634, 2011.

16 G. Barles, E. Chasseigne, and C. Imbert. Hölder continuity of solutions of second-order non-linear elliptic integro-differential equations. *Journal of the European Mathematical Society*, 13:1–26, 2011.

17 D. Bates. Jumps and stochastic volatility: the exchange rate processes implicit in deutschemark options. *The Review of Financial Studies*, 9:69–107, 1996.

18 M.P. Beccar-Varela, H. Gonzalez-Huizar, M.C. Mariani, and O.K. Tweneboah. Use of wavelets techniques to discriminate between explosions and natural earthquakes. *Physica A: Statistical Mechanics and Its Applications*, 457:42–51, 2016.

19 M.P. Beccar-Varela, M.C. Mariani, O.K. Tweneboah, and I. Florescu. Analysis of the lehman brothers collapse and the flash crash event by applying wavelets methodologies. *Physica A: Statistical Mechanics and Its Applications*, 474:162–171, 2017.

20 N.A. Beliaeva and S.K. Nawalkha. A simple approach to pricing american options under the heston stochastic volatility model. *The Journal of Derivatives*, 17(4):25–43, 2010.

21 J. Bertoin. *Lévy Processes*. Cambridge University Press, 1996.

22 D. Bertsimas and J. Tsitsiklis. Simulated annealing. *Statistical Science*, 8:10–15, 1993.

23 C. Bingham, M.D. Godfrey, and J.W. Tukey. Modern techniques of power spectrum estimation. *IEEE Transactions on Audio Electroacoustic*, AU-15:55–66, 1967.

24 F. Black and M. Scholes. The valuation of options and corporate liability. *Journal of Political Economy*, 81:637–654, 1973.

25 S. Bochner. Diffusion equation and stochastic processes. *Proceedings of the National Academy of Sciences of the United States of America*, 35(7):368370, 1949.

26 K. Boukhetala and A. Guidoum. A package for simulation of diffusion processes in R., ffhal-00629841f, 2011.

27 P.P. Boyle and T. Vorst. Option replication in discrete time with transaction costs. *The Journal of Finance*, 47:271, 1992.

28 E.O. Brigham. *The Fast Fourier Transform*. Prentice-Hall, Inc., 1974.

29 D. Brigo and F. Mercurio. *Interest Rate Models – Theory and Practice: With Smile, Inflation and Credit*. Springer, 2nd edition, 2006.

30 C.G. Bucher. Adaptive sampling – An iterative fast Monte Carlo procedure. *Structural Safety*, 5:119–128, 1988.
31 S.V. Buldyrev, A.L. Goldberger, S. Havlin, R.N. Mantegna, M.E. Matsa, C.K. Peng, M. Simons, and H.E. Stanley. Long-range correlation properties of coding and noncoding dna sequences: Genbank analysis. *Physical Review E*, 51:5084–5091, 1995.
32 R.E. Caflisch and S. Chaudhary. Monte carlo Simulation for American Options. Springer, 116, 2004.
33 G. Casella and R.L. Berger. *Statistical Inference*. Duxbury advanced series in statistics and decision sciences. Thomson Learning, 2002.
34 A.L. Cauchy. *Analyse Algébrique*. Imprimerie Royale, 1821.
35 N. Chen, R. Roll, and S.A. Ross. Economic forces and the stock market. *The Journal of Business*, 59(3):383–403, 1986.
36 M. Chesney and L. Scott. Pricing European currency options: a comparison of the modified black-scholes model and a random variance model. *Journal of Financial and Quantitative Analysis*, 24:267–284, 1989.
37 P. Cizeau, Y.H. Liu, M. Meyer, C.K. Peng, and H.E. Stanley. Volatility distribution in the s&p500 stock index. *Physica A: Statistical Mechanics and Its Applications*, 245:441–445, 1997.
38 L. Clewlow and C. Strickland. *Implementing Derivatives Models*. Wiley, 1998.
39 V. Connor. The three types of factor models: A comparison of their explanatory power. *Financial Analysts Journal*, 51(3):42–46, 1995.
40 R. Cont and P. Tankov. *Financial Modelling with Jumps Processes*. CRC Financial mathematics series. Chapman & Hall, 2004.
41 R. Cont and E. Voltchkova. Integro-differential equations for option prices in exponential lévy models. *Finance and Stochastic*, 9:299–325, 2005.
42 D.R. Cox. Some statistical methods connected with series of events. *Journal of the Royal Statistical Society*, 17(2):129–164, 1955.
43 J. Cox. Notes on option pricing i: Constant elasticity of variance diffusions. Working paper, Stanford University, 1975.
44 J. Cox. Notes on option pricing i: Constant elasticity of variance diffusions. *Journal of Portfolio Management*, 22:15–17, 1996.
45 J. Cox and S. Ross. The valuation of options for alternative stochastic processes. *Journal of Financial Economics*, 3:145–166, 1976.
46 J.C. Cox, J.E. Ingersoll, and S.A. Ross. A theory of the term structure of interest rates. *Econometrica*, (53):385–407, 1985.
47 J. Cox, J.E. Ingersoll Jr,, and S.A. Ross. A theory of the term structure of interest rates. *Econometrica*, 53(2):385–408, 1985.
48 J.C. Cox, S.A. Ross, and M. Rubinstein. Option pricing: A simplified approach. *Journal of Financial Econometrics*, 7:229–263, 1979.

49 S. Dajcman. Interdependence between some major european stock markets-a wavelet lead/lag analysis. *Prague Economic Papers*, pages 28–49, 2013.

50 M.H.A. Davis, G.P. Vassilios, and T. Zariphopoulou. European option pricing with transaction costs. *SIAM Journal on Control and Optimization*, *31*:470–493, 1993.

51 F. Delbaen and W. Schachermayer. A general version of the fundamental theorem of asset pricing. *Mathematische Annalen*, *300*:463–520, 1994.

52 P. Del Moral. *Feynman-Kac Formulae. Genealogical and Interacting Particle Systems with Applications*. Probability and Its Applications (New York). Springer-Verlag, 2004.

53 P. Del Moral, J. Jacod, and P. Protter. The monte-carlo method for filtering with discrete time observations. *Probability Theory and Related Fields*, *120*:346–368, 2001.

54 A. Dembo. Lecture notes in probability. http://statweb.stanford.edu/~adembo/stat-310b/lnotes.pdf

55 J.N. Dewynne A.E. Whalley and P. Wilmott. Path-dependent options and transaction costs. *Philosophical Transactions: Physical Sciences and Engineering 347* (1684):517–5291994

56 A. Doucet, N. DeFreitas, and N. Gordon. An introduction to sequential monte carlo methods. *Sequential Monte Carlo Methods in Practice*, Springer–Verlag, *214*, 2001.

57 D. Duffie. *Dynamic Asset Pricing Theory*. Princeton University Press, 2001.

58 D. Duffie and N. Garleanu. Risk and valuation of collateralized debt obligations. *Financial Analysts Journal*, *57*:41–62, 2001.

59 B. Dupire. Pricing with a smile. *Risk*, *7*: 18–20, 1994.

60 B. Dupire. *Pricing and Hedging with Smiles, in Mathematics of Derivative Securities*. Cambridge University Press, 1997.

61 B. Engelmann, M. Fengler, and P. Schwendner. Better than its reputation: An empirical hedging analysis of the local volatility model for barrier options. *Journal of Risk*, *12*:53–77, 2005.

62 S.N. Ethier and T.G. Kurtz. *Markov Processes: Characterization and Convergence*. Wiley, 2005.

63 L.C. Evans. *Partial Differential Equations*. American Mathematical Society, 1998.

64 F. Fang and C. Oosterlee. A novel pricing method for european options based on fourier-cosine series expansions. *SIAM Journal on Scientific Computing*, *31*(2):826–848, 2008.

65 H. Federer. A note on the gauss-green theorem. *Proceedings of the American Mathematical Society*, *9*(3):447–451, 1958.

66 P.J. Van Fleet. *Discrete Wavelet Tranformations: An Elementary Approach with Applications*. Wiley, 2008.

67 I. Florescu. *Probability and Stochastic Processes*. John Wiley & Sons, 2014.
68 I. Florescu and M.C. Mariani. Solutions to an integro-differential parabolic problem arising in the pricing of financial options in a levy market. *Electronic Journal of Differential Equations*, *62*:1–10, 2010.
69 I. Florescu and C.G. Păsărică. A study about the existence of the leverage effect in stochastic volatility models. *Physica A: Statistical Mechanics and Its Applications*, *388*(4):419–432, 2009.
70 I. Florescu and C.A. Tudor. *Handbook of Probability*. Wiley, 2014.
71 I. Florescu and F. Viens. Stochastic volatility: Option pricing using a multinomial recombining tree. *Applied Mathematical Finance*, *15*(2):151–181, 2008.
72 I. Florescu, R. Liu, and M.C. Mariani. Numerical solutions to an integro-differential parabolic problem arising in the pricing of financial options in a levy market. *Electronic Journal of Differential Equations*, *2012*(231):1–12, 2012.
73 I. Florescu, M.C. Mariani, and G. Sewell. Numerical solutions to an integro-differential parabolic problem arising in the pricing of financial options in a levy market. *Quantitative Finance*, *14*(8):1445–1452, 2011.
74 G.B. Folland. *Introduction to Partial Differential Equations*. Princeton University Press, 2nd edition, 1995.
75 E. Foufoula-Georgiou and P. Kumar. *Wavelet Analysis and Its Applications*. Academic Press, Inc, 1994.
76 J.E. Fowler. The redundant discrete wavelet transform and additive noise. *IEEE Signal Processing Letters*, *12*(9):629–632, 2005.
77 A. Friedman. *Variational Principles and Free Boundary Problems*. Wiley, 1982.
78 G. Fusai and A. Roncoroni. *Implementing Models in Quantitative Finance: Methods and Cases*. Springer Science & Business Media, 2007.
79 C. Genest and L.P. Rivest. Statistical inference procedures for bivariate archimedean copulas. *Journal of the American Statistical Association*, *55*:698–707, 1993.
80 T. Gerstner and P.E. Kloeden. *Recent Developments in Computational Finance: Foundations, Algorithms and Applications*. World Scientific – Business & Economics, 2013.
81 P. Glasserman. *Monte Carlo Methods in Financial Engineering.*, Volume 53 of Stochastic Modelling and Applied Probability. Springer New York, 2010, 2003.
82 A. Graps. An introduction to wavelets. *IEEE Computational Science and Engineering*, *2*:50–61, 1995.
83 A.C. Guidoum and K. Boukhetala. Estimation of stochastic differential equations with sim.diffproc package version 3.0, 2015.

84 X. Guo. Information and option pricings. *Quantitative Finance*, 1(1):38–44, 2000.

85 T.J. Haahtela. Recombining trinomial tree for real option valuation with changing volatility. A http://ssrn.com/abstract=1932411, 2010.

86 P. Hagan, D. Kumar, A. Lesniewski, and D. Woodward. *Managing smile risk*. Wilmott Magazine, 2002.

87 M. Hanke and E.R. Osler. Computation of local volatilities from regularized dupire equations. *International Journal of Theoretical and Applied Finance*, 8(2):207–222, 2005.

88 J.M. Harrison and S.R. Pliska. Martingales and stochastic integrals in the theory of continuous trading. *Stochastic Processes and Their Applications*, 11(3):215–260, 1981.

89 P. Hartman and G. Stampacchia. On some nonlinear elliptic differential functional equations. *Acta Mathematica*, 115:271–310, 1966.

90 J.G. Hayes. Numerical methods for curve and surface fitting. *Bulletin Institute of Mathematics and Its Applications*, 10:144–152, 1974.

91 J.G. Hayes and J. Halliday. The least-squares fitting of cubic spline surfaces to general data sets. *Journal of the Institute of Mathematics and Its Applications*, 14:89–103, 1974.

92 J.H. He. Homotopy perturbation technique. *Computer Methods in Applied Mechanics and Engineering*, 178:257–262, 1999.

93 D.C. Heath and G. Swindle. *Introduction to Mathematical Finance: American Mathematical Society Short Course*. American Mathematical Society, Providence, R.I., 1999.

94 D. Heath, R. Morton, and A. Morton. Bond pricing and the term structure of interest rates: A new methodology for contingent claims valuation. *Econometrica, Econometric Society*, 60(1):77–105, 1992.

95 S.L. Heston. A closed-form solution for options with stochastic volatility with applications to bond and currency options. *Review of Financial Studies*, 6(2):327–43, 1993.

96 S.D. Hodges and A. Neuberger. Optimal replication of contingent claims under transaction costs. *Review Futures Market*, 8:222–239, 1993.

97 T. Hoggard, A.E. Whalley, and P. Wilmott. Hedging option portfolios in the presence of transaction costs. *Advance Futures Optics Research*, 7:21, 1994.

98 R. Howard. The gronwall inequality, 1998. http://www.math.sc.edu/howard/Notes/gronwall.pdf.

99 Y.L. Hsu, T.I. Lin, and C.F. Lee. Constant elasticity of variance (cev) option pricing model: Integration and detailed derivations. *Mathematics and Computers in Simulation*, 79:60–71, 2008.

100 R. Huang. Valuation of mortgage-backed securities. Master of Arts Dissertation, Simon Fraser University, 2006.

101 J.C. Hull. *Options, Futures and Other Derivatives*. Prentice Hall, 7th edition, 2008.
102 J.C. Hull and A.D. White. The pricing of options on assets with stochastic volatilities. *Journal of Finance*, 42(2):281–300, 1987.
103 H.E. Hurst. Long term storage of reservoirs. *Transactions of the American Society of Civil Engineers*, 116:770–808, 1950.
104 N. Ikeda. *Stochastic Differential Equations and Diffusion Processes*. North Holland, 2nd edition, 1989.
105 H. Imai, N. Ishimura, I. Mottate, and M. Nakamura. On the Hoggard–Whalley–Wilmott equation for the pricing of options with transaction costs. *Asia Pacific Financial Markets*, 13(4):315–326, 2006.
106 Investopedia. Bond etfs: A viable alternative, 2009. Investopedia.com.
107 Investopedia. Commodity etf, 2017. Investopedia.com.
108 Investopedia. Introduction to exchange-traded funds, 2017. Investopedia.com.
109 Investopedia. Leverage etf, 2017. Investopedia.com.
110 K. Ivanova and M. Ausloos. Application of the detrended fluctuation analysis (dfa) method for describing cloud breaking. *Physica A: Statistical Mechanics and Its Applications*, 274:349–354, 1999.
111 K. Ito. On stochastic differential equations. *Memoirs of the American Mathematical Society*, 4:1–51, 1951.
112 S. Janson. Stable distributions. arXiv :1112.0220 [math], 2001.
113 R. Jarrow and A. Rudd. *Option Pricing*. Dow Jones-Irwin, 1983.
114 Y.M. Kabanov and M.M. Safarian. On leland's strategy of option pricing with transaction costs. *Finance Stochast*, 1:239–250, 1997.
115 R.V.D. Kamp. Local volatility modeling, 2009. Master's thesis, University of Twente, The Netherlands.
116 J.W. Kantelhardt, R. Berkovits, S. Havlin, and A. Bunde. Are the phases in the anderson model long-range correlated? *Physica A: Statistical Mechanics and Its Applications*, 266:461–464, 1999.
117 I. Karatzas and S.E. Shreve. *Brownian Motion and Stochastic Calculus*. Springer, 2nd edition, 1991.
118 I. Karatzas and S.E. Shreve. *Methods of Mathematical Finance*. Springer, 1998.
119 S. Karlin and H.M. Taylor. *A First Course in Stochastic Processes*. Academic Press, 2nd edition, 1975.
120 M. Kessler. Estimation of an ergodic diffusion from discrete observations. *Scandinavian Journal of Statistics*, 24:211–229, 1997.
121 A.Ya. Khintchine and P. Lévy. Sur les lois stables. *Comptes Rendus de l'Académie des Sciences*, 202:374, 1936.
122 D. Kincaid and W. Cheney. *Numerical Analysis: Mathematics of Scientific Computing*. American Mathematical Society, 3rd edition, 2002.

123 P.E. Kloeden and E. Platen. *Numerical Solution of Stochastic Differential Equations*. Stochastic Modelling and Applied Probability. Springer, 1992.

124 I. Koponen. Analytic Approach to the Problem of Convergence of Truncated Lévy Flights toward the Gaussian Stochastic Process. *Physical Review E*, *52*:1197, 1995.

125 E. Koscienly-Bunde, A. Bunde, S. Havlin, H.E. Roman, Y. Goldreich, and H.J. Schellnhuber. Indication of universal persistence law governing atmospheric variability. *Physical Review Letters*, *81*:729–732, 1998.

126 E. Koscienly-Bunde, H.E. Roman, A. Bunde, S. Havlin, and H.J. Schellnhuber. Long-range power-law correlations in local daily temperature fluctuations. *Philosophical Magazine B*, *77*:1331–1339, 1998.

127 S.G. Kou. A jump diffusion model for option pricing. *Management Science*, *48*:1086–1101, 2002.

128 S.G. Kou and H. Wang. Option pricing under a double exponential jump diffusion model. Preprint, *Columbia University and Brown University*.

129 N.V. Krylov. *Lectures on Elliptic and Parabolic Equations in Hölder Spaces, volume 12*. American Mathematical Society, 1996.

130 O.A. Ladyzenskaja, V.A. Solonikov, and N.N. Ural'ceva. *Linear and Quasilinear Equations of Parabolic Type*. American Mathematical Society, Providence, RI, 1964.

131 B. Lee and Y.S. Tarng. Application of the discrete wavelet transform to the monitoring of tool failure in end milling using the spindle motor current. *International Journal of Advanced Manufacturing Technology*, *15*(4):238243, 1999.

132 H.E. Leland. Option pricing and replication with transaction costs. *Journal of Finance*, *40*:1283–301, 1985.

133 A. Lesniewski. Short rate models. https://www.math.nyu.edu/alberts/spring07/Lecture5.pdf, 2008.

134 G.M. Lieberman. *Second Order Parabolic Differential Equations*. World Scientific, 1996.

135 R.H. Liu. Regime-switching recombining tree for option pricing. *International Journal of Theoretical and Applied Finance*, *13*:479–499, 2010.

136 R.H. Liu, Q. Zhang, and G. Yin. Option pricing in a regime switching model using the fast fourier transform. *Journal of Applied Mathematics and Stochastic Analysis*, *22*:1–22, 2006.

137 Y.H. Liu, P. Cizeau, M. Meyer, C.K. Peng, and H.E. Stanley. Quantification of correlations in economic time series. *Physica A: Statistical Mechanics and its Applications*, *245*:437–440, 1997.

138 A. Lo. *Hedge Funds: An Analytic Perspectives*. Princeton University Press, 2008.

139 A. Lo. *Structured Finance & Collaterlaized Debt Obligations*. Wiley & Sons, 2008.
140 F.A. Longstaff and E.S. Schwartz. Valuing american options by simulation: a simple least-squares approach. *The Review of Financial Studies*, 14(1):113–147, 2001.
141 B.B. Mandelbrot. *The Fractal Geometry of Nature*. Freeman and Co., 1982.
142 B.B. Mandelbrot and J.W. Van Ness. Fractional brownian motions, fractional noises and applications. *SIAM Review*, 10:422–437, 1968.
143 R.N. Mantegna and H.E. Stanley. Stochastic process with ultraslow convergence to a Gaussian: The truncated Lévy flight. *Physical Review Letters*, 73:2946, 1994.
144 M.C. Mariani, I. Florescu, M.P. Beccar Varela, and E. Ncheuguim. Long correlations and lévy models applied to the study of memory effects in high frequency (tick) data. *Physica A: Statistical Mechanics and Its Applications*, 388(8):1659–1664, 2009.
145 M.C. Mariani, I. SenGupta, and P. Bezdek. Numerical solutions for option pricing models including transaction costs and stochastic volatility. *Acta Applicandae Mathematicae*, 118:203–220, 2012.
146 M.C. Mariani and O.K. Tweneboah. Stochastic differential equations applied to the study of geophysical and financial time series. *Physica A: Statistical Mechanics and Its Applications*, 443:170–178, 2016.
147 H. Markowitz. *Portfolio Selection: Efficient Diversification of Investments*. John Wiley & Sons, Inc, 1959.
148 G. Marsaglia and W.W. Tsang. The ziggurat method for generating random variables. *Journal of Statistical Software*, 5:1–7, 2000.
149 D. Maxey. Fidelity the latest to caution on etfs, 2009. Online.wsj.com.
150 R.C. Merton. An intertemporal capital asset pricing model. *Econometrica*, 7:867–887, 1973.
151 R.C. Merton. Option pricing when the underlying stock returns are discontinuous. *Journal of Financial Econometrics*, 3:115–144, 1976.
152 R.C. Merton. *Continuous-Time Finance*. Wiley-Blackwell, 1992.
153 D.E. Muller. A method for solving algebraic equations using an automatic computer. *Mathematical Tables and Other Aids to Computation*, 10:208–215, 1956.
154 R.B. Nelsen. *An Introduction to Copulas*. Springer, 1999.
155 R.B. Nelsen. Properties and applications of copulas: a brief survey. *Proceedings of the First Brazilian Conference on Statistical Modeling in Insurance and Finance*, University Press USP, 10–28, 2003.
156 R.B. Nelsen. *An Introduction to Copulas*. Springer, 2006.
157 D.E. Newland. Harmonic wavelet analysis. *Proceedings of the Royal Society of London, Series A (Mathematical and Physical Sciences)*, 443(1917):203–225, 1993.

158 B. Øksendal. *Stochastic Differential Equations*. Springer Verlag, 5th edition, 2003.

159 B. Øksendal. *Stochastic Differential Equations: An Introduction with Applications*. Springer, 10th edition, 2010.

160 A.V. Ovanesova and L.E. Surez. Applications of wavelet transforms to damage detection in frame structures. *Engineering Structures*, 26:39–49, 2004.

161 L. Paul. *Calcul des Probabilits*. Gauthier-Villars, 1925.

162 W. Paul. *Introduction to Quantitative Finance*, volume 2. John Wiley & Sons Ltd. 2007.

163 C.K. Peng, S. Havlin, H.E. Stanley, and A.L. Goldberger. Quantification of scaling exponents and crossover phenomena in nonstationary heartbeat time series. *Chaos*, 5:82–87, 1995.

164 C.K. Peng, S.V. Buldyrev, A.L. Goldberger, R.N. Mantegna, M. Simons, and H.E. Stanley. Statistical properties of dna sequences. *Physica A: Statistical Mechanics and Its Applications*, 221:180–192, 1995.

165 C.K. Peng, S.V. Buldyrev, S. Havlin, M. Simons, H.E. Stanley, and A.L. Goldberger. Mosaic organization of dna nucleotides. *Physical Review E*, 49:1685–1689, 1994.

166 C.K. Peng, J. Mietus, J.M. Hausdorff, S. Havlin, H.E. Stanley, and A.L. Goldberger. Long-range anticorrelations and non-gaussian behavior of the heartbeat. *Physical Review Letters*, 70:1343–1346, 1993.

167 M. Pilz and S. Parolai. Tapering of windowed time series. *New Manual of Seismological Observatory Practice 2 (NMSOP-2)*, Deutsches Geo-ForschungsZentrum GFZ, 1–4. 2012.

168 L.D. Pitt. Two problems in markov processes. Extending the life span of markov processes. Products of markovian semi-groups. PhD thesis, Princeton University, 1967.

169 L.C.G. Rogers. Monte carlo valuation of american options. *Mathematical Finance*, 12:271–286, 2002.

170 F.D. Rouah. The sabr model. http://www.volopta.com.

171 R. Rowland. Etf statistics for june 2012: Actively managed assets less than, 2012.

172 H. Royden. *Real Analysis*. Prentice Hall, 3rd edition, 1988.

173 D.B. Rubin. Using the sir algorithm to simulate posterior distributions. *Bayesian Statistics 3*. Oxford University Press, 1998.

174 H. Rubin and B.C. Johnson. Efficient generation of exponential and normal deviates. *Journal of Statistical Computation and Simulation*, 76(6):509–518, 2006.

175 RYDEX. Rydex etf trust (form: 485apos), 2006. https://www.sec.gov/Archives/edgar/data/1208211/000119312512085741/d274410d485bpos.htm

176 G. Samorodnitsky and M.S. Taqqu. *Stable Non-Gaussian Random processes: Stochastic Models with Infinite Variance.* Chapman and Hall, 1994.
177 K. Sato. *Lévy Processes and Infinitely Divisible Distributions.* Cambridge University Press, 1999.
178 N. Scafetta and P. Grigolini. Scaling detection in time series: Diffusion entropy analysis. *Physical Review E,* 66:036–130, 2002.
179 W. Schoutens. *Lévy Processes in Finance: Pricing Financial Derivatives.* Wiley Series in Probability and Stochastics, 2003.
180 W. Schoutens and J. Cariboni. *Lévy Processes in Credit Risk.* John Wiley & Sons, 2009.
181 L.O. Scott. Option pricing when the variance changes randomly: Theory, estimation, and an application. *The Journal of Financial and Quantitative Analysis,* 22(4):419–438, 1987.
182 SEC. Actively managed exchange-traded funds, 2001. SEC Release No. IC-25258, 66 Fed. Reg. 57614.
183 SEC. Exchange-traded funds (etfs), 2013. U.S. Securities and Exchange Commission.
184 I.W. Selesnick, R.G. Baraniuk, and N.C. Kingsbury. The dual-tree complex wavelet transform. *IEEE Signal Processing Magazine,* 22(6):123–151, 2005.
185 I. SenGupta. Differential operator related to the generalized superradiance integral equation. *Journal of Mathematical Analysis and Applications,* 369:101–111, 2010.
186 W.F. Sharpe. Capital asset prices: A theory of market equilibrium under conditions of risk. *Journal of Finance,* 19(3):425–442, 1964.
187 A.N. Shiryaev. *Essentials of the Stochastic Finance.* World Scientific, 2008.
188 S.E. Shreve. *Stochastic Calculus for Finance II.* Springer, 2004.
189 R.H. Shumway and D.S. Stoffer. *Time Series Analysis and Its Applications With R Examples.* Springer, 2010.
190 K. Simmons. The diagonal argument and the liar. *Journal of Philosophical Logic,* 19:277–303, 1990.
191 D.R. Smart. *Fixed Point Theorem.* Cambridge University Press, 1980.
192 H.M. Soner, S.E. Shreve, and J. Cvitanic. There is no nontrivial hedging portfolio for option pricing with transaction costs. *The Annals of Applied Probability,* 5(2):327–355, 1995.
193 I. Stakgold. *Green's Functions and Boundary Value Problems.* Wiley, 1979.
194 E.M. Stein and J.C. Stein. Stock price distributions with stochastic volatility: An analytic approach. *Review of Financial Studies,* 4(4):727–52, 1991.
195 L. Stentoft. Convergence of the least-squares monte carlo approach to american option valuation. Center for Analytical Finance 16, Working Paper Series 113, University of Aarhus School of Business, 2002.

196 W.J. Stewart. *Probability, Markov Chains, Queues, and Simulation: The Mathematical Basis of Performance Modeling*. Princeton University Press, 2009.

197 V. Stojanovic, M. Stankovic, and I. Radovanovic. Application of wavelet analysis to seismic signals. *4th International Conference on Telecommunications in Modern Satellite, Cable and Broadcasting Services. TELSIKS'99 (Cat. No.99EX365)*. Nis, Yugoslavia (13–15 October 1999). IEEE, 13–15, 1999.

198 D.W. Stroock and S.R.S. Varadhan. *Multidimensional Diffusion Processes*. Springer, 1979.

199 L. Trigeorgis. A log-transformed binomial numerical analysis method for valuing complex multi-option investments. *Journal of Financial and Quantitative Analysis*, 26(3):309–326, 1991.

200 O.K. Tweneboah. Stochastic differential equation applied to high frequency data arising in geophysics and other disciplines. ETD Collection for University of Texas, El Paso, Paper AAI1600353, 2015.

201 N. Vandewalle, M. Ausloos, M. Houssa, P.W. Mertens, and M.M. Heyns. Non-gaussian behavior and anticorrelations in ultrathin gate oxides after soft breakdown. *Applied Physics Letters*, 74:1579–1581, 1999.

202 O. Vasicek. An equilibrium characterisation of the term structure. *Journal of Financial Economics*, (5):177188, 1977.

203 T.A. Vuorenmaa. A wavelet analysis of scaling laws and long-memory in stock market volatility. *Proceeding of the SPIE 5848, Noise and Fluctuations in Econophysics and Finance*, 27:39–54, 2005.

204 W. Wang and M.T. Wells. Model selection and semiparametric inference for bivariate failure-time data. *Journal of the American Statistical Association*, 95:62–76, 2000.

205 C. Wang, Z. Wu, and J. Yin. *Elliptic and Parabolic Equations*. World Scientific Publishing, 2006.

206 A.E. Whaley and P. Wilmott. An asymptotic analysis of an optimal hedging model for option pricing with transaction costs. *Mathematical Finance*, 7(3):307–324, 1997.

207 R.L. Wheeden and A. Zygmund. *Measure and Integral: An Introduction to Real Analysis*. Marcel Dekker Inc., 1977.

208 M.J. Wichura. Algorithm as 241: The percentage points of the normal distribution. *Journal of the Royal Statistical Society. Series C (Applied Statistics)*, 37(3):477–484, 1988.

209 J.B. Wiggins. Option values under stochastic volatility: Theory and empirical estimates. *Journal of Financial Economics*, 19(2):351–372, 1987.

210 P. Wilmott. *The Mathematics of Financial Derivatives*, volume 2, Cambridge University Press. 1999.

211 P. Wilmott, S. Howison, and J. Dewynne. *The Mathematics of Financial Derivatives a Student Introduction*. Cambridge University Press, 1995.

212 H. Wong, Z. Hie W. IP, and X. Lui. Modelling and forecasting by wavelets, and the application to exchange rate. *Journal of Applied Statistics*, 30(5):537–553, 2003.

213 H.M. Yin. A uniqueness theorem for a class of non-classical parabolic equations. *Applicable Analysis*, 34:67–78, 1989.

214 H. Zhao, R. Chatterjee, T. Lonon, and I. Florescu. Pricing bermudan variance swaptions using multinomial trees. *The Journal of Derivatives*, 26(3): 22–34, 2018.

215 H. Zhao, Z. Zhao, R. Chatterjee, T. Lonon, and I. Florescu. Pricing variance, gamma, and corridor swaps using multinomial trees. *The Journal of Derivatives*, 25(2):7–21, 2017.

216 L. Zhengyan and B. Zhidong. *Probability Inequalities*. Springer, 2011.

217 E. Zivot. *Unit root tests, Modeling Financial Time Series with S-PLUS*. Springer, 2006.

218 V. Zolotarev. One-dimensional stable distributions. *American Mathematical Society*, 195, 1990.

Index

a

abscissa nodes 26
accept–reject method 224–235
actual volatility 70
adapted stochastic process 6
adaptive importance sampling technique 241–242
additive tree 123
ADF *see* Augmented Dickey Fuller Test (ADF)
affine term structure 427–428
American Call option 36
American Call prices 37
American option contract 36
American option pricing
  binomial tree 126–127
  finite difference methods 199–201
  Monte Carlo simulation method
    LSM 210–216
    martingale optimization 210
    risk-neutral valuation 209
American Put option 127
American Put prices 37
antithetic variates method 205–206
approximate likelihood estimation 24
arbitrage 33–35
arbitrage-free payoff 34
Archimedean copulas 411–412
ARCH processes *see* autoregressive conditionally heteroscedastic (ARCH) processes

Arzelà–Ascoli theorem 374
Augmented Dickey Fuller Test (ADF) 335
autoregressive conditionally heteroscedastic (ARCH) processes 9–10

b

Banach spaces 86, 371
BARRA approach *see* Bar Rosenberg (BARRA) approach
barrier options 139–140
Bar Rosenberg (BARRA) approach 408
basket credit default swaps 439–440
Bayesian statistics 217
Bernoulli process 14–16
binary credit default swaps 439
binomial hedging 39
binomial tree
  American option pricing 126–131
  Brownian motion 132–135
  compute hedge sensitivities 128
  European Call 125–126
  one-step 122–125
  pricing path-dependent options 127–128
bisection method 65–66
Black–Karasinski model 264
Black–Scholes–Merton derivation 49–50
Black–Scholes–Merton model 44, 60

Quantitative Finance, First Edition. Maria C. Mariani and Ionut Florescu.
© 2020 John Wiley & Sons, Inc. Published 2020 by John Wiley & Sons, Inc.

460 Index

Black–Scholes–Merton PDE 156, 179
Black–Scholes (B–S) model 49
   Call option 58–60
   cap, floor, and collar 420
   continuous hedging 50
   Euler–Milstein scheme 276
   European Call option in 51–54, 58
   European Put options 54–57
   finite difference methods 178
   with jumps 305–310
   PDEs 61, 96–98
   SABR model 65
   swaption 420
   trinomial trees 140
   zero coupon bond options 420
bond ETFs 433
bond market
   coupon bond option 417
   forward contract on the zero coupon bond 416
   forward rate 416
   instantaneous forward rate 416–417
   options on bonds 417
   short rate 415–416
   spot rate associated with a zero coupon bond 416
   swap rate calculation 417–419
   swaption 417
   term structure consistent models 422–426
   yield curve inversion 426–428
   zero coupon and vanilla options 419–422
   zero coupon bond price 416
Borel—Cantelli lemmas 19–20
boundary conditions (BC)
   Heston PDE 189–190
   for PDE 84–85
bounded domains 372–385
Box-Muller method 221–222
Brent's method 68
Brownian motion 14, 17–19, 42, 122, 132–135, 257–258
B–S model *see* Black–Scholes (B–S) model

c

Call option 35, 36, 58–60
Capital Asset Pricing Model (CAPM) 405
CAPM single factor model 407
Cauchy's functional equation 11
Cauchy stable distribution 340
CBO *see* collateralized bond obligation (CBO)
CDO *see* collateralized debt obligation (CDO)
CDS *see* credit default swap (CDS)
central limit theorems (CLT) 16, 19, 20, 203, 204
Chapman–Kolmogorov equation 144, 146
characteristic function 326
Cholesky decomposition 270, 282–283
CIR models *see* Cox–Ingersoll–Ross (CIR) models
classical mean-reverting Ørnstein–Uhlenbeck (O–U) process 294
collar 419
collateralized bond obligation (CBO) 442–443
collateralized debt obligation (CDO)
   collateralized bond obligation 442–443
   collateralized loan obligation 442
   collateralized mortgage obligation 441–442
collateralized loan obligation (CLO) 442
collateralized mortgage obligation (CMO) 441–442
commodity ETFs 433–434
complete filtration 5

compound Poisson process 304–305
constant elasticity of variance (CEV)
  models 291–292
continuous hedging 50
continuous time model 41–45
continuous time SV models 290–291
  CEV models 291–292
  Heston model 295–296
  Hull–White model 292–293
  mean-reverting processes 296–301
  SABR model 293–294
  Scott model 294–295
  Stein and Stein model 295
contraction 149
control variates method 206–208
convergence
  Crank–Nicolson finite difference
    methods 183–184
  explicit finite difference method
    180
  implicit finite difference method
    182–183
copula models 409–412
coupon bond option 417
covariance matrix 278
covariance stationary 9
Cox–Ingersoll–Ross (CIR) models 23,
    63–64, 421
  analysis of 263–265
  moments calculation for the
    265–267
  parameter estimation for the
    267–268
Cox–Ross–Rubinstein (CRR) tree 124
Crank–Nicolson finite difference
  methods
  number of nodes 185–186
  stability and convergence 183–184
credit default swap (CDS) 436–437
  basket credit default swaps
    439–440
  binary credit default swaps 439
  example of 437
  recovery rate estimates 439

valuation of 437–439
credit event 436
cross-sectional regression 404–406
cumulative distribution function (cdf)
  218–220
currency ETFs 434–435

d

Daniell kernel transforms 103–104
Darboux sum 12
Daubechies wavelets 115–116
delivery price 34
density estimation 163
detrended fluctuation analysis (DFA)
  332–334, 357, 362–364
diffusion equation 84
Dirichlet distribution 279–280
Dirichlet-type conditions 84, 85
discrete Fourier transform (DFT)
  102, 104–106
discrete Markov chain 157
discrete-time Fourier transform (DTFT)
  104
discrete-time stochastic variance
  processes 10
discrete wavelet transform (DWT)
  111–112
  Daubechies wavelets 115–116
  Haar functions 113
  Haar wavelets 112–114
dividend-paying assets
  continuous dividends
    135–136
  discrete cash dividend
    136–137
  discrete proportional dividends
    136
  time-varying volatility 137–139
Dow Jones Industrial Average (DJIA)
  index 352, 356
Dupire formula 76
Dupire's equation 74–77
DWT see discrete wavelet transform
  (DWT)

dynamic Fourier analysis
  DFT   104–106
  FFT   106–109
  spectral density with Daniell Kernel   103–104
  tapering   102–103
dynamic model   157

e

ETFs *see* exchange traded funds (ETFs)
Eugene Fama and Kenneth French (FAMA-FRENCH) approach   408
Euler method   273–277
Euler–Milstein method   274–276
Euler's algorithm   94
European Call option   36, 37, 51–54, 68, 125–126, 129
European option contract   35
European Put options   36, 37, 54–57
exchange traded funds (ETFs)   431–432
  bond   433
  commodity   433–434
  currency   434–435
  index   432–433
  inverse   435
  leverage   435–436
  stock   433
expected returns   406–407
explicit finite difference methods
  Black–Scholes–Merton PDE   179
  number of nodes   184–185
  stability and convergence   180
exponential Lévy models   330–331

f

factor model   403–404
  cross-sectional regression   404–406
  expected returns   406–407
  fundamental factor models   408
  macroeconomic factor models   407
  statistical factor models   408–409
fast Fourier transforms (FFT)
  algorithms   106–107
"fat tailed" distribution   344
Feller semigroup   151–152
Feynman–Kaç theorem   271, 366
FFT algorithms *see* fast Fourier transforms (FFT) algorithms
filtration   5–6
finance *see* mathematical finance; quantitative finance
finite difference methods
  B–S PDE   178
  Crank–Nicolson   183–184
  explicit   179–180
  free boundaries
    American Option Valuations   192
    LCP   193–196
    obstacle problem   196–199
  implicit   180–183
  pricing American options   199–201
finite dimensional distribution   8
Fisher, R. A.   3
Fokker–Planck equation   23
Fong model   264
forward contract on the zero coupon bond   416
forward price   34
forward rate   416
Fourier series approximation method   93–96
Fourier transform
  dynamic Fourier analysis
    DFT   104–106
    FFT   106–109
    spectral density with Daniell Kernel   103–104
    tapering   102–103
  Lévy flight models   342–343
Frobenius norm   148
Fubini's theorem   265
functional approximation methods
  Dupire's equation   74–77

local volatility model   74
numerical solution techniques
    78–79
pricing surface   79
spline approximation   77–78
fundamental factor models   408

g

Gamma process   329–330
Gauss–Green theorem   380–385
Gaussian copulas   411
Gaussian distributions, prescribed
    covariance structure   281–283
Gaussian random variable   339
Gaussian vectors   281–283
Gaussian white noise process   9
Gauss–Jordan algorithm   186, 187
Gauss–Seidel method   197–198
general diffusion process   152–156
generalized Black–Scholes equation
    366
generalized Erlang distribution
    349–350
general parabolic integro-differential
    problem   370–372
general PIDE system
    Black–Scholes equation   314, 315
    method of upper and lower solutions
        316–322
generating random variables
    importance sampling   236–242
    normally distributed random
        numbers
        Box–Muller method   221–222
        polar rejection method   222–224
    one-dimensional random variable,
        cdf   218–220
    rejection sampling method
        224–235
Girsanov's theorem   122, 125–126,
    135
Green's function   376
Gronwall inequality   382

h

Haar functions   112–114
Haar transform matrix   114
Haar wavelets   112–114
Hall, Philip   223
H-a relationship for the truncated Lévy
    flight   346–347
hedging   39–40
Heston differential equation   61–63
Heston model   60–63, 156, 295–296
Heston PDE
    boundary conditions   189–190
    nonuniform grid   191
    uniform grid   190–191
Hölder space   87, 376
homogeneous Markov process   144,
    146
homotopy perturbation method   309
Hull–White model   25, 292–293
Hurst analysis   332–334
Hurst methods   362–364
hypoexponential distribution
    349–350

i

implicit finite difference methods
    Black–Scholes–Merton PDE   180
    number of nodes   185
    Put option   181
    stability and convergence   182–183
implied volatility
    bisection method   65–66
    hedging with   71–73
    Newton's method   66–69
    pricing surface   79
    Put–Call parity   69–70
    secant method   67–68
importance sampling   236–242
increments   10–11
independent increments   10–11
index ETFs   432–433
index of dependence   332
index of long-range dependence   332

infinite divisibility of Lévy processes 327
infinitely divisible distribution 326
infinitesimal operator of semigroup 150–151
instantaneous forward rate 416–417
integro-differential parabolic problems
  assumptions 370–371
  Black–Scholes model 365–366
  construction of the solution in the whole domain 385–386
  Feynman-Kac lemma 366
  generalized tree process 367
  initial-boundary value problem 370
  Lévy-like stochastic process 367
  partial integro-differential equations 366
  solutions in bounded domains 372–385
  statement of the problem 368–370
interest rate cap 419
interest rate floor 419
In type barrier option 139, 140
inverse ETFs 435
inverse Gaussian (IG) process 330
inverse quadratic interpolation 67
Irwin–Hall distribution 223
Irwin, Joseph Oscar 223
Itô integral 246
Itô integral construction 249–251
Itô lemma 42–44, 254–256
Itô process 254

j

Jacobi method 197
Jarrow–Rudd tree 124
joint distribution function 9
jump diffusion models
  Black–Scholes models 305–310
  compound Poisson process 304–305
  Poisson process 303–304
  solutions to partial-integral differential systems 310–322

k

Kessler method 25

l

lagrange interpolating polynomial 28–29
Laplace equations 84
Laplace transform 91
  Black–Scholes PDE 96–98
  moment generating function 91–93
least squares Monte Carlo (LSM) 208, 210–216
Lebesgue's dominated convergence theorem 319
left continuous filtrations 6
leverage effect 294
leverage ETFs 435–436
Lévy alpha-stable distribution 337
Lévy flight models
  Cauchy stable distribution 340
  characteristic functions 339
  Fourier transform 342–343
  H-α relationship for the truncated Lévy flight 346–347
  kurtosis 343–344
  normal distribution with density 340
  one-sided stable distribution 340
  self-similarity 345–346
  stability exponent 340
  symmetric Lévy distribution 341
  truncated Lévy flight 341
Lévy flight parameter 352
  for City Group 352–353
  for the Dow Jones Industrial Average (DJIA) index 352, 356
  for Google 357, 359
  for IBM 357, 358
  for JPMorgan 352, 354
  for Microsoft Corporation 352, 355

for Walmart 357, 360
for Walt Disney Company 357, 361
Levy–Khintchine formula 106, 327
Lévy measure 328–329
Lévy processes 154
  characteristic function 326
  detrended fluctuation analysis 334–335
  exponential Lévy models 330–331
  Gamma process 329–330
  infinite divisibility of 327
  infinitely divisible distribution 326
  inverse Gaussian (IG) process 330
  Lévy measure 328–329
  rescaled range analysis 332–334
  stationarity and unit root test 335–336
  stochastic continuity 326
  stochastic process 325–326
  subordination of 331–332
Lévy stochastic variables 347
  sum of exponential random variables 348–350
  sum of Lévy random variables 351–352
linear complementarity problems (LCP) 192–196
linear regression 404
linear systems of SDEs 268–271
Lipschitz condition 258
local volatility model 74
log-likelihood function 268
Longstaff model 264
Lorentzian random variable 338

m

macroeconomic factor models 407
Markov Chain Monte Carlo (MCMC) methods 145, 217
Markov process 14, 19, 132, 133
  definition 143
  linear space/vector space 147
  norm 148–149
  semigroup
    definition 149
    Feller semigroup 151–152
    infinitesimal operator of 150–151
    operator 146–149
    strongly continuous 149–150
    transition function 143–146
Marsaglia's polar rejection method 222–224
Marsaglia's Ziggurat Method 226–235
martingale optimization method 210
martingales 14
mathematical finance
  arbitrage 33–35
  continuous time model 41–45
  options
    Call type 35
    definition 35
    hedging 39–40
    Put–Call Parity 36–38
    Put type 35
    vanilla 35–36
  stock return model 40–41
maximum likelihood estimator (MLE) 23
mean-reverting O–U process 263–264, 296–301
Merton 154
metatheorem 422–426
Metropolis–Hastings method 217
MGF *see* moment generating function (MGF)
midpoint rule 28
modified Hull–White process 393
moment generating function (MGF)
  definition 91
  Fourier series approximation method 93–96
  Laplace transform 91–93
Monte Carlo simulation method
  American option pricing 208–216
  integrals 205
  nonstandard 216–217

Monte Carlo simulation method (*contd.*)
   plain vanilla   203–204
   variance reduction
      antithetic variates   205–206
      control variates   206–208
mortgage backed securities (MBS)   440–441
Muller method   67
multifactor models   407
multiplicative tree   123
multivariate (Gaussian) distributions with prescribed covariance structure   281–283
Multivariate normal distribution   280–281

n

Newman-type BC   85
Newton's method   66–67
noise   245
nonlinear PDE   395–397
nonstandard Monte Carlo methods
   MCMC   217
   SMC methods   216–217
normal distribution with density   340
number of nodes
   Crank–Nicolson finite difference method   185–186
   explicit finite difference method   184–185
   implicit finite difference method   185

o

obstacle problem
   Gauss–Seidel method   197–198
   Jacobi method   197
   SOR technique   198–199
one-dimensional Brownian motion   146
one-dimensional standard Brownian motion   273

one-sided stable distribution   340
one-step binomial tree   122–125
option pricing
   PDE derivation   393–400
   stochastic volatility model   392–393
   with transaction costs and stochastic volatility   389–390
   valuation in the geometric Brownian motion case   390–392
options
   Call type   35
   definition   35
   hedging   39–40
   Put–Call Parity   36–38
   Put type   35
   vanilla   35–36
options on bonds   417
ordinary differential equations (ODE)   88
Ornstein–Uhlenbeck (O–U) process   45, 63, 262–263
Out type barrier option   139
Ozaki method   24–25

p

partial differential equations (PDEs)   155, 393–395
   Black–Scholes   96–98
   boundary conditions   84–85
   definition   83
   derivation of the option value   397–400
   finite difference methods (*see* finite difference methods)
   functional spaces   85–87
   Heston PDE   188–191
   MGF (*see* moment generating function (MGF))
   nonlinear PDE   395–397
   vs. SDE   271–272
   separation of variables   88–91
   two-dimensional linear   83–84
partial-integral differential systems   310–311

general PIDE system 314–322
option pricing problem 313–314
regime-switching jump diffusion model 312–313
suitability of the stochastic model postulated 311–312
particle filter construction 159–163
payoff 35
PDEs *see* partial differential equations (PDEs)
plain vanilla Monte Carlo method 203–204
p-norm 148
point processes 13
Poisson process 153, 303–304
polar rejection method 222–224
portfolio rebalancing 391
positive definite matrix 278–279
positive semidefinite matrix 278–279
pricing path-dependent options 127–128, 139–140
pricing surface 79
pseudo maximum likelihood estimation 24
Put–Call parity
  American Options 37–38
  binomial tree 129, 130
  definition 37
  In type barrier options 140

q

quadratic variation of stochastic process 11–13
quadrature methods
  definition 26
  midpoint rule 28
  rectangle rule 27
  Simpson's rule 28–29
  trapezoid rule 28
quadrature nodes 26
quadrinomial interpolating polynomial 28–29
quadrinomial tree approximation
  advantage 164

construction of
  multiperiod model 170–173
  one-period model 164–170
quantitative finance
  Black–Scholes–Merton derivation 49–50
B-S model
  Call option 58–60
  continuous hedging 50
  European Call option in 51–54, 58
  European Put options 54–57
CIR model 63–64
functional approximation methods
  Dupire's equation 74–77
  local volatility model 74
  numerical solution techniques 78–79
  pricing surface 79
  spline approximation 77–78
hedging using volatility 70–73
Heston model 60–63
implied volatility
  bisection method 65–66
  Newton's method 66–69
  Put–Call parity 69–70
  secant method 67–68
SABR model 64–65

r

random differential equation 21
random vectors 277–281
recovery rate estimates 439
rectangle rule 27
reference entity 436
regime-switching jump diffusion model 311
regressand/dependent variable 405
regression coefficients 405
regressors/independent variables 405
rejection sampling method 224–235
rescaled range analysis 332–334, 357, 362–364
Riemann–Stieltjes integral 253

right continuous filtrations 6
risk-free interest rate 392

S
SABR model *see* stochastic alpha beta rho (SABR) model
Schaefer's fixed point theorem 371–272
Schwartz model 264
score function 24
Scott model 294–295
SDE *see* stochastic differential equation (SDE)
self-similarity 345–346
semigroup theory
 definition 149
 Feller semigroup 151–152
 infinitesimal operator of 150–151
 operator 146–149
 strongly continuous 149–150
seminorm 148
sequential Monte Carlo (SMC) methods 216–217
Shoji–Ozaki method 25
short rate 415–416
short rate models 420–422
sigma algebras 5, 6
simple random walk 14–16
Simpson's rule 28–29
Sklar's theorem 410–411
Sobolev spaces 86
SOR method *see* Successive Over Relaxation (SOR) method
spot rate associated with a zero coupon bond 416
stability
 Crank–Nicolson finite difference methods 183–184
 explicit finite difference method 180
 implicit finite difference method 182–183
stable distributions 337–339
standard Brownian motion 365, 366

standard filtration 6
standard risk-neutral pricing principle 310, 313
stationarity and unit root test 335–336
stationary increments 10–11
stationary process 9, 10
statistical factor models 408–409
Stein and Stein model 295
stochastic alpha beta rho (SABR) model 64–65, 293–294
stochastic calculus 246
stochastic continuity 326
stochastic differential equation (SDE) 20–21, 245–246
 Brownian motion on the unit circle 257–258
 CIR model 63
 construction of the stochastic integral 246–252
 Euler method for approximating 273–277
 examples of 260–268
 Itô lemma 45, 254–256
 linear systems of 268–271
 Markov process 144
 multivariate (Gaussian) distributions with prescribed covariance structure 281–283
 O–U process 262–263
 vs. partial differential equations 271–272
 properties of the stochastic integral 253–254
 random vectors 277–281
 solution methods for 259–260
 stock process 123
 violating the linear growth condition 258–259
stochastic integral 21–22, 246
 construction of the 246–252
 properties of the 253–254
stochastic Lévy processes *see* Lévy processes

stochastic process 121, 325–326
  adaptiveness 6
  Bernoulli process 14–16
  Borel—Cantelli lemmas 19–20
  Brownian motion 17–19
  central limit theorems 20
  definition 3
  filtration 5–6
  finite dimensional distribution of 8
  independent components 9
  independent increments 10–11
  Markov processes 14
  Martingales 14
  maximization and parameter calibration of
    Kessler method 25
    method 23–24
    Ozaki method 24–25
    pseudo maximum likelihood estimation 24
    Shoji–Ozaki method 25
  pathwise realizations 7–8
  point processes 13
  quadrature methods
    definition 26
    midpoint rule 28
    rectangle rule 27
    Simpson's rule 28–29
    trapezoid rule 28
  SDE 20–21
  set I indexes 4
  standard filtration 6–7
  state space $\mathcal{F}$ 4–5
  stationary increments 10–11
  stationary process 9–10
  stochastic integral 21–22
  variation and quadratic variation of 11–13
stochastic version 44
stochastic volatility models 264, 289–290
  mean-reverting processes 296–301
  types of continuous time SV models 290–296

stock dynamics 45
stock ETFs 433
strictly stationary process 9, 10
strongly continuous semigroup 149, 150
strongly unique solution 259
strong Markov Process 145–146
strong solution 259
subordination of Lévy processes 331–332
subordinator 331–332
Successive Over Relaxation (SOR) method 198–199
swap rate calculation 417–419
swaption 417

t

taper 102–103
term structure consistent models 422–426
three-dimensional Brownian motion 256
traded volatility index 394
transaction costs
  option price valuation in the geometric Brownian motion case 390–392
  stochastic volatility model 392–393
  and stochastic volatility, option pricing with 389–390
transition density function 143
trapezoid rule 28
tree methods
  approximate general diffusion processes 156–159
  binomial tree (see binomial tree)
  dividend-paying assets
    continuous dividends 135–136
    discrete cash dividend 136–137
    discrete proportional dividends 136
    time-varying volatility 137–139
  general diffusion process 152–156
  Markov process (see Markov process)

tree methods (contd.)
  particle filter construction   159–163
  pricing path-dependent options
    139–140
  quadrinomial tree approximation
    advantage   164
    multiperiod model construction of
      170–173
    one-period model construction of
      164–170
  trinomial   140–142
tridiagonal matrix
  algorithm for solving   187–188
  invert   186–187
Trigeorgis tree   124, 136
trinomial trees   140–142
truncated Lévy flight   346–347
truncated Lévy models   352–361
two-dimensional linear PDE   83–84
two factor models   421–422

v

vanilla options   35–36
variance reduction technique, Monte
  Carlo methods
  antithetic variates   205–206
  control variates   206–208
variational inequalities   192
variation of stochastic process
  11–13
Vasicek model   264, 421

volatility   41–42
volatility index (VIX)   393

w

Wald's martingale   14
wavelets theory
  damage detection in frame structures
    118
  definition   109–110
  DWT   112–116
  finance   116–117
  image compression   117
  modeling and forecasting   117
  seismic signals   117–118
  and time series   110–112
weakly unique solution   259
weak solution   259
weak stationary   9
white noise process   9, 245
Wiener process   17–19,
  132–135

y

yield curve inversion   426–428

z

zero coupon and vanilla options
  Black model   419–420
  short rate models   420–422
zero coupon bond price   416
Ziggurat algorithm   224

WILEY SERIES IN STATISTICS IN PRACTICE

Advisory Editor, Marian Scott, University of Glasgow, Scotland, UK

Founding Editor, Vic Barnett, Nottingham Trent University, UK

Human and Biological Sciences
  Brown and Prescott · Applied Mixed Models in Medicine
  Ellenberg, Fleming and DeMets · Data Monitoring Committees
    in Clinical Trials:
  A Practical Perspective
  Lawson, Browne and Vidal Rodeiro · Disease Mapping With WinBUGS and
    MLwiN
  Lui · Statistical Estimation of Epidemiological Risk
  *Marubini and Valsecchi · Analysing Survival Data from Clinical Trials and
    Observation Studies
  Parmigiani · Modeling in Medical Decision Making: A Bayesian Approach
  Senn · Cross-over Trials in Clinical Research, Second Edition
  Senn · Statistical Issues in Drug Development
  Spiegelhalter, Abrams and Myles · Bayesian Approaches to Clinical Trials
    and Health-Care Evaluation
  Turner · New Drug Development: Design, Methodology, and Analysis
  Whitehead · Design and Analysis of Sequential Clinical Trials, Revised
    Second Edition
  Whitehead · Meta-Analysis of Controlled Clinical Trials
  Zhou, Zhou, Liu and Ding · Applied Missing Data Analysis in the
    Health Sciences

Earth and Environmental Sciences
  Buck, Cavanagh and Litton · Bayesian Approach to Interpreting
    Archaeological Data
  Cooke · Uncertainty Modeling in Dose Response: Bench Testing
    Environmental Toxicity
  Gibbons, Bhaumik, and Aryal · Statistical Methods for Groundwater
    Monitoring, Second Edition
  Glasbey and Horgan · Image Analysis in the Biological Sciences
  Helsel · Nondetects and Data Analysis: Statistics for Censored
    Environmental Data
  Helsel · Statistics for Censored Environmental Data Using Minitab® and R,
    Second Edition
  McBride · Using Statistical Methods for Water Quality Management: Issues,
    Problems and Solutions

Ofungwu · Statistical Applications for Environmental Analysis and
  Risk Assessment
Webster and Oliver · Geostatistics for Environmental Scientists

Industry, Commerce and Finance
  Aitken and Taroni · Statistics and the Evaluation of Evidence for
    Forensic Scientists, Second Edition
  Brandimarte · Numerical Methods in Finance and Economics:
    A MATLAB-Based Introduction, Second Edition
  Brandimarte and Zotteri · Introduction to Distribution Logistics
  Chan and Wong · Simulation Techniques in Financial Risk Management,
    Second Edition
  Jank · Statistical Methods in eCommerce Research
  Jank and Shmueli · Modeling Online Auctions
  Lehtonen and Pahkinen · Practical Methods for Design and Analysis
    of Complex Surveys, Second Edition
  Lloyd · Data Driven Business Decisions
  Ohser and Mücklich · Statistical Analysis of Microstructures in Materials
    Science
  Rausand · Risk Assessment: Theory, Methods, and Applications

Printed and bound by CPI Group (UK) Ltd, Croydon, CR0 4YY